Lecture Notes in Comp

Founding Editors

Gerhard Goos
Juris Hartmanis

Editorial Board Members

Elisa Bertino, *Purdue University, West Lafayette, IN, USA*
Wen Gao, *Peking University, Beijing, China*
Bernhard Steffen, *TU Dortmund University, Dortmund, Germany*
Moti Yung, *Columbia University, New York, NY, USA*

The series Lecture Notes in Computer Science (LNCS), including its subseries Lecture Notes in Artificial Intelligence (LNAI) and Lecture Notes in Bioinformatics (LNBI), has established itself as a medium for the publication of new developments in computer science and information technology research, teaching, and education.

LNCS enjoys close cooperation with the computer science R & D community, the series counts many renowned academics among its volume editors and paper authors, and collaborates with prestigious societies. Its mission is to serve this international community by providing an invaluable service, mainly focused on the publication of conference and workshop proceedings and postproceedings. LNCS commenced publication in 1973.

Apostolos Antonacopoulos ·
Subhasis Chaudhuri · Rama Chellappa ·
Cheng-Lin Liu · Saumik Bhattacharya ·
Umapada Pal
Editors

Pattern Recognition

27th International Conference, ICPR 2024
Kolkata, India, December 1–5, 2024
Proceedings, Part XII

Editors
Apostolos Antonacopoulos
University of Salford
Salford, UK

Subhasis Chaudhuri
Indian Institute of Technology Bombay
Mumbai, India

Rama Chellappa
Johns Hopkins University
Baltimore, MD, USA

Cheng-Lin Liu
Chinese Academy of Sciences
Beijing, China

Saumik Bhattacharya
IIT Kharagpur
Kharagpur, India

Umapada Pal
Indian Statistical Institute Kolkata
Kolkata, India

ISSN 0302-9743 ISSN 1611-3349 (electronic)
Lecture Notes in Computer Science
ISBN 978-3-031-78197-1 ISBN 978-3-031-78198-8 (eBook)
https://doi.org/10.1007/978-3-031-78198-8

© The Editor(s) (if applicable) and The Author(s), under exclusive license to Springer Nature Switzerland AG 2025

This work is subject to copyright. All rights are solely and exclusively licensed by the Publisher, whether the whole or part of the material is concerned, specifically the rights of translation, reprinting, reuse of illustrations, recitation, broadcasting, reproduction on microfilms or in any other physical way, and transmission or information storage and retrieval, electronic adaptation, computer software, or by similar or dissimilar methodology now known or hereafter developed.
The use of general descriptive names, registered names, trademarks, service marks, etc. in this publication does not imply, even in the absence of a specific statement, that such names are exempt from the relevant protective laws and regulations and therefore free for general use.
The publisher, the authors and the editors are safe to assume that the advice and information in this book are believed to be true and accurate at the date of publication. Neither the publisher nor the authors or the editors give a warranty, expressed or implied, with respect to the material contained herein or for any errors or omissions that may have been made. The publisher remains neutral with regard to jurisdictional claims in published maps and institutional affiliations.

This Springer imprint is published by the registered company Springer Nature Switzerland AG
The registered company address is: Gewerbestrasse 11, 6330 Cham, Switzerland

If disposing of this product, please recycle the paper.

President's Address

On behalf of the Executive Committee of the International Association for Pattern Recognition (IAPR), I am pleased to welcome you to the 27th International Conference on Pattern Recognition (ICPR 2024), the main scientific event of the IAPR.

After a completely digital ICPR in the middle of the COVID pandemic and the first hybrid version in 2022, we can now enjoy a fully back-to-normal ICPR this year. I look forward to hearing inspirational talks and keynotes, catching up with colleagues during the breaks and making new contacts in an informal way. At the same time, the conference landscape has changed. Hybrid meetings have made their entrance and will continue. It is exciting to experience how this will influence the conference. Planning for a major event like ICPR must take place over a period of several years. This means many decisions had to be made under a cloud of uncertainty, adding to the already large effort needed to produce a successful conference. It is with enormous gratitude, then, that we must thank the team of organizers for their hard work, flexibility, and creativity in organizing this ICPR. ICPR always provides a wonderful opportunity for the community to gather together. I can think of no better location than Kolkata to renew the bonds of our international research community.

Each ICPR is a bit different owing to the vision of its organizing committee. For 2024, the conference has six different tracks reflecting major themes in pattern recognition: Artificial Intelligence, Pattern Recognition and Machine Learning; Computer and Robot Vision; Image, Speech, Signal and Video Processing; Biometrics and Human Computer Interaction; Document Analysis and Recognition; and Biomedical Imaging and Bioinformatics. This reflects the richness of our field. ICPR 2024 also features two dozen workshops, seven tutorials, and 15 competitions; there is something for everyone. Many thanks to those who are leading these activities, which together add significant value to attending ICPR, whether in person or virtually. Because it is important for ICPR to be as accessible as possible to colleagues from all around the world, we are pleased that the IAPR, working with the ICPR organizers, is continuing our practice of awarding travel stipends to a number of early-career authors who demonstrate financial need. Last but not least, we are thankful to the Springer LNCS team for their effort to publish these proceedings.

Among the presentations from distinguished keynote speakers, we are looking forward to the three IAPR Prize Lectures at ICPR 2024. This year we honor the achievements of Tin Kam Ho (IBM Research) with the IAPR's most prestigious King-Sun Fu Prize "for pioneering contributions to multi-classifier systems, random decision forests, and data complexity analysis". The King-Sun Fu Prize is given in recognition of an outstanding technical contribution to the field of pattern recognition. It honors the memory of Professor King-Sun Fu who was instrumental in the founding of IAPR, served as its first president, and is widely recognized for his extensive contributions to the field of pattern recognition.

The Maria Petrou Prize is given to a living female scientist/engineer who has made substantial contributions to the field of Pattern Recognition and whose past contributions, current research activity and future potential may be regarded as a model to both aspiring and established researchers. It honours the memory of Professor Maria Petrou as a scientist of the first rank, and particularly her role as a pioneer for women researchers. This year, the Maria Petrou Prize is given to Guoying Zhao (University of Oulu), "for contributions to video analysis for facial micro-behavior recognition and remote biosignal reading (RPPG) for heart rate analysis and face anti-spoofing".

The J.K. Aggarwal Prize is given to a young scientist who has brought a substantial contribution to a field that is relevant to the IAPR community and whose research work has had a major impact on the field. Professor Aggarwal is widely recognized for his extensive contributions to the field of pattern recognition and for his participation in IAPR's activities. This year, the J.K. Aggarwal Prize goes to Xiaolong Wang (UC San Diego) "for groundbreaking contributions to advancing visual representation learning, utilizing self-supervised and attention-based models to establish fundamental frameworks for creating versatile, general-purpose pattern recognition systems".

During the conference we will also recognize 21 new IAPR Fellows selected from a field of very strong candidates. In addition, a number of Best Scientific Paper and Best Student Paper awards will be presented, along with the Best Industry Related Paper Award and the Piero Zamperoni Best Student Paper Award. Congratulations to the recipients of these very well-deserved awards!

I would like to close by again thanking everyone involved in making ICPR 2024 a tremendous success; your hard work is deeply appreciated. These thanks extend to all who chaired the various aspects of the conference and the associated workshops, my ExCo colleagues, and the IAPR Standing and Technical Committees. Linda O'Gorman, the IAPR Secretariat, deserves special recognition for her experience, historical perspective, and attention to detail when it comes to supporting many of the IAPR's most important activities. Her tasks became so numerous that she recently got support from Carolyn Buckley (layout, newsletter), Ugur Halici (ICPR matters), and Rosemary Stramka (secretariat). The IAPR website got a completely new design. Ed Sobczak has taken care of our web presence for so many years already. A big thank you to all of you!

This is, of course, the 27th ICPR conference. Knowing that ICPR is organized every two years, and that the first conference in the series (1973!) pre-dated the formal founding of the IAPR by a few years, it is also exciting to consider that we are celebrating over 50 years of ICPR and at the same time approaching the official IAPR 50th anniversary in 2028: you'll get all information you need at ICPR 2024. In the meantime, I offer my thanks and my best wishes to all who are involved in supporting the IAPR throughout the world.

September 2024

Arjan Kuijper
President of the IAPR

Preface

It is our great pleasure to welcome you to the proceedings of the 27th International Conference on Pattern Recognition (ICPR 2024), held in Kolkata, India. The city, formerly known as 'Calcutta', is the home of the fabled Indian Statistical Institute (ISI), which has been at the forefront of statistical pattern recognition for almost a century. Concepts like the Mahalanobis distance, Bhattacharyya bound, Cramer–Rao bound, and Fisher–Rao metric were invented by pioneers associated with ISI. The first ICPR (called IJCPR then) was held in 1973, and the second in 1974. Subsequently, ICPR has been held every other year. The International Association for Pattern Recognition (IAPR) was founded in 1978 and became the sponsor of the ICPR series. Over the past 50 years, ICPR has attracted huge numbers of scientists, engineers and students from all over the world and contributed to advancing research, development and applications in pattern recognition technology.

ICPR 2024 was held at the Biswa Bangla Convention Centre, one of the largest such facilities in South Asia, situated just 7 kilometers from Kolkata Airport (CCU). According to ChatGPT "Kolkata is often called the 'Cultural Capital of India'. The city has a deep connection to literature, music, theater, and art. It was home to Nobel laureate Rabindranath Tagore, and the Bengali film industry has produced globally renowned filmmakers like Satyajit Ray. The city boasts remarkable colonial architecture, with landmarks like Victoria Memorial, Howrah Bridge, and the Indian Museum (the oldest and largest museum in India). Kolkata's streets are dotted with old mansions and buildings that tell stories of its colonial past. Walking through the city can feel like stepping back into a different era. Finally, Kolkata is also known for its street food."

ICPR 2024 followed a two-round paper submission format. We received a total of 2135 papers (1501 papers in round-1 submissions, and 634 papers in round-2 submissions). Each paper, on average, received 2.84 reviews, in single-blind mode. For the first-round papers we had a rebuttal option available to authors.

In total, 945 papers (669 from round-1 and 276 from round-2) were accepted for presentation, resulting in an acceptance rate of 44.26%, which is consistent with previous ICPR events. At ICPR 2024 the papers were categorized into six tracks: Artificial Intelligence, Machine Learning for Pattern Analysis; Computer Vision and Robotic Perception; Image, Video, Speech, and Signal Analysis; Biometrics and Human-Machine Interaction; Document and Media Analysis; and Biomedical Image Analysis and Informatics.

The main conference ran over December 2–5, 2024. The main program included the presentation of 188 oral papers (19.89% of the accepted papers), 757 poster papers and 12 competition papers (out of 15 submitted). A total 10 oral sessions were held concurrently in four meeting rooms with a total of 40 oral sessions. In total 24 workshops and 7 tutorials were held on December 1, 2024.

The plenary sessions included three prize lectures and three invited presentations. The prize lectures were delivered by Tin Kam Ho (IBM Research, USA; King Sun

Fu Prize winner), Xiaolong Wang (University of California, San Diego, USA; J.K. Aggarwal Prize winner), and Guoying Zhao (University of Oulu, Finland; Maria Petrou Prize winner). The invited speakers were Timothy Hospedales (University of Edinburgh, UK), Venu Govindaraju (University at Buffalo, USA), and Shuicheng Yan (Skywork AI, Singapore).

Several best paper awards were presented in ICPR: the Piero Zamperoni Award for the best paper authored by a student, the BIRPA Best Industry Related Paper Award, and the Best Paper Awards and Best Student Paper Awards for each of the six tracks of ICPR 2024.

The organization of such a large conference would not be possible without the help of many volunteers. Our special gratitude goes to the Program Chairs (Apostolos Antonacopoulos, Subhasis Chaudhuri, Rama Chellappa and Cheng-Lin Liu), for their leadership in organizing the program. Thanks to our Publication Chairs (Ananda S. Chowdhury and Wataru Ohyama) for handling the overwhelming workload of publishing the conference proceedings. We also thank our Competition Chairs (Richard Zanibbi, Lianwen Jin and Laurence Likforman-Sulem) for arranging 12 important competitions as part of ICPR 2024. We are thankful to our Workshop Chairs (P. Shivakumara, Stephanie Schuckers, Jean-Marc Ogier and Prabir Bhattacharya) and Tutorial Chairs (B.B. Chaudhuri, Michael R. Jenkin and Guoying Zhao) for arranging the workshops and tutorials on emerging topics. ICPR 2024, for the first time, held a Doctoral Consortium. We would like to thank our Doctoral Consortium Chairs (Véronique Eglin, Dan Lopresti and Mayank Vatsa) for organizing it.

Thanks go to the Track Chairs and the meta reviewers who devoted significant time to the review process and preparation of the program. We also sincerely thank the reviewers who provided valuable feedback to the authors.

Finally, we acknowledge the work of other conference committee members, like the Organizing Chairs and Organizing Committee Members, Finance Chairs, Award Chair, Sponsorship Chairs, and Exhibition and Demonstration Chairs, Visa Chair, Publicity Chairs, and Women in ICPR Chairs, whose efforts made this event successful. We also thank our event manager Alpcord Network for their help.

We hope that all the participants found the technical program informative and enjoyed the sights, culture and cuisine of Kolkata.

October 2024

Umapada Pal
Josef Kittler
Anil Jain

Organization

General Chairs

Umapada Pal Indian Statistical Institute, Kolkata, India
Josef Kittler University of Surrey, UK
Anil Jain Michigan State University, USA

Program Chairs

Apostolos Antonacopoulos University of Salford, UK
Subhasis Chaudhuri Indian Institute of Technology, Bombay, India
Rama Chellappa Johns Hopkins University, USA
Cheng-Lin Liu Institute of Automation, Chinese Academy of Sciences, China

Publication Chairs

Ananda S. Chowdhury Jadavpur University, India
Wataru Ohyama Tokyo Denki University, Japan

Competition Chairs

Richard Zanibbi Rochester Institute of Technology, USA
Lianwen Jin South China University of Technology, China
Laurence Likforman-Sulem Télécom Paris, France

Workshop Chairs

P. Shivakumara University of Salford, UK
Stephanie Schuckers Clarkson University, USA
Jean-Marc Ogier Université de la Rochelle, France
Prabir Bhattacharya Concordia University, Canada

Tutorial Chairs

B. B. Chaudhuri Indian Statistical Institute, Kolkata, India
Michael R. Jenkin York University, Canada
Guoying Zhao University of Oulu, Finland

Doctoral Consortium Chairs

Véronique Eglin CNRS, France
Daniel P. Lopresti Lehigh University, USA
Mayank Vatsa Indian Institute of Technology, Jodhpur, India

Organizing Chairs

Saumik Bhattacharya Indian Institute of Technology, Kharagpur, India
Palash Ghosal Sikkim Manipal University, India

Organizing Committee

Santanu Phadikar West Bengal University of Technology, India
SK Md Obaidullah Aliah University, India
Sayantari Ghosh National Institute of Technology Durgapur, India
Himadri Mukherjee West Bengal State University, India
Nilamadhaba Tripathy Clarivate Analytics, USA
Chayan Halder West Bengal State University, India
Shibaprasad Sen Techno Main Salt Lake, India

Finance Chairs

Kaushik Roy West Bengal State University, India
Michael Blumenstein University of Technology Sydney, Australia

Awards Committee Chair

Arpan Pal Tata Consultancy Services, India

Sponsorship Chairs

P. J. Narayanan   Indian Institute of Technology, Hyderabad, India
Yasushi Yagi   Osaka University, Japan
Venu Govindaraju   University at Buffalo, USA
Alberto Bel Bimbo   Università di Firenze, Italy

Exhibition and Demonstration Chairs

Arjun Jain   FastCode AI, India
Agnimitra Biswas   National Institute of Technology, Silchar, India

International Liaison, Visa Chair

Balasubramanian Raman   Indian Institute of Technology, Roorkee, India

Publicity Chairs

Dipti Prasad Mukherjee   Indian Statistical Institute, Kolkata, India
Bob Fisher   University of Edinburgh, UK
Xiaojun Wu   Jiangnan University, China

Women in ICPR Chairs

Ingela Nystrom   Uppsala University, Sweden
Alexandra B. Albu   University of Victoria, Canada
Jing Dong   Institute of Automation, Chinese Academy of
                             Sciences, China
Sarbani Palit   Indian Statistical Institute, Kolkata, India

Event Manager

Alpcord Network

Track Chairs – Artificial Intelligence, Machine Learning for Pattern Analysis

Larry O'Gorman   Nokia Bell Labs, USA
Dacheng Tao   University of Sydney, Australia
Petia Radeva   University of Barcelona, Spain
Susmita Mitra   Indian Statistical Institute, Kolkata, India
Jiliang Tang   Michigan State University, USA

Track Chairs – Computer and Robot Vision

C. V. Jawahar   International Institute of Information Technology (IIIT), Hyderabad, India
João Paulo Papa   São Paulo State University, Brazil
Maja Pantic   Imperial College London, UK
Gang Hua   Dolby Laboratories, USA
Junwei Han   Northwestern Polytechnical University, China

Track Chairs – Image, Speech, Signal and Video Processing

P. K. Biswas   Indian Institute of Technology, Kharagpur, India
Shang-Hong Lai   National Tsing Hua University, Taiwan
Hugo Jair Escalante   INAOE, CINVESTAV, Mexico
Sergio Escalera   Universitat de Barcelona, Spain
Prem Natarajan   University of Southern California, USA

Track Chairs – Biometrics and Human Computer Interaction

Richa Singh   Indian Institute of Technology, Jodhpur, India
Massimo Tistarelli   University of Sassari, Italy
Vishal Patel   Johns Hopkins University, USA
Wei-Shi Zheng   Sun Yat-sen University, China
Jian Wang   Snap, USA

Track Chairs – Document Analysis and Recognition

Xiang Bai	Huazhong University of Science and Technology, China
David Doermann	University at Buffalo, USA
Josep Llados	Universitat Autònoma de Barcelona, Spain
Mita Nasipuri	Jadavpur University, India

Track Chairs – Biomedical Imaging and Bioinformatics

Jayanta Mukhopadhyay	Indian Institute of Technology, Kharagpur, India
Xiaoyi Jiang	Universität Münster, Germany
Seong-Whan Lee	Korea University, Korea

Metareviewers (Conference Papers and Competition Papers)

Wael Abd-Almageed	University of Southern California, USA
Maya Aghaei	NHL Stenden University, Netherlands
Alireza Alaei	Southern Cross University, Australia
Rajagopalan N. Ambasamudram	Indian Institute of Technology, Madras, India
Suyash P. Awate	Indian Institute of Technology, Bombay, India
Inci M. Baytas	Bogazici University, Turkey
Aparna Bharati	Lehigh University, USA
Brojeshwar Bhowmick	Tata Consultancy Services, India
Jean-Christophe Burie	University of La Rochelle, France
Gustavo Carneiro	University of Surrey, UK
Chee Seng Chan	Universiti Malaya, Malaysia
Sumohana S. Channappayya	Indian Institute of Technology, Hyderabad, India
Dongdong Chen	Microsoft, USA
Shengyong Chen	Tianjin University of Technology, China
Jun Cheng	Institute for Infocomm Research, A*STAR, Singapore
Albert Clapés	University of Barcelona, Spain
Oscar Dalmau	Center for Research in Mathematics, Mexico

Tyler Derr	Vanderbilt University, USA
Abhinav Dhall	Indian Institute of Technology, Ropar, India
Bo Du	Wuhan University, China
Yuxuan Du	University of Sydney, Australia
Ayman S. El-Baz	University of Louisville, USA
Francisco Escolano	University of Alicante, Spain
Siamac Fazli	Nazarbayev University, Kazakhstan
Jianjiang Feng	Tsinghua University, China
Gernot A. Fink	TU Dortmund University, Germany
Alicia Fornes	CVC, Spain
Junbin Gao	University of Sydney, Australia
Yan Gao	Amazon, USA
Yongsheng Gao	Griffith University, Australia
Caren Han	University of Melbourne, Australia
Ran He	Institute of Automation, Chinese Academy of Sciences, China
Tin Kam Ho	IBM, USA
Di Huang	Beihang University, China
Kaizhu Huang	Duke Kunshan University, China
Donato Impedovo	University of Bari, Italy
Julio Jacques	University of Barcelona and Computer Vision Center, Spain
Lianwen Jin	South China University of Technology, China
Wei Jin	Emory University, USA
Danilo Samuel Jodas	São Paulo State University, Brazil
Manjunath V. Joshi	DA-IICT, India
Jayashree Kalpathy-Cramer	Massachusetts General Hospital, USA
Dimosthenis Karatzas	Computer Vision Centre, Spain
Hamid Karimi	Utah State University, USA
Baiying Lei	Shenzhen University, China
Guoqi Li	Chinese Academy of Sciences, and Peng Cheng Lab, China
Laurence Likforman-Sulem	Institut Polytechnique de Paris/Télécom Paris, France
Aishan Liu	Beihang University, China
Bo Liu	Bytedance, USA
Chen Liu	Clarkson University, USA
Cheng-Lin Liu	Institute of Automation, Chinese Academy of Sciences, China
Hongmin Liu	University of Science and Technology Beijing, China
Hui Liu	Michigan State University, USA

Jing Liu	Institute of Automation, Chinese Academy of Sciences, China
Li Liu	University of Oulu, Finland
Qingshan Liu	Nanjing University of Posts and Telecommunications, China
Adrian P. Lopez-Monroy	Centro de Investigacion en Matematicas AC, Mexico
Daniel P. Lopresti	Lehigh University, USA
Shijian Lu	Nanyang Technological University, Singapore
Yong Luo	Wuhan University, China
Andreas K. Maier	FAU Erlangen-Nuremberg, Germany
Davide Maltoni	University of Bologna, Italy
Hong Man	Stevens Institute of Technology, USA
Lingtong Min	Northwestern Polytechnical University, China
Paolo Napoletano	University of Milano-Bicocca, Italy
Kamal Nasrollahi	Milestone Systems, Aalborg University, Denmark
Marcos Ortega	University of A Coruña, Spain
Shivakumara Palaiahnakote	University of Salford, UK
P. Jonathon Phillips	NIST, USA
Filiberto Pla	University Jaume I, Spain
Ajit Rajwade	Indian Institute of Technology, Bombay, India
Shanmuganathan Raman	Indian Institute of Technology, Gandhinagar, India
Imran Razzak	UNSW, Australia
Beatriz Remeseiro	University of Oviedo, Spain
Gustavo Rohde	University of Virginia, USA
Partha Pratim Roy	Indian Institute of Technology, Roorkee, India
Sanjoy K. Saha	Jadavpur University, India
Joan Andreu Sánchez	Universitat Politècnica de València, Spain
Claudio F. Santos	UFSCar, Brazil
Shin'ichi Satoh	National Institute of Informatics, Japan
Stephanie Schuckers	Clarkson University, USA
Srirangaraj Setlur	University at Buffalo, SUNY, USA
Debdoot Sheet	Indian Institute of Technology, Kharagpur, India
Jun Shen	University of Wollongong, Australia
Li Shen	JD Explore Academy, China
Chen Shengyong	Zhejiang University of Technology and Tianjin University of Technology, China
Andy Song	RMIT University, Australia
Akihiro Sugimoto	National Institute of Informatics, Japan
Qianru Sun	Singapore Management University, Singapore
Arijit Sur	Indian Institute of Technology, Guwahati, India
Estefania Talavera	University of Twente, Netherlands

Wei Tang	University of Illinois at Chicago, USA
Joao M. Tavares	Universidade do Porto, Portugal
Jun Wan	NLPR, CASIA, China
Le Wang	Xi'an Jiaotong University, China
Lei Wang	Australian National University, Australia
Xiaoyang Wang	Tencent AI Lab, USA
Xinggang Wang	Huazhong University of Science and Technology, China
Xiao-Jun Wu	Jiangnan University, China
Yiding Yang	Bytedance, China
Xiwen Yao	Northwestern Polytechnical University, China
Xu-Cheng Yin	University of Science and Technology Beijing, China
Baosheng Yu	University of Sydney, Australia
Shiqi Yu	Southern University of Science and Technology, China
Xin Yuan	Westlake University, China
Yibing Zhan	JD Explore Academy, China
Jing Zhang	University of Sydney, Australia
Lefei Zhang	Wuhan University, China
Min-Ling Zhang	Southeast University, China
Wenbin Zhang	Florida International University, USA
Jiahuan Zhou	Peking University, China
Sanping Zhou	Xi'an Jiaotong University, China
Tianyi Zhou	University of Maryland, USA
Lei Zhu	Shandong Normal University, China
Pengfei Zhu	Tianjin University, China
Wangmeng Zuo	Harbin Institute of Technology, China

Reviewers (Competition Papers)

Liangcai Gao
Mingxin Huang
Lei Kang
Wenhui Liao
Yuliang Liu
Yongxin Shi

Da-Han Wang
Yang Xue
Wentao Yang
Jiaxin Zhang
Yiwu Zhong

Reviewers (Conference Papers)

Aakanksha Aakanksha
Aayush Singla
Abdul Muqeet
Abhay Yadav
Abhijeet Vijay Nandedkar
Abhimanyu Sahu
Abhinav Rajvanshi
Abhisek Ray
Abhishek Shrivastava
Abhra Chaudhuri
Aditi Roy
Adriano Simonetto
Adrien Maglo
Ahmed Abdulkadir
Ahmed Boudissa
Ahmed Hamdi
Ahmed Rida Sekkat
Ahmed Sharafeldeen
Aiman Farooq
Aishwarya Venkataramanan
Ajay Kumar
Ajay Kumar Reddy Poreddy
Ajita Rattani
Ajoy Mondal
Akbar K.
Akbar Telikani
Akshay Agarwal
Akshit Jindal
Al Zadid Sultan Bin Habib
Albert Clapés
Alceu Britto
Alejandro Peña
Alessandro Ortis
Alessia Auriemma Citarella
Alexandre Stenger
Alexandros Sopasakis
Alexia Toumpa
Ali Khan
Alik Pramanick
Alireza Alaei
Alper Yilmaz
Aman Verma
Amit Bhardwaj
Amit More
Amit Nandedkar
Amitava Chatterjee
Amos L. Abbott
Amrita Mohan
Anand Mishra
Ananda S. Chowdhury
Anastasia Zakharova
Anastasios L. Kesidis
Andras Horvath
Andre Gustavo Hochuli
André P. Kelm
Andre Wyzykowski
Andrea Bottino
Andrea Lagorio
Andrea Torsello
Andreas Fischer
Andreas K. Maier
Andreu Girbau Xalabarder
Andrew Beng Jin Teoh
Andrew Shin
Andy J. Ma
Aneesh S. Chivukula
Ángela Casado-García
Anh Quoc Nguyen
Anindya Sen
Anirban Saha
Anjali Gautam
Ankan Bhattacharyya
Ankit Jha
Anna Scius-Bertrand
Annalisa Franco
Antoine Doucet
Antonino Staiano
Antonio Fernández
Antonio Parziale
Anu Singha
Anustup Choudhury
Anwesan Pal
Anwesha Sengupta
Archisman Adhikary
Arjan Kuijper
Arnab Kumar Das

Arnav Bhavsar
Arnav Varma
Arpita Dutta
Arshad Jamal
Artur Jordao
Arunkumar Chinnaswamy
Aryan Jadon
Aryaz Baradarani
Ashima Anand
Ashis Dhara
Ashish Phophalia
Ashok K. Bhateja
Ashutosh Vaish
Ashwani Kumar
Asifuzzaman Lasker
Atefeh Khoshkhahtinat
Athira Nambiar
Attilio Fiandrotti
Avandra S. Hemachandra
Avik Hati
Avinash Sharma
B. H. Shekar
B. Uma Shankar
Bala Krishna Thunakala
Balaji Tk
Balázs Pálffy
Banafsheh Adami
Bang-Dang Pham
Baochang Zhang
Baodi Liu
Bashirul Azam Biswas
Beiduo Chen
Benedikt Kottler
Beomseok Oh
Berkay Aydin
Berlin S. Shaheema
Bertrand Kerautret
Bettina Finzel
Bhavana Singh
Bibhas C. Dhara
Bilge Gunsel
Bin Chen
Bin Li
Bin Liu
Bin Yao

Bin-Bin Jia
Binbin Yong
Bindita Chaudhuri
Bindu Madhavi Tummala
Binh M. Le
Bi-Ru Dai
Bo Huang
Bo Jiang
Bob Zhang
Bowen Liu
Bowen Zhang
Boyang Zhang
Boyu Diao
Boyun Li
Brian M. Sadler
Bruce A. Maxwell
Bryan Bo Cao
Buddhika L. Semage
Bushra Jalil
Byeong-Seok Shin
Byung-Gyu Kim
Caihua Liu
Cairong Zhao
Camille Kurtz
Carlos A. Caetano
Carlos D. Martã-Nez-Hinarejos
Ce Wang
Cevahir Cigla
Chakravarthy Bhagvati
Chandrakanth Vipparla
Changchun Zhang
Changde Du
Changkun Ye
Changxu Cheng
Chao Fan
Chao Guo
Chao Qu
Chao Wen
Chayan Halder
Che-Jui Chang
Chen Feng
Chenan Wang
Cheng Yu
Chenghao Qian
Cheng-Lin Liu

Chengxu Liu
Chenru Jiang
Chensheng Peng
Chetan Ralekar
Chih-Wei Lin
Chih-Yi Chiu
Chinmay Sahu
Chintan Patel
Chintan Shah
Chiranjoy Chattopadhyay
Chong Wang
Choudhary Shyam Prakash
Christophe Charrier
Christos Smailis
Chuanwei Zhou
Chun-Ming Tsai
Chunpeng Wang
Ciro Russo
Claudio De Stefano
Claudio F. Santos
Claudio Marrocco
Connor Levenson
Constantine Dovrolis
Constantine Kotropoulos
Dai Shi
Dakshina Ranjan Kisku
Dan Anitei
Dandan Zhu
Daniela Pamplona
Danli Wang
Danqing Huang
Daoan Zhang
Daqing Hou
David A. Clausi
David Freire Obregon
David Münch
David Pujol Perich
Davide Marelli
De Zhang
Debalina Barik
Debapriya Roy (Kundu)
Debashis Das
Debashis Das Chakladar
Debi Prasad Dogra
Debraj D. Basu

Decheng Liu
Deen Dayal Mohan
Deep A. Patel
Deepak Kumar
Dengpan Liu
Denis Coquenet
Désiré Sidibé
Devesh Walawalkar
Dewan Md. Farid
Di Ming
Di Qiu
Di Yuan
Dian Jia
Dianmo Sheng
Diego Thomas
Diganta Saha
Dimitri Bulatov
Dimpy Varshni
Dingcheng Yang
Dipanjan Das
Dipanjyoti Paul
Divya Biligere Shivanna
Divya Saxena
Divya Sharma
Dmitrii Matveichev
Dmitry Minskiy
Dmitry V. Sorokin
Dong Zhang
Donghua Wang
Donglin Zhang
Dongming Wu
Dongqiangzi Ye
Dongqing Zou
Dongrui Liu
Dongyang Zhang
Dongzhan Zhou
Douglas Rodrigues
Duarte Folgado
Duc Minh Vo
Duoxuan Pei
Durai Arun Pannir Selvam
Durga Bhavani S.
Eckart Michaelsen
Elena Goyanes
Élodie Puybareau

Emanuele Vivoli
Emna Ghorbel
Enrique Naredo
Enyu Cai
Eric Patterson
Ernest Valveny
Eva Blanco-Mallo
Eva Breznik
Evangelos Sartinas
Fabio Solari
Fabiola De Marco
Fan Wang
Fangda Li
Fangyuan Lei
Fangzhou Lin
Fangzhou Luo
Fares Bougourzi
Farman Ali
Fatiha Mokdad
Fei Shen
Fei Teng
Fei Zhu
Feiyan Hu
Felipe Gomes Oliveira
Feng Li
Fengbei Liu
Fenghua Zhu
Fillipe D. M. De Souza
Flavio Piccoli
Flavio Prieto
Florian Kleber
Francesc Serratosa
Francesco Bianconi
Francesco Castro
Francesco Ponzio
Francisco Javier Hernández López
Frédéric Rayar
Furkan Osman Kar
Fushuo Huo
Fuxiao Liu
Fu-Zhao Ou
Gabriel Turinici
Gabrielle Flood
Gajjala Viswanatha Reddy
Gaku Nakano

Galal Binamakhashen
Ganesh Krishnasamy
Gang Pan
Gangyan Zeng
Gani Rahmon
Gaurav Harit
Gennaro Vessio
Genoveffa Tortora
George Azzopardi
Gerard Ortega
Gerardo E. Altamirano-Gomez
Gernot A. Fink
Gibran Benitez-Garcia
Gil Ben-Artzi
Gilbert Lim
Giorgia Minello
Giorgio Fumera
Giovanna Castellano
Giovanni Puglisi
Giulia Orrù
Giuliana Ramella
Gökçe Uludoğan
Gopi Ramena
Gorthi Rama Krishna Sai Subrahmanyam
Gourav Datta
Gowri Srinivasa
Gozde Sahin
Gregory Randall
Guanjie Huang
Guanjun Li
Guanwen Zhang
Guanyu Xu
Guanyu Yang
Guanzhou Ke
Guhnoo Yun
Guido Borghi
Guilherme Brandão Martins
Guillaume Caron
Guillaume Tochon
Guocai Du
Guohao Li
Guoqiang Zhong
Guorong Li
Guotao Li
Gurman Gill

Haechang Lee
Haichao Zhang
Haidong Xie
Haifeng Zhao
Haimei Zhao
Hainan Cui
Haixia Wang
Haiyan Guo
Hakime Ozturk
Hamid Kazemi
Han Gao
Hang Zou
Hanjia Lyu
Hanjoo Cho
Hanqing Zhao
Hanyuan Liu
Hanzhou Wu
Hao Li
Hao Meng
Hao Sun
Hao Wang
Hao Xing
Hao Zhao
Haoan Feng
Haodi Feng
Haofeng Li
Haoji Hu
Haojie Hao
Haojun Ai
Haopeng Zhang
Haoran Li
Haoran Wang
Haorui Ji
Haoxiang Ma
Haoyu Chen
Haoyue Shi
Harald Koestler
Harbinder Singh
Harris V. Georgiou
Hasan F. Ates
Hasan S. M. Al-Khaffaf
Hatef Otroshi Shahreza
Hebeizi Li
Heng Zhang
Hengli Wang
Hengyue Liu
Hertog Nugroho
Hieyong Jeong
Himadri Mukherjee
Hoai Ngo
Hoda Mohaghegh
Hong Liu
Hong Man
Hongcheng Wang
Hongjian Zhan
Hongxi Wei
Hongyu Hu
Hoseong Kim
Hossein Ebrahimnezhad
Hossein Malekmohamadi
Hrishav Bakul Barua
Hsueh-Yi Sean Lin
Hua Wei
Huafeng Li
Huali Xu
Huaming Chen
Huan Wang
Huang Chen
Huanran Chen
Hua-Wen Chang
Huawen Liu
Huayi Zhan
Hugo Jair Escalante
Hui Chen
Hui Li
Huichen Yang
Huiqiang Jiang
Huiyuan Yang
Huizi Yu
Hung T. Nguyen
Hyeongyu Kim
Hyeonjeong Park
Hyeonjun Lee
Hymalai Bello
Hyung-Gun Chi
Hyunsoo Kim
I-Chen Lin
Ik Hyun Lee
Ilan Shimshoni
Imad Eddine Toubal

Imran Sarker
Inderjot Singh Saggu
Indrani Mukherjee
Indranil Sur
Ines Rieger
Ioannis Pierros
Irina Rabaev
Ivan V. Medri
J. Rafid Siddiqui
Jacek Komorowski
Jacopo Bonato
Jacson Rodrigues Correia-Silva
Jaekoo Lee
Jaime Cardoso
Jakob Gawlikowski
Jakub Nalepa
James L. Wayman
Jan Čech
Jangho Lee
Jani Boutellier
Javier Gurrola-Ramos
Javier Lorenzo-Navarro
Jayasree Saha
Jean Lee
Jean Paul Barddal
Jean-Bernard Hayet
Jean-Philippe G. Tarel
Jean-Yves Ramel
Jenny Benois-Pineau
Jens Bayer
Jerin Geo James
Jesús Miguel García-Gorrostieta
Jia Qu
Jiahong Chen
Jiaji Wang
Jian Hou
Jian Liang
Jian Xu
Jian Zhu
Jianfeng Lu
Jianfeng Ren
Jiangfan Liu
Jianguo Wang
Jiangyan Yi
Jiangyong Duan

Jianhua Yang
Jianhua Zhang
Jianhui Chen
Jianjia Wang
Jianli Xiao
Jianqiang Xiao
Jianwu Wang
Jianxin Zhang
Jianxiong Gao
Jianxiong Zhou
Jianyu Wang
Jianzhong Wang
Jiaru Zhang
Jiashu Liao
Jiaxin Chen
Jiaxin Lu
Jiaxing Ye
Jiaxuan Chen
Jiaxuan Li
Jiayi He
Jiayin Lin
Jie Ou
Jiehua Zhang
Jiejie Zhao
Jignesh S. Bhatt
Jin Gao
Jin Hou
Jin Hu
Jin Shang
Jing Tian
Jing Yu Chen
Jingfeng Yao
Jinglun Feng
Jingtong Yue
Jingwei Guo
Jingwen Xu
Jingyuan Xia
Jingzhe Ma
Jinhong Wang
Jinjia Wang
Jinlai Zhang
Jinlong Fan
Jinming Su
Jinrong He
Jintao Huang

Jinwoo Ahn
Jinwoo Choi
Jinyang Liu
Jinyu Tian
Jionghao Lin
Jiuding Duan
Jiwei Shen
Jiyan Pan
Jiyoun Kim
João Papa
Johan Debayle
John Atanbori
John Wilson
John Zhang
Jónathan Heras
Joohi Chauhan
Jorge Calvo-Zaragoza
Jorge Figueroa
Jorma Laaksonen
José Joaquim De Moura Ramos
Jose Vicent
Joseph Damilola Akinyemi
Josiane Zerubia
Juan Wen
Judit Szücs
Juepeng Zheng
Juha Roning
Jumana H. Alsubhi
Jun Cheng
Jun Ni
Jun Wan
Junghyun Cho
Junjie Liang
Junjie Ye
Junlin Hu
Juntong Ni
Junxin Lu
Junxuan Li
Junyaup Kim
Junyeong Kim
Jürgen Seiler
Jushang Qiu
Juyang Weng
Jyostna Devi Bodapati
Jyoti Singh Kirar

Kai Jiang
Kaiqiang Song
Kalidas Yeturu
Kalle Åström
Kamalakar Vijay Thakare
Kang Gu
Kang Ma
Kanji Tanaka
Karthik Seemakurthy
Kaushik Roy
Kavisha Jayathunge
Kazuki Uehara
Ke Shi
Keigo Kimura
Keiji Yanai
Kelton A. P. Costa
Kenneth Camilleri
Kenny Davila
Ketan Atul Bapat
Ketan Kotwal
Kevin Desai
Keyu Long
Khadiga Mohamed Ali
Khakon Das
Khan Muhammad
Kilho Son
Kim-Ngan Nguyen
Kishan Kc
Kishor P. Upla
Klaas Dijkstra
Komal Bharti
Konstantinos Triaridis
Kostas Ioannidis
Koyel Ghosh
Kripabandhu Ghosh
Krishnendu Ghosh
Kshitij S. Jadhav
Kuan Yan
Kun Ding
Kun Xia
Kun Zeng
Kunal Banerjee
Kunal Biswas
Kunchi Li
Kurban Ubul

Lahiru N. Wijayasingha
Laines Schmalwasser
Lakshman Mahto
Lala Shakti Swarup Ray
Lale Akarun
Lan Yan
Lawrence Amadi
Lee Kang Il
Lei Fan
Lei Shi
Lei Wang
Leonardo Rossi
Lequan Lin
Levente Tamas
Li Bing
Li Li
Li Ma
Li Song
Lia Morra
Liang Xie
Liang Zhao
Lianwen Jin
Libing Zeng
Lidia Sánchez-González
Lidong Zeng
Lijun Li
Likang Wang
Lili Zhao
Lin Chen
Lin Huang
Linfei Wang
Ling Lo
Lingchen Meng
Lingheng Meng
Lingxiao Li
Lingzhong Fan
Liqi Yan
Liqiang Jing
Lisa Gutzeit
Liu Ziyi
Liushuai Shi
Liviu-Daniel Stefan
Liyuan Ma
Liyun Zhu
Lizuo Jin

Longteng Guo
Lorena Álvarez Rodríguez
Lorenzo Putzu
Lu Leng
Lu Pang
Lu Wang
Luan Pham
Luc Brun
Luca Guarnera
Luca Piano
Lucas Alexandre Ramos
Lucas Goncalves
Lucas M. Gago
Luigi Celona
Luis C. S. Afonso
Luis Gerardo De La Fraga
Luis S. Luevano
Luis Teixeira
Lunke Fei
M. Hassaballah
Maddimsetti Srinivas
Mahendran N.
Mahesh Mohan M. R.
Maiko Lie
Mainak Singha
Makoto Hirose
Malay Bhattacharyya
Mamadou Dian Bah
Man Yao
Manali J. Patel
Manav Prabhakar
Manikandan V. M.
Manish Bhatt
Manjunath Shantharamu
Manuel Curado
Manuel Günther
Manuel Marques
Marc A. Kastner
Marc Chaumont
Marc Cheong
Marc Lalonde
Marco Cotogni
Marcos C. Santana
Mario Molinara
Mariofanna Milanova

Markus Bauer
Marlon Becker
Mårten Wadenbäck
Martin G. Ljungqvist
Martin Kampel
Martina Pastorino
Marwan Torki
Masashi Nishiyama
Masayuki Tanaka
Massimo O. Spata
Matteo Ferrara
Matthew D. Dawkins
Matthew Gadd
Matthew S. Watson
Maura Pintor
Max Ehrlich
Maxim Popov
Mayukh Das
Md Baharul Islam
Md Sajid
Meghna Kapoor
Meghna P. Ayyar
Mei Wang
Meiqi Wu
Melissa L. Tijink
Meng Li
Meng Liu
Meng-Luen Wu
Mengnan Liu
Mengxi China Guo
Mengya Han
Michaël Clément
Michal Kawulok
Mickael Coustaty
Miguel Domingo
Milind G. Padalkar
Ming Liu
Ming Ma
Mingchen Feng
Mingde Yao
Minghao Li
Mingjie Sun
Ming-Kuang Daniel Wu
Mingle Xu
Mingyong Li

Mingyuan Jiu
Minh P. Nguyen
Minh Q. Tran
Minheng Ni
Minsu Kim
Minyi Zhao
Mirko Paolo Barbato
Mo Zhou
Modesto Castrillón-Santana
Mohamed Amine Mezghich
Mohamed Dahmane
Mohamed Elsharkawy
Mohamed Yousuf
Mohammad Hashemi
Mohammad Khalooei
Mohammad Khateri
Mohammad Mahdi Dehshibi
Mohammad Sadil Khan
Mohammed Mahmoud
Moises Diaz
Monalisha Mahapatra
Monidipa Das
Mostafa Kamali Tabrizi
Mridul Ghosh
Mrinal Kanti Bhowmik
Muchao Ye
Mugalodi Ramesha Rakesh
Muhammad Rameez Ur Rahman
Muhammad Suhaib Kanroo
Muming Zhao
Munender Varshney
Munsif Ali
Na Lv
Nader Karimi
Nagabhushan Somraj
Nakkwan Choi
Nakul Agarwal
Nan Pu
Nan Zhou
Nancy Mehta
Nand Kumar Yadav
Nandakishor Nandakishor
Nandyala Hemachandra
Nanfeng Jiang
Narayan Hegde

Narayan Ji Mishra
Narayan Vetrekar
Narendra D. Londhe
Nathalie Girard
Nati Ofir
Naval Kishore Mehta
Nazmul Shahadat
Neeti Narayan
Neha Bhargava
Nemanja Djuric
Newlin Shebiah R.
Ngo Ba Hung
Nhat-Tan Bui
Niaz Ahmad
Nick Theisen
Nicolas Passat
Nicolas Ragot
Nicolas Sidere
Nikolaos Mitianoudis
Nikolas Ebert
Nilah Ravi Nair
Nilesh A. Ahuja
Nilkanta Sahu
Nils Murrugarra-Llerena
Nina S. T. Hirata
Ninad Aithal
Ning Xu
Ningzhi Wang
Niraj Kumar
Nirmal S. Punjabi
Nisha Varghese
Norio Tagawa
Obaidullah Md Sk
Oguzhan Ulucan
Olfa Mechi
Oliver Tüselmann
Orazio Pontorno
Oriol Ramos Terrades
Osman Akin
Ouadi Beya
Ozge Mercanoglu Sincan
Pabitra Mitra
Padmanabha Reddy Y. C. A.
Palaash Agrawal
Palaiahnakote Shivakumara

Palash Ghosal
Pallav Dutta
Paolo Rota
Paramanand Chandramouli
Paria Mehrani
Parth Agrawal
Partha Basuchowdhuri
Patrick Horain
Pavan Kumar
Pavan Kumar Anasosalu Vasu
Pedro Castro
Peipei Li
Peipei Yang
Peisong Shen
Peiyu Li
Peng Li
Pengfei He
Pengrui Quan
Pengxin Zeng
Pengyu Yan
Peter Eisert
Petra Gomez-Krämer
Pierrick Bruneau
Ping Cao
Pingping Zhang
Pintu Kumar
Pooja Kumari
Pooja Sahani
Prabhu Prasad Dev
Pradeep Kumar
Pradeep Singh
Pranjal Sahu
Prasun Roy
Prateek Keserwani
Prateek Mittal
Praveen Kumar Chandaliya
Praveen Tirupattur
Pravin Nair
Preeti Gopal
Preety Singh
Prem Shanker Yadav
Prerana Mukherjee
Prerna A. Mishra
Prianka Dey
Priyanka Mudgal

Qc Kha Ng
Qi Li
Qi Ming
Qi Wang
Qi Zuo
Qian Li
Qiang Gan
Qiang He
Qiang Wu
Qiangqiang Zhou
Qianli Zhao
Qiansen Hong
Qiao Wang
Qidong Huang
Qihua Dong
Qin Yuke
Qing Guo
Qingbei Guo
Qingchao Zhang
Qingjie Liu
Qinhong Yang
Qiushi Shi
Qixiang Chen
Quan Gan
Quanlong Guan
Rachit Chhaya
Radu Tudor Ionescu
Rafal Zdunek
Raghavendra Ramachandra
Rahimul I. Mazumdar
Rahul Kumar Ray
Rajib Dutta
Rajib Ghosh
Rakesh Kumar
Rakesh Paul
Rama Chellappa
Rami O. Skaik
Ramon Aranda
Ran Wei
Ranga Raju Vatsavai
Ranganath Krishnan
Rasha Friji
Rashmi S.
Razaib Tariq
Rémi Giraud

René Schuster
Renlong Hang
Renrong Shao
Renu Sharma
Reza Sadeghian
Richard Zanibbi
Rimon Elias
Rishabh Shukla
Rita Delussu
Riya Verma
Robert J. Ravier
Robert Sablatnig
Robin Strand
Rocco Pietrini
Rocio Diaz Martin
Rocio Gonzalez-Diaz
Rohit Venkata Sai Dulam
Romain Giot
Romi Banerjee
Ru Wang
Ruben Machucho
Ruddy Théodose
Ruggero Pintus
Rui Deng
Rui P. Paiva
Rui Zhao
Ruifan Li
Ruigang Fu
Ruikun Li
Ruirui Li
Ruixiang Jiang
Ruowei Jiang
Rushi Lan
Rustam Zhumagambetov
S. Amutha
S. Divakar Bhat
Sagar Goyal
Sahar Siddiqui
Sahbi Bahroun
Sai Karthikeya Vemuri
Saibal Dutta
Saihui Hou
Sajad Ahmad Rather
Saksham Aggarwal
Sakthi U.

Salimeh Sekeh
Samar Bouazizi
Samia Boukir
Samir F. Harb
Samit Biswas
Samrat Mukhopadhyay
Samriddha Sanyal
Sandika Biswas
Sandip Purnapatra
Sanghyun Jo
Sangwoo Cho
Sanjay Kumar
Sankaran Iyer
Sanket Biswas
Santanu Roy
Santosh D. Pandure
Santosh Ku Behera
Santosh Nanabhau Palaskar
Santosh Prakash Chouhan
Sarah S. Alotaibi
Sasanka Katreddi
Sathyanarayanan N. Aakur
Saurabh Yadav
Sayan Rakshit
Scott McCloskey
Sebastian Bunda
Sejuti Rahman
Selim Aksoy
Sen Wang
Seraj A. Mostafa
Shanmuganathan Raman
Shao-Yuan Lo
Shaoyuan Xu
Sharia Arfin Tanim
Shehreen Azad
Sheng Wan
Shengdong Zhang
Shengwei Qin
Shenyuan Gao
Sherry X. Chen
Shibaprasad Sen
Shigeaki Namiki
Shiguang Liu
Shijie Ma
Shikun Li

Shinichiro Omachi
Shirley David
Shishir Shah
Shiv Ram Dubey
Shiva Baghel
Shivanand S. Gornale
Shogo Sato
Shotaro Miwa
Shreya Ghosh
Shreya Goyal
Shuai Su
Shuai Wang
Shuai Zheng
Shuaifeng Zhi
Shuang Qiu
Shuhei Tarashima
Shujing Lyu
Shuliang Wang
Shun Zhang
Shunming Li
Shunxin Wang
Shuping Zhao
Shuquan Ye
Shuwei Huo
Shuyue Lan
Shyi-Chyi Cheng
Si Chen
Siddarth Ravichandran
Sihan Chen
Siladittya Manna
Silambarasan Elkana Ebinazer
Simon Benaïchouche
Simon S. Woo
Simone Caldarella
Simone Milani
Simone Zini
Sina Lotfian
Sitao Luan
Sivaselvan B.
Siwei Li
Siwei Wang
Siwen Luo
Siyu Chen
Sk Aziz Ali
Sk Md Obaidullah

Sneha Shukla
Snehasis Banerjee
Snehasis Mukherjee
Snigdha Sen
Sofia Casarin
Soheila Farokhi
Soma Bandyopadhyay
Son Minh Nguyen
Son Xuan Ha
Sonal Kumar
Sonam Gupta
Sonam Nahar
Song Ouyang
Sotiris Kotsiantis
Souhaila Djaffal
Soumen Biswas
Soumen Sinha
Soumitri Chattopadhyay
Souvik Sengupta
Spiros Kostopoulos
Sreeraj Ramachandran
Sreya Banerjee
Srikanta Pal
Srinivas Arukonda
Stephane A. Guinard
Su O. Ruan
Subhadip Basu
Subhajit Paul
Subhankar Ghosh
Subhankar Mishra
Subhankar Roy
Subhash Chandra Pal
Subhayu Ghosh
Sudip Das
Sudipta Banerjee
Suhas Pillai
Sujit Das
Sukalpa Chanda
Sukhendu Das
Suklav Ghosh
Suman K. Ghosh
Suman Samui
Sumit Mishra
Sungho Suh
Sunny Gupta

Suraj Kumar Pandey
Surendrabikram Thapa
Suresh Sundaram
Sushil Bhattacharjee
Susmita Ghosh
Swakkhar Shatabda
Syed Ms Islam
Syed Tousiful Haque
Taegyeong Lee
Taihui Li
Takashi Shibata
Takeshi Oishi
Talha Ahmad Siddiqui
Tanguy Gernot
Tangwen Qian
Tanima Bhowmik
Tanpia Tasnim
Tao Dai
Tao Hu
Tao Sun
Taoran Yi
Tapan Shah
Taveena Lotey
Teng Huang
Tengqi Ye
Teresa Alarcon
Tetsuji Ogawa
Thanh Phuong Nguyen
Thanh Tuan Nguyen
Thattapon Surasak
Thibault Napolão
Thierry Bouwmans
Thinh Truong Huynh Nguyen
Thomas De Min
Thomas E. K. Zielke
Thomas Swearingen
Tianatahina Jimmy Francky Randrianasoa
Tianheng Cheng
Tianjiao He
Tianyi Wei
Tianyuan Zhang
Tianyue Zheng
Tiecheng Song
Tilottama Goswami
Tim Büchner

Tim H. Langer
Tim Raven
Tingkai Liu
Tingting Yao
Tobias Meisen
Toby P. Breckon
Tong Chen
Tonghua Su
Tran Tuan Anh
Tri-Cong Pham
Trishna Saikia
Trung Quang Truong
Tuan T. Nguyen
Tuan Vo Van
Tushar Shinde
Ujjwal Karn
Ukrit Watchareeruetai
Uma Mudenagudi
Umarani Jayaraman
V. S. Malemath
Vallidevi Krishnamurthy
Ved Prakash
Venkata Krishna Kishore Kolli
Venkata R. Vavilthota
Venkatesh Thirugnana Sambandham
Verónica Maria Vasconcelos
Véronique Ve Eglin
Víctor E. Alonso-Pérez
Vinay Palakkode
Vinayak S. Nageli
Vincent J. Whannou De Dravo
Vincenzo Conti
Vincenzo Gattulli
Vineet Padmanabhan
Vishakha Pareek
Viswanath Gopalakrishnan
Vivek Singh Baghel
Vivekraj K.
Vladimir V. Arlazarov
Vu-Hoang Tran
W. Sylvia Lilly Jebarani
Wachirawit Ponghiran
Wafa Khlif
Wang An-Zhi
Wanli Xue

Wataru Ohyama
Wee Kheng Leow
Wei Chen
Wei Cheng
Wei Hua
Wei Lu
Wei Pan
Wei Tian
Wei Wang
Wei Wei
Wei Zhou
Weidi Liu
Weidong Yang
Weijun Tan
Weimin Lyu
Weinan Guan
Weining Wang
Weiqiang Wang
Weiwei Guo
Weixia Zhang
Wei-Xuan Bao
Weizhong Jiang
Wen Xie
Wenbin Qian
Wenbin Tian
Wenbin Wang
Wenbo Zheng
Wenhan Luo
Wenhao Wang
Wen-Hung Liao
Wenjie Li
Wenkui Yang
Wenwen Si
Wenwen Yu
Wenwen Zhang
Wenwu Yang
Wenxi Li
Wenxi Yue
Wenxue Cui
Wenzhuo Liu
Widhiyo Sudiyono
Willem Dijkstra
Wolfgang Fuhl
Xi Zhang
Xia Yuan

Xianda Zhang
Xiang Zhang
Xiangdong Su
Xiang-Ru Yu
Xiangtai Li
Xiangyu Xu
Xiao Guo
Xiao Hu
Xiao Wu
Xiao Yang
Xiaofeng Zhang
Xiaogang Du
Xiaoguang Zhao
Xiaoheng Jiang
Xiaohong Zhang
Xiaohua Huang
Xiaohua Li
Xiao-Hui Li
Xiaolong Sun
Xiaosong Li
Xiaotian Li
Xiaoting Wu
Xiaotong Luo
Xiaoyan Li
Xiaoyang Kang
Xiaoyi Dong
Xin Guo
Xin Lin
Xin Ma
Xinchi Zhou
Xingguang Zhang
Xingjian Leng
Xingpeng Zhang
Xingzheng Lyu
Xinjian Huang
Xinqi Fan
Xinqi Liu
Xinqiao Zhang
Xinrui Cui
Xizhan Gao
Xu Cao
Xu Ouyang
Xu Zhao
Xuan Shen
Xuan Zhou

Xuchen Li
Xuejing Lei
Xuelu Feng
Xueting Liu
Xuewei Li
Xueyi X. Wang
Xugong Qin
Xu-Qian Fan
Xuxu Liu
Xu-Yao Zhang
Yan Huang
Yan Li
Yan Wang
Yan Xia
Yan Zhuang
Yanan Li
Yanan Zhang
Yang Hou
Yang Jiao
Yang Liping
Yang Liu
Yang Qian
Yang Yang
Yang Zhao
Yangbin Chen
Yangfan Zhou
Yanhui Guo
Yanjia Huang
Yanjun Zhu
Yanming Zhang
Yanqing Shen
Yaoming Cai
Yaoxin Zhuo
Yaoyan Zheng
Yaping Zhang
Yaqian Liang
Yarong Feng
Yasmina Benmabrouk
Yasufumi Sakai
Yasutomo Kawanishi
Yazeed Alzahrani
Ye Du
Ye Duan
Yechao Zhang
Yeong-Jun Cho

Yi Huo
Yi Shi
Yi Yu
Yi Zhang
Yibo Liu
Yibo Wang
Yi-Chieh Wu
Yifan Chen
Yifei Huang
Yihao Ding
Yijie Tang
Yikun Bai
Yimin Wen
Yinan Yang
Yin-Dong Zheng
Yinfeng Yu
Ying Dai
Yingbo Li
Yiqiao Li
Yiqing Huang
Yisheng Lv
Yisong Xiao
Yite Wang
Yizhe Li
Yong Wang
Yonghao Dong
Yong-Hyuk Moon
Yongjie Li
Yongqian Li
Yongqiang Mao
Yongxu Liu
Yongyu Wang
Yongzhi Li
Youngha Hwang
Yousri Kessentini
Yu Wang
Yu Zhou
Yuan Tian
Yuan Zhang
Yuanbo Wen
Yuanxin Wang
Yubin Hu
Yubo Huang
Yuchen Ren
Yucheng Xing

Yuchong Yao
Yuecong Min
Yuewei Yang
Yufei Zhang
Yufeng Yin
Yugen Yi
Yuhang Ming
Yujia Zhang
Yujun Ma
Yukiko Kenmochi
Yun Hoyeoung
Yun Liu
Yunhe Feng
Yunxiao Shi
Yuru Wang
Yushun Tang
Yusuf Osmanlioglu
Yusuke Fujita
Yuta Nakashima
Yuwei Yang
Yuwu Lu
Yuxi Liu
Yuya Obinata
Yuyao Yan
Yuzhi Guo
Zaipeng Xie
Zander W. Blasingame
Zedong Wang
Zeliang Zhang
Zexin Ji
Zhanxiang Feng
Zhaofei Yu
Zhe Chen
Zhe Cui
Zhe Liu
Zhe Wang
Zhekun Luo
Zhen Yang
Zhenbo Li
Zhenchun Lei
Zhenfei Zhang
Zheng Liu
Zheng Wang
Zhengming Yu
Zhengyin Du

Zhengyun Cheng
Zhenshen Qu
Zhenwei Shi
Zhenzhong Kuang
Zhi Cai
Zhi Chen
Zhibo Chu
Zhicun Yin
Zhida Huang
Zhida Zhang
Zhifan Gao
Zhihang Ren
Zhihang Yuan
Zhihao Wang
Zhihua Xie
Zhihui Wang
Zhikang Zhang
Zhiming Zou
Zhiqi Shao
Zhiwei Dong
Zhiwei Qi
Zhixiang Wang
Zhixuan Li
Zhiyu Jiang
Zhiyuan Yan
Zhiyuan Yu
Zhiyuan Zhang
Zhong Chen
Zhongwei Teng
Zhongzhan Huang
Zhongzhi Yu
Zhuan Han
Zhuangzhuang Chen
Zhuo Liu
Zhuo Su
Zhuojun Zou
Zhuoyue Wang
Ziang Song
Zicheng Zhang
Zied Mnasri
Zifan Chen
Žiga Babnik
Zijing Chen
Zikai Zhang
Ziling Huang
Zilong Du
Ziqi Cai
Ziqi Zhou
Zi-Rui Wang
Zirui Zhou
Ziwen He
Ziyao Zeng
Ziyi Zhang
Ziyue Xiang
Zonglei Jing
Zongyi Xu

Contents – Part XII

AD-Lite Net: A Lightweight and Concatenated CNN Model
for Alzheimer's Detection from MRI Images 1
    Santanu Roy, Archit Gupta, Shubhi Tiwari, and Palak Sahu

Leveraging Persistent Homology for Differential Diagnosis of Mild
Cognitive Impairment 17
    *Ninad Aithal, Debanjali Bhattacharya, Neelam Sinha,
and Thomas Gregor Issac*

Medical Image Classification Attack Based on Texture Manipulation 33
    Yunrui Gu, Cong Kong, Zhaoxia Yin, Yan Wang, and Qingli Li

An Attention Transformer-Based Method for the Modelling of Functional
Connectivity and the Diagnosis of Autism Spectrum Disorder 49
    *Ge Yang, Linbo Qing, Yanteng Zhang, Feng Gao, Li Gao, Xiaohai He,
and Yonghong Peng*

A Fast Domain-Inspired Unsupervised Method to Compute COVID-19
Severity Scores from Lung CT 60
    *Samiran Dey, Bijon Kundu, Partha Basuchowdhuri,
Sanjoy Kumar Saha, and Tapabrata Chakraborti*

Privacy-Preserving Tabular Data Generation: Application to Sepsis
Detection 75
    *Eric Macias-Fassio, Aythami Morales, Cristina Pruenza,
and Julian Fierrez*

SVD-Grad-CAM: Singular Value Decomposition filtered Gradient
Weighted Class Activation Map 90
    Gokaramaiah Thota, K. Nagaraju, and Sathya Babu Korra

Applying Layer-Wise Relevance Propagation on U-Net Architectures 106
    *Patrick Weinberger, Bernhard Fröhler, Anja Heim, Alexander Gall,
Ulrich Bodenhofer, and Sascha Senck*

Visualizing Dynamics of Federated Medical Models via Conversational
Memory Elements 122
    *Sanidhya Kumar, Varun Vilvadrinath, Jignesh S. Bhatt,
and Ashish Phophalia*

Generating Counterfactual Trajectories with Latent Diffusion Models
for Concept Discovery 138
    Payal Varshney, Adriano Lucieri, Christoph Balada, Andreas Dengel,
    and Sheraz Ahmed

Fusing Forces: Deep-Human-Guided Refinement of Segmentation Masks 154
    Rafael Sterzinger, Christian Stippel, and Robert Sablatnig

Specular Region Detection and Covariant Feature Extraction 170
    D. M. Bappy, Donghwa Kang, Jinkyu Lee, Youngmoon Lee,
    Minsuk Koo, and Hyeongboo Baek

Low-Rank Adaptation of Segment Anything Model for Surgical Scene
Segmentation 187
    Jay N. Paranjape, Shameema Sikder, S. Swaroop Vedula,
    and Vishal M. Patel

DRIVPocket: A Dual-stream Rotation Invariance in Feature Sampling
and Voxel Fusion Approach for Protein Binding Site Prediction 203
    Bowen Deng, Yang Hua, Wenjie Zhang, Xiaoning Song, and Xiao-jun Wu

TotalCT-SAM: A Whole-Body CT Segment Anything Model
with Memorizing Transformer 220
    Zhiwei Zhang and Yiqing Shen

Multi-modal Multitask Learning Model for Simultaneous Classification
of Two Epilepsy Biomarkers 235
    Nawara Mahmood Broti, Masaki Sawada, Yutaro Takayama,
    Keiya Iijima, Masaki Iwasaki, and Yumie Ono

Learning Neural Networks for Multi-label Medical Image Retrieval Using
Hamming Distance Fabricated with Jaccard Similarity Coefficient 251
    Asim Manna and Debdoot Sheet

Adaptive Cross-Modal Representation Learning for Heterogeneous Data
Types in Alzheimer Disease Progression Prediction with Missing Time
Point and Modalities 267
    S. P. Dhivyaa, Duy-Phuong Dao, Hyung-Jeong Yang, and Jahae Kim

Enhancing Medical Image Analysis with MA-DTNet: A Dual Task
Network Guided by Morphological Attention 283
    Susmita Ghosh and Swagatam Das

TransNetOCT: An Efficient Transformer-Based Model for 3D-OCT
Segmentation Using Prior Shape 301
    Mohamed Elsharkawy, Ibrahim Abdelhalim, Mohammed Ghazal,
    Mohammad Z. Haq, Rayan Haq, Ali Mahmoud, Harpal S. Sandhu,
    Aristomenis Thanos, and Ayman El-Baz

On the Importance of Local and Global Feature Learning for Automated
Measurable Residual Disease Detection in Flow Cytometry Data 316
    Lisa Weijler, Michael Reiter, Pedro Hermosilla,
    Margarita Maurer-Granofszky, and Michael Dworzak

Self-supervised Siamese Network Using Vision Transformer for Depth
Estimation in Endoscopic Surgeries 332
    Snigdha Agarwal and Neelam Sinha

Enhanced 3D Dense U-Net with Two Independent Teachers for Infant
Brain Image Segmentation 345
    Afifa Khaled and Ahmed Elazab

IDQCE: Instance Discrimination Learning Through Quantized Contextual
Embeddings for Medical Images 360
    Azad Singh and Deepak Mishra

Superpixel-Based Sparse Labeling for Efficient and Certain Medical
Image Annotation 376
    Somayeh Rezaei and Xiaoyi Jiang

Attention Seekers U-Net with Mamba for Sub-cellular Segmentation 391
    Pratik Sinha and Arif Ahmed Sekh

Cross-Domain Multi-contrast MR Image Synthesis via Generative
Adversarial Network 408
    Guowen Wang, Silei Wang, Lu Wang, Congbo Cai, Shuhui Cai,
    and Zhong Chen

Fusion of Machine Learning and Deep Neural Networks for Pulmonary
Arteries and Veins Segmentation in Lung Cancer Surgery Planning 422
    Hongyu Cheng, Limin Zheng, Zeyu Yan, Haoran Zhang, Bo Meng,
    and Xiaowei Xu

SEANet: Rethinking Skip-Connections Design in Encoder-Decoder
Networks via Synergistic Spatial-Spectral Fusion for LDCT Denoising 439
    Abhijit Das, Vandan Gorade, Dwarikanath Mahapatra, and Sudipta Roy

Extracting Vitals from ICU Monitor Images: An Insight from Analysis of 10K Patient Data 455
   Akshat Rampuria, Kushagra Khare, Ayush Soni, and Debi Prosad Dogra

Author Index 473

AD-Lite Net: A Lightweight and Concatenated CNN Model for Alzheimer's Detection from MRI Images

Santanu Roy[1]( ), Archit Gupta[2], Shubhi Tiwari[2], and Palak Sahu[2]

[1] Department of CSE, Pandit Deendayal Energy University (PDEU), Gandhinagar, India
santanuroy35@gmail.com
[2] Department of CSE, NIIT University, Neemrana, Rajasthan, India
palak.sahu20@st.niituniversity.in
http://www.springer.com/gp/computer-science/lncs

Abstract. Alzheimer's Disease (AD) is a non-curable progressive neurodegenerative disorder that affects the human brain, leading to a decline in memory, cognitive abilities, and eventually, the ability to carry out daily tasks. Manual diagnosis of Alzheimer's disease from MRI images is fraught with less sensitivity and it is a very tedious process for neurologists. Therefore, there is a need for an automatic Computer Assisted Diagnosis (CAD) system, which can detect AD at early stages with higher accuracy. Until now, numerous researchers have proposed several deep-learning models to detect AD efficiently from MRI datasets. However, most of their methods have deployed lots of pre-processing and image-processing techniques, which yields a lack of generalization in the model performance. In this research, we have proposed a novel AD-Lite Net model (trained from scratch), that could alleviate the aforementioned problem. The novelties we bring here in this research are, (I) We have proposed a very lightweight CNN model by incorporating Depth Wise Separable Convolutional (DWSC) layers and Global Average Pooling (GAP) layers. (II) We have leveraged a "parallel concatenation block" (pcb), in the proposed AD-Lite Net model. This pcb consists of a Transformation layer (Tx-layer), followed by two convolutional layers, which are thereby concatenated with the original base model. This Tx-layer converts the features into very distinct kind of features, which are imperative for the Alzheimer's disease. As a consequence, the proposed AD-Lite Net model with "parallel concatenation" converges faster and automatically mitigates the class imbalance problem from the MRI datasets in a very generalized way. For the validity of our proposed model, we have implemented it on three different MRI datasets. Furthermore, we have combined the ADNI and AD datasets and subsequently performed a 10-fold cross-validation experiment to verify the model's generalization ability. Extensive experimental results showed that our proposed model has outperformed all the existing CNN models, and one recent trend Vision Transformer (ViT) model by a significant margin.

Supported by organization x.

© The Author(s), under exclusive license to Springer Nature Switzerland AG 2025
A. Antonacopoulos et al. (Eds.): ICPR 2024, LNCS 15312, pp. 1–16, 2025.
https://doi.org/10.1007/978-3-031-78198-8_1

Keywords: Alzheimer's Disease Detection · Magnetic Resonance Imaging (MRI) Images · Convolutional Neural Network (CNN) · Attention-based Models · Vision Transformer (ViT)

1 Introduction

Alzheimer's Disease (AD) is a severe, and fatal neurodegenerative disease [1] that usually targets older individuals. The early signs of Alzheimer's are forgetting recent events, language issues, having problems with reasoning and gradually it leads to loss of one's ability to perform everyday tasks. AD occurs due to abnormal protein accumulation including beta amyloid plaques and tau tangles in the brain. These changes cause mental deterioration since nerve cells are lost gradually and the connections between brain cells and communication get disrupted. Alzheimer's Disease International (ADI) has estimated that dementia affects more than 50 million people across the world [2], which is a term that refers to symptoms of brain impairment. AD is the leading cause of dementia and accounts for 60-80% of cases. AD affects particular structures within the brain, the hippocampus [3], which is one of those first attacked by AD. Neural changes in the hippocampus's anatomy can be identified by measuring its volume and form, as well as that of gray matter substance with highly advanced imaging techniques such as Computed Tomography (CT), Positron Emission Tomography (PET) and Magnetic Resonance Imaging (MRI). Out of all these image-acquiring techniques, MRI is the most frequently employed. Because it is noninvasive and easily available, moreover, it causes less radiation to the human body. Examining the alterations in the Cerebrospinal Fluid System (CFS) [4] aids in identifying the phase of AD. As this disease progresses, there is an enlargement in CFS region and reduction of the cerebral cortex and hippocampus. At present, there is no effective treatment available for Alzheimer's disease (AD), and the only way to prevent it is through early detection, as modern methods can only delay the course of progression. However, manually extracting and interpreting the features of Alzheimer's disease and furthermore, classifying them into different grades (from MRI images), is a very tedious and complex task for Neurologists. Hence, there is a need for an automatic CAD system, in order to detect AD efficiently from MRI images.

Various deep learning models have been widely employed recently by numerous researchers, in order to develop an automatic CAD system of AD detection from MRI images. Modupe Odusam et al. [5] proposed a pre-trained ResNet-18 which detects Alzheimer's disease from MRI images at an accuracy of around 98-99%. However, they considered any two classes, thus, their classification problem (binary) was slightly lesser complex than the multi-class classification. Hadeer A. Helaly et al. [6] proposed a CNN model E2AD2C (trained from scratch) which is comprised of 3 convolutional layers, 2 Fully Connected (FC) layers, and 1 output layer. Their model architecture had less number of hyper-parameters (to train) and was inspired by the standard VGG-16 model. Nevertheless, they have deployed many pre-processing techniques, for example, over-sampling and under-sampling methods, data-augmentation, MRI filtering and normalization etc. prior to feeding the data into a classifier. Shakarami et al. [7] proposed an AlexNet-SVM model in order to predict Alzheimer's disease from PET images. Their method encompasses four different steps. (I) First, 3D PET images are converted into 2D slices (or, images), (II) The pixels (in 2D slices) have values more than 150 are only passed through, otherwise avoiding all other pixels. (III) AlexNet-SVM model is utilized for the feature extraction part, and (IV) the final classification is done by the majority voting on slices. Although their method seems like a reasonable method, after converting 3D images into 2D slices, it may lose some important information, thus, it is not so feasible. K.G. Achilleos et al. [8] proposed a manual feature extraction method, in which they had computed Haralick texture [9] features for hippocampal atrophy which is the most vital part for predicting AD from MRI images. Moreover, they combined these hippocampal textures with their volume and subsequently, they applied all these features to a 10-fold cross-validation Decision Tree (for a 4-class classification task). Another potential direction of approaching this imbalanced MRI datasets is to deploy Weighted Categorical Cross Entropy (WCCE) [10] which assigns weights for every class which is inversely proportional to the number of images in that class. M. Masud et al. [11] have employed similar WCCE on top of a lightweight CNN model in order to resolve the issue of class imbalance from MRI datasets for AD detection. Besides that, many more related research works can be found in [12]-[15].

Another valid direction of this research could be leveraging new recent trends, that is, self-supervised models [16] or, attention-based models, in order to alleviate class imbalance problem from these MRI datasets. Numerous self-attention transformer models have been widely popular and proposed in the domain of NLP [17]. However, their equivalent model, i.e., Vision Transformer (ViT) [18], still is not an automatic choice for researchers in the domain of computer vision or image classification. The reason why still CNN outperforms ViT is that, ViT needs larger data in order to generalize well, however, in most of the medical image diagnoses, we have weakly supervised data or very limited imbalanced data. Moreover, unlike CNN model, ViT does not leverage a multi-scale hierarchical structure [16] which has a special significance for image classification. Therefore, numerous researchers [19],[20] come up with the idea of integrating both of the notions of ViT and CNN simultaneously. Recently, Byeongho Heo et al. [20] have proposed a Pooling-based ViT (PiT), which incorporates pooling layers in the ViT model. This leverages a multi-scale hierarchical architecture in the ViT, moreover, due to utilizing many pooling layers the number of hyperparameters in PiT has been drastically reduced. Numerous researchers also tried to incorporate equivalent channel attention [21–23] named Squeeze Attention or, Swin Transformer [24] on top of CNN model, in order to improve the efficacy of AD detection from MRI images. Jiayi Zhu et al. [21] proposed a Sparse self-attention block in order to detect Alzheimer's disease at early stages, from MRI images. This "Sparse self-attention block" can reduce the elements (by $logN$) that can represent the overall features N. Therefore, overall, the computational complexity of their model (called BraInf) has been considerably reduced. Z. Liu et al. [22] have proposed a novel Multi-Scale Convolutional Network (MSC-Net) comprising four parallel concatenations of convolutional layers with varying dilation rates. Additionally, they have integrated an attention module "SE-Net" into their MSC-Net to enhance channel independence.

We have observed that most of the aforementioned state-of-the-art models [5–14] struggle to generalize across different MRI datasets for Alzheimer's detection. These models particularly exhibit overfitting when dealing with imbalanced and small datasets. Researchers utilize image processing techniques as pre-processing methods [6,7] to augment datasets in order to improve the efficacy of the deep learning model. However, while these techniques may work well on a specific dataset, they do not ensure effective generalization across diverse datasets. Furthermore, several attention modules [21–24] proposed for AD detection could not directly address the issue of class imbalance. Therefore, in this research, we aim to develop a lightweight CNN model (trained from scratch), specifically designed for Alzheimer's detection, such that it can alleviate the class imbalance problem and generalize well across diverse MRI datasets.

1.1 MRI Images Dataset and Its Challenges

For extensive experimentation, we have employed 3 MRI datasets which are readily available on Kaggle. The first dataset of Alzheimer's Disease [25] contains a total of 5000 images which are labeled further into 4 classes - Mild-Demented, Moderate-Demented, Non-Demented, and Very Mild-Demented. We call this dataset "Alzheimer's Detection (AD) dataset". Here, Moderate-Demented is severely demented and is analogous to Alzheimer's Disease (AD). Whereas, Mild-Demented and very Mild-Demented are early stages of Alzheimer's Disease. The number of images in Mild-Demented, Moderate-Demented, Non-Demented, and Very Mild-Demented are 717, 52, 2560, and 1792 respectively. A Second dataset, named "ADNI-Extracted-Axial", consists of 2D axial images extracted from the Nifti ADNI from ADNI website [26]. This ADNI dataset is the most authentic MRI dataset for Alzheimer's disease, followed by numerous researchers. This ADNI contains 5000 images which are further divided into 3 classes - Alzheimer's Disease (AD), Mild Cognitive Impaired (CI), and Common Normal (CN). A third dataset OASIS [26] of four classes, is also utilized in this research. The number of images in Mild Dementia, Moderate Dementia, Non- Dementia, and Very mild Dementia are 5002, 488, 67200 and 13725 respectively. Hence, this is a huge class imbalance problem and conventional CNN models' efficacy may suffer due to the lack of generalizing ability in the minor classes.

1.2 Contributions

The contributions of this paper are as follows:

1. A very lightweight CNN model, AD-Lite Net, has been proposed as a base model for detecting Alzheimer's disease efficiently, from MRI images dataset.

2. A "parallel concatenation block" is incorporated on top of this base model in order to alleviate the class imbalance problem and to increase generalization ability of the model. In this "parallel concatenation block" (pcb), one Transformation layer (Tx-layer) is employed which enables the model to extract distinct and complementary features which were essential for Alzheimer's detection.
3. A mathematical analysis of the proposed model AD-Lite Net is presented in this research. In this analysis, one new *lemma* has also been proposed.
4. For validity purpose, the proposed AD-Lite Net has been implemented on three different MRI image datasets. Moreover, we merged the ADNI and AD datasets and subsequently conducted a 10-fold cross-validation experiment to test the model's generalization ability.

2 Methodology

This methodology section can be further divided into two parts: (a) Alzheimer's Detection Lite Network (AD-Lite Net), (b) Mathematical Analysis of AD-Lite Net.

2.1 Alzheimer's Detection-Lite Network (AD-Lite Net)

The proposed AD-Lite Net model is explored in Fig.1. The proposed model is comprised of main two parts: (I) Main backbone CNN model (which is a very lightweight model or base model), (II) One parallel concatenation block is leveraged into this backbone CNN model in order to increase the generalization ability of the model. Overall, in the proposed framework, a total of 7 convolutional layers and two Depth-wise Separable Convolutional (DWSC) layers [27] are employed, as shown in Fig.1. The number of filters deployed in the backbone model are 16, 32, 64, 96, and 128 from the 1st to 5th convolutional layer respectively. Every convolution layer has the same kernel size 3×3 (except the 1st one having kernel size 5×5) with zero padding "same". ReLU activation function is employed in all the convolutional layers, whereas, SoftMax activation function is incorporated in the output of the CNN model. Each convolutional layer is followed by a Max-pooling layer, which down-samples the image size by half, because of using stride 2. Subsequently, a batch normalization layer is also incorporated after every Max-pooling layer or convolution layer, in the model. This batch normalization layer converts the scattered 2D tensor input (after convolution) into a normalized distribution having mean 0 and standard deviation 1. It ensures a smooth gradient flow throughout the network and hence, reduces the over-fitting problem, to a certain extent.

The "parallel concatenation" block starts from a transformation layer (or, Tx-layer) which converts the tensor output (coming from the 3rd convolutional block) into a very different kind of image (i.e., negative image). This is further shown in Fig.2. This tx-layer is further followed by 2 convolutional layers and 2

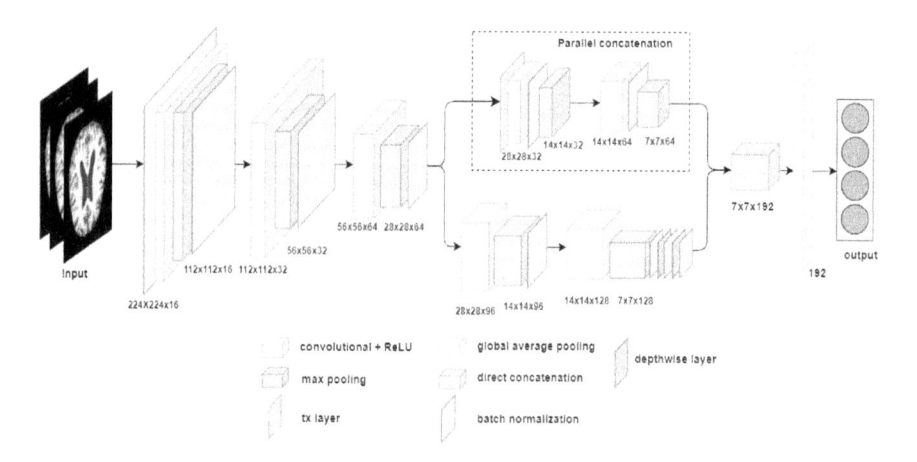

Fig. 1. Block diagram of the proposed model AD-Lite Net

Max-pooling layers. These two back-to-back convolutional layers have the number of filters 32 and 64 respectively. These numbers are chosen empirically, which is further explored in an ablation study in Supplementary material. This parallel concatenation block (pcb) can work like like an equivalent 'Attention block' in the CNN model, which is further exploited in the next subsection. Thereafter, these two parallel blocks are concatenated by a concatenation block which is followed by Global Average Pooling (GAP) Layer [28] and output layer, as shown in Fig.1. This is to clarify that DWSC layers and GAP layers (instead of flatten layer) have reduced the computational complexity of the AD-Lite Net considerably. Moreover, due to avoiding the entire dense layer part, the number of hyper-parameters of this AD-Lite Net is reduced to only 2.3 lakhs (approximately), hence, the proposed framework can work efficiently even on a very small and imbalanced dataset without being affected much by overfitting.

2.2 Mathematical Analysis of AD Lite-Net

A mathematical analysis of AD Lite-Net is presented in this section, in order to understand the credibility of the proposed research with much clarity.

The convoluted tensor output (after any convolutional layer) in our proposed model, can be represented by

$$O_i(f)_{w\times w} = ReLU((\sum_{j=1}^{p_i} O_j(z)_{3\times 3} * I(f))_{w\times w} + b) \tag{1}$$

Here in equation (1), p_i is the number of filters in the current convolutional layer, $I(f)$ is the original image having size w×w, $O_j(z)$ is the convolutional filter, having kernel size 3×3 or 5×5, and the same stride=1, with zero padding "same". Thus, the size of the convoluted output will be also the same, i.e., $w \times w$, b is the bias, '∗' in equation (1) indicates convolution operation.

The number of hyper-parameters $h_{c,i}$ in this i^{th} convolutional layer can be computed by the following equation.

$$h_{c,i} = (3^2 . p_{i-1} + 1).p_i \qquad (2)$$

Here, in equation (2), p_{i-1} is the number of filters in the previous layer, '.' indicates point-wise multiplication.

On the other hand, the number of hyper-parameters $h_{D,i}$ in this i^{th} DWCS layer is represented in equation (3). Comparing equation (2) and (3), we can conclude that $h_{D,i} \ll h_{c,i}$ if p_i is higher, because DWSC utilizes only one 3×3 convolution layer followed by 1×1 layers [27] that does point-wise multiplication.

$$h_{D,i} = (3^2 . 1 + 1).p_{i-1} = 10 p_{i-1} \qquad (3)$$

We have employed 2 such DWSC layers at the last block (as shown in Fig.1) such that the number of hyper-parameters will not be raised significantly.

The Max-pooling with stride 2 (and pool size 2×2), is a down-sampling operation [18] that would reduce the original image size to its half. After utilizing a total n number of Max-pooling layers, the tensor output will be

$$(Max^n(I(f)))_{w \times w}{}_{2 \times 2 | 2} = (O_n(f))_{(w/2^n . w/2^n)} \qquad (4)$$

Here, in our proposed model, $n = 5$. Thus, the spatial dimension of the output will be $224/2^5$ x $224/2^5$ = 7 × 7. The spectral dimension in this last block is $(64 + 128) = 192$, shown in the Fig.1. This last layer is passed through the GAP layer, instead of flatten layer. This GAP layer [28] takes an average in the spatial dimension, thus, the number of neurons in this GAP is reduced to 192 only. Whereas, the number of neurons in the flatten layer would be 7 × 7 × 192. Thus, the number of neurons has decreased considerably, after leveraging GAP in the proposed model. This will have a significant impact on the total number of hyper-parameters in the model. Hence, this can be concluded that the proposed CNN model is indeed a very lightweight model, and it has a very less number of hyper-parameters (2.3 lakhs only), as compared to other existing CNN models (trained from scratch).

The parallel concatenation block (pcb) is one of the novelties of our research, shown in Fig.1. This pcb starts from a transformation layer (tx-layer), output of this tx-layer $I_o(f)$ is given in equation (5). This tx-layer is followed another 2 convolutional layers and two Max-Pooling layers, as shown in Fig.1.

$$I_o(f)_{w \times w} = m * (255 - I(f)_{w \times w}) \qquad (5)$$

where, $I_o(f)$ is the output of that transformation layer, $I(f)$ is the input tensor coming to the transformation layer, m here is a real constant whose value is supposed to be $0 < m < 1$, empirically we have chosen the value of $m = 0.8$ in this research. The purpose of this layer is to present the MRI images in such a format that it can highlight some hidden features which was not so prominent previously in the input tensor. In other words, it converts the original images into its negative version, such that it can extract additional essential features

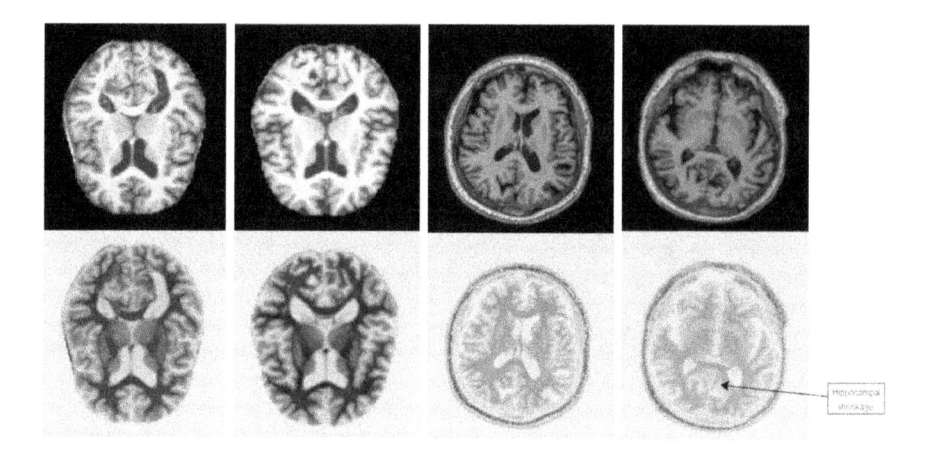

Fig. 2. First row represents the original MRI images, 2nd row represents the images after passing it through Tx-layer

for Alzheimer's detection. We have further ensured that with medical hospital doctors. For instance, this is evident from Fig.2 (first two images) that the gray matter substance in hippocampus's anatomy [3] of the original MRI image is more prominent after passing it through this Tx-layer. Similarly, in the last two images, in Fig.2 it has been highlighted that hippocampus shrinking [4] is more clear in the 2nd row. Moreover, abnormal levels of beta-amyloid [4] and widespread deposits of this protein becomes more visible after passing the MRI images through this Tx-layer, according to the neurologists. These are significant features of AD that get more highlighted after utilizing the Tx-layer.

The significance of Tx-layer in pcb, is explained in the following:

1. It can be observed from Fig.2 that the regions in the original image which were white, become more prominent and clear after passing through the Tx-layer. In contrast, areas in the transformed images that have changed to white (previously it was black in the original) become less prominent. Hence, it can be concluded that these two pairs of images (original and Tx-layer images) possess kind of complementary features. After consulting with neurologists, we came to know that this complementary features also carry some important information for AD detection. Therefore, incorporating both combinations of these features, enables the CNN model to learn more distinct and essential feature maps (for AD detection) than previous.
2. Moreover, it is evident from Fig.2 that the overall statistics in the original image and the processed image (i.e., after passing it through Tx-layer), differ significantly, thus, pcb may work like an efficient data augmenter inside the model. According to the research in [29], an efficient data-augmenter must generate synthetic images which have slightly different statistics compared to original images, otherwise, it induces overfitting in the model performance.

3. Numerous researchers [21–23] proposed attention module in the form of parallel concatenation in their CNN framework. However, none of their techniques deployed transformation layers before, hence, there is a possibility that redundant features (or, very similar features) might have been extracted in those parallel concatenation blocks, leading to overfitting in the model performance. Our proposed framework first time introduced the concept of the Tx-layer (through pcb), which automatically transforms original feature maps into its complementary version. Thus, proposed pcb works like an efficient data (or, feature) augmenter inside the model, to the best of our knowledge. As a consequence, the proposed pcb block automatically increases the generalization ability of the model, thus, mitigating the class imbalance problem to a certain extent.

We propose a new kind of *lemma* of CNN model in this research, in a very generalized way which is as follows:

Lemma1: If a CNN model, comprised of two parallel connections, extracts distinct features (in both such connections) that are essential for the final classification task, then that makes the model more stable than a series connection. Furthermore, extra distinct (or, complementary) features extracted in parallel concatenation, enable the network to generalize better for minor classes and thus, automatically alleviating the class imbalance problem efficiently.

This is to clarify that, the idea of parallel concatenation is not exactly new. Previously Cornia Marcella et al. [30] pointed out one of the limitations of a Deep CNN model (having a large number of layers) that, the features that were extracted earlier at the beginning layers (of CNN), are mostly forgotten at the final decision of classification. Thus, many researchers suggested making a parallel concatenation to fuse those features from previous layers to the output layer. Later it becomes trends while numerous researchers [21–24] started employing attention module through parallel connection. In this research, we have furthermore extended that concept into a generalized concept that any CNN model, having those parallel concatenation layers, if extracting a bit distinct kinds of features, automatically resolves the class imbalance problem in a generalized way. For example, MobileNet-V2 [28], and Xception [27] models have already utilized similar kinds of parallel concatenation in their model architecture, therefore, they have decent performances on these imbalanced MRI datasets, despite having higher complexity of their architecture.

3 Results and Analysis

The results and analysis section can be further summarized into two, (a) Training specification, (b) Experimental results comparisons and analysis.

3.1 Training Specifications

The training specifications of all of the models are given below:

1. The model was built using TensorFlow and Keras sequential API and the experiments were run on T4 GPU(Colab) environment as well as GPU P100(Kaggle). Colab environment provided a RAM of 25GB and Kaggle provided 100GB of RAM for the experiments.
2. All the datasets were randomly split into 80-20% ratio in a stratified way which is more feasible for class imbalance problem. This random splitting of train-test is the most authentic way of data splitting [31] so far for deep learning model. The train set was further partitioned into 80-20% split (random) for creating the validation dataset.
3. All the images in the entire dataset were resized to 224 x 224 prior to splitting the dataset.
4. A Batch size of 64 was employed throughout all experiments to train all the CNN models.
5. A learning rate (lr) of 0.00095 was chosen empirically, for Adams optimizer.
6. For "AD Dataset", the model was trained for 18 epochs and moreover, an adaptive learning rate (alr) of 5% decaying rate, is deployed after 8 epochs.
7. For "ADNI dataset", we have not employed any alr, which means we train it for a fixed lr of 0.00095 for 15 epochs, because we have found ADNI (Axial) dataset is a very simple dataset and loss was converging much smoother way, without having any fluctuation.
8. For "OASIS" dataset, we employed a total of 7 epochs only, with alr (5% decaying rate) employed after 4 epochs.
9. This is to clarify that, we have not employed any early stopping criteria for model training, because we noticed that for a model (trained from scratch) early stopping often stops the training too earlier than expected.
10. We have also implemented a pre-trained Pool-based Vision Transformer (PiT) model, on all three datasets. First, we have implemented it with the same training framework i.e., total 18 epochs with alr after 8 epochs. However, we observed that their model does not have the capability to learn very fast (in only 15 or 18 epochs). Thus, especially for PiT model, we also implemented the model for 50 epochs on all MRI datasets.

3.2 Experimental results comparisons and analysis

We have implemented numerous pre-trained CNN models VGG-16, Xception, DenseNet-121, MobileNet etc. (which are 100% fine-tuned from ImageNet dataset) on all three MRI datasets. Along with it, we have also implemented two existing CNN model, (I) 2D-M2IC (proposed by Helaly et al. [6]), and (II) MSC-Net (proposed by Liu, Z. et al. [22]) (trained-from-scratch) which were for AD detection. Furthermore, we have compared the efficacy of the proposed framework with a recent trend Pooling-based Vision Transformer (PiT) model [20]. Experimental results in Table 1, reveal that the proposed "AD-Lite Net" (trained from scratch) has consistently outperformed all the CNNs and PiT models by a substantial margin on all three MRI datasets. Furthermore, a comparison of the classification reports of the proposed AD-Lite Net model and

Table 1. Comparisons of several existing CNN models with the proposed framework (AD Lite-Net) on testing, for all three MRI datasets (Weighted Average)

Model/ Methods	AD-Dataset Accur-acy	AD-Dataset F1score	AD-Dataset secs/ep	ADNI Dataset Accur-acy	ADNI Dataset F1score	ADNI Dataset secs/ep	OASIS Dataset Accur-acy	OASIS Dataset F1score	OASIS Dataset secs/ep	No. of param (lakhs)
DenseNet-121 (fine tuning)	0.500	0.500	48	0.739	0.739	35	0.962	0.962	76	71.54
VGG-16 (fine tuning)	0.648	0.627	56	0.502	0.327	44	0.251	0.167	87	147.17
Xception (fine tuning)	0.892	0.891	70	0.997	0.997	54	0.951	0.951	112	208.15
MobileNet-V2 (fine tuning)	0.938	0.938	15	0.994	0.994	12	0.973	0.973	30	32.11
Pooling-based ViT (PiT) [20]	0.581	0.584	20	0.621	0.618	18	0.313	0.234	30	45.91
Pooling-based ViT (PiT) [20] with 50 epochs	0.917	0.917	20	0.925	0.926	18	0.285	0.267	30	45.91
2D-M2IC [6] (train-from-scratch)	0.882	0.881	2	0.996	0.996	1	0.937	0.937	3	8.19
MSC-Net+SE-Net [22] (train-from-scratch)	0.893	0.901	81	0.530	0.51	64	0.877	0.877	113	144.58
AD-Lite Net (train-from-scratch) proposed	**0.982**	**0.981**	**5**	**0.999**	**0.999**	**4**	**0.996**	**0.996**	**12**	**2.32**

AD-Lite without parallel concatenation, is presented in Table 2. This is to clarify that accuracy can not be counted on a specific class, it is always the overall accuracy of the model, thus, in Table 2 only one value of "Accuracy" is presented in one column. The results in Table 1 and Table 2 further strengthen and verify our proposed theory which was proposed in Section 2.2. Furthermore, the quality metrics along with their graphs, and confusion matrices of all these experiments (mentioned in table-1) are available in a Github link: **https://github.com/ArchitGupta16/Alzheimer-Detection/tree/main**.

An ablation study of the proposed AD-Lite model is also available in that link and this is further explored in a supplementary material.

From Table-1 this is evident that the efficacy of the VGG-16 and Dense-Net are relatively lesser than that of other pre-trained CNN models. VGG-16 [32] usually does not deal well with the class imbalance problem, due to the lack of feature extraction in both spatial and spectral domains. Moreover, due to utilizing back-to-back convolutional layers (both in VGG-16 and DenseNet), the number of hyper-parameters in their model increased significantly, thus, over-fitting is inevitable in their model performances for small datasets. The most imbalanced dataset was the Oasis dataset, in which this is evident that VGG-16 suffers considerably to achieve higher accuracy and F1 score. Moreover, DenseNet-121 model suffers from very poor accuracy both in AD-Dataset and ADNI dataset. On the other hand, MobileNet-V2, Xception models have performed way better than VGG-16 and DenseNet-121, because of their lightweight framework. Xception is a modified version of Inception-V3 and the first time they incorporated Depth-Wise Separable Convolutional (DWSC) layers in their model, explored in Section 2.2. Whereas, MobileNet-V2 utilizes both DWSC layers and convolutional layers in its model, additionally, it leverages GAP layer instead of flatten layer. Due to utilizing these components in their model, both of these models avoid overfitting and as a consequence, they have decent performances throughout all these (small) MRI datasets.

Table 2. Comparisons of Classification Reports of the proposed AD-Lite Net model with and without Parallel Concatenation, on the "AD-Dataset"

Classes	AD-Lite Net without Parallel Concatenation				AD-Lite Net with Parallel Concatenation			
	Precision	Recall	F1-Score	Accuracy	Precision	Recall	F1-score	Accuracy
Very-Mild Demented	0.97	0.94	0.95	0.96	1.00	0.96	0.98	**0.98**
Mild Demented	0.99	0.94	0.97		0.99	0.98	0.99	
Moderate Demented	1.00	0.92	0.96		1.00	1.00	1.00	
Non-Demented	0.95	0.99	0.97		0.97	1.00	0.98	
Macro-Average	0.98	0.95	0.96		**0.99**	**0.98**	**0.99**	
Weighted Average	0.96	0.96	0.96		**0.98**	**0.98**	**0.98**	

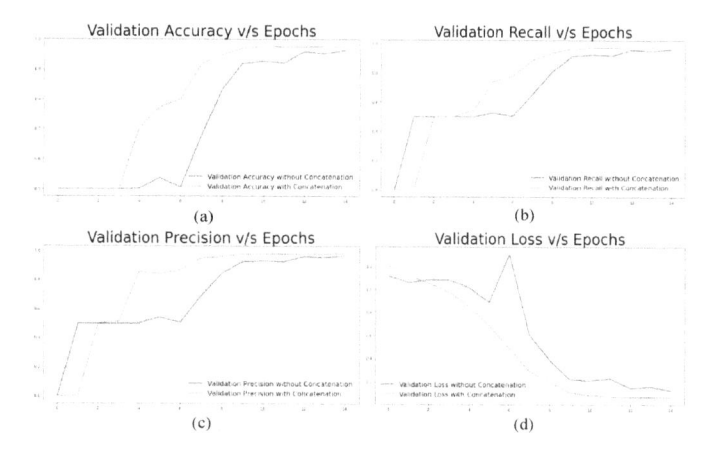

Fig. 3. Validation graph comparison of proposed AD-Lite Net with vs AD-Lite Net without pcb, blue line indicates performance of AD-Lite without pcb and green line indicates performance of AD-Lite with pcb; (a) Accuracy vs epochs, (b) Recall vs epochs, (c) Precision vs epochs, (d) Loss vs Epochs

We have also implemented one of the recent trend models, Pooling based ViT (PiT) [20], on all three MRI datasets. Conventional ViT models can not be implemented on these small datasets, due to the complexity in their model architecture. Therefore, we have implemented PiT instead of ViT. From Table-1 , this is apparent that the PiT model with 50 epochs, has achieved decent efficacy in both AD and ADNI datasets, however, their model has struggled to generalize in minor classes, for OASIS dataset. A recently proposed 2D CNN model (2D-M2IC) [6] is also implemented in this study, which is trained from scratch. The number of hyper-parameters in 2D-M2IC is considerably lesser (8.19 lakhs) than in other models. Table-1 shows that 2D-M2IC achieves good accuracy, and F1 score both in ADNI and OASIS datasets, however, it struggles to generalize the same in AD Dataset. Additionally, we have implemented a recently proposed model for AD detection, that is, "MSC-Net," along with SE-Net attention block [22] on all three datasets. This model was trained from scratch with the same specification as the proposed model. Experimental results suggest that MSC-Net (with the SE-Net attention block) has achieved a commendable accuracy of 89.3% and 87.7% for AD and Oasis dataset respectively, nevertheless, it severely failed on ADNI dataset. Due to employing higher number of hyper-parameters (144.6 lakhs) it exhibited over-fitting for small dataset.

Overall, Table 1 reveals that some models performed occasionally well on particular datasets, however, most of them failed to generalize on all three MRI datasets. Only MobileNet-V2 [29], and the proposed AD-Lite Net model have obtained decent accuracy and F1 score more than 90% consistently, over all three MRI datasets. Furthermore, this can be observed from Table 1 and Table 2 that the proposed AD-Lite Net has achieved the best accuracy, precision, recall, and F1 score (so far) on all three MRI datasets. This is also apparent from the graph in Fig.3 that the proposed "AD-Lite Net" has converged to higher accuracy and precision much faster after integrating the "parallel concatenation block (pcb)". This also reveals that by utilizing this pcb, the proposed framework generalizes much more effectively than previous and the validation graph becomes more stable. Furthermore, from Table 2, this is evident that the macro-averages of precision, recall, and F1-score have been boosted by 1-3%, after leveraging pcb on the AD-Lite Net. This is a significant improvement, which justifies the necessity of incorporating "pcb" in the proposed framework. Hence, these experimental results support our proposed theory and *Lemma1* which were proposed in Section 2.2.

We have also conducted a 10-fold cross validation experiment by combining two datasets. 'AD dataset' and 'ADNI dataset', which had dis-similar statistics. This merging is done after labelling the 'Mild Demented' and 'Very Mild Demented' classes in AD dataset into a single class Mild-Demented class. The idea was to blend diverse statistical images from these two datasets to create a challenging dataset. By this 10-fold cross-validation experiment, we effectively created the equivalent of 10 different datasets (we call them fold1-to-fold10 in Table-3), where each dataset has distinct testing set, having different statistics compared to the same of other 9 datasets. The results of this 10-fold cross-validation, with mean and standard deviation values, have been presented in Table-3 and also available in the aforementioned GitHub repository. These results demonstrate that the proposed "AD-Lite Net" is capable of achieving 98.3-99.7% (Mean 99%) accuracy consistently, in this challenging 10-fold cross-validation experiment as well. Furthermore, the standard deviation of accuracy, precision, recall and F1 score across these 10 folds is significantly low, that is 0.4% only. This also indicates that the performance of the proposed model has been remarkably stable and it indeed resolved the class imbalance issue in a very generalized way. Hence, this experiment validates the generalization capability of the proposed model in a highly efficient way.

Table 3. Testing results for 10-fold cross validation on merged dataset

folds	Accuracy	Precision	Recall	F1score	AUC	secs/ ep
fold1	0.995	0.995	0.994	0.994	0.999	7
fold2	0.984	0.984	0.984	0.984	0.998	6
fold3	0.992	0.992	0.992	0.992	0.999	7
fold4	0.991	0.991	0.991	0.991	0.999	6
fold5	0.983	0.983	0.983	0.983	0.999	6
fold6	0.992	0.992	0.992	0.992	0.999	6
fold7	0.994	0.994	0.994	0.994	0.999	6
fold8	0.988	0.988	0.988	0.988	0.999	7
fold9	0.997	0.997	0.997	0.997	0.999	6
fold10	0.985	0.985	0.985	0.985	0.997	7
Mean± Std dev	0.990± 0.004	0.990± 0.004	0.990± 0.004	0.990± 0.004	0.999± 0.0008	6.4

4 Conclusion and Future Work

One lightweight and concatenated CNN model (train from scratch) was proposed for automatic Alzheimer's detection from MRI images. "Parallel concatenation block", incorporated into the base model, leveraged a novel Tx-layer which extracted unique salient features for Alzheimer's disease, thus, automatically mitigating the class imbalance problem in a generalized way. Experimental results on three different MRI datasets showed that there was a lack of generalization of all the existing and pre-trained CNN models. The AD-Lite Net model with concatenation block, not only generalized well for all three MRI datasets, but also, achieved the best accuracy, precision, recall, F1 score for all three datasets. Furthermore, the proposed framework outperformed one recent trends model, Pooling-based Vision Transformer (PiT), by a significant margin. Hence, this can be concluded that the proposed AD-Lite Net successfully alleviated all the challenges for AD detection from MRI datasets, and this proposed framework can perform well uniformly for any MRI dataset. A 10-fold cross-validation experiment also demonstrated the strong generalization capability of the proposed "AD-Lite Net".

This is to clarify that, until now, we worked with MRI datasets that did not include subject-specific images. Moving forward, our goal is to extend this project to predicting Alzheimer's disease at different subjects instantly which will be a more challenging and valid direction from the perspective of medical experts. In order to deal with more practical (noisy) data taken from a hospital, we are also planning to incorporate one extra attention module in our model.

References

1. Sweeney, Melanie D., Abhay P. Sagare, and Berislav V. Zlokovic.: Blood-brain barrier breakdown in Alzheimer disease and other neurodegenerative disorders. Nature Reviews Neurology 14(3), 133-150 (2018)

2. Prince, Martin, et al. World Alzheimer Report 2015. The Global Impact of Dementia: An analysis of prevalence, incidence, cost and trends. Diss. Alzheimer's Disease International, (2015)
3. Rao, Y. Lakshmisha, et al.: Hippocampus and its involvement in Alzheimer's disease: a review. 3 Biotech 12(2), 55 (2022)
4. Tarawneh, Rawan, et al.: Cerebrospinal fluid markers of neurodegeneration and rates of brain atrophy in early Alzheimer disease. JAMA neurology 72(6), 656-665 (2015)
5. Odusami, Modupe, et al.: Analysis of features of Alzheimer's disease: Detection of early stage from functional brain changes in magnetic resonance images using a finetuned ResNet18 network. Diagnostics 11(6), 1071 (2021)
6. Helaly, Hadeer A., Mahmoud Badawy, and Amira Y. Haikal.: Deep learning approach for early detection of Alzheimer's disease. Cognitive computation 14(5), 1711-1727 (2022)
7. Shakarami, Ashkan, Hadis Tarrah, and Ali Mahdavi-Hormat.: A CAD system for diagnosing Alzheimer's disease using 2D slices and an improved AlexNet-SVM method. Optik 212, 164237 (2020)
8. Achilleos, K. G., Leandrou, S., Prentzas, N., Kyriacou, P. A., Kakas, A. C., and Pattichis, C. S.: Extracting explainable assessments of Alzheimer's disease via machine learning on brain MRI imaging data. In 2020 IEEE 20th International Conference on Bioinformatics and Bioengineering (BIBE), pp. 1036-1041. IEEE, (2020)
9. Textural features for image classification: Haralick, Robert M., Karthikeyan Shanmugam, and Its' Hak Dinstein. IEEE Trans. Syst. Man Cybern. **6**, 610–621 (1973)
10. Tyagi, M. et al.: Custom Weighted Balanced Loss function for Covid 19 Detection from an Imbalanced CXR Dataset. In 26th International Conference on Pattern Recognition (ICPR), pp. 2707-2713, IEEE (2022)
11. Masud, M., Almars, A.M., Rokaya, M.B., Meshref, H., Gad, I., Atlam, E.S.: A Novel Light-Weight Convolutional Neural Network Model to Predict Alzheimer's Disease Applying Weighted Loss Function. Journal of Disability Research **3**(4), 20240042 (2024)
12. Silva, J., Bispo, B.C., Rodrigues, P.M.: Structural MRI texture analysis for detecting Alzheimer's disease. Journal of Medical and Biological Engineering **43**(3), 227–238 (2023)
13. Khan, Afreen, and Swaleha Zubair.: An improved multi-modal based machine learning approach for the prognosis of Alzheimer's disease. Journal of King Saud University-Computer and Information Sciences, 34(6), 2688-2706, (2022)
14. Zhu, Wenyong, et al.: Dual attention multi-instance deep learning for Alzheimer's disease diagnosis with structural MRI: IEEE Transactions on Medical Imaging 40(9), 2354-2366, (2021)
15. Zhang, Qiongmin, et al.: Lightweight neural network for Alzheimer's disease classification using multi-slice sMRI. Magnetic Resonance Imaging 107, 164-170 (2024)
16. Hong, Y., Wu, Q., Qi, Y., Rodriguez-Opazo, C., Gould, S.: Vln bert: A recurrent vision-and-language bert for navigation. In: Proceedings of the IEEE/CVF conference on Computer Vision and Pattern Recognition. pp. 1643-1653 (2021)
17. Ranftl, R., Bochkovskiy, A., Koltun, V.: Vision transformers for dense prediction. In: Proceedings of the IEEE/CVF international conference on computer vision, pp. 12179-12188 (2021)
18. Bronstein, M.M., Bruna, J., LeCun, Y., Szlam, A., Vandergheynst, P.: Geometric deep learning: going beyond euclidean data. IEEE Signal Process. Mag. **34**(4), 18–42 (2017)

19. Xie, Z., Lin, Y., Yao, Z., Zhang, Z., Dai, Q., Cao, Y., Hu, H.: Self-supervised learning with swin transformers. arXiv preprint arXiv:2105.04553 (2021)
20. Heo, B., Yun, S., Han, D., Chun, S., Choe, J., Oh, S.J.: Rethinking spatial dimensions of vision transformers. In: Proceedings of the IEEE/CVF International Conference on Computer Vision. pp. 11936-11945 (2021)
21. Zhu, Jiayi, et al.: Efficient self-attention mechanism and structural distilling model for Alzheimer's disease diagnosis. Computers in Biology and Medicine, 147, 105737 (2022)
22. Liu, Z., Lu, H., Pan, X., Xu, M., Lan, R., Luo, X.: Diagnosis of Alzheimer's disease via an attention-based multi-scale convolutional neural network. Knowl.-Based Syst. **238**, 107942 (2022)
23. Ji, Huanhuan, et al.: Early diagnosis of Alzheimer's disease based on selective kernel network with spatial attention. Asian Conference on Pattern Recognition (ACPR). Cham: Springer International Publishing, pp. 503-515 (2019)
24. Hu, Zhentao, et al.: VGG-TSwinformer: Transformer-based deep learning model for early Alzheimer's disease prediction. Computer Methods and Programs in Biomedicine 229, 107291 (2023)
25. https://www.kaggle.com/datasets/tourist55/alzheimers-dataset-4-class-of-images
26. Popuri, Karteek, et al.: Using machine learning to quantify structural MRI neurodegeneration patterns of Alzheimer's disease into dementia score: Independent validation on 8,834 images from ADNI, AIBL, OASIS, and MIRIAD databases. Human Brain Mapping 41(14), 4127-4147 (2020)
27. Chollet, F.: Xception: Deep learning with depthwise separable convolutions. In: Proceedings of the IEEE conference on computer vision and pattern recognition, pp. 1251-1258 (2017)
28. Sandler, M., Howard, A., Zhu, M., Zhmoginov, A., Chen, L.C.: Mobilenetv2: Inverted residuals and linear bottlenecks. In: Proceedings of the IEEE conference on computer vision and pattern recognition. pp. 4510-4520 (2018)
29. Roy, Santanu, et al.: Svd-clahe boosting and balanced loss function for covid-19 detection from an imbalanced chest x-ray dataset. Computers in Biology and Medicine, 150, 106092 (2022)
30. Cornia, Marcella, et al.: A deep multi-level network for saliency prediction. 2016 23rd International Conference on Pattern Recognition (ICPR), pp. 3488-3493, IEEE (2016)
31. S. Song, Congzheng, Thomas Ristenpart, and Vitaly Shmatikov. "Machine learning models that remember too much." In Proceedings of the 2017 ACM SIGSAC Conference on computer and communications security, 587-601 (2017)
32. Alippi, C., Disabato, S., Roveri, M.: Moving convolutional neural networks to embedded systems: the alexnet and VGG-16 case. In 2018 17th ACM/IEEE International Conference on Information Processing in Sensor Networks (IPSN), pp. 212-223, IEEE (2018)

Leveraging Persistent Homology for Differential Diagnosis of Mild Cognitive Impairment

Ninad Aithal[✉], Debanjali Bhattacharya, Neelam Sinha, and Thomas Gregor Issac

Center for Brain Research, Indian Institute of Science (IISc), Bengaluru, India
reachninadaithal@gmail.com

Abstract. Mild cognitive impairment (MCI) is characterized by subtle changes in cognitive functions, often associated with disruptions in brain connectivity. The present study introduces a novel fine-grained analysis to examine topological alterations in neurodegeneration pertaining to six different brain networks of MCI subjects (Early/Late MCI). To achieve this, fMRI time series from two distinct populations are investigated: (i) the publicly accessible ADNI dataset and (ii) our in-house dataset. The study utilizes sliding window embedding to convert each fMRI time series into a sequence of 3-dimensional vectors, facilitating the assessment of changes in regional brain topology. Distinct persistence diagrams are computed for Betti descriptors of dimension-0, 1, and 2. Wasserstein distance metric is used to quantify differences in topological characteristics. We have examined both (i) ROI-specific inter-subject interactions and (ii) subject-specific inter-ROI interactions. Further, a new deep learning model is proposed for classification, achieving a maximum classification accuracy of 95% for the ADNI dataset and 85% for the in-house dataset. This methodology is further adapted for the differential diagnosis of MCI sub-types, resulting in a peak accuracy of 76.5%, 91.1% and 80% in classifying HC Vs. EMCI, HC Vs. LMCI and EMCI Vs. LMCI, respectively. We showed that the proposed approach surpasses current state-of-the-art techniques designed for classifying MCI and its sub-types using fMRI.

Keywords: fMRI time series · Sliding window embedding · Persistent homology · Wasserstein distance · Deep learning

1 Introduction

Mild Cognitive Impairment (MCI) stands as a crucial stage bridging normal cognitive aging and dementia, often serving as a precursor to conditions like Alzheimer's disease (AD) and other neurodegenerative disorders [1,2,18]. Research indicates that individuals with MCI progress to AD at a rate of approximately 10-15% per year [13], making MCI as the most challenging group for early detection and diagnosis of AD. Based on the extent of episodic memory impairment, MCI can be primarily categorized into Early Mild Cognitive Impairment

© The Author(s), under exclusive license to Springer Nature Switzerland AG 2025
A. Antonacopoulos et al. (Eds.): ICPR 2024, LNCS 15312, pp. 17–32, 2025.
https://doi.org/10.1007/978-3-031-78198-8_2

(EMCI) and Late Mild Cognitive Impairment (LMCI). Notably, the risk of LMCI transitioning to AD surpasses that of EMCI. However, detecting EMCI remains clinically challenging due to subtle alteration from healthy controls (HC). Additionally, the classification of EMCI and LMCI based solely on memory scores may lead to low specificity and misclassifications. The search for sensitive biomarkers that change alongside disease progression offers hope in refining disease staging, potentially decreasing the prevalence of AD through early intervention. To be more specific, EMCI is particularly an important sub-type of MCI for implementing interventions, aimed at potentially modifying the progression of the condition. Hence, there is a growing emphasis on delineating the neurobiological alterations associated with EMCI and LMCI, vital for early diagnosis, prognosis, and intervention strategies [27]. The identification of potentially high-sensitivity diagnostic markers evolving alongside disease progression can significantly aid physicians in making accurate diagnoses. Recent improvements in brain imaging, especially with fMRI, offer valuable insights into how MCI affects brain function. By studying changes in various brain networks, we gain a better understanding of how MCI disrupts communication within the brain. These disruptions are key to understanding the disease early on, which could help delay or reverse cognitive decline associated with MCI and its sub-types. In the current study, rather than relying on state-of-the-art machine learning and deep learning methods which are commonly used for classification of MCI and its sub-types using various MRI modalities[1,2], we have reported completely different approach. We have utilized *persistent homology*- an advanced tool in computational topology, in order to explore potential differences in topology between MCI sub-types and compared them to HC. This novel approach diverges from conventional feature engineering techniques and offers a unique perspective on understanding the subtle yet significant variations associated between MCI sub-types and HC.

Persistent homology is a powerful topological data analysis (TDA) approach falling under the branch of algebraic topology. It provides a robust framework for analyzing the topological features of data, particularly in the context of shape and structure. At its core, persistent homology aims to capture the evolution of topological features across different spatial scales by constructing a sequence of topological spaces based on the input data. It reveals how specific topological characteristics remain consistent or evolve as we observe these spaces at different levels of detail. Persistent homology finds its application in various domains. In the context of medical image analysis, it is used for analysis of endoscopy [11], breast cancer [30], analysis of brain networks for differentiating various types of brain disorders [24,29], detecting transition between states in EEG [26], identifying epileptic seizures [8] and distinguishing between male and female brain networks [9]. Notably, persistent homology of time series emerges as a rapidly growing field with several compelling applications like computing stability of dynamic systems [23], to quantify periodicity [28] and differences in visual brain networks [5].

In this study, we analyze fMRI time series data derived from 160 Dosenbach Regions of Interest (ROIs), which are selected from six classical brain networks.

Each ROI corresponds to a distinct time series, and our focus is on leveraging persistent homology to analyze these fMRI time series for the classification of MCI and its sub-types. The present study offers two significant contributions:
(i) The introduction of a novel paradigm that employs persistent homology on fMRI time series to quantify topological changes in functional connectivity patterns between MCI sub-types.
(ii) A comprehensive statistical analysis of topological features between HC and MCI sub-types is conducted to detect critical brain regions to examine distinct pattern linked with different stages of MCI across each homology dimension. To the best knowledge of authors, no other research has examined the efficacy of persistent homology of fMRI time series for differential diagnosis of MCI.

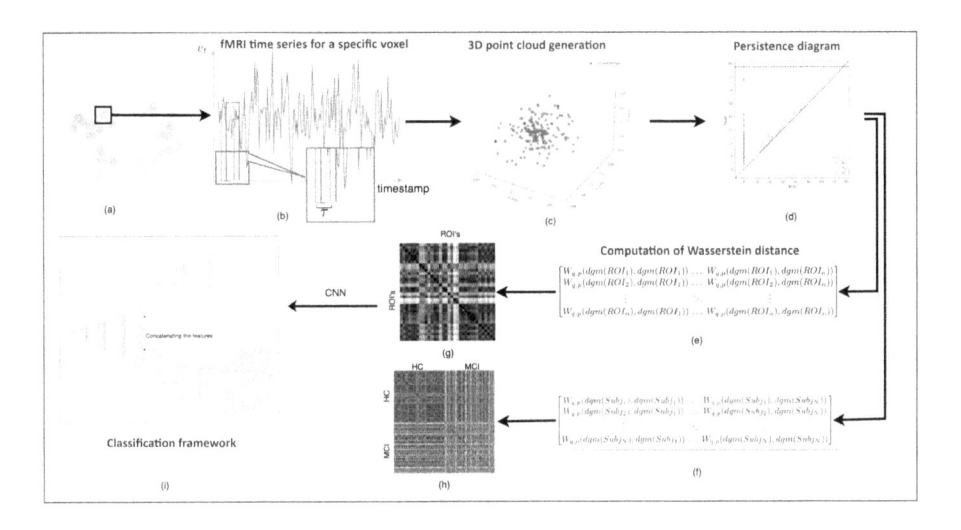

Fig. 1. Block schematic of the proposed methodology

2 Dataset Description

The study utilizes fMRI images from cohorts representing two distinct populations. The baseline study incorporates subjects from the publicly available Alzheimer's Disease Neuroimaging Initiative (ADNI) dataset [17], with a repetition time (TR) of 3000 ms and an echo time (TE) of 30 ms. This is complemented by our in-house TLSA (TATA Longitudinal Study for Aging) cohort, which has a TR of 3200 ms and a TE of 30 ms. The TLSA is an urban cohort investigation aimed at accumulating long-term data to discern risk and protective factors associated with dementia in India. Both cohorts feature images acquired in the sagittal acquisition plane with a 3D acquisition type. The subjects included from ADNI are: MCI ($N = 50$), EMCI ($N = 163$), LMCI ($N = 141$) and HC ($N = 179$). The efficacy of the proposed methodology is further verified on our in-house MCI ($N = 50$) and HC ($N = 50$) cohort. In our in-house MCI cohorts, individuals diagnosed with MCI met specific criteria, having clinical dementia rating or CDR value (the current gold standard for assessing the stages of patients diagnosed having dementia) of 0.5. All fMRI images underwent the same preprocessing pipeline, which included motion correction, adjustment for slice timing, normalization to the standard MNI space, and regression to account for nuisance variables. These preprocessing steps were performed using FSL (FMRIB Software Library) version 6.0.6 [19].

3 Proposed Methodology

The block diagram of the proposed methodology is shown in Figure 1. Our method involves converting the 1D fMRI time series data into a 3D point cloud, from which persistent diagrams are obtained. We utilize the Wasserstein distance metric to compare all possible pairs of persistent diagrams, enabling quantification of topological alterations. These alterations are examined in two contexts: (i)*across subjects for a given ROI* (ii)*across ROIs for a given subject*. The in-house data will be made available to the research community after procedural formalities by the administration at the center. [1]

3.1 Extraction of Time Series

The fMRI time series captures the temporal evolution of brain activity, providing a dynamic view of how different brain regions interact over time. The study carefully selects relevant brain regions from Dosenbach's ROIs to extract fMRI time series for identifying meaningful patterns and differences between MCI and its sub-types. The Dosenbach's ROIs[10] ($n = 160$) are divided into six distinct brain networks: cerebellum ($n=18$), cingulo-opercular ($n=32$), default mode ($n=34$), fronto-parietal ($n=21$), occipital ($n=22$), and sensorimotor ($n=33$) networks, encompassing various interconnected brain regions. Each of these brain network is linked to specific cognitive, sensory, and motor functions and may show unique disruption patterns at various stages of MCI. Therefore, by analyzing all six networks, the study offers a comprehensive assessment of network-specific changes, which could serve as distinct biomarkers for different stages or types of cognitive impairment. A representative of 5mm radius sphere drawn at each voxel location contributed as a time series $\{v_t, t = 1, 2, \ldots, N\}$.

3.2 Point Cloud Construction from Time Series

Creating an efficient point cloud representation from 1D time series data is a crucial step for computing persistent homology [14, 28]. In this step, the goal is to construct a richer feature space that enables the analysis of changes in the

---

[1] The codes of this analysis are available at https://github.com/blackpearl006/ICPR-2024

intrinsic topological properties of MCI. Each point in the time series is mapped to a vector in 3D space, in order to capture the complex temporal dynamics (Figure 1.C). For this, the study employs sliding window embedding ($\mathcal{SW}$) with embedding dimension $M = 2$ and time lag $\tau = 1$ in order to convert the fMRI time series v_t into point clouds $\mathcal{S}$. The choice of length of the sliding window is set to 3 to minimise noise interference and better interpretability.

The sliding window embedding of a function f based at $t \in \mathbb{R}$ into $\mathbb{R}^{M+1}$ is represented as follows (Equation 1):

$$\mathcal{SW}_{M,\tau}f : \mathbb{R} \to \mathbb{R}^{(M+1)}, \quad t \to \begin{bmatrix} f(t) \\ f(t+\tau) \\ f(t+2\tau) \end{bmatrix} \tag{1}$$

Thus, selecting different values of t results in a set of points, known as a sliding window point cloud for the function f. Multiple literature sources point to the efficiency of sliding window methods [15] in capturing dynamic functional connectivity in resting-state fMRI. In this study, for embedding dimension $M = 2$, the point cloud of the fMRI time series is represented by Equation 2:

$$\mathcal{S} = \{v_i : i = 1, \ldots, N, \ v_i \in \mathbb{R}^3\} \tag{2}$$

This representation ensures each point v_i in the point cloud $\mathcal{S}$ is a vector in $\mathbb{R}^3$, maintaining consistency with the spatial dimensions of the fMRI data.

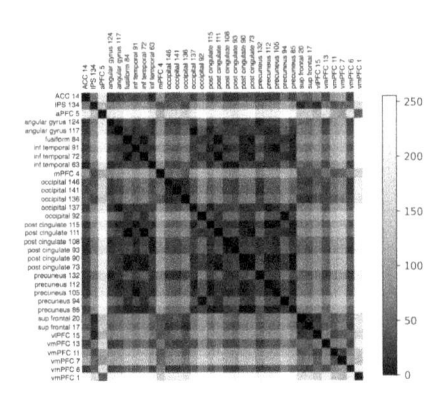

Fig. 2. Illustrating the Wasserstein distance as derived from the persistence diagram of fMRI time series for homology dimension-0, showing interactions among ROIs within the DMN for one representative subject. The 34 ROIs within the DMN exhibit a consistent spatial arrangement, wherein nearby brain regions are grouped together. The arrangement of these 34 ROIs remains consistent across all subjects. A low value of Wasserstein distance suggests synchronized neural activity and coordination between brain regions, while a high value of distance indicates distinct activity patterns and potential functional independence.

3.3 Persistent Homology

This step involves computing persistent homology from the generated 3D point clouds to extract topological features of varying dimensions. This is accomplished by constructing a series of simplicial complexes using Vietoris-Rips filtration and calculating their homological features. These features capture the underlying topological structure in the data, identifying meaningful patterns and differences between MCI and its sub-types. We exploit the information encoded in *persistence diagram* to analyze the differences in topology of brain networks of individuals with MCI from HC. Persistence diagram encodes the persistence features in data across the filtration parameter range as a collection of points in the two-dimensional Euclidean space $\mathbb{R}^2$. A common approach for constructing a filtration from a point cloud is through the Vietoris-Rips complex. This complex is generated from the point cloud by connecting any subset of points whose pairwise distances fall within a specified threshold, creating a simplex. Thus, filtration is a collection $\mathcal{F} = \{F_\epsilon\}_{\epsilon \geq 0}$ of spaces with $F_\epsilon \subset F_{\epsilon'}$ continuous $\forall \epsilon \leq \epsilon'$. The ith persistence diagram of $\mathcal{F}$ is a multiset $dgm_i(\mathcal{F}) \subset \{(p,q) \in [0,\infty] \times [0,\infty] \mid 0 \leq p < q\}$ where each pair $(a,b) \in dgm_i(\mathcal{F})$ encodes a i-dimensional topological feature, in other words Betti descriptors[2] associated with a simplicial complex that born at F_b and dies at F_d. The quantity $(d-b)$ is the persistence of the feature, and typically measures significance across the filtration. In our study, given a time series (V_t) the sliding window point cloud $\mathcal{SW}_{M,\tau}f$ is computed which is in a metric space (X, M_X). The Rips filtration $\mathcal{VR}(X, M_X)$ is derived from the Vietoris-Rips complex $VR_\epsilon(X, M_X)$, computed at each scale $\epsilon \geq 0$. The mathematical expression for computing Rips filtration is depicted in Equation 3

$$\mathcal{VR}(X, M_X) := \{VR_\epsilon(X, M_X)\}_{\epsilon \geq 0}, where$$
$$VR_\epsilon(X, M_X) := \{\{x_0, ..., x_n\} \in X \mid \max_{0 \leq i,j \leq n} M_X(x_i, x_j) < \epsilon, n \in \mathbb{N} \quad (3)$$

The birth-death pairs (b,d) in the Rips persistence diagrams $dgm_i^{\mathcal{VR}}(X) := dgm_i \mathcal{VR}(X, M_X)$ reveal the underlying topology of space X. The points (b,d) in $dgm_i^{\mathcal{VR}}(X)$ with large persistence values $(d-b)$ suggest the most persistent topological features of the continuous space where X is concentrated.

The paper employs the Wasserstein distance to measure the dissimilarity between the persistence diagrams. This metric quantifies the differences in the topological features between subjects, providing a robust basis for distinguishing between different MCI sub-types. Thus, it enhances the ability to compare topological changes in brain activity. In this particular step, the paper contribution lies in analysing both inter-subject and inter-ROI Wasserstein distance

---

[2] In algebraic topology, the topological features of a space are represented as *holes* or *cycles* in various dimensions. The number of k-dimensional holes in a d-dimensional simplicial complex (with $k \leq d$) is denoted by the Betti number β_k or H_k. Thus, 0-dimensional holes ($\beta_0 or H_0$) correspond to connected components, 1-dimensional holes ($\beta_1 or H_1$) represent tunnels (or loops) and 2-dimensional holes ($\beta_2 or H_2$) are voids.

measures to highlight the specific brain regions where changes are pronounced in MCI sub-types. In context of fMRI time series analysis using topological persistence diagram, the Wasserstein distance between fMRI time series of two brain regions captures the degree of similarity in their patterns of neural activity. A low Wasserstein distance indicates that the two fMRI time series exhibit similar patterns of neural activity over time. In other words, it suggests that the two brain regions are functionally synchronized and are likely engaged in coordinated activity. On the contrary, a high Wasserstein distance suggests that the fMRI time series of two brain regions have distinct patterns of neural activity, may be functionally dissociated or independent from each other. High Wasserstein distances may also indicate abnormalities in functional connectivity between the two brain regions, which could be indicative of any neurological or psychiatric disorders. Figure 2 illustrates the variability in Wasserstein distance among all ROI pairs for a single representative subject. In computational topology, Bottleneck distance and Wasserstain distance are the two widely used measures for quantifying the dissimilarity between two persistence diagrams [12]. Suppose, f_1 and f_2 are two different filtrations and let $X = dgm_p(f_1)$ and $Y = dgm_p(f_2)$ denote the pth persistence diagrams corresponding to f_1 and f_2. The Wasserstein and Bottleneck distance metrices are used to quantify the dissimilarity between these two multisets X and Y. Let $L_\infty(f_1, f_2) = \|f_1 - f_2\|_\infty$ denote the supremum distance between f_1 and f_2, and η denotes a bijection of $X \to Y$, then, the q–Wasserstein distance between two persistence diagrams X and Y is defined as

$$W_{q,p}(X,Y) = \left[\inf_{\eta:X\to Y} \sum_{x\in X} \|x - \eta(x)\|_\infty^q\right]^{\frac{1}{q}} \tag{4}$$

To compute the distance elements of X and Y one-to-one (bijection η) are matched. It is usually done in the following way: first for each pair of elements, $x \in X$ and $y = \eta(x) \in Y$, the difference between them (the cost function) is calculated using $\|x - \eta(x)\|_\infty$ that is basically L_∞ norm. Adding up the qth degrees $\|.\|_\infty^q$, we get a notion of the difference between the whole multisets X and Y under the matching $\eta : X \to Y$. Taking the infimum over all possible bijections η, we get the difference between multisets X and Y under the best matching possible, effectively removing η from further consideration. The bottleneck distance is the Wasserstein distance, with parameter $q \to \infty$. Hence, one drawback of the bottleneck distance is its insensitivity to details of the bijection beyond the furthest pair of corresponding points. Due to this, the present study considers Wasserstein distance for quantification.

In this study, the persistence diagram is computed for two different scenarios: (i) *ROI-specific-* across all ROIs of a particular subject, and (ii) *Subject-specific-* across all considered subjects of HC and MCI for a particular ROI. Therefore, Wasserstein distance $W_{q,p}(X,Y)$ in Equation 4 is computed to measure the pairwise distance between all ROIs as well as between all subjects. This is described in the next sub-sections (Section 3.4 and Section 3.5).

3.4 *ROI-Specific* Inter-Subject Interactions

ROI-specific analysis for all brain networks is conducted to understand the dissimilarity in topological patterns across all considered subjects for each of the three Betti descriptors (H_0, H_1, and H_2). This is represented in the pairwise-subject distance matrix (PS) of dimensions $N \times N$, where N is the number of subjects. The mathematical representation of matrix PS is shown in Equation 5. Each element $PS(i,j)$ indicates the Wasserstein distance between the persistence diagrams of subject i and subject j for a given ROI of a specific brain network. The sample images as obtained from PS for one specific ROI of each brain network are shown in Figure 4.

$$PS = \begin{bmatrix} W_{q,p}(dgm(Subj_1), dgm(Subj_1)) & \dots & W_{q,p}(dgm(Subj_1), dgm(Subj_N)) \\ W_{q,p}(dgm(Subj_2), dgm(Subj_1)) & \dots & W_{q,p}(dgm(Subj_2), dgm(Subj_N)) \\ \vdots & \ddots & \vdots \\ W_{q,p}(dgm(Subj_N), dgm(Subj_1)) & \dots & W_{q,p}(dgm(Subj_N), dgm(Subj_N)) \end{bmatrix}$$
(5)

3.5 *Subject-Specific* Inter-ROI Interactions

Here, for each of the three Betti descriptors (H_0, H_1, H_2), we compute the pairwise-ROI distance matrix (PR) with dimensions $n \times n$, where n is the number of ROIs of a specific brain network. The mathematical representation of the matrix PR is shown in Equation 6. The Wasserstein distance in matrix PR represents the interaction between different ROIs. Each element in $PR(i,j)$ indicates the distance between the persistence diagrams of ROI i and ROI j. However, computing persistent homology and Wasserstein distance matrices (PR and PS) are computationally very expensive, for which we utilised high-performance computing (HPC) resources, specifically an Intel(R) Xeon(R) Gold 6240 CPU @ 2.60GHz with dual CPUs and 192 GB of memory, were utilized.

$$PR = \begin{bmatrix} W_{q,p}(dgm(ROI_1), dgm(ROI_1)) & \dots & W_{q,p}(dgm(ROI_1), dgm(ROI_n)) \\ W_{q,p}(dgm(ROI_2), dgm(ROI_1)) & \dots & W_{q,p}(dgm(ROI_2), dgm(ROI_n)) \\ \vdots & \ddots & \vdots \\ W_{q,p}(dgm(ROI_n), dgm(ROI_1)) & \dots & W_{q,p}(dgm(ROI_n), dgm(ROI_n)) \end{bmatrix}$$
(6)

3.6 Classification

The study integrates the 1D and 2D features from Wasserstein distances that captures the inter-ROI interaction for each subject into a classification framework using conventional CNN. The classification results demonstrate the effectiveness of the proposed method in distinguishing between different stages of MCI, showcasing the practical applicability of persistent homology in medical diagnostics. To perform classification, the subject-specific PR matrix $(n \times n)$ is utilized. The proposed CNN architecture used for classification purpose is depicted in Figure 3. This model integrates 1D features extracted from each ROI pair of the PR matrix with the 2D CNN features. Thus, the proposed classification model includes two steps. First, the Wasserstein distance matrix is flattened to create 1D features, focusing on pairwise relationships. In the second step, 2D features are extracted from the distance matrix through CNN layers to capture local patterns and spatial hierarchies (Figure 2). The features from the flattened matrix and the CNN layers are concatenated, creating a unified feature vector that combines information from both linear and convolutional layers. The concatenated features are then passed through several dense layers with dropout for regularization, reducing the risk of overfitting. Combining 1D and 2D features allows the model to create a richer and more diverse feature space that can capture different aspects of the data. 1D features can represent sequential or linear relationships, while 2D features can capture spatial or topological relationships. Hence, integrating both 1D and 2D features offer several advantages including a comprehensive and richer feature space for classification, enhanced learning from different perspectives, helps in mitigating the impact of noise and artifacts, and the ability to capture both local and global patterns. For the 2D features, our model comprises three CNN layers with 16, 32, and 64 filters respectively. This is followed by a max-pooling layer and another convolutional block with two CNN layers having 128 and 256 filters. We use kernel size of 3 for all CNN layers. Subsequently, a global average pooling layer condenses each feature map into a single value, forming a linear feature vector. ReLU activation functions are employed throughout the model. Simultaneously, the 1D features of the PR matrix $(n^2 \times 1)$, are processed through a linear layer, reducing them to 256 features. The resulting 256-dimensional feature vector from this linear layer and the 256-dimensional feature vector from the 2D CNN are concatenated and fed as an input to a series of fully connected layers of size 128, 64, and 32, each incorporating dropout set to 0.2. A softmax activation function is applied at the final layer for classification. Each Betti descriptor (H_0, H_1, and H_2) from every brain network is independently analyzed. The model is trained over 100 epochs using the Adam optimizer with a learning rate of 0.001 and a train-test split of 80-20. Cross-Entropy loss is employed for training. The model is trained to classify the following scenarios: (i) MCI versus HC (for both ADNI and our in-house dataset), (ii) EMCI versus HC (ADNI), (iii) LMCI versus HC (ADNI), and (iv) EMCI versus LMCI (ADNI). The classification part of this experiment is conducted using Kaggle notebooks with 2 × 16GB NVIDIA Tesla T4 GPU and the PyTorch deep learning framework.

4 Results and Discussion

The current study examines alterations in brain network topology between HC and MCI sub-types using persistent homology of fMRI time series. Numerous studies have been conducted over past decades to classify MCI from HC, with

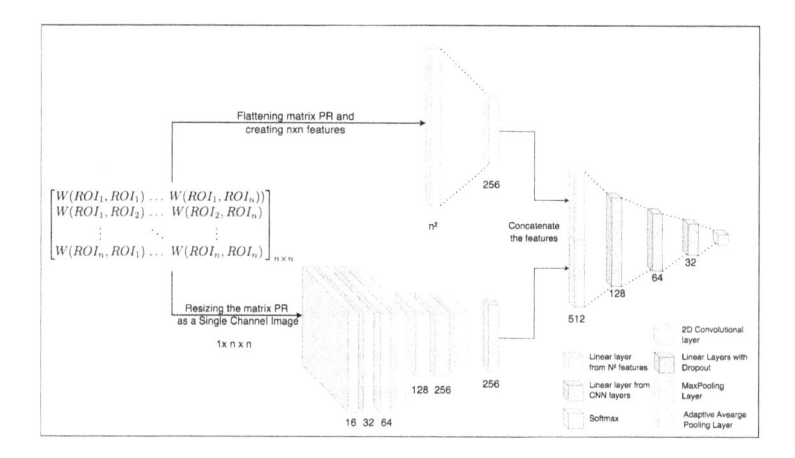

Fig. 3. The proposed deep learning architecture

several groups reporting fair classification performance. However the novelty in the proposed method lies in incorporation of an innovative methodology that combines sliding window embedding of fMRI time series with a tool originating from the emerging field of computational topology. Furthermore, an extensive statistical analysis on topological features comparing HC and MCI sub-types is carried out, aiming to identify statistically significant ROI-pairs in each functional brain network for investigating unique neurobiological patterns associated with various stages of MCI across different homology dimensions.

Persistence homology is computed for dimension-0, 1, and 2 on each point cloud using Vietoris-Rips filtration. The computation of persistent homology is performed using Ripser software [3]. The filtrations provide a basis for computing the persistent topological features that exist within the point cloud. The persistence of topological features is then encoded in persistence diagram. For three different homology dimensions H_0, H_1 and H_2, the derived persistence diagram for a point cloud is shown in Figure 1(d). As described in methodology section, in the persistence diagram, every point corresponds to a specific topological feature. The magnitude of the difference between the "birth" (b) and "death" (d) values indicates the life-span or persistence of the topological descriptors. The Wasserstein distance metric is used to compute the dissimilarity between two persistence diagrams. Figure 2 and Figure 4 capture this dissimilarity across all ROIs and across all subjects, respectively.

From persistence diagram both inter-subject (PS) and inter ROI (PR) Wasserstein distance is computed for each homology dimension as shown in Figure 1.(g) and Figure 1.(h). The Wilcoxon rank-sum test is conducted at 99% C.I on PR matrix to identify key ROI-pairs within brain and their associated patterns in order to distingusish EMCI and LMCI. As discussed previously, since the Wasserstein distance between fMRI time series of two brain regions captures the degree of topological similarity in their neural activation pattern, our study hypothesis seeks to find ROI-pairs for which the inter-ROI Wasserstein distance between (i)

Leveraging Persistent Homology for Differential Diagnosis 27

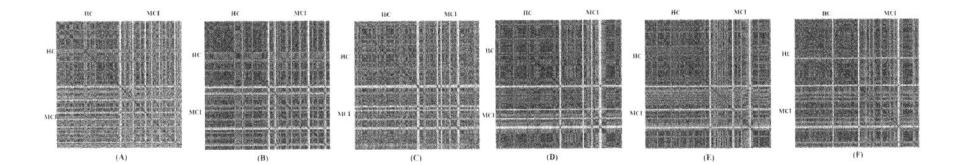

Fig. 4. ROI-specific Wasserstein distance (dim-0) across all subjects of HC ($n=50$) and MCI ($n=50$) from each considered brain network- showing the visible difference in pattern of Wasserstein distance between HC and MCI subjects. These ROIs are (A) Post cingulate 108 (DMN), (B) inf cerebellum 121 (CB), (C) IPL 96 (FP), (D) Occipital1 106 (OP), (E) Post cingulate 80 (CO) and (F) Pre-SMA 41 (SM).

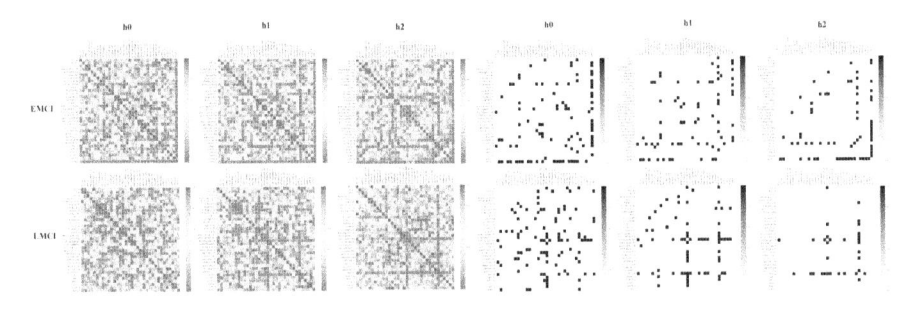

Fig. 5. Visualization of P-plot at 99% C.I for Default mode network for EMCI (*top row*) and LMCI *bottom row*. The *column 1, 2 and 3* show the P-plot for homology dimension 0, 1 and 2, respectively. *Column 4, 5, and 6* highlight the ROI-pair that showed significant differences ($p < 0.01$) in topology across all subjects of between (i) HC and EMCI, (ii) HC and LMCI. The visualization clearly depicts significant ROIs which are seen to be more concentrated in fewer brain regions as homology dimension increases in case of LMCI as compared to EMCI.

HC Vs. EMCI, (ii) HC Vs. LMCI groups is statistically significant ($p < 0.01$). Identification of such ROI-pairs in EMCI and LMCI may aid to uncover distinct neurobiological signatures associated with different stages of MCI. The visualization of the p-value plot which we refer as *P-plot*, as obtained for each ROI pair of DMN is shown in Figure 5. The P-plot visualization shows clear disparity in ROI pairs between EMCI and LMCI for each homology dimension-0 (*column-4*), 1 (*column-5*) and 2 (*column-6*). It is seen that significant ROI pairs tend to concentrate within fewer regions of DMN in the case of LMCI as compared to EMCI. This pattern is seen to be consistent for all six brain networks and pronounced particularly for homology dimension 2, followed by dimension 1. For example, as seen from Figure 5, in LMCI, significant dissimilarities in brain activity with other ROIs are primarily concentrated in two regions of DMN: post-cingulate-108 (PC108) and Ventrolateral Prefrontal Cortex (vlPFC). Conversely, in EMCI, significant dissimilarities in brain activity with other ROIs are extended to additional regions such as vmPFC, occipital, and inf temporal, in addition to vlPFC and PC108, for homology dimension-2. For dimension-1 also this spread in pattern across DMN

Table 1. The classification accuracy as obtained across six distinct brain networks.

Comparison	Dataset	Dim.	DMN	FP	OP	SM	CO	CB
HC vs MCI	ADNI	H_0	70.0	90.0	85.0	85.0	**95.0**	85.0
		H_1	65.0	90.0	**90.0**	75.0	75.0	75.0
		H_2	75.0	80.0	85.0	85.0	75.0	80.0
HC vs MCI	In-house TLSA	H_0	82.4	71.4	57.8	**85.0**	70.6	63.0
		H_1	76.5	52.9	68.4	72.2	52.9	63.2
		H_2	70.6	64.7	73.8	55.0	76.5	57.9
HC vs EMCI	ADNI	H_0	**76.5**	60.3	60.3	58.8	66.2	64.7
		H_1	61.8	69.1	75	57.4	52.9	60.3
		H_2	66.2	50	69.1	55.9	61.8	55.9
HC vs LMCI	ADNI	H_0	**91.1**	76.8	84.1	80.1	74.6	71.4
		H_1	85.7	55.6	79.3	74.6	73	61.9
		H_2	63.5	65.1	58.7	66.6	53.9	60.3
EMCI vs LMCI	ADNI	H_0	**80.0**	65.0	71.7	76.7	**80.0**	71.7
		H_1	75.0	66.7	68.3	70.0	61.6	68.3
		H_2	56.7	53.3	53.3	56.7	56.6	60.0

Table 2. Comparison of peak classification accuracy of the proposed CNN architecture with the conventional Densenet-121 and also with random forest ensemble classifier to check the efficacy of the proposed model in classifying HC and MCI.

Brain Network	Densenet-121 ADNI	Densenet-121 In-house TLSA	Ensemble Classifier ADNI	Ensemble Classifier In-house TLSA	**Proposed Model** ADNI	**Proposed Model** In-house TLSA
DMN	80.0%	82.0%	67.0%	72.48%	**91.1%**	**82.4%**
FP	89.6%	71.4%	83.0%	61.67%	**90.0%**	**71.4%**
OP	74.4%	72.9%	84.0%	70.67%	**90.0%**	**73.8%**
SM	80.0%	72.2%	80.0%	69.3%	**85.0%**	**85.0%**
CO	90.0%	75.0%	86.0%	68.19%	**95.0%**	**76.5%**
CB	85.0%	62.6%	82.0%	63.44%	**85.0%**	**63.2%**

regions is more noticeable for EMCI than LMCI cases. Post cingulate cortex plays a central role in various cognitive functions, including memory retrieval, attention, and self-referential processing. The vlPFC is portion of the prefrontal cortex which is located on the inferior frontal gyrus, involved in higher-order cognitive functions and executive control such as decision-making, response inhibition, working memory, and goal-directed behavior. Thus, the significant topological dissimilarities in activation pattern the post cingulate regions and vlPFC with several other brain regions of DMN in LMCI as compared to EMCI likely reflects the progressive neurodegenerative changes associated with advanced stages of cognitive impairment. These findings shed light on differences in the evolving patterns of neural activity and functional connectivity within specific brain regions, offering

Table 3. Comparative analysis with recent state-of-the-art techniques that used fMRI data to differentiate disease sub-types as well as to distinguish MCI and its sub-types.

Reference	Year	Modality (dataset)	Subjects	Accuracy
[16]	2022	fMRI (local)	HC Vs. MCI	65.14%
[7]	2023	fMRI (ADNI)	HC Vs. MCI	87%
[20]	2018	fMRI (ADNI)	HC Vs. MCI	82.6%
			EMCI Vs. LMCI	74.3%
[33]	2021	fMRI (ADNI)	HC Vs. MCI	82.8%
[6]	2020	fMRI (ADNI)	HC Vs. MCI	80%
[4]	2024	fMRI (ADNI)	HC Vs. MCI	89.47%
[22]	2018	fMRI (ADNI)	HC Vs. EMCI	74.23%
[21]	2020	fMRI (ADNI)	HC Vs. EMCI	76.07%
[25]	2021	fMRI (ADNI)	HC Vs. EMCI	74.42%
[32]	2021	fMRI (ADNI)	HC Vs. LMCI	87.2%
[31]	2021	fMRI (ADNI)	EMCI Vs. LMCI	79.36%
Proposed method	2024	fMRI (ADNI)	HC Vs. MCI	**95%** (H_0)
		fMRI (in-house TLSA)	HC Vs. MCI	**85%** (H_0)
		fMRI (ADNI)	HC Vs. EMCI	**76.5%** (H_0)
			HC Vs. LMCI	**91.1%** (H_0)
			EMCI Vs. LMCI	**80%** (H_0)

potential biomarkers for differential diagnosis. Furthermore, the significant dissimilarities in Wasserstein distance within fewer regions particularly in the PCC and vlPFC of DMN, in LMCI compared to EMCI implies a progression towards more localized disruptions in brain activity as cognitive impairment advances. These insights deepen our understanding of the topological changes occurring in neural networks as cognitive impairment progresses, offering new directions for tailored diagnostic and therapeutic interventions.

The performance of the proposed model in classifying MCI and its sub-types is tabulated in Table 1. The classification using the proposed CNN model yields the highest accuracy of 95% and 85% for ADNI and in-house TLSA MCI cohort, respectively. In case of MCI sub-types classification from HC, the accuracy is decreased to 76.5% to classify EMCI from HC. However, in classifying LMCI from HC, highest accuracy of 91.1% is obtained. While distinguishing disease sub-types EMCI Vs. LMCI, 80% accuracy is obtained. The variability in classification accuracy within the DMN can be attributed to the degree of cognitive impairment that influence the model's ability to distinguish between MCI sub-types. While comparing the peak accuracy across all networks, it is found that 0-dimensional topological features (H_0) perform the best. This is because H_0 captures the most fundamental topological aspect of data which is the number of connected components. This translates to identifying how functional connec-

Fig. 6. Illustrating the variation in six standard clinical dementia rating features between two distinct population: (i) ADNI cohort and (ii) in-house TLSA cohort.

tivity among different regions in a specific brain network evolve over time. In neurodegenerative disease like MCI, initial changes often manifest as alterations in basic connectivity pattern rather than more complex topological structures like loops (H_1) or voids (H_2). This makes H_0 a reliable measure for detecting such changes, leading to better classification performance. Moreover, H_0 features are simpler and less prone to noise as compared to higher-dimensional features. These complex features might introduce variability that does not contribute meaningfully to classification performance. To validate the efficacy of the proposed CNN model, its results are compared with those of the classical DenseNet-121 architecture and the random forest, which has shown superior performance among other random forest ensemble classifiers. It has been observed that the deep learning model generally outperforms traditional machine learning classifiers. On the ADNI dataset, the proposed CNN model surpasses both the DenseNet-121 and the ensemble classifier. For the TLSA dataset, the performance of the proposed CNN model is comparable to that of the DenseNet-121 model across all brain networks. This is shown in Table 2. However, the notable differences in classification accuracy are observed between the two distinct population. This can arise due to several factors. Population from different regions or ethnicity may have distinct demographic characteristics, genetic backgrounds, lifestyle factors, cultural practices, socioeconomic status, education levels, environmental exposures and prevalence rates of certain diseases. These differences can contribute to variations in brain structure, function, and connectivity patterns, affecting the results of fMRI analyses. Figure 6 shows the clear differences in six standard CDR features as obtained for the two distinct population. Moreover, differences in data acquisition protocols and imaging parameters may result in variations in the functional connectivity patterns between datasets of distinct population. Hence, the present study reports the classification performance separately on two distinct populations to ensure that the data is as comparable as possible and not confounded by site differences. This approach enhances the generalization ability and reliability of the proposed CNN methodology. It is seen that the proposed methodology outperforms the recent state-of-the art techniques that utilized fMRI to study MCI. The table comparing the proposed method with SOTA is shown in Table 3. The clinical relevance of the study lies in its ability to localize critical regions in the brain network whose activity patterns and connectivity are altered across different stages of MCI. By identifying these key regions in brain network and their associated topological patterns (Figure 5), the study provides valuable insights into the progression of cognitive impairment and the underlying neurobiological changes. The findings from the current study suggest the potential utility of employing persistent homology for differential diagnosis of MCI. Nevertheless, further research is required in order to validate these findings across diverse cohorts with more samples for conclusive inferences.

Acknowledgement. We thank the Director, Dr. K.V.S. Hari and the administration of Centre for Brain Research, IISc, for the support provided throughout the study.

References

1. Aithal, N., Pradeep, C.S., Sinha, N.: Mci detection using fmri time series embeddings of recurrence plots. In: 2024 IEEE International Symposium on Biomedical Imaging (ISBI). pp. 1–4 (2024). https://doi.org/10.1109/ISBI56570.2024.10635716
2. Ammu, R., Sinha, N.: Analysis of mild cognitive impairment utilizing covariance matrices of brain regions. In: 2023 IEEE 33rd International Workshop on Machine Learning for Signal Processing (MLSP). pp. 1–6. IEEE (2023)
3. Bauer, U.: Ripser: efficient computation of vietoris-rips persistence barcodes. Journal of Applied and Computational Topology **5**(3), 391–423 (2021)
4. Bhattacharya, D., , Sinha, N.e.: Multi-scale fmri time series analysis for understanding neurodegeneration in mci. arXiv preprint arXiv:2402.02811 (2024)
5. Bhattacharya, D., Sinha, N., Chattopadhyay, A., et al.: Image complexity based fmri-bold visual network categorization across visual datasets using topological descriptors and deep-hybrid learning. arXiv preprint arXiv:2311.08417 (2023)
6. Bi, X.a., Hu, X.e.: Multimodal data analysis of alzheimer's disease based on clustering evolutionary random forest. IEEE Journal of Biomedical and Health Informatics **24**(10), 2973–2983 (2020)
7. Bolla, G., Berente, D.B.e.: Comparison of the diagnostic accuracy of resting-state fmri driven machine learning algorithms in the detection of mild cognitive impairment. Scientific Reports **13**(1), 22285 (2023)
8. Caputi, L., Pidnebesna, A., Hlinka, J.: Promises and pitfalls of topological data analysis for brain connectivity analysis. Neuroimage **238**, 118245 (2021)
9. Das, S., Anand, D.V., Chung, M.K.: Topological data analysis of human brain networks through order statistics. PLoS ONE **18**(3), e0276419 (2023)
10. Dosenbach, N.U., Nardos, B.e.: Prediction of individual brain maturity using fmri. Science **329**(5997), 1358–1361 (2010)
11. The classification of endoscopy images with persistent homology: Dunaeva, O., Edelsbrunner, H.e. Pattern Recogn. Lett. **83**, 13–22 (2016)
12. Edelsbrunner, H., Harer, J.L.: Computational topology: an introduction. American Mathematical Society (2022)

13. Farias, S.T., Mungas, D., Reed, B.R., Harvey, D., DeCarli, C.: Progression of mild cognitive impairment to dementia in clinic-vs community-based cohorts. Arch. Neurol. **66**(9), 1151–1157 (2009)
14. Gakhar, H., Perea, J.A.: Sliding window persistence of quasiperiodic functions. Journal of Applied and Computational Topology pp. 1–38 (2023)
15. Hindriks, R., Adhikari, M.H.e.: Can sliding-window correlations reveal dynamic functional connectivity in resting-state fmri? Neuroimage **127**, 242–256 (2016)
16. Hu, M., Yu, Y.e.: Classification and interpretability of mild cognitive impairment based on resting-state functional magnetic resonance and ensemble learning. Computational intelligence and neuroscience **2022** (2022)
17. Jack, J., Clifford, R.e.: The alzheimer's disease neuroimaging initiative (adni): Mri methods. Journal of Magnetic Resonance Imaging **27**(4), 685–691 (2008)
18. Janoutová, J., Serỳ, O.e.: Is mild cognitive impairment a precursor of alzheimer's disease? short review. Central European journal of public health **23**(4), 365 (2015)
19. Jenkinson, M., Beckmann, C.F.e.: Fsl. Neuroimage **62**(2), 782–790 (2012)
20. Jie, B., Liu, M.e.: Sub-network kernels for measuring similarity of brain connectivity networks in disease diagnosis. IEEE Transactions on Image Processing **27**(5), 2340–2353 (2018)
21. Kam, T.E., Zhang, H., Jiao, Z.e.a.: Deep learning of static and dynamic brain functional networks for early mci detection. IEEE Trans. Med. Imaging pp. 39:478–87 (2020)
22. Kam, T.E., Zhang, H., Shen, D.: A novel deep learning framework on brain functional networks for early mci diagnosis. Medical image computing and computer assisted intervention – (MICCAI 2018). pp. 293–301 (2018)
23. Khasawneh, F.A., Munch, E.: Chatter detection in turning using persistent homology. Mech. Syst. Signal Process. **70**, 527–541 (2016)
24. Lee, H., Kang, H.e.: Persistent brain network homology from the perspective of dendrogram. IEEE transactions on medical imaging **31**(12), 2267–2277 (2012)
25. Lee, J., KoW, Kang, E.e.a.: A unified framework for personalized regions selection and functional relation modeling for early mci identification. Neuroimage (2021)
26. Merelli, E., Piangerelli, M.e.: A topological approach for multivariate time series characterization: the epileptic brain. In: In Proc. EAI International Conference on Bio-inspired Information and Communications Technologies. pp. 201–204 (2016)
27. Morley, J.E.: Anticholinergic medications and cognition. J. Am. Med. Dir. Assoc. **12**(8), 543–543 (2011)
28. Perea, J.A., Harer, J.: Sliding windows and persistence: An application of topological methods to signal analysis. Found. Comput. Math. **15**, 799–838 (2015)
29. Stolz, B.J., Emerson, T.e.: Topological data analysis of task-based fmri data from experiments on schizophrenia. Journal of Physics: Complexity **2**(3), 035006 (2021)
30. Wang, F., Kapse, S., Liu, S., Prasanna, P., Chen, C.: Topotxr: A topological biomarker for predicting treatment response in breast cancer (05 2021)
31. Wang, M., Lian, C., Yao, D.e.a.: Spatial-temporal dependency modeling and network hub detection for functional mri analysis via convolutional-recurrent network. IEEE Trans Biomed Eng pp. 2241–2252 (2020)
32. Yang, P.e.a.: Fused sparse network learning for longitudinal analysis of mild cognitive impairment. IEEE Transactions on Cybernetics p. 233–246 (2021)
33. Yang, P., Zhou, F.e.: Fused sparse network learning for longitudinal analysis of mild cognitive impairment. IEEE transactions on cybernetics **51**(1), 233–246 (2019)

Medical Image Classification Attack Based on Texture Manipulation

Yunrui Gu[1,2], Cong Kong[1,2], Zhaoxia Yin[1,2(✉)], Yan Wang[1,2], and Qingli Li[1,2]

[1] School of Communication and Electronic Engineering,
East China Normal University, Shanghai 200241, China
zxyin@cee.ecnu.edu.cn
[2] Shanghai Key Laboratory of Multidimensional Information Processing,
East China Normal University, Shanghai 200241, China

Abstract. The security of artificial intelligence systems has received great attention, especially in the field of smart medical diagnosis in over the past few years. In order to enhance the security of smart medical systems, it is important to study adversarial attack methods to increase defense performance, and the central aspect of adversarial attacks lies in crafting effective strategies that can integrate covert malicious behaviors within the system. However, due to the diversity of medical imaging modes and dimensions, creating a unified attack approach that produces imperceptible examples with high content similarity and applies them across various medical image classification systems presents significant challenges. Most existing attack methods aim at attacking natural image classification models, which inevitably add global noise to the image and make the attack more visible, simultaneously does not taking into account that medical image classification task considers texture information more. To address this issue, we propose a new adversarial attack method based on changing texture information that utilizes the CycleGAN approach, while also incorporating AdvGAN to ensure the attack success rate. Our method can provide attacks in various medical image classification tasks. Our experiment includes two public medical image datasets, including chest X-Ray image dataset and melanoma dermoscopy dataset, which contain different imaging modes and dimensions. The results indicate that our model has superior performance in attacking medical image classification tasks in different imaging modes and dimensions compared to other state-of-the-art adversarial attack methods.

Keywords: Medical diagnosis · Adversarial attack · Texture

1 Introduction

Deep neural networks(DNNs) have demonstrated excellent performance in tasks involving natural images, such as image classification[10,24,25], object detection[32,34], and image segmentation[8]. With their success in natural image tasks, deep neural networks have also shown strong performance in medical image tasks[1,9], not only in the aforementioned tasks but also in various datasets including dermoscopy and X-Rays. However, research indicates that even well-trained DNNs are susceptible to adversarial attacks[2,7], both in natural images and medical images, e.g., small perturbations applied to the input images can deceive DNNs to have wrong conclusions. The global medical pressure is increasing currently and an increasing number of networks are being introduced to assist doctors in addressing clinical issues while the vulnerability exhibited by DNNs in the face of adversarial attacks can prove to be extremely detrimental[17]. Adversarial examples are likely to result in misdiagnoses and various social disturbances, thus limiting the applicability of deep learning in both safety- and security-critical environments[19]. Medical misdiagnosis not only leads to unnecessary waste of medical resources but also poses a serious threat to patient safety and well-being.

Studying adversarial attacks on medical images can help identify and expose vulnerabilities in medical diagnostic systems[17], thereby assisting researchers in enhancing their robustness against adversarial attacks more effectively. The existing methods for generating adversarial examples on natural images to attack DNNs mostly involve adding global noise to the image to disrupt the DNN's classification decisions[21]. This adversarial attack method is very effective, but it can also result in adversarial example not matching the original image, making it easy to detect. While in the classification task of medical images, researches have found that compared to natural images, DNNs tend to classify medical images more through texture rather than shape[12]. So attacking the texture information of medical images will be a more effective way to realize attacks.

Most of the existing adversarial attacks on medical images are based on the transfer of attack methods on natural images. Although recent attack methods have taken into account the characteristics of medical images in the feature space, their methods have too many limitations on features, resulting in suboptimal visual effects of the generated adversarial examples.

In this paper, we propose a simple but effective attack method that can fool deep medical diagnosis systems working with different medical datasets. Using CycleGAN[33] can extract malignant texture information from deep features and generate it on benign images, while also ensuring that the transformed images maintain a high degree of similarity to the original benign images. By incorporating the classification results of a white-box model into the generation process of the GAN network, we can further guide the network to generate adversarial examples in the wrong direction, increasing the attack success rate of our method.

Our main contribution in this paper is summarized as follows:

- Our method can achieve a good success rate in attacks compared to existing adversarial attack methods, nearly 100%.
- Our method better utilizes the texture information of medical images. Since DNNs tend to prioritize texture in medical image classification, our method offers better transferability compared to others.

- Our method put emphasis on textures and generates fewer perturbations. Consequently, our method generates adversarial examples with less visual discrepancies and better imperceptibility compared to existing methods.

2 Related works

2.1 Adversarial Attack

In Nature Images. To attack a given a pretrained DNN model f and a normal example x with a corresponding class label y, adversarial attack method aims to maximize the classification error of the DNN model while keeping x_{adv} within a small ϵ-ball centered on the original example $x(||x_{adv} - x||_p \leq \epsilon, p = 2$ or $p = \infty)$, where p is the L_p-norm.

Goodfellow introduces the Fast Gradient Sign Method (FGSM)[7], a pioneering approach that generates adversarial examples by approximating the loss function using a first-order derivative. FGSM remains one of the most widely used and effective attack techniques. Building on this, the Basic Iterative Method (BIM)[15] and Projected Gradient Descent (PGD)[18] follow. These methods extend single-step perturbations into multi-step processes, incorporating smaller perturbations and random initializations using a uniform distribution. They significantly enhance attack effectiveness, with PGD often recognized as the most robust first-order attack.

Another category of attacks focuses on optimization-based techniques, such as the Carlini-Wagner(CW) method[3], which optimizes both the perturbation magnitude and the misclassification loss of adversarial examples. However, both gradient-based and optimization-based approaches face limitations. They typically modify one example at a time, which can be time-consuming due to the need for multiple optimization iterations. Drawing inspiration from Generative Adversarial Networks (GANs)[6], Xiao proposes AdvGAN[29], which utilizes adversarial loss and an image-to-image framework to map original images to perturbed versions. AdvGAN produces adversarial examples that are both highly effective against target networks and visually indistinguishable. Further developments, such as AdvGAN++[11], aim to generate more precisely targeted adversarial examples.

In Medical Images. The majority of adversarial attack studies in medical image analysis predominantly concentrate on the white-box scenario. Specifically, these studies mainly address the vulnerability of computer-aided diagnosis models across different medical imaging tasks, leveraging comprehensive understanding of medical deep neural networks (DNNs). During adversary generation, the attacker may regard the target diagnosis DNN as a locally deployed model.

Several researchers dedicate significant efforts to crafting tailored adversarial attack techniques [14,22,26,31] specifically suited for medical imaging tasks. Yao [31] delves into the vulnerability of medical image representations, devising the Hierarchical Feature Constraint(HFC) to cloak adversarial representations within the clean feature domain. This HFC module serves as supplementary guidance, seamlessly integrating into existing attack frameworks to mitigate detection risks. Functionally, HFC incentivizes adversarial features to align with high-density regions of normal feature distributions by maximizing adversarial feature log-likelihood. Addressing diverse medical image modalities, Qi et al. [22] introduce a Stablized Medical Image Attack(SMIA). They optimize an objective function comprising deviation and stabilization loss terms. The deviation term widens the prediction gap between adversarial outputs and ground truth, while the stabilization term acts as regularization, confining adversarial perturbations to low variance. This regularization mitigates instances of local optima during optimization, often induced by instance-wise image noise.

Wang et al. [22] introduce Feature-Space-Restricted Attention Attack (FSRAA) to generate adversarial examples across various medical modalities with minimal visual interference. This method imposes feature-level constraints to ensure that the adversarial examples remain near the decision boundaries within the feature space. Furthermore, they employ an attention mechanism to manage image-level perturbations, directing them specifically to the diseased region by integrating class-specific attention information into the process of generating adversarial perturbations.

Chen et al.[4] introduce a novel adversarial attack approach that leverages frequency constraints, allowing it to be effective across different medical image classification tasks. This approach involves adding perturbations primarily to high-frequency components while preserving the integrity of low-frequency content, thereby maintaining the overall similarity of the image.

Recently, Yao et al.[30] also make certain improvements and supplements to HFC. They provide a comprehensive proof of the theorem regarding feature vulnerability within binary classification and extended this theorem to apply to multi-class settings. Their work offers both empirical and theoretical insights, particularly concerning out-of-distribution adversarial features in these multi-class scenarios.

Although these existing methods consciously attack the feature representation of medical images, especially HFC and FSRAA, they have excessive limitations on the feature space, resulting in more details being lost in adversarial examples. SMIA, although having relatively better transferability, lacks consideration of features, resulting in lower attack success rate.

2.2 Texture Transfer

Texture Networks by Ulyanov et al.[27] leverage convolutional neural networks (CNNs) to perform texture generation and style transfer tasks in a feed-forward manner, yielding impressive results with reduced computational overhead. Additionally, advancements in convolutional neural network architectures, such as the VGG network[24], have facilitated the development of more sophisticated style transfer methods.

Beyond texture synthesis and style transfer, other relevant techniques like photorealistic image stylization[16] have contributed to the broader realm of image manipulation and synthesis. These methodologies offer valuable insights and alternative avenues for achieving realistic and visually compelling outcomes in texture synthesis and transfer tasks.

Unlike previous approaches reliant on paired training data, CycleGAN[33] introduces an unsupervised learning framework, eliminating the need for direct correspondences between images from different domains. This innovation is underpinned by its novel use of cycle-consistency loss, ensuring fidelity in both content preservation and texture/style translation. Prior methodologies such as pix2pix and dualGAN, while pioneering in their own right, are constrained by their dependence on paired examples. CycleGAN's dual-generator and discriminator architecture further enhances its efficacy, facilitating high-fidelity texture conversions through adversarial training dynamics.

3 Method

Figure 1 illustrates the overall architecture of CycleAdvGAN including two mappings, which mainly consist of five parts: the generator G_M, the generator G_B, the discriminator D_M of domain $Malignant$, the discriminator D_B of domain $Benign$, and the target neural network F. Figure 1(a) represents the mapping process of transforming malignant images into benign adversarial examples, with G_M playing the main role in the texture transfer generation, while figure 1(b) is the opposite, representing the mapping process of transforming benign images into malignant ones, with G_B playing the main role in the texture transfer generation. Two mapping processes share two generators G_M, G_B and two discriminators D_M, D_B respectively. The process of cyclic learning through GAN allows G_M to learn features of benign images and transfer them on malignant images, and G_B does the same in reverse.

Specifically, in figure 1(a), here the generator G_M takes the original image $Malignant$ as its input and generates $Benign_{fake}$ as $G_M(m)$. Then, the generated adversarial example $Benign_{fake}$ will be sent to the discriminator D_B, which

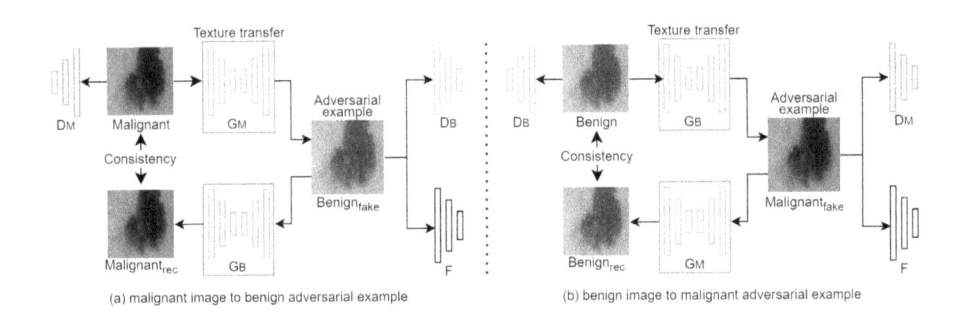

Fig. 1. The overall architecture of CycleAdvGAN. (a) represents the process of transforming malignant images into benign adversarial example, while (b) does the opposite, i.e., transforming benign images into malignant ones.

distinguishes between the adversarial data and the original $Benign$ example. D_B guides G_M to transform $Malignant$ images into outputs indistinguishable from domain $Benign$. As shown in Figure 1, the two parts of the architecture are symmetrical. Therefore, in figure 1(b) the generator G_B takes the original image example as its input and generates $Malignant_{fake}$ as $G_B(b)$. Then, the generated adversarial example $Malignant_{fake}$ will be sent to the discriminator D_M, which differentiates between the adversarial data and the original $Benign$ example. D_M instructs G_B to convert $Benign$ images into outputs indistinguishable from $Malignant$ domain. To achieve the goal of deceiving the learning model, we execute a white-box attack, where the target model is F. F receives $Benign_{fake}$ and $Malignant_{fake}$ as input, with its loss L_{adv} representing the distance between the prediction and the ground truth class.

3.1 Texture Transfer

Adversarial Loss for G_M. We apply adversarial losses to both mapping functions. For the mapping function G_M: $Malignant$ to $Benign$ and its discriminator D_B, we express the objective as

$$L_{G_M}(G_M, D_B) = \mathbb{E}_{b \sim P_B}\left[\log\left(D_B(b)\right)\right] \\ + \mathbb{E}_{m \sim P_M}\left[\log\left(1 - D_B\left(G_M(m)\right)\right)\right], \quad (1)$$

where generator G_M aims to generate imperceptible adversarial example $Benign_{fake}$ as $G_M(m)$ that is looking similar to original $Benign$ images, while D_B aims to distinguish between generated adversarial example and original $Benign$ example.

Adversarial Loss for G_B. We apply a similar adversarial loss for the mapping function G_B:$Benign$ to $Malignant$, and its discriminator D_M. The loss is defined as

$$L_{G_B}(G_B, D_M) = \mathbb{E}_{m \sim P_M}\left[\log\left(D_M(m)\right)\right] \\ + \mathbb{E}_{b \sim P_B}\left[\log\left(1 - D_M\left(G_B(b)\right)\right)\right], \quad (2)$$

where generator G_B aims to recover adversarial example $Benign_{fake}$ to clean example $Malignant_{rec}$, while D_M aims to distinguish between adversarial example and generated clean example.

Cycle Consistency Loss. After engaging in the min-max game with discriminator D_B, the generator G_M is expected to produce visually convincing adversarial examples. Nevertheless, due to the absence of pairwise supervision, the reconstructed image might not preserve the content information from the original clean image. Drawing inspiration from CycleGAN, which demonstrates the concept of cycle consistency constraints in style transfer, we incorporate different forms of cycle consistency losses to facilitate the seamless integration of adversarial attacks. The adversarial loss guides the generator's output, after perturbation, to resemble the corresponding instance in the target domain. However, relying solely on adversarial losses may not ensure that the trained function can accurately map an input classified as *Malignant* to the desired *Benign* output. To further refine the perturbations, we introduce a cycle consistency loss into the architecture. The loss is defined as

$$L_{cycle} = \mathbb{E}_{m \sim P_M}\left[\| G_B(G_M(m)) - m \|_1\right] \\ + \mathbb{E}_{b \sim P_B}\left[\| G_M(G_B(b)) - b \|_1\right]. \quad (3)$$

3.2 Adversarial attack

Adversarial Loss for F. The loss for fooling the target model F in an untargeted attack is

$$L_{adv} = \mathbb{E}_{m \sim P_M}\left[\mathcal{L}_{\mathcal{F}}(b_{fake}, l_c)\right] \\ + \mathbb{E}_{b \sim P_B}\left[\mathcal{L}_{\mathcal{F}}(m_{fake}, l_c)\right], \quad (4)$$

where l_c represents the true class. Meanwhile, $\mathcal{L}_{\mathcal{F}}$ denotes the loss function (e.g., cross-entropy loss) used to train the original model F. The L_{adv} loss is designed to cause the perturbed image to be misclassified while ensuring that the recovered instance is correctly classified.

3.3 Total Loss

We optimize a objective function defined as

$$L = (1-\lambda)[L_{G_M} + L_{G_B} + L_{cycle}] + \lambda L_{adv}, \quad (5)$$

where λ is a hyperparameter used to control the degree to which the loss tends towards CycleGAN or AdvGAN. The smaller the value of λ, the more biased the loss will be towards CycleGAN, resulting in better visual effects of adversarial examples and more covert attacks; The larger the value of λ, the more biased the loss will be towards AdvGAN. Although it can increase the attack success rate, the attack will also be more noticeable.

4 Experiments

4.1 Experiment Setting

Datasets. Chest X-Ray and Dermoscopy are selected as benchmarks. These tasks represent some of the most notable achievements in deep learning for medical imaging, partly due to the availability of standard public datasets[5].

For the Chest X-Ray task, the goal is to classify X-rays into two categories to identify pneumonia. We utilize a specific Chest X-Ray dataset[13], which contains 5,232 chest X-ray images. Unlike the Chest X-Ray 14 dataset[28] that includes multiple labels, this dataset provides only two labels for all images:

3,883 labeled as pneumonia and 1,349 as normal. This binary labeling helps reduce variability caused by differing examples.

For Dermoscopy, the International Skin Imaging Collaboration (ISIC) 2020 dataset[23] is used, chosen for its validation set availability and the task at hand. This dataset includes over 30,000 dermoscopy images and is organized according to the categories of "benign" and "malignant" melanoma[17].

Comparison Methods. We have selected classic adversarial attack methods such as FGSM and PGD, which have been designed and migrated from natural images to medical images, and have also achieved excellent attack results. At the same time, we also select SMIA and HFC, two adversarial attack methods specifically designed for medical images, as a comparison, to demonstrate that our method have good results when applied to medical images.

Implementation Details. ResNet-50, Inception-V3 and DenseNet-121 serve as the backbone architectures, with their final layers replaced by two newly added fully connected layers for classification. The network weights, excluding the final layers, are initialized using pre-trained weights from ImageNet instead of random initialization. The Adam optimizer operates with a base learning rate set at 1×10^{-5}. Given the unique characteristics of medical images, several preprocessing steps are included in the data loading process. Initially, all images convert to the RGB format, and then the images undergo random resizing to a fixed scale of 224×224 from their original dimensions. During our experiments, attacks proceed with a perturbation magnitude of 0.03 for both FGSM and PGD attacks. For PGD attacks, the iterative step size is set to $\alpha = 0.01$. For the transferability experiments, we utilize a robust vision transformer, MedViT-s[20], as the validation model to effectively assess the transferability of our method. The parameter settings remain consistent with those described in the original paper.

Evaluation Metrics. The impact of the attack is assessed from multiple perspectives. Firstly, the effectiveness of the adversarial attack on DNNs is quantified using the Attack Success Rate(ASR), which measures the proportion of successful attacks out of all attempts. A higher ASR reflects a greater effectiveness of the attack method and indicates the vulnerability of the targeted system. Additionally, the similarity between the original and adversarial examples is evaluated using the Structural Similarity(SSIM), which considers three aspects of similarity: luminance, contrast, and structure. Higher SSIM values suggest that the original and adversarial images are more alike in terms of their structural and content features.

4.2 Effects of Hyper-Parameters

We also study the hyperparameters λ in Equation (5), the impact is shown in the Figure 2. As previously described, as λ increases, the loss will shift from CycleGAN to AdvGAN, which means that the attack success rate will continue to increase, while the SSIM maintained by relying on cycle consistency loss will continue to decrease. In order to achieve a balance between ASR and SSIM, we seet hyperparameters λ to 0.7, so that we can ensure that both ASR and SSIM are greater than 0.9 at the same time.

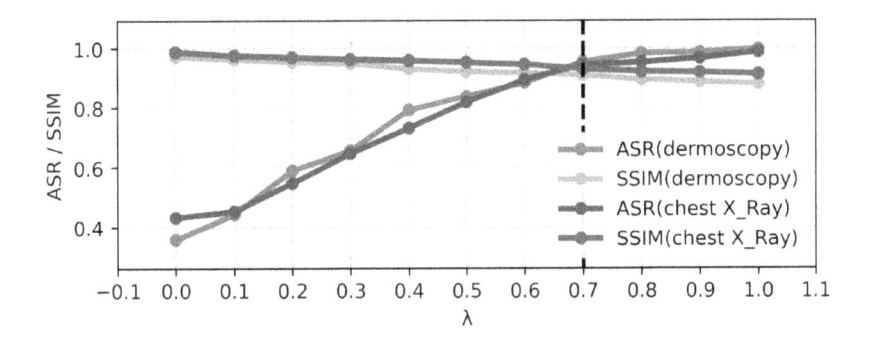

Fig. 2. ASR and SSIM under the influence of different hyperparameters λ.

4.3 Attack Visualization

To illustrate the result and advantage of our method, we visualize several examples in Figure 3. Compared with previous works, the noise generated by our method attached to the original image is more inclined towards high-frequency texture content rather than global noise, which ensures the invisibility of our attack method. According to the previous description, DNN tends to rely more on texture in medical image classification tasks.Through the cyclic learning of GAN, the generators G_M and G_B respectively generate texture features of benign and malignant images on each other, which can make perturbations more biased towards texture features, and as a result the method proposed in this article can also improve the attack success rate.

4.4 Attack Effectiveness

Table 1 and Table 2 shows the result of five attack methods against different classification models (i.e., ResNet50, Inception-V3, DenseNet121) on two medical image datasets. With the parameter information of the white-box model brought by AdvGAN, our method can achieve high attack success rates for all datasets and all backbone networks, nearly 100%. Additionally, the SSIM value serves as

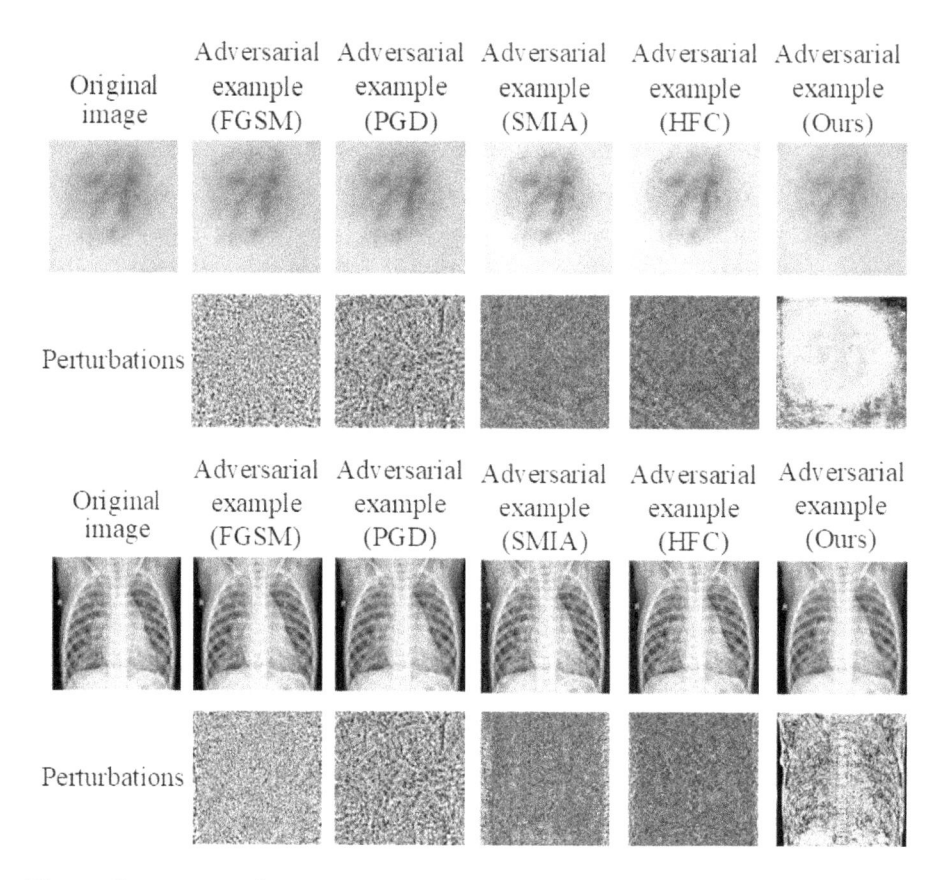

Fig. 3. Comparison of the original and adversarial examples generated by different attack methods. The first line is the original image and the adversarial examples under different attack methods, while the second line is the difference between the adversarial examples and the original image.

Table 1. Results of attack success rate(ASR), structural similarity(SSIM) values by five attack methods on Dermoscopy dataset against respective white-box model. Both ASR and SSIM are better as they are higher.

Attack Method	Resnet50 ASR(↑)	Resnet50 SSIM(↑)	Densenet121 ASR(↑)	Densenet121 SSIM(↑)	Inception-V3 ASR(↑)	Inception-V3 SSIM(↑)
FGSM[7]	0.8973	0.7965	0.8864	0.8165	0.9012	0.7983
PGD[18]	1.0000	0.8398	1.0000	0.8263	1.0000	0.8204
SMIA[22]	0.8968	0.8835	0.8234	0.8879	0.9194	0.8991
HFC[30]	0.9532	0.8979	0.9627	0.8825	0.9367	0.8817
Ours	0.9961	0.9126	0.9895	0.9038	0.9529	0.9021

a dependable metric to gauge the extent of human perceptibility in adversarial examples. By incorporating the cycle consistency loss, our method achieves the highest SSIM values, around 90%, surpassing other methods. These findings indicate that our attack method is effective across multiple medical datasets with varying modalities.

Table 2. Results of attack success rate(ASR), structural similarity(SSIM) values by five attack methods on Chest X-Ray dataset against respective white-box model. Both ASR and SSIM are better as they are higher.

Attack Method	Resnet50 ASR(↑)	SSIM(↑)	Densenet121 ASR(↑)	SSIM(↑)	Inception-V3 ASR(↑)	SSIM(↑)
FGSM[7]	0.9032	0.8043	0.9124	0.8139	0.9046	0.8231
PGD[18]	0.9954	0.8264	0.9961	0.8157	0.9929	0.8167
SMIA[22]	0.9652	0.8863	0.9531	0.8755	0.9643	0.8924
HFC[30]	0.9583	0.8753	0.9724	0.8861	0.9514	0.8961
Ours	0.9918	0.8937	0.9986	0.8924	0.9962	0.9034

4.5 Transferability

We conduct untargeted attack experiments on two datasets and four models to assess the transferability of our methods. We first choose a white-box model to generate adversarial examples and then applied them to other black-box models to verify the transportability

For the dermoscopy dataset, Table 3 presents the transferability results of five attack methods-FGSM, PGD, SMIA, HFC, and our proposed method-across various classification models, including ResNet50, DenseNet121, Inception-V3, and MedViT. According to Table 1, our attack method achieves a nearly 100% success rate in a white-box setting.

Next, we analyze how well our method transfers to target models under black-box conditions. Our approach matches the white-box success rate of other methods while outperforming them in terms of transportability, achieving a significantly higher attack success rate, with an improvement of about 20%. Additionally, our method outperforms SMIA, which shows a 3% improvement in transferability. We also utilize MedViT to further verify transferability. Our white-box attacks on commonly used CNN models successfully transfer to black-box MedViT models, while white-box MedViT attacks are equally effective against other models.

To verify the effectiveness of our approach across different datasets, we perform experiments using the chest X-Ray dataset. The same attack methods are applied within the context of the Dermoscopy dataset. The experimental outcomes are shown in Table 4. Consistent with the data presented in Table 3, the transferability on the chest X-Ray dataset is approximately 3% higher than the best results obtained from previous methods.

These findings highlight the superior performance and enhanced transferability of our approach in comparison to other attack methods. This can be attributed to our method's focus on semantic elements such as texture and boundaries, which aligns with DNN's tendency to emphasize texture in classification tasks. Consequently, our method achieves better transferability across other DNN classification models.

Table 3. Transferability results reflected by attack success rate(ASR). The first column is the source white-box model, while the second column is the used attack method, followed by the ASR results on the target black-box model. The best results are highlighted in bold.

Source	Attack	Densenet121	Inception-V3	MedViT
Resnet50	FGSM[7]	0.2471	0.3127	0.2785
	PGD[18]	0.1143	0.1851	0.1624
	SMIA[22]	0.4067	0.4216	0.4367
	HFC[30]	0.3461	0.3954	0.3789
	Ours	**0.4651**	**0.4867**	**0.4965**
		Resnet50	Inception-V3	MedViT
Densenet121	FGSM[7]	0.3251	0.1249	0.2157
	PGD[18]	0.1934	0.1047	0.1753
	SMIA[22]	0.3968	0.3564	0.3852
	HFC[30]	0.3579	0.3327	0.3445
	Ours	**0.4142**	**0.3649**	**0.4621**
		Resnet50	Densenet121	MedViT
Inception-V3	FGSM[7]	0.1368	0.1594	0.1954
	PGD[18]	0.0997	0.1162	0.1462
	SMIA[22]	0.3849	0.4249	0.4368
	HFC[30]	0.3958	0.4398	0.4523
	Ours	**0.4176**	**0.4517**	**0.4784**
		Resnet50	Densenet121	Inception-V3
MedViT	FGSM[7]	0.3145	0.2714	0.2214
	PGD[18]	0.1623	0.1945	0.1842
	SMIA[22]	0.4372	0.4185	0.4321
	HFC[30]	0.3894	0.3956	0.4489
	Ours	**0.5123**	**0.4697**	**0.4723**

Table 4. Transferability results reflected by attack success rate(ASR). The first column is the source white-box model, while the second column is the used attack method, followed by the ASR results on the target black-box model. The best results are highlighted in bold.

Source	Attack	Densenet121	Inception-V3	MedViT
Resnet50	FGSM[7]	0.1567	0.1038	0.2785
	PGD[18]	0.1134	0.0917	0.3624
	SMIA[22]	0.4356	0.3846	0.4967
	HFC[30]	0.4678	0.3819	0.2789
	Ours	**0.4791**	**0.3961**	**0.5185**
		Resnet50	Inception-V3	MedViT
Densenet121	FGSM[7]	0.1681	0.1237	0.3157
	PGD[18]	0.1169	0.0867	0.2753
	SMIA[22]	0.4319	0.3961	**0.4852**
	HFC[30]	0.3967	0.3754	0.3445
	Ours	**0.4293**	**0.4029**	0.4721
		Resnet50	Densenet121	MedViT
Inception-V3	FGSM[7]	0.1394	0.1469	0.2954
	PGD[18]	0.1063	0.1087	0.2462
	SMIA[22]	0.4268	0.4397	**0.5368**
	HFC[30]	0.4073	0.4469	0.3523
	Ours	**0.4381**	**0.4483**	0.4781
		Resnet50	Densenet121	Inception-V3
MedViT	FGSM[7]	0.3745	0.3214	0.2214
	PGD[18]	0.2623	0.3945	0.1842
	SMIA[22]	0.4372	**0.5185**	0.4321
	HFC[30]	0.2894	0.2956	0.4489
	Ours	**0.5123**	0.4697	**0.4723**

5 Conclusion

This study proposes a new attack method targeting multiple medical image datasets. Our method transforms benign and malignant images into adversarial examples through a CycleGAN network, while ensuring their structural similarity to enhance invisibility. Additionally, AdvGAN guides the network to learn towards the incorrect classification direction of DNN. The method proposed in this article can achieve both high ASR and the best SSIM. Although the proposed method demonstrates excellent performance, there remain several areas that warrant further exploration and are recommended for future research. We can further restrict perturbations more strictly to texture and boundary regions through frequency domain or attention mechanisms, thereby achieving better results. We anticipate that our research will encourage more focused studies on medical images.

Acknowledgments. This research work is partly supported by National Natural Science Foundation of China No.62472177, No.62172001, and Science and Technology Commission of Shanghai Municipality (Grant No. 22DZ2229004).

References

1. Abbas, A., Abdelsamea, M.M., Gaber, M.M.: Classification of covid-19 in chest x-ray images using detrac deep convolutional neural network. Appl. Intell. **51**, 854–864 (2021)
2. Akhtar, N., Mian, A.: Threat of adversarial attacks on deep learning in computer vision: A survey. Ieee Access **6**, 14410–14430 (2018)
3. Carlini, N., Wagner, D.: Towards evaluating the robustness of neural networks. In: 2017 ieee symposium on security and privacy (sp). pp. 39–57. Ieee (2017)
4. Chen, F., Wang, J., Liu, H., Kong, W., Zhao, Z., Ma, L., Liao, H., Zhang, D.: Frequency constraint-based adversarial attack on deep neural networks for medical image classification. Comput. Biol. Med. **164**, 107248 (2023)
5. Finlayson, S.G., Chung, H.W., Kohane, I.S., Beam, A.L.: Adversarial attacks against medical deep learning systems. arXiv preprint arXiv:1804.05296 (2018)
6. Goodfellow, I., Pouget-Abadie, J., Mirza, M., Xu, B., Warde-Farley, D., Ozair, S., Courville, A., Bengio, Y.: Generative adversarial nets. Advances in neural information processing systems **27** (2014)
7. Goodfellow, I.J., Shlens, J., Szegedy, C.: Explaining and harnessing adversarial examples (2015)
8. Guo, Y., Liu, Y., Georgiou, T., Lew, M.S.: A review of semantic segmentation using deep neural networks. International journal of multimedia information retrieval **7**, 87–93 (2018)
9. Hatamizadeh, A., Tang, Y., Nath, V., Yang, D., Myronenko, A., Landman, B., Roth, H.R., Xu, D.: Unetr: Transformers for 3d medical image segmentation. In: Proceedings of the IEEE/CVF winter conference on applications of computer vision. pp. 574–584 (2022)
10. He, K., Zhang, X., Ren, S., Sun, J.: Deep residual learning for image recognition. In: Proceedings of the IEEE conference on computer vision and pattern recognition. pp. 770–778 (2016)
11. Jandial, S., Mangla, P., Varshney, S., Balasubramanian, V.: Advgan++: Harnessing latent layers for adversary generation. In: Proceedings of the IEEE/CVF International Conference on Computer Vision Workshops. pp. 0–0 (2019)
12. Kaviani, S., Han, K.J., Sohn, I.: Adversarial attacks and defenses on ai in medical imaging informatics: A survey. Expert Syst. Appl. **198**, 116815 (2022)
13. Kermany, D., Zhang, K., Goldbaum, M., et al.: Labeled optical coherence tomography (oct) and chest x-ray images for classification. Mendeley data **2**(2), 651 (2018)
14. Kulkarni, Y., Bhambani, K.: Kryptonite: An adversarial attack using regional focus. In: Applied Cryptography and Network Security Workshops: ACNS 2021 Satellite Workshops, AIBlock, AIHWS, AIoTS, CIMSS, Cloud S&P, SCI, SecMT, and SiMLA, Kamakura, Japan, June 21–24, 2021, Proceedings. pp. 463–481. Springer (2021)
15. Kurakin, A., Goodfellow, I.J., Bengio, S.: Adversarial examples in the physical world. In: Artificial intelligence safety and security, pp. 99–112. Chapman and Hall/CRC (2018)

16. Luan, F., Paris, S., Shechtman, E., Bala, K.: Deep photo style transfer. In: Proceedings of the IEEE conference on computer vision and pattern recognition. pp. 4990–4998 (2017)
17. Ma, X., Niu, Y., Gu, L., Wang, Y., Zhao, Y., Bailey, J., Lu, F.: Understanding adversarial attacks on deep learning based medical image analysis systems. Pattern Recogn. **110**, 107332 (2021)
18. Madry, A., Makelov, A., Schmidt, L., Tsipras, D., Vladu, A.: Towards deep learning models resistant to adversarial attacks. arXiv preprint arXiv:1706.06083 (2017)
19. Mangaokar, N., Pu, J., Bhattacharya, P., Reddy, C.K., Viswanath, B.: Jekyll: Attacking medical image diagnostics using deep generative models. In: 2020 IEEE European Symposium on Security and Privacy (EuroS&P). pp. 139–157. IEEE (2020)
20. Manzari, O.N., Ahmadabadi, H., Kashiani, H., Shokouhi, S.B., Ayatollahi, A.: Medvit: a robust vision transformer for generalized medical image classification. Comput. Biol. Med. **157**, 106791 (2023)
21. Ortiz-Jiménez, G., Modas, A., Moosavi-Dezfooli, S.M., Frossard, P.: Optimism in the face of adversity: Understanding and improving deep learning through adversarial robustness. Proc. IEEE **109**(5), 635–659 (2021)
22. Qi, G., GONG, L., Song, Y., Ma, K., Zheng, Y.: Stabilized medical image attacks. In: International Conference on Learning Representations (2021), https://openreview.net/forum?id=QfTXQiGYudJ
23. Rotemberg, V., Kurtansky, N., Betz-Stablein, B., Caffery, L., Chousakos, E., Codella, N., Combalia, M., Dusza, S., Guitera, P., Gutman, D., et al.: A patient-centric dataset of images and metadata for identifying melanomas using clinical context. Scientific data **8**(1), 34 (2021)
24. Simonyan, K., Zisserman, A.: Very deep convolutional networks for large-scale image recognition. arXiv preprint arXiv:1409.1556 (2014)
25. Szegedy, C., Vanhoucke, V., Ioffe, S., Shlens, J., Wojna, Z.: Rethinking the inception architecture for computer vision. In: Proceedings of the IEEE conference on computer vision and pattern recognition. pp. 2818–2826 (2016)
26. Tian, B., Guo, Q., Juefei-Xu, F., Le Chan, W., Cheng, Y., Li, X., Xie, X., Qin, S.: Bias field poses a threat to dnn-based x-ray recognition. In: 2021 IEEE international conference on multimedia and expo (ICME). pp. 1–6. IEEE (2021)
27. Ulyanov, D., Lebedev, V., Vedaldi, A., Lempitsky, V.: Texture networks: Feed-forward synthesis of textures and stylized images. arXiv preprint arXiv:1603.03417 (2016)
28. Wang, X., Peng, Y., Lu, L., Lu, Z., Bagheri, M., Summers, R.M.: Chestx-ray8: Hospital-scale chest x-ray database and benchmarks on weakly-supervised classification and localization of common thorax diseases. In: Proceedings of the IEEE conference on computer vision and pattern recognition. pp. 2097–2106 (2017)
29. Xiao, C., Li, B., Zhu, J., He, W., Liu, M., Song, D.: Generating adversarial examples with adversarial networks. In: Lang, J. (ed.) Proceedings of the Twenty-Seventh International Joint Conference on Artificial Intelligence, IJCAI 2018, July 13-19, 2018, Stockholm, Sweden. pp. 3905–3911. ijcai.org (2018). https://doi.org/10.24963/IJCAI.2018/543, https://doi.org/10.24963/ijcai.2018/543
30. Yao, Q., He, Z., Li, Y., Lin, Y., Ma, K., Zheng, Y., Zhou, S.K.: Adversarial medical image with hierarchical feature hiding. IEEE Transactions on Medical Imaging (2023)
31. Yao, Q., He, Z., Lin, Y., Ma, K., Zheng, Y., Zhou, S.K.: A hierarchical feature constraint to camouflage medical adversarial attacks. In: Medical Image Computing

and Computer Assisted Intervention–MICCAI 2021: 24th International Conference, Strasbourg, France, September 27–October 1, 2021, Proceedings, Part III 24. pp. 36–47. Springer (2021)
32. Zhao, Z.Q., Zheng, P., Xu, S.t., Wu, X.: Object detection with deep learning: A review. IEEE transactions on neural networks and learning systems **30**(11), 3212–3232 (2019)
33. Zhu, J.Y., Park, T., Isola, P., Efros, A.A.: Unpaired image-to-image translation using cycle-consistent adversarial networks. In: Proceedings of the IEEE international conference on computer vision. pp. 2223–2232 (2017)
34. Zou, Z., Chen, K., Shi, Z., Guo, Y., Ye, J.: Object detection in 20 years: A survey. Proc. IEEE **111**(3), 257–276 (2023)

An Attention Transformer-Based Method for the Modelling of Functional Connectivity and the Diagnosis of Autism Spectrum Disorder

Ge Yang[1], Linbo Qing[1(✉)], Yanteng Zhang[1], Feng Gao[1,2], Li Gao[3], Xiaohai He[1], and Yonghong Peng[4]

[1] College of Electronics and Information Engineering, Sichuan University, Chengdu 610064, Sichuan, China
qing_lb@scu.edu.cn
[2] National Interdisciplinary Institute on Aging (NIIA), Southwest Jiaotong University, Chengdu 611756, Sichuan, China
[3] The Third People's Hospital of Chengdu, Chengdu 610014, Sichuan, China
[4] Faculty of Science and Engineering, Anglia Ruskin University, East Road, Cambridge CB1 1PT, UK

Abstract. Autism Spectrum Disorder (ASD), as a developmental disorder of brain, affects the ability of individuals to express themselves verbally, participate in social activities and perform normal behaviors. Multi-site dataset inevitably introduces experimental and environmental variability in data acquisition and processing, which is not disease-related. For the purpose of reducing the impact of site effects and utilizing the connection between different functional community of the brain, a harmonization method is used to process the feature matrix and the brain topology metric nodal local efficiency is introduced as a weighting coefficient for feature enhancement in this paper, based on which a transformer architecture is developed to incorporate a community-interaction module. The result shows that our method achieves an accuracy of 73.4%, an AUROC of 79.97%, a sensitivity of 68.1% and a specificity of 78.63% in the binary classification task of ASD identification on the Autism Brain Imaging Data Exchange (ABIDE) dataset.

Keywords: ASD · Transformer · fMRI · computer-aided diagnosis

1 Introduction

Autism Spectrum Disorder (ASD), is a condition related to brain development that affects the way a person perceives and socializes with others, causing problems in social interaction and communication. Limited and repetitive patterns of behavior are also characteristic of this disorder. Over time, the prevalence of ASD has increased in many countries, with a median prevalence of 65/10,0007 in a 2021 survey [1], placing a burden on society and on the families with ASD individuals. Currently, the mainstream diagnostic approach for ASD is clinician assessment and diagnosis by clinicians based on interaction with the patient and the patient's symptoms, which introduces some bias due to subjective factors [2, 3]. Functional Magnetic Resonance Imaging (fMRI) is an imaging technique used to study human brain activities and indirectly reflects neural activities by detecting and measuring changes in oxygen levels in the blood. The various features constructed from fMRI data can reflect the differences in the brain due to the disease, which is one of the determining factors in the performance of computerized diagnostic classification and now is widely used to determine the ASD-related biomarkers [4, 5]. The functional connectivity (FC) matrix is extracted from the fMRI data, and it is commonly constructed to describe neural connections in terms of Pearson correlation coefficients in regions of interest (ROIs). Brain connectivity analysis based on FC matrix is a useful tool for studying the connections between ROIs and their correlations with cognitive processes, which commonly used in the diagnosis and treatment of brain disorders including ASD [6, 7]. Several existing models based on brain functional connectivity using transformer to classify disease samples have performed well, but the publicly available ABIDE dataset, which is commonly used to study ASD, collects different data from different locations, so statistical effectiveness and generalizability of results would be affected and it may limit the ability to detect real phenomenon and may lead to spurious findings [8]. To reduce this negative impact known as site effects and to stabilize feature extraction, a harmonization method needs to be taken to preserve biologically significant variability while dealing with this non-biological variability [9], and adding a weighting coefficient will be helpful for feature enhancement.

The topological metrics of brain functional networks are a set of mathematical measures used to quantify and describe how the brain is functionally organized and connected. In some studies, it has been shown that the local efficiency of patients with Alzheimer's disease have reduced in functional brain networks [10]; the topology of the functional brain networks of patients with major depression is disrupted, and both the local efficiency and the efficiency of the overall brain network are reduced [11]; the local efficiency of the brain network can also be used to discriminate between patients with Parkinson's disease and normal subjects [12]. Normal subjects in the ASD study showed tighter functional network organization, manifested in a higher local efficiency [13, 14].

Thus, we introduce nodal local efficiency as a weighting coefficient, proposing a deep learning method that reduces site effects and enhances feature based on the transformer structure [15]. In addition, an interaction module has been presented to learn the connection between different functional community of the brain. Specifically, the nodal local efficiency as a weighting coefficient of the FC matrix is computed based on the harmonized FC matrix by Combat, then the weighted FC matrix is rearranged according to the functional communities and sequentially fed into the first transformer encoder layer with initialized community prompt tokens. The obtained node feature embedding of each community is interacted with the community prompt token embedding in order of attention weights, then pass their combination to the second transformer encoder layer to finally obtain the prediction results through the readout layer. Our main contributions are as follows:

1. We use the method Combat to reduce site effects and propose to use the nodal local efficiency as a weighting factor for FC matrix.

2. We propose an interaction module that the community node features output interacts with the community prompt tokens output from the first transformer encoder layer.

3. We achieve better result on the ABIDE dataset in binary classification task of ASD identification.

2 Related Work

Deep learning plays an important role in fMRI-based brain network connectivity analysis. Classical convolutional neural network (CNN) is capable of extracting disease-related features. For example, [16] uses multiple 2D CNNs of different sizes to extract features from FC matrix for classification. The transformer model, a further developed neural network model, was initially applied in natural language processing, and nowadays there have been tremendous advances in various fields such as computer vision, and the use of transformer for analyzing brain connectome is one of the examples. In the ASD classification task, Brain Network Transformer (BNT) feeds FC matrix into the transformer encoder layer and performs cluster readout of features through the orthogonal clustering readout layer [6]; Community-Aware Transformer (Com-TF) proposes a local-global transformer architecture based on BNT considering the communication within different functional communities of brain, while embedding the learnable prompt tokens of the communities [7]; [17] brings up a spatial–temporal multi-headed attention unit to obtain the spatial and temporal representation of fMRI data in the distinguishing ASD subjects. Some of the above methods take into account the relationship between time and space, and some take into account the relationship between the local and the whole, but the consideration of the data sources and the relationship between functional communities is missing.

3 Materials and Methods

3.1 Dataset and Experimental Setup

The Autism Brain Imaging Data Exchange (ABIDE) is a dataset sharing fMRI data from 17 international sites. This dataset contains fMRI images of 1009 subjects, 516 ASD subjects (51.14%) and 493 healthy controls (48.86%). The nodes of the brain connectivity network, i.e. the brain regions of interest, were defined based on the Craddock 200 atlas [18]. Images were preprocessed using the Configurable Pipeline for the Analysis of Connectomes (CPAC) pipeline and time series correlations were calculated using the Pearson correlation coefficient [19]. We implement our model in PyTorch and train on a GeForce RTX 2080 Ti with 11 GB of memory. We split the train\validating\test set with a ratio of 7:1:2, and we set the batch size as 64. The training set is fixed and repeated five times to randomly divide the validation and test sets, and each time the model is trained for 50 epochs. The performance evaluation metrics are AUROC, accuracy, sensitivity and specificity of ASD vs. NC classification predictions on the test set. The epoch for comparison on the test set is selected based on the epoch with the largest AUC on the validation set. All reported performances are the average of 5 random runs on the test set with the standard deviation. During the training process, we use CrossEntropyLoss as the loss function, using the adam optimisation algorithm to optimize.

3.2 Methods

During the preprocessing of the ABIDE dataset, the brain region is firstly divided into N ROI regions according to the selected brain atlas, and each ROI can be regarded as a node. The time series of each ROI are extracted and correlations are calculated using the Pearson correlation coefficient to construct a symmetric functional connectivity (FC) matrix $X \in \mathbb{R}^{N*N}$. Combat algorithm is used to reduce site effects of the FC matrix data, then nodal local efficiency as a weighting coefficient computed based on the processed FC matrix is multiplied by the FC matrix. Rearrangement according to the functional communities of brain is performed, bundling regions with similar patterns of functional connectivity. The brain communities are in order from the Yeo 7 network template [20].

The community node features are combined with the randomly initialized community prompt tokens and fed sequentially into the first transformer encoder layer, so connections within these functional communities can be obtained. We set eight heads for multi-heads attention mechanism of transformer encoder as there are eight functional communities. Since the connection between functional communities is also very important for ASD prediction [21], the encoded community node features are interacted with the encoded community prompt tokens from the first transformer encoder layer in the interaction module. The encoded community node features with high attention weight are combined with the encoded community prompt token with low attention weight, so that communities that receive less attention are compensated and information can be obtained from their connection. Then rearranging the whole encoded node feature matrix again in the order of community attention weights from the largest to the smallest first, and the combination with encoded community prompt token is fed into the second transformer encoder layer, and the output is put into the orthogonal clustering readout layer for the prediction classification. The structure of the proposed model structure is shown in Fig. 1.

Correcting for site effects in multi-site fMRI data. In computer-aided diagnosis, a large amount of data is needed to increase the generalizability of results and to find features that can discriminate diseased from normal subjects, but site effects can be introduced as confounder in the training process that reduce the accuracy. Combat is designed to correct for the batch effects in genomic studies [22–24], and it determines the magnitude of the batch effect by estimating the difference between the expression levels of each feature across batches through a statistical model and adjusting the data for each sample; it is a batch effect correction method based on an empirical Bayesian framework. The Combat model can be written as:

$$y_{ijm} = \alpha + \gamma_{im} + \delta_{im}\varepsilon_{ijm} \qquad (1)$$

where y_{ijm} represents the m th value (m $\in$ {1,…,19900}) in the upper delta of the functional connectivity matrix for the j th subject at site i, γ_{im} and δ_{im} denote the estimated additive and multiplicative site effects of the selected feature M at site I, and ε_{ijm} is a normally distributed error term with zero mean and variance [9].

Nodal local efficiency. The nodal local efficiency of a node measures communication efficiency between its neighbors when the node is removed. If a node's local efficiency is low, it means that the node is of low importance in the brain network, and the brain network can complete the information exchange and processing tasks without it; if

An Attention Transformer-Based Method   53

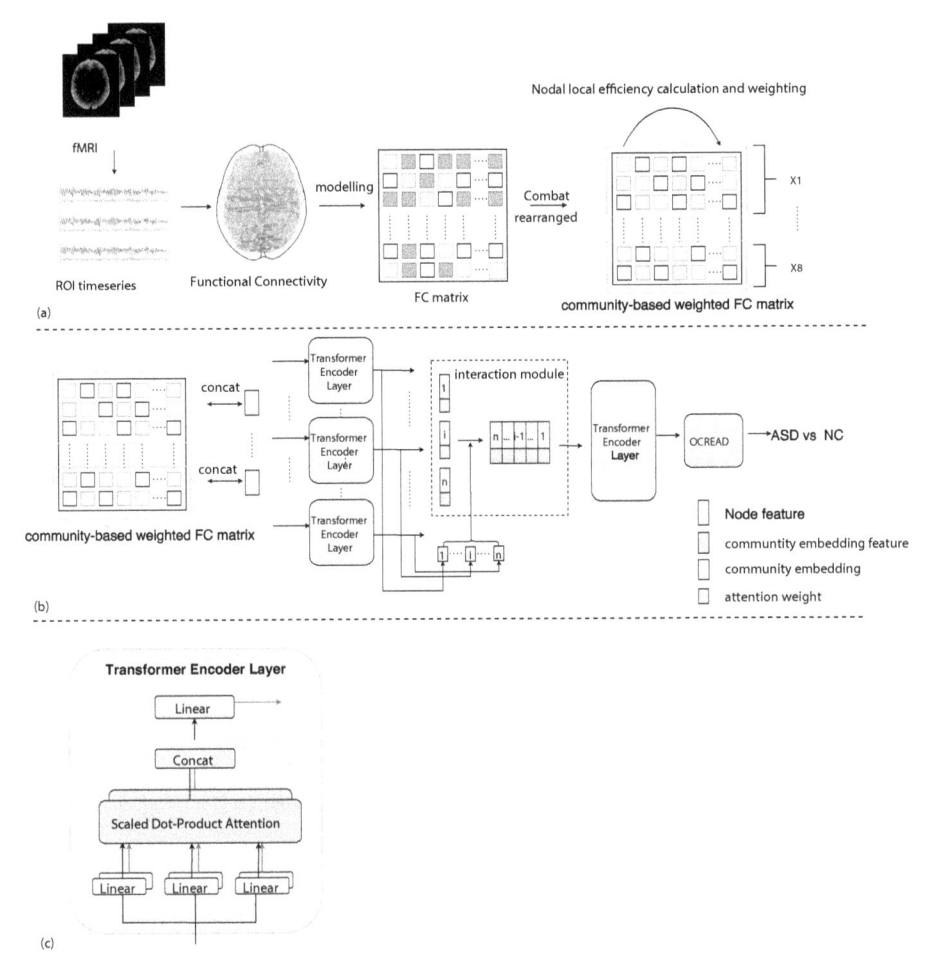

Fig. 1. (a) FC matrix modelled and processed by Combat, weighted with nodal local efficiency, then rearranged based on functional community (b) The proposed model structure (c) Transformer Encoder Layer

a node's local efficiency is high, it means that the node is significant in the brain's collaborative functional work. The node local efficiency is defined as

$$[E_{local}(v) = \frac{1}{N_v(N_v-1)} \sum_{i \neq j \in N(v)} \frac{1}{d_{ij}}] \tag{2}$$

where N_v is the number of neighbors of node v, $N(v)$ represents the set of neighbors of node v, d_{ij} is the shortest path length between node i and node j, , and the summation traverses all the pairs of neighboring nodes (i,j) of node v, with $i \neq j$.

Transformer encoder layer. In the model, Transformer encoder layer is used as the feature extraction layer and the encoder consists of a multi-head attention mechanism:

$$h_i = MHSA(X_i') \tag{3}$$

$$h_i = (\|_{m=1}^{M} \text{softmax}\left(\frac{Q^m(K^m)^T}{\sqrt{d_k^m}}\right)V^m)W_O \tag{4}$$

where $Q^m = W_Q X_i'$, $K^m = W_K X_i'$, $V^m = W_V X_i'$, $X_i' = E_{local}(V_i)X_i$, $X_i \in R^{N_i \times N}$, M is the number of attention headers, W_Q, W_K, W_V, W_O are the learnable rearrange weight parameters, and $\|$ stands for the splicing operation, $i \in \{1,2,...,8\}$.

Reordering of weighted functional connectivity matrix and initialized community prompt tokens are performed on nodes according to the functional community they belong to. The weighted functional connectivity matrix and tokens are fed into the local transformer encoder layer.

$$p_i', H_i = Transformer\ encoder\ layer([p_i, X_i']) \tag{5}$$

The obtained node feature embeddings are rearranged based on the order of attention weights, while the community prompt tokens are spliced together with the node feature embeddings in reverse order, so as to perform the inter-community information interaction to exploit the connection between different functional regions in the functional network. In order to obtain the overall inter-community node interrelationships, the spliced features are sent into the second transformer encoder layer, which allows the model to better learn the disease-related features:

$$p_{global} = Concat(rearrange(p_i', p_1' \ldots p_K')) \tag{6}$$

$$H_{global} = Concat(rearrange((H_k, H_2, \ldots, H_i))) \tag{7}$$

$$p', Z^L = Transformer encoder layer([p_{total}, H_{total}]) \tag{8}$$

Readout Layer. The OCREAD layer is used for the aggregation of whole brain feature embeddings. OCREAD initializes the orthogonal clustering centres and softly assigns the feature embeddings to these centres, and subsequently obtains the graph-level embedding representation Z_G, which is linearised and passed to the MLP layer for the graph-level prediction [6]:

$$P_{ik} = \frac{e^{<Z_i^L, E_k>}}{\sum_{K\prime}^{K} e^{<Z_i^L, E_{k\prime}>}} \tag{9}$$

where $(\bullet)$ represents the inner product operation, Z_L is the node feature matrix output from the transformer encoder layer, P_{ik} denotes the probability of assigning node i to the k th cluster, and finally Z_L is aggregated under the guidance of the clustering probability:

$$Z_G = P^T Z_L \tag{10}$$

4 Experimental Results

4.1 Performance Comparison

Baseline method comparison: we compare with (i) FBNETGEN; (ii) BrainNetCNN; (iii) BNT; and (iv) Com-TF. The comparison results are shown in Table 1.

Table 1. Performance comparison with baselines (%).

Model	Accuracy	AUROC	Sensitivity	Specificity
FBNETGEN	65.6 ± 8.5	72.68 ± 8.9	66.7 ± 8.0	64.53 ± 9.2
BrainNetCNN	69.2 ± 4.7	76.15 ± 4.2	66.56 ± 12.6	71.99 ± 7.6
BNT	69.8 ± 3.0	77.78 ± 1.9	65.13 ± 20.2	75.56 ± 18.3
Com-TF	70 ± 5.8	77.76 ± 4.4	75.31 ± 13.8	63.95 ± 10.5
Ours	73.4 ± 1.8	79.97 ± 2.6	68.1 ± 11.5	78.63 ± 13.3

4.2 Ablation Experiments

Role of weighting coefficient and node features interacting with the prompt token. The site NYU contains the largest number of samples, the site UCLA_2 contains the smallest number of samples, and the site containing the median number of samples of the 17 sites is STANFORD. Figure 2 shows the mean brain/DMN/SMN functional connectivity data distribution of the samples of the three sites. Figure 3 shows nodes with significant alteration in nodal local efficiency. Figure 4 illustrates the attention weights change of the second transformer encoder layer before and after performing the interaction. Table 2 shows comparison with weighting coefficient and interaction.

Table 2. Performance comparison with weighting coefficient and interaction (%).

Model	Accuracy	AUROC	Sensitivity	Specificity
With Nle-weighted	70.4 ± 6.4	79.38 ± 4.7	76.48 ± 8.4	64.73 ± 17.7
With interaction	71 ± 5.3	79.53 ± 4.8	60.58 ± 14.2	80.21 ± 9.1
proposed	73.4 ± 1.8	79.97 ± 2.6	68.1 ± 11.5	78.63 ± 13.3

4.3 Discussion

Figure 2 shows the FC matrix data distribution of the original abide, the FC matrix data distribution processed by Combat. The data distribution of different sites obtained after Combat is more approximate, which can retain the differences between different nodes while avoiding the emergence of a few extreme values, so that the model can more effectively give different attention to different nodes to reduce the impact of deviation values on the model. Most of the nodes with significant changes in nodal local efficiency are in the DMN and SMN in Fig. 3. Figure 4 shows that more attention is paid to the disease-related regions after FC matrix has been processed through Combat and multiplied by a weighting coefficient. From Table 1 Performance comparison with baselines, we can see that the model we proposed have better performance on Accuracy and AUROC. Sensitivity and specificity often show opposite trends during the classification task, and the performance achieved by our proposed method compared to other models is between the performance of these models. And from Table 2 it can be seen that there is some improvement in the metrics of the result after we have processed the features more.

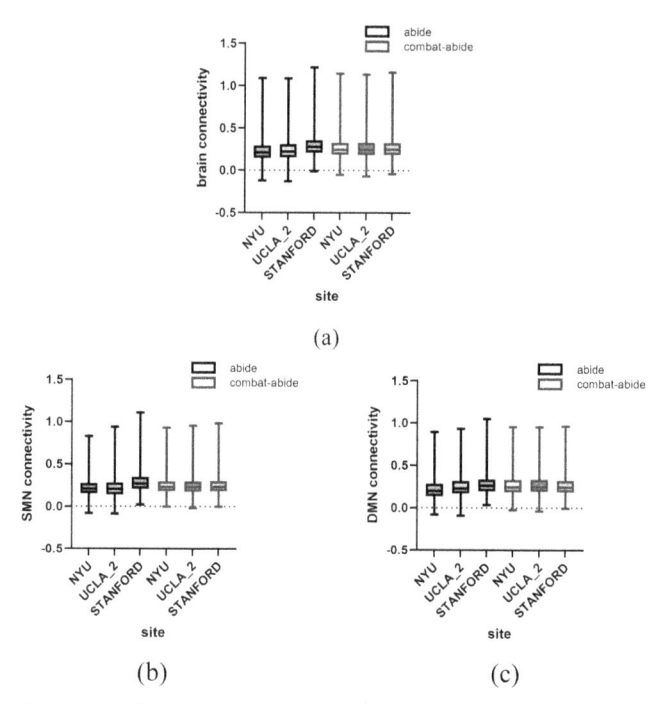

Fig. 2. (a) Brain functional connectivity data distribution (b) DMN functional connectivity data distribution (c) SMN functional connectivity data distribution.

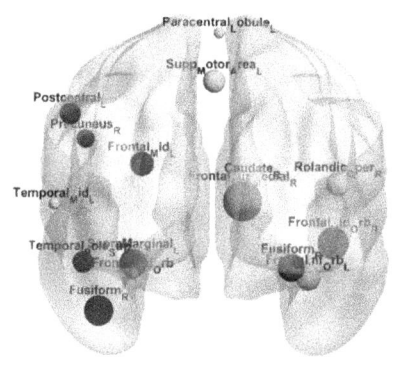

Fig. 3. The nodes with nodal local efficiency show a significant (P ≤ 0.05) alteration in subjects with ASD compared to NC subjects.

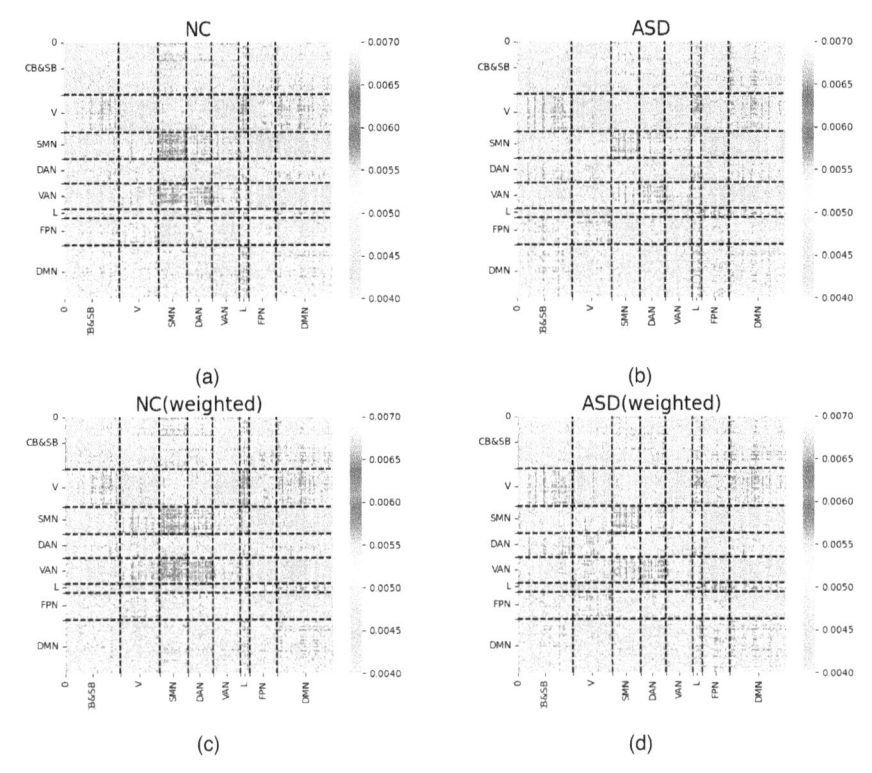

Fig. 4. (a) NC's average attention weights from the second transformer encoder layer (b) ASD's average attention weights from the second transformer encoder layer (c) NC's average attention weights from the second transformer encoder layer with Nle (d) ASD's average attention weights from the second transformer encoder layer with Nle.

5 Conclusion

In this work, we propose a transformer architecture with an interaction module for ASD analysis. Our model reduces site effects, adds a weighting factor to enhance the extracted data features. It also optimizes the ability of feature learning while learning node embeddings within and between brain functional network communities to deepen the information interaction between different communities for ASD classification tasks. The proposed model is trained and tested on the ABIDE dataset. The experimental results show the effectiveness of our method in ASD diagnosis, which can achieve better results for the classification of the whole dataset. Our future work is to investigate the impact of different harmonization methods on the model and the effect of different atlases on the model performance to gain a comprehensive understanding of how site effects can be reduced.

Acknowledgments. . This work was supported by the Key Research and Development Program of Sichuan, China under Grant 2023YFS0195.

References

1. Zeidan, J., Fombonne, E., Scorah, J., Ibrahim, A., Durkin, M.S., Saxena, S., Yusuf, A., Shih, A., Elsabbagh, M.: Global prevalence of autism: A systematic review update. Autism research **15**(5), 778–790 (2022)
2. Mandell, D.S., Ittenbach, R.F., Levy, S.E., Pinto-Martin, J.A.: Disparities in diagnoses received prior to a diagnosis of autism spectrum disorder. Journal of autism and developmental disorders **37**, 1795–1802 (2007)
3. Yahata, N., Morimoto, J., Hashimoto, R., Lisi, G., Shibata, K., Kawakubo, Y., Kuwabara, H., Kuroda, M., Yamada, T., Megumi, F., et al.: A small number of abnormal brain connections predicts adult autism spectrum disorder. Nature communications **7**(1), 11254 (2016)
4. Zhang, J., Feng, F., Han, T., Gong, X., Duan, F.: Detection of autism spectrum disorder using fmri functional connectivity with feature selection and deep learning. Cognitive Computation **15**(4), 1106–1117 (2023)
5. Almuqhim, F., Saeed, F.: Asd-grestm: Deep learning framework for asd classification using gramian angular field. In: 2023 IEEE International Conference on Bioinformatics and Biomedicine (BIBM). pp. 2837–2843. IEEE (2023)
6. Kan, X., Dai, W., Cui, H., Zhang, Z., Guo, Y., Yang, C.: Brain network transformer. Advances in Neural Information Processing Systems **35**, 25586–25599 (2022)
7. Bannadabhavi, A., Lee, S., Deng, W., Ying, R., Li, X.: Community-aware transformer for autism prediction in fmri connectome. In: International Conference on Medical Image Computing and Computer-Assisted Intervention. pp. 287–297. Springer (2023)
8. Abraham, A., Milham, M.P., Di Martino, A., Craddock, R.C., Samaras, D., Thirion, B., Varoquaux, G.: Deriving reproducible biomarkers from multi-site resting-state data: An autism-based example. NeuroImage **147**, 736–745 (2017)
9. Fortin, J.P., Parker, D., Tunç, B., Watanabe, T., Elliott, M.A., Ruparel, K., Roalf, D.R., Satterthwaite, T.D., Gur, R.C., Gur, R.E., et al.: Harmonization of multi-site diffusion tensor imaging data. Neuroimage **161**, 149–170 (2017)
10. Reijmer, Y.D., Leemans, A., Caeyenberghs, K., Heringa, S.M., Koek, H.L., Biessels, G.J., Group, U.V.C.I.S.: Disruption of cerebral networks and cognitive impairmentin alzheimer disease. Neurology **80**(15), 1370–1377 (2013)
11. Yang, H., Chen, X., Chen, Z.B., Li, L., Li, X.Y., Castellanos, F.X., Bai, T.J., Bo, Q.J., Cao, J., Chang, Z.K., et al.: Disrupted intrinsic functional brain topology in patients with major depressive disorder. Molecular psychiatry **26**(12), 7363–7371 (2021)
12. Zhang, D., Wang, J., Liu, X., Chen, J., Liu, B.: Aberrant brain network efficiency in parkinson's disease patients with tremor: A multi-modality study. Frontiers in aging neuroscience **7**, 169 (2015)
13. Alaerts, K., Geerlings, F., Herremans, L., Swinnen, S.P., Verhoeven, J., Sunaert, S., Wenderoth, N.: Functional organization of the action observation network in autism: a graph theory approach. PloS one **10**(8), e0137020 (2015
14. Qin, B., Wang, L.: Enhanced topological network efficiency in preschool autism spectrum disorder: a diffusion tensor imaging study. Frontiers in Psychiatry **9**, 365939 (2018)
15. Vaswani, A., Shazeer, N., Parmar, N., Uszkoreit, J., Jones, L., Gomez, A.N., Kaiser, Ł., Polosukhin, I.: Attention is all you need. Advances in neural information processing systems **30** (2017)
16. Sherkatghanad, Z., Akhondzadeh, M., Salari, S., Zomorodi-Moghadam, M., Abdar, M., Acharya, U.R., Khosrowabadi, R., Salari, V.: Automated detection of autism spectrum disorder using a convolutional neural network. Frontiers in neuroscience **13**, 1325 (2020)
17. Bedel, H.A., Sivgin, I., Dalmaz, O., Dar, S.U., Çukur, T.: Bolt: Fused window transformers for fmri time series analysis. Medical Image Analysis **88**, 102841 (2023)

18. Craddock, R.C., James, G.A., Holtzheimer III, P.E., Hu, X.P., Mayberg, H.S.: A whole brain fmri atlas generated via spatially constrained spectral clustering. Human brain mapping **33**(8), 1914–1928 (2012)
19. Craddock, C., Benhajali, Y., Chu, C., Chouinard, F., Evans, A., Jakab, A., Khundrakpam, B.S., Lewis, J.D., Li, Q., Milham, M., et al.: The neuro bureau preprocessing initiative: open sharing of preprocessed neuroimaging data and derivatives. Frontiers in Neuroinformatics **7**(27), 5 (2013)
20. Yeo, B.T., Krienen, F.M., Sepulcre, J., Sabuncu, M.R., Lashkari, D., Hollinshead, M., Roffman, J.L., Smoller, J.W., Zöllei, L., Polimeni, J.R., et al.: The organization of the human cerebral cortex estimated by intrinsic functional connectivity. Journal of neurophysiology (2011)
21. Padmanabhan, A., Lynch, C.J., Schaer, M., Menon, V.: The default mode network in autism. Biological Psychiatry: Cognitive Neuroscience and Neuroimaging **2**(6), 476–486 (2017)
22. Fortin, J.P., Labbe, A., Lemire, M., Zanke, B.W., Hudson, T.J., Fertig, E.J., Greenwood, C.M., Hansen, K.D.: Functional normalization of 450k methylation array data improves replication in large cancer studies. Genome biology **15**(11), 1–17 (2014)
23. Yan, C.G., Craddock, R.C., Zuo, X.N., Zang, Y.F., Milham, M.P.: Standardizing the intrinsic brain: towards robust measurement of inter-individual variation in 1000 functional connectomes. Neuroimage **80**, 246–262 (2013)
24. Johnson, W.E., Li, C., Rabinovic, A.: Adjusting batch effects in microarray expression data using empirical bayes methods. Biostatistics **8**(1), 118–127 (2007)

A Fast Domain-Inspired Unsupervised Method to Compute COVID-19 Severity Scores from Lung CT

Samiran Dey[1]($\boxtimes$), Bijon Kundu[2], Partha Basuchowdhuri[1], Sanjoy Kumar Saha[3], and Tapabrata Chakraborti[4]($\boxtimes$)

[1] Indian Association for the Cultivation of Science, Kolkata, India
Smcssd2661@iacs.res.in
[2] EKO X Ray and Imaging Institute, Kolkata, India
[3] Department of Computer Science, Jadavpur University, Kolkata, India
[4] The Alan Turing Institute and University College London, London, UK
tchakraborty@turing.ac.uk, t.chakraborty@ucl.ac.uk

Abstract. There has been a deluge of data-driven deep learning approaches to detect COVID-19 from computed tomography (CT) images over the pandemic, most of which use ad-hoc deep learning black boxes of little to no relevance to the actual process clinicians use and hence have not seen translation to real-life practical settings. Radiologists use a clinically established process of estimating the percentage of the affected area of the lung to grade the severity of infection out of a score of 0-25 from lung CT scans. Hence any computer-automated process that has aspirations of being adopted in the clinic to alleviate the workload of radiologists while being trustworthy and safe, needs to follow this clearly defined clinical process religiously. Keeping this in mind, we propose a simple yet effective methodology that uses explainable mechanistic modelling using classical image processing and pattern recognition techniques. The proposed pipeline has no learning element and hence is fast. It mimics the clinical process and hence is transparent. We collaborate with an experienced radiologist to enhance an existing benchmark COVID-19 lung CT dataset by adding the grading labels, which is another contribution of this paper, along with the methodology which has a higher potential of becoming a clinical decision support system (CDSS) due to its rapid and explainable nature. The radiologist gradations and the code is available at https://github.com/Samiran-Dey/explainable_seg.

Keywords: COVID-19 severity · unsupervised biomedical image segmentation · lung computed tomography (CT) · ground glass detection

1 Introduction

To estimate the severity of infection caused by COVID-19, lung computed tomography (CT) scans have been strongly recommended by clinicians because of the

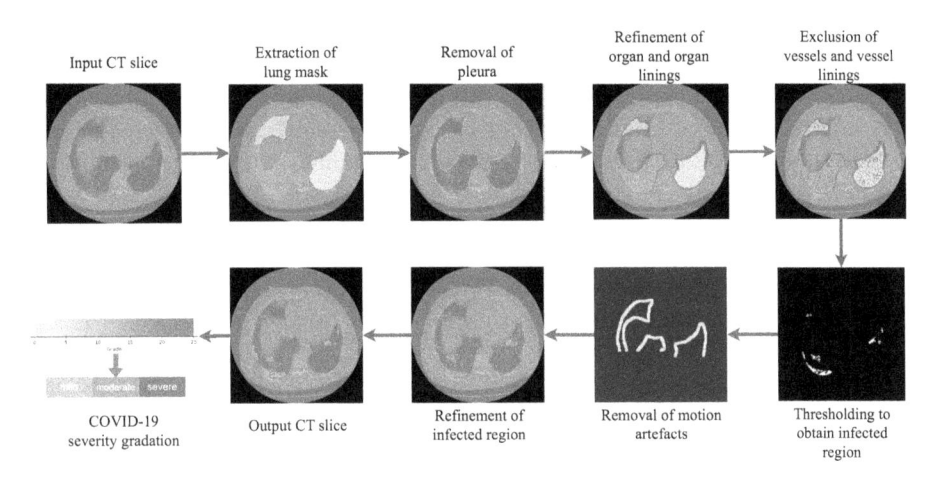

Fig. 1. Pipeline for the proposed methodology. The infection region is segmented using classical image processing and pattern recognition techniques. The percentage involvement of each lobe of the lung is hence computed to grade the severity of COVID-19.

primary involvement of the respiratory system [1]. Radiologists manually perform COVID-19 severity gradation from lung CT scans using eye estimates of the volume of the lung lobes that the virus has infected [11]. Infections of the lung caused by COVID-19 are mainly identified by ground-glass opacities, vascular enlargement, bilateral abnormalities, lower lobe involvement, and posterior predilection [2]. Some other abnormalities observed in chest CTs include consolidation, linear opacity, septal thickening and/or reticulation, crazy-paving pattern, air bronchogram, pleural thickening, etc [2]. Besides morphological and structural differences, these infection regions of the lungs may be discriminated from normal regions using pixel values of the lung CT slices measured in the scale of Hounsfield Unit (HU) in which the radiodensity of distilled water and air in standard temperature pressure (STP) is 0 HU -1000 HU respectively [6].

There have been several works on predicting COVID-19 and segmenting COVID-19-affected regions in lung CT scans. But most of them use deep learning methodologies [16, 19–22, 24–26] and are thus black boxes. Because of this, they cannot be adopted in real clinical scenarios [27], as patient safety is of primary concern. For AI-automated assistive models to be adopted in clinical settings to alleviate the workload of radiologists, these methods need to follow the clinical workflow faithfully and transparently, without which clinical translation of AI models will continue to be abysmally low. Thus, we propose a methodology inspired by the clinical procedure of COVID-19 severity gradation followed by the clinical practitioners and is completely explainable. Since there is no training involved, no ground-truth annotation is required, the time required is less and in the absence of randomness, there is no uncertainty in prediction. Moreover, our methodology may be extended to a clinical decision support system (CDSS) for guided estimation of lung involvement and act as a suggestion to the eye estimates that radiologists make. The proposed methodology pipeline is shown in Fig. 1. The contributions of our work are as follows.

1. A novel unsupervised methodology based on classical image processing and pattern recognition methods for segmentation of infected regions in lung CT scans, that closely mimics the clinical workflow in a transparent and interpretable manner.
2. Contribution to Medical Imaging Data Resource Center - RSNA International COVID-19 Open Radiology Database - Release 1a (MIDRC-RICORD-1a) dataset by providing ground truth COVID-19 severity grades for the CT scans in the dataset, scored by an experienced radiologist. The gradations are available at https://github.com/Samiran-Dey/explainable_seg.
3. Since the method is unsupervised, there is no training needed, hence it is lightweight, fast and easily deployable in computational resource-limited clinical settings. Also because of its unsupervised nature, our method does not need any costly ground-truth annotations and labels from human experts.
4. An analysis of algorithmic fairness is also presented for the MIDRC-RICORD-1a dataset with respect to sex and age, from the point of view of health equity for different patient sub-groups.

In Sect. 1.1, we describe the related works for COVID-19 infected region segmentation and detection. In Sect. 2, we explain our proposed methodology with illustrative images of intermediate results at each step. In Sect. 3, we provide the details of the dataset used followed by the qualitative and quantitative results of our experiments. We also provide an analysis of the fairness offered by our methodology for the given dataset, in Sect. 3. And finally in Sect. 4, we state the conclusions of our paper.

1.1 Related Works

Since the advent of the COVID-19 pandemic, there have many works for the detection and segmentation of COVID-19 infection from lung CT scans. Most of the benchmark models used U-Net as the base framework. Ahmed et al.[21] applied spatial, colour and noise augmentation, and used soft attention in U-Net layers for Covid-19 lesion segmentation. Whereas in COVID TV-Unet [24], a modified loss function with 2D-anisotropic total variation was used for the connectivity-promoting regularization of a U-Net model. A hierarchical model involving the extraction of semantic data with two cascaded residual attention inception U-Net was proposed by Punn et. al. [22] for segmentation of infection region. In SD-UNet [26], a modified U-Net framework, was introduced with the squeeze-and-attention (SA) and dense atrous spatial pyramid pooling (Dense ASPP) modules for segmenting ground glass opacity and consolidation lesion in lung CT images. In CARes-UNet [25], the content-aware upsampling module residual UNet was used for improving the segmentation performance. Antar et al. [16] used a density heat map tool to colour lung images and used three colour channels separately, after image inverse and histogram equalization, to obtain segmentation results by feeding them through three separate U-Nets with the same architecture. The segmentation results were combined and run through a convolution layer one by one to get the detection.

Many works used ResNet-based frameworks as well for the segmentation and detection of COVID-19 lesions in lung CT scans. Aleem et al. [20] used ResNet50 and ResNet101-based Mask R-CNN to segment the area of the COVID-19 infection and provide the ratio of the infected area in the lungs. Enshaei et al. [19] proposed an encoder-decoder-based network comprising an encoding path, transition layers, context perception boosting module, and decoding path for segmenting the ground glass opacity and consolidation infection regions. Sarkisov [23] proposed a variation of the Mask R-CNN model based on ResNet, Feature Pyramid Net (FPN) and Region Proposal Net (RPN) to detect and segment COVID lesions and create the bounding boxes for the detected ROIs. Sailunaz et al. [17] proposed a web-based application framework for COVID-19 lung infection detection and segmentation with a feedback mechanism for self-learning and tuning using Mask R-CNN, U-Net, and U-Net++. Oulefki et al. [18] proposed a Covid-19 lung infection segmentation and measurement methodology that uses a new masking algorithm containing multiple thresholding, filtering and entropy calculations on the image histogram to generate masks for the infected regions of the lungs.

2 Methodology

Our methodology for COVID-19 severity gradation is a clinically explainable approach involving classical image processing and pattern recognition techniques. The methodology pipeline is illustrated in Fig.1. In this section, we discuss in detail each of the steps.

2.1 Extraction of Lung Mask

We use pre-trained Unet_R231_LTRCLobes by Hofmanninger [3] for the segmentation of lungs and their five lobes from the CT slices as illustrated in Fig.2.b. The segmentation of lung lobes is required to compute the percentage involvement of each lobe for CT severity gradation.

The pixel values of the CT slices are in the Hounsfield Unit (HU) scale. It can be observed by studying CT slices, that the infection regions have an intensity between -700 HU to +10 HU approximately, with pleural effusion having an intensity greater than +15 HU [7]. Thus, once the lung is extracted, the infection region may be obtained by thresholding based on the HU values. However, further observation has shown that several other anatomical structures have values within a similar range of HU. Moreover, due to the scattering of the radiations the surrounding region of different structures having a higher value in HU scale have values within the specified range [6]. Thus, before thresholding for obtaining the infected regions, these anomalous regions need to be removed. In the following steps, we proceed with the removal of anatomical structures and their surrounding regions that are present in the extracted lungs having HU values within the observed intensity range of the infected region. We consciously attempt to exclude all possible false positive regions for patient safety.

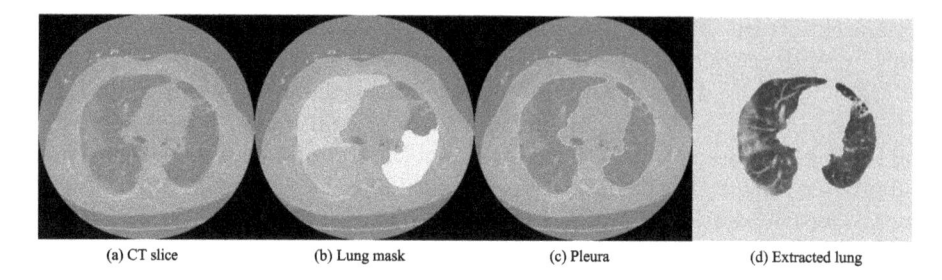

Fig. 2. Extraction of the lung from CT slices. (a) CT slice (b) Lung mask generated using pre-trained Unet_R231_LTRCLobes. The different colours of the mask indicate different lobes of the lungs. (c) The pleura region obtained by eroding the lung mask generated in (b), marked in red. (d) The extracted lung after removal of the pleura.

2.2 Removal of pleura

We begin with the removal of the pleura and its surrounding region. The pleura is a thin layer of tissue that covers the lungs and lines the interior wall of the chest cavity. Thus, we erode the lung mask obtained in section 2.1 by a small amount to remove the outer lining of the lung, illustrated in Fig.2.c. Hence we obtain the lungs where the pleura and its surroundings have been removed as illustrated in Fig.2.d.

2.3 Removal of organ linings

As illustrated in Fig.3, as we move from top to bottom of the lungs using the axial CT slices, other organs anatomically located beneath the lungs such as the liver, pancreas, etc. gradually start appearing in the CT scan. However, because of the scattering of radiations their surrounding region has values in the HU scale that intersect with the range of HU values for the infected region. Thus, the surrounding regions need to be excluded from the lung mask before thresholding to obtain infected region segmentation. In Fig.3.b., an organ appears in the right lung at slice number 42, and a small surrounding region of the organ in slices 42 and 43 appears to have ground glass opacity indicative of an infection which in this case however is because of scattering of radiation as evident from the images. A similar thing is noticed for Fig.3.a. in CT slices 37, 38 and 39. However, in slice 37 it gets merged with an actual infection region and hence is more critical. As we intend to have no false-positive regions in the infection mask, we exclude a considerable surrounding region with ground glass opacity in such cases. We have also observed that in CT scans with more frequent slices that is where the slice thickness is less, there happens to be a slice just above the slice in which the organ appears where there is a region of ground glass opacity suggestive of the organ in the next slice and not an infection. This is illustrated in Fig.3.b. with a red arrow in slice 41. Such regions also require to be excluded.

To exclude these surrounding regions, we first attempt to find out the slice where the organ starts appearing in the right lung by counting the number of

A Fast Domain-Inspired Unsupervised Method 65

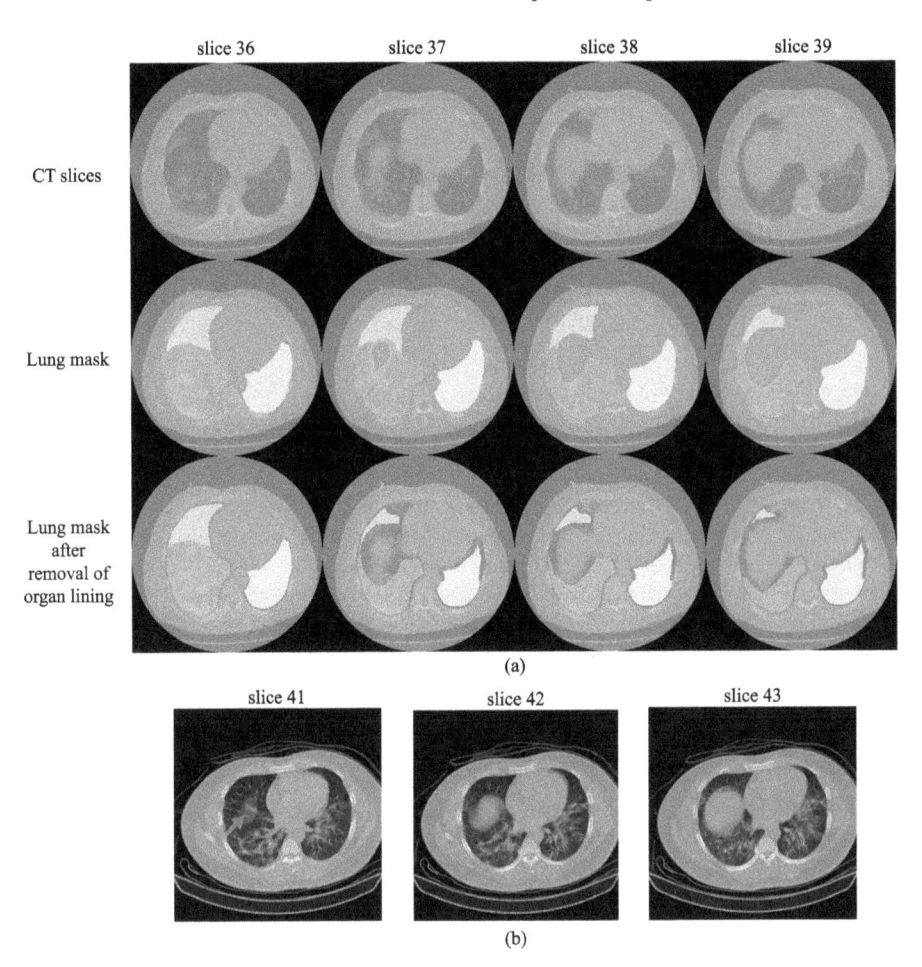

Fig. 3. Removal of surrounding regions of other organs like liver, pancreas, etc. where intensity values are similar to the range of HU value of the infected region because of scattering of radiation. A sequence of CT slices is shown where an organ gradually appears as we proceed. (a) CT scan with slices of greater thickness. (b) CT scan with slices of less thickness. The red arrow in (b), slice 41, shows a region of ground glass opacification which appears due to the presence of an organ in slice 42 and not because of infection. Our method efficiently removes the surrounding regions of the organ where the intensity values are such that might be mistaken as an infected region. Different colours in the lung mask indicate different lobes of the lung. It is to be noted that the lungs in the axial CT slices are laterally inverted.

connected components in the lung mask. It is seen that for the particular slice where the organ first appears there happens to be 4 connected components in the lung mask - the background, the left lung, the right lung and the organ as evident in Fig.3.a. slice 37. Once we have the organ mask separated, we iterate through the prior slices for a total thickness of 4mm from the slice where the

(a) Blood vessel exclusion

(b) Removal of motion artefacts

(c) Segmentation of infected region

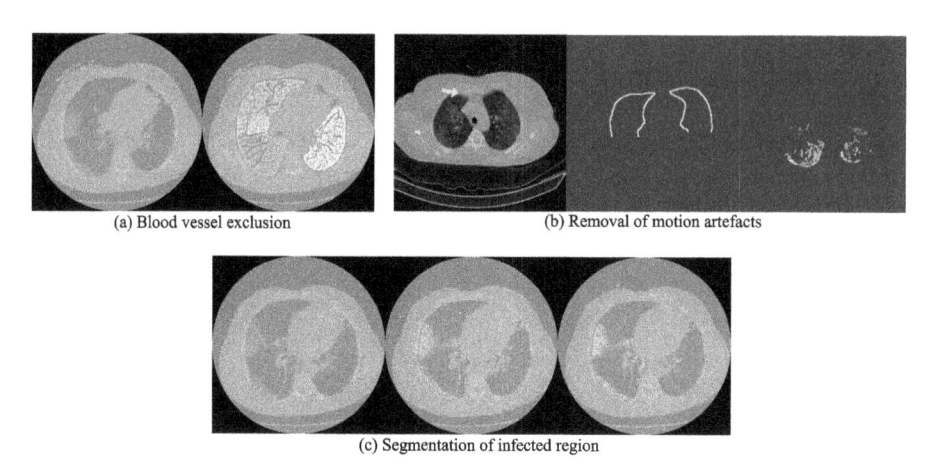

Fig. 4. Illustration of different steps of our proposed methodology. (a) Removal of blood vessels and vessel linings from the lung mask. Different colours correspond to different lobes of the lung. (b) Removal of motion artefacts. The yellow arrow points to a region of motion artefact. (c) Segmentation of infected regions by thresholding the modified lung mask. The red regions comprise the infected region mask which is further refined in the third image. (Color figure online)

organ appears while eroding the organ mask and removing the region of the organ mask from the lung mask. Thus, for instances like Fig.3.b., the region indicative of the organ's appearance in some following slice is removed. Next, we iterate through the slices following the slice where the organ first appears for a total thickness of 12mm while dilating the organ mask and excluding the corresponding area from the lung mask. To remove the surrounding region of other organs we find the upper edge of the lung mask separately for the left and right lungs, dilate the edge, and then remove the corresponding area from the lung mask for all slices following the slice where the organ first appears, as seen in the left lung in slices 37, 38 and 39 of Fig.3.a. Our methodology effectively removes all surrounding regions of other organs which have intensity values similar to that of the infected regions as evident from the third row of Fig.3.a.

2.4 Removal of blood vessels and vessel linings

Next, we proceed to remove the blood vessel and vessel linings as their observed intensity value intersects with the range of HU values corresponding to the infected regions. We use the MultiScaleHessianBasedMeasureImageFilter [4] of the ITK library for extracting the vessels. To remove the vessel linings we dilate the vessel mask thus obtained, by a small amount. We refine the vessel mask by further dilating the blood vessels that occupy an area of greater than 460 pixels, as broader vessels have more scattering in the surrounding regions. The area corresponding to the vessel mask is thus removed from the lung mask, as illustrated in Fig.4.a.

2.5 Removal of motion artefacts

Motion artefact [8] is a patient-based artefact that occurs with voluntary or involuntary patient movement during image acquisition. Misregistration artefacts, which appear as blurring, streaking, or shading, are caused by patient movement during a CT scan. A region of motion artefact is illustrated in Fig.4.b. using a yellow arrow. It is observed that such artefacts appear in lung CTs for COVID-19 patients in the upper region of the lung in axial CT slices. Thus, the edge of the upper region of the lungs is obtained as shown in the second image of Fig.4.b where motion artefacts are most probable. After thresholding the lung mask for infection regions, we check for infection regions having their centroid within the upper edge region previously considered and remove them from the infection mask. As seen in the third image of Fig.4, though motion artefacts are present in the CT slice, the infection mask has them removed.

2.6 Segmentation of infection regions

The lung mask obtained in section 2.1 is modified using the methodology described in section 2.2 to section 2.5 by removing all such regions that have intensity values that intersect with the range of HU value of infection regions. The lung region considered in the lung mask is now thresholded based on the HU value for infection regions. We perform experiments and choose a threshold of -595 HU to -1 HU that corresponds best to infection regions. It is to be noted that a region with a density of -600 HU contains an average of 60% air and 40% "tissue" [6]. These tissues comprise soft tissue of the lungs, tissues of fluids like blood and tissues arising because of infection [9]. The infection mask so obtained is further refined by removing all small regions comprising 30 pixels. Then all small holes of 30 pixels are filled to make the regions continuous. In Fig.4, the middle image shows the infection mask and the third image shows the refined infection mask, thus proving the relevance of refinement. The false positive regions are reduced while the true positive regions are filled.

2.7 COVID-19 severity gradation

CT severity score (CTSS) is used to quantify lung involvement in COVID-19 [10]. To compute CTSS, the volume of infected regions is estimated and the percentage of involvement is computed for each lobe of the lung. We use the methodology proposed by Li et. al. [11] to grade the severity. Each lobe is awarded a CT score between 0 to 5, depending on the percentage of the lobe involvement [12]: score 0 - for 0% involvement; score 1 - for less than 5% involvement; score 2 - for 5% to 25% involvement; score 3 - for 26% to 49% involvement; score 4 - for 50% to 75% involvement; and score 5 - for greater than 75% involvement. The overall

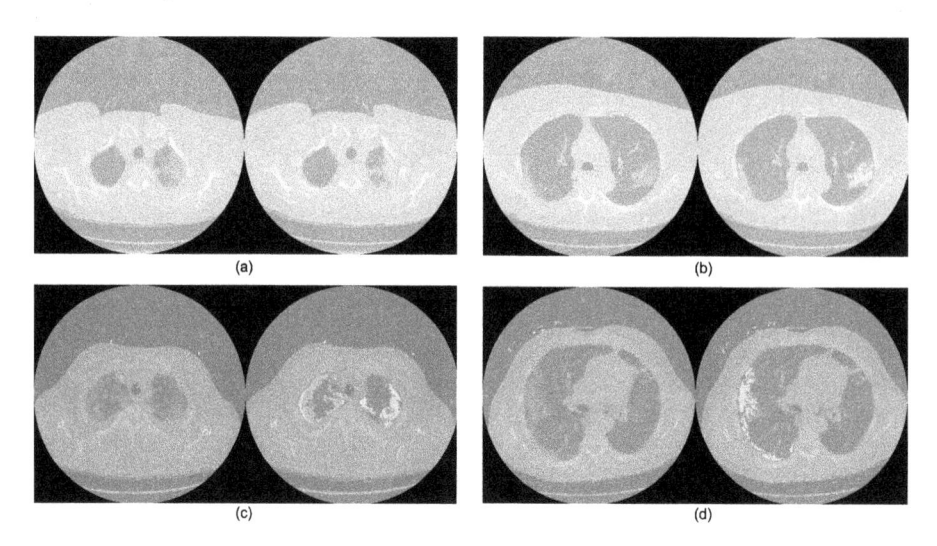

Fig. 5. Qualitative results for segmentation of infected regions. On the left are the images of CT slides and on the right are the CT images with overlapped segmentation maps. Blue denotes dataset annotations, green denotes infected region segmentation mask obtained using our methodology and red denotes intersection areas of dataset annotations and our methodology. (Color figure online)

CT score is the sum of the scores from each lobe and ranges from 0 to 25. The cases are further classified into mild, moderate and severe based on the scores thus computed [13]: mild - for a score <8, moderate - for a score from 8 to 15 and severe - for a score > 15.

3 Experiments and results

In this section, we first give the details of the data used in our experiments. Then we illustrate the result of the segmentation of infected regions followed by the result of COVID-19 severity gradation. We also perform an analysis of coverage and fairness offered by our methodology.

3.1 Dataset

Lung CT scans from the Medical Imaging Data Resource Center - RSNA International COVID-19 Open Radiology Database - Release 1a (MIDRC-RICORD-1a) [14] of The Cancer Imaging Archive (TCIA) are used for our experiments. The data comes from patients at least 18 years of age receiving positive diagnoses for COVID-19. In addition to lung CT scans, the dataset contains clinical data of the patients and a rough annotation of the infected regions. We contribute to the benchmark dataset by providing the COVID-19 severity gradations scored by an experienced radiologist corresponding to each patient. The radiologist gradations are available at https://github.com/Samiran-Dey/explainable_seg. We

Table 1. Table of results for automated gradation of COVID-19 score and severity. QWK stands for quadratic weighted kappa. α denotes the range of grades considered for evaluating the coverage. Best scores are marked in bold.

ROI segmentation	Severity	Gradation						
	QWK	QWK	Coverage					
			$\alpha=0$	$\alpha=1$	$\alpha=2$	$\alpha=3$	$\alpha=4$	$\alpha=5$
Dataset annotations	0.43860	0.52245	0.19	**0.45**	0.58	0.69	0.74	0.81
Our methodology	**0.73492**	**0.81312**	**0.21**	0.38	**0.65**	**0.79**	**0.87**	**0.94**

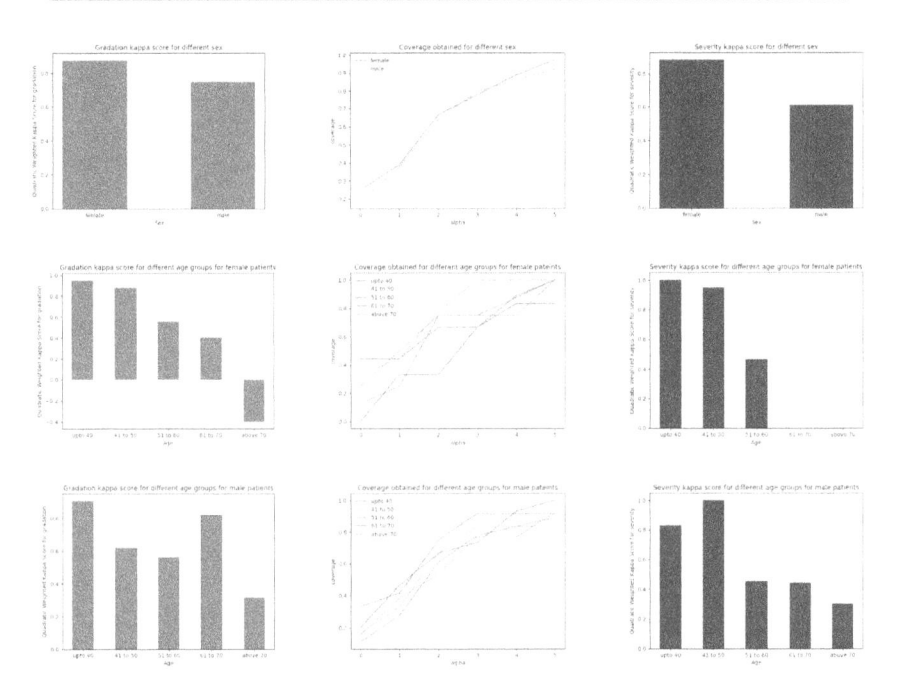

Fig. 6. Plots for the result of fairness evaluation of COVID-19 severity score prediction by our methodology for different sex and age groups. The first column shows bar graphs of quadratic weighted kappa scores for COVID-19 gradation, the second column shows line graphs for the coverage obtained for different values of α for COVID-19 gradation and the third column shows bar graphs of quadratic weighted kappa for severity estimation. The first row corresponds to different sexes, the second row corresponds to different age groups of sex females and the third row corresponds to different age groups of sex males.

used 100 CT scans from the dataset for our experiments. The rest of the cases were not used because of the presence of different exceptional patterns like cystic broncho-ectatic changes in the right lung with bilateral pleural effusion, only mosaic perfusions in both lungs, only scars in both lower lobes, only mosaic perfusions bilaterally, only nodular and streaky scars, etc. which makes the process of COVID-19 severity gradation far more complex.

3.2 Qualitative results for segmentation of infected regions

Illustrative images for the qualitative results of the segmentation of infected regions are provided in Fig.5, for different cases. In Fig. 5.a, we obtain the closest match with the dataset annotations. However, the segmentation map produced by our methodology marks the region of ground glass opacity more precisely than the dataset annotations, as can be observed from the CT image on the left. In Fig. 5.b, it is observed that the dataset annotations are partly missing, which our methodology can effectively provide. A similar observation is made for Fig. 5.c. In both the images annotations for the left lung are missing and in Fig. 5.d, dataset annotations are completely missing. However, since our methodology is unsupervised and based on classical image processing and pattern recognition techniques, we can easily predict the infected regions for the cases where dataset annotations are missing. Also in Fig. 5.c, the dataset annotation marks the entire right lung as infected, whereas using our methodology we can obtain a more precise segmentation of the infected regions that matches the observed ground glass opacity for the specific CT slice. Since the dataset annotations are missing, we do not analyse the segmentation results quantitatively. Also, since data with missing annotations cannot be used to train supervised deep learning models, we do not compare our methodology with any deep learning methods for the given dataset. Thus, it may be concluded that our unsupervised clinically inclined approach is capable of segmenting the COVID-19-infected lung regions from CT slices effectively and accurately.

3.3 Quantitative results for COVID-19 severity gradation

To evaluate the predicted COVID-19 severity scores we use quadratic weighted kappa (QWK) [15] as a metric. QWK measures the degree of agreement of predicted severity scores with the ground truth gradation. While a value of 1 denotes complete agreement, -1 denotes complete disagreement. Our methodology can obtain a QWK of 0.81 for the gradation of COVID-19 and a QWK of 0.73 for severity estimation, as seen in Table 1. However, on using the dataset annotations of infected regions for obtaining the percentage involvement of each lobe, a QWK of 0.52 is obtained for COVID-19 gradation and 0.44 for severity estimation, thus showing less agreement with the radiologist gradations. This is because our methodology has pixel-level precision when deciding on the infected regions. Whereas, in the manual annotations provided with the dataset a wider area is roughly marked around the infection region as evident in Fig 5. Also, for some slices annotations are missing. As our methodology has greater agreement with the ground-truth gradations, it suggests that our methodology may be used as an assistance tool by radiologists for precise annotation of COVID-19 infection regions.

Table 2. Table of results for evaluation of the fairness of COVID-19 severity prediction by our proposed methodology for different sex and age groups. QWK stands for quadratic weighted kappa. α denotes the range of grades considered for evaluating the coverage.

Sex	Age	Severity	Gradation						
		QWK	QWK	Coverage					
				$\alpha=0$	$\alpha=1$	$\alpha=2$	$\alpha=3$	$\alpha=4$	$\alpha=5$
Female	-	0.88053	0.87315	0.25	0.39	0.67	0.78	0.89	0.97
Male	-	0.61181	0.74746	0.19	0.38	0.64	0.80	0.86	0.92
Female	up to 40	1.00000	0.95000	0.44	0.44	0.67	0.67	0.89	1.00
Female	41 to 50	0.94915	0.87861	0.33	0.44	0.78	1.00	1.00	1.00
Female	51 to 60	0.46667	0.55932	0.13	0.25	0.75	0.75	0.88	1.00
Female	61 to 70	0.00000	0.40496	0.00	0.33	0.33	0.67	0.83	0.83
Female	above 70	0.00000	-0.40000	0.25	0.5	0.75	0.75	0.75	1.00
Male	up to 40	0.83019	0.90789	0.33	0.42	0.75	0.92	0.92	0.92
Male	41 to 50	1.00000	0.61878	0.17	0.33	0.67	0.83	0.83	0.83
Male	51 to 60	0.45455	0.56288	0.11	0.28	0.61	0.78	0.83	0.89
Male	61 to 70	0.44444	0.82326	0.20	0.47	0.67	0.73	0.93	1.00
Make	above 70	0.30357	0.31484	0.15	0.38	0.54	0.77	0.77	0.92

3.4 Coverage and Fairness evaluation

We also evaluate the coverage offered by our methodology for different ranges of error grades, denoted by α, and provide the results in Table 1. Coverage evaluates whether the true grade is present in the range given by predicted grade $\pm \alpha$. Our methodology obtains a coverage of 0.21 for $\alpha = 0$, which means that 21% of the predictions are an exact match to the true grades and gradually goes up to a coverage of 0.94 for $\alpha = 5$. Our methodology of segmenting infection regions obtains a higher coverage than dataset annotations in all cases, except for $\alpha = 5$.

Further, we attempt to evaluate the fairness of predictions by our methodology for different sex and age groups. The results are given in Table 2 and Fig. 6. It is observed that QWK for female patients is 0.87 for COVID-19 gradation and 0.88 for severity estimation whereas for male patients, QWK is 0.75 for COVID-19 gradation and 0.61 for severity estimation. The coverage obtained for female patients is also more than that obtained for male patients for a particular value of α, the only exception being $\alpha = 3$. Thus, for the MIDRC-RICORD-1a dataset, our methodology performs better for female patients than male patients. On studying the performance for different age groups, it is observed that the QWK score decreases for female patients as age increases for both COVID-19 gradation and severity estimate, with a negative QWK value for the age group above 70 for COVID-19 gradation and a 0 value for age of above 60 for severity estimation. On observing the coverage values for different age groups a relatable trend is found for female patients. However, for different age groups of male patients, no such trend can be observed for QWK scores and coverage. Thus, it may be concluded that for the given dataset, the performance of our methodology decreases as the age group increases for female patients but no such conclusion can be drawn for male patients.

4 Conclusion

In this paper, we propose a domain-inspired unsupervised approach for the prediction of COVID-19 severity grades from lung CT scans using classical image processing and pattern recognition techniques based on the clinical procedure of gradation. Radiologists estimate by eye the percentage involvement of each lobe of the lung which is then added to give a score between 0 and 25. We propose a computer-automated assistance tool that mimics this eye estimation to provide pixel-level segmentation of infected regions of the lung with high precision followed by gradation of COVID-19 severity scores and act as a clinical decision support system (CDSS). Our method is purely mechanistic and hence provides several advantages: 1) it is unsupervised and hence does not need costly annotations and labels from human experts for training (only for validation), 2) since there is no learning stage, it is fast and lightweight, and hence can be easily deployed to a remote clinical setting; 3) it is interpretable and trustworthy as it follows the clinical process in a transparent way as opposed to deep learning black boxes. We achieve high-fidelity results on a popular benchmark dataset and even add new resources to the same through severity scores generated by our contributing senior radiologist. Finally, our method can be easily generalised to other similar tasks like pulmonary oedema, etc. The code and the COVID-19 CT severity gradations by our experienced radiologist are available at https://github.com/Samiran-Dey/explainable_seg.

Acknowledgements and Declarations. We acknowledge the Indo-Swedish DBT-Vinnova project, BT/PR41025/Swdn/135/9/2020, for supporting this research. S Dey is funded by AI4ICPS Chanakya PhD Fellowship. TC is funded by the Turing-Roche Strategic Partnership. The authors have no conflict of interest.

References

1. Jin YH, Cai L, Cheng ZS, Cheng H, Deng T, Fan YP, Fang C, Huang D, Huang LQ, Huang Q, Han Y. A rapid advice guideline for the diagnosis and treatment of 2019 novel coronavirus (2019-nCoV) infected pneumonia (standard version) Mil Med Res. 2020;7(1):1-23. https://doi.org/10.1186/s40779-020-0233-6
2. Kwee, T.C., Kwee, R.M.: Chest CT in COVID-19: What the Radiologist Needs to Know. Radiographics **40**(7), 1848–1865 (2020). https://doi.org/10.1148/rg.2020200159
3. Hofmanninger, J., Prayer, F., Pan, J., et al.: Automatic lung segmentation in routine imaging is primarily a data diversity problem, not a methodology problem. Eur Radiol Exp **4**, 50 (2020). https://doi.org/10.1186/s41747-020-00173-2

4. Antiga L. "Generalizing vesselness with respect to dimensionality and shape". The Insight Journal. 2007 Aug. http://hdl.handle.net/1926/576
5. Abdel-Tawab M, Basha MAA, Mohamed IAI, Ibrahim HM. A simple chest CT score for assessing the severity of pulmonary involvement in COVID-19. Egypt J Radiol Nucl Med. 2021;52(1):149. https://doi.org/10.1186/s43055-021-00525-x.Epub 2021 Jun 18. PMCID: PMC8211934
6. Simon BA, Christensen GE, Low DA, Reinhardt JM. Computed tomography studies of lung mechanics. Proc Am Thorac Soc. 2005;2(6):517-21, 506-7. https://doi.org/10.1513/pats.200507-076DS.PMID: 16352757; PMCID: PMC2713339
7. Rupanagudi, Vijay A. et al. CAN PLEURAL FLUID DENSITY MEASURED BY HOUNSFIELD UNITS(HU) ON CHEST CT BE USED TO DIFFERENTIATE BETWEEN TRANSUDATE AND EXUDATE? CHEST, Volume 128, Issue 4, 361S https://doi.org/10.1378/chest.128.4_MeetingAbstracts.361S.
8. Murphy A, Hacking C, Iflaq P, et al. Motion artifact. Reference article, Radiopaedia.org (Accessed on 17 Mar 2024) https://doi.org/10.53347/rID-48589
9. Schaller MA, Sharma Y, Dupee Z, Nguyen D, Urueña J, Smolchek R, Loeb JC, Machuca TN, Lednicky JA, Odde DJ, Campbell RF, Sawyer WG, Mehrad B. Ex vivo SARS-CoV-2 infection of human lung reveals heterogeneous host defense and therapeutic responses. JCI Insight. 2021 Sep 22;6(18):e148003. https://doi.org/10.1172/jci.insight.148003.PMID: 34357881
10. Almasi Nokiani A, Shahnazari R, Abbasi MA, Divsalar F, Bayazidi M, Sadatnaseri A. CT severity score in COVID-19 patients, assessment of performance in triage and outcome prediction: a comparative study of different methods. Egypt J Radiol Nucl Med. 2022;53(1):116. https://doi.org/10.1186/s43055-022-00781-5.Epub 2022 May 18
11. Li, K., Wu, J., Wu, F., et al.: The clinical and chest CT features associated with severe and critical COVID-19 pneumonia. Invest. Radiol. (2020). https://doi.org/10.1097/RLI.0000000000000672
12. Wasilewski PG, Mruk B, Mazur S, Półtorak-Szymczak G, Sklinda K, Walecki J. COVID-19 severity scoring systems in radiological imaging - a review. Pol J Radiol. 2020 Jul 17;85:e361-e368. https://doi.org/10.5114/pjr.2020.98009.PMID: 32817769;
13. Sharma S, Aggarwal A, Sharma RK, Patras E, Singhal A. Correlation of chest CT severity score with clinical parameters in COVID-19 pulmonary disease in a tertiary care hospital in Delhi during the pandemic period. Egypt J Radiol Nucl Med. 2022;53(1):166. https://doi.org/10.1186/s43055-022-00832-x.Epub 2022 Jul 28
14. Tsai, E., Simpson, S., Lungren, M.P., Hershman, M., Roshkovan, L., Colak, E., Erickson, B.J., Shih, G., Stein, A., Kalpathy-Cramer, J., Shen, J., Hafez, M.A.F., John, S., Rajiah, P., Pogatchnik, B.P., Mongan, J.T., Altinmakas, E., Ranschaert, E., Kitamura, F.C., Topff, L., Moy, L., Kanne, J.P., Wu, C.: Data from the Medical Imaging Data Resource Center - RSNA International COVID Radiology Database Release 1a - Chest CT Covid+ (MIDRC-RICORD-1A). The Cancer Imaging Archive (2020). https://doi.org/10.7937/VTW4-X588
15. Doewes, A., Kurdhi, N., Saxena, A. (2023). Evaluating Quadratic Weighted Kappa as the Standard Performance Metric for Automated Essay Scoring. In Proceedings of the 16th International Conference on Educational Data Mining (pp. 103-113). International Educational Data Mining Society (IEDMS) https://doi.org/10.5281/zenodo.8115784

16. Antar, S., Abd El-Sattar, H.K.H., Abdel-Rahman, M.H. et al. COVID-19 infection segmentation using hybrid deep learning and image processing techniques. Sci Rep 13, 22737 (2023) https://doi.org/10.1038/s41598-023-49337-1
17. Sailunaz K, Bestepe D, Özyer T, Rokne J, Alhajj R. Interactive framework for Covid-19 detection and segmentation with feedback facility for dynamically improved accuracy and trust. PLoS One. 2022 Dec 22;17(12):e0278487. https://doi.org/10.1371/journal.pone.0278487.PMID: 36548288; PMCID: PMC9778629
18. Oulefki A, Agaian S, Trongtirakul T, Kassah Laouar A. Automatic COVID-19 lung infected region segmentation and measurement using CT-scans images. Pattern Recognit. 2021 Jun;114:107747. https://doi.org/10.1016/j.patcog.2020.107747.Epub 2020 Nov 2. PMID: 33162612; PMCID: PMC7605758
19. Enshaei N, Oikonomou A, Rafiee MJ, Afshar P, Heidarian S, Mohammadi A, Plataniotis KN, Naderkhani F. COVID-rate: an automated framework for segmentation of COVID-19 lesions from chest CT images. Sci Rep. 2022 Feb 25;12(1):3212. https://doi.org/10.1038/s41598-022-06854-9.PMID: 35217712; PMCID: PMC8881477
20. Aleem M, Raj R, Khan A. Comparative performance analysis of the resnet backbones of mask rcnn to segment the signs of covid-19 in chest ct scans. arXiv preprint arXiv:2008.09713. 2020 Aug 21
21. Ahmed, I., Chehri, A., Jeon, G.: A Sustainable Deep Learning-Based Framework for Automated Segmentation of COVID-19 Infected Regions: Using U-Net with an Attention Mechanism and Boundary Loss Function. Electronics 11, 2296 (2022). https://doi.org/10.3390/electronics11152296
22. Punn NS, Agarwal S. CHS-Net: A Deep Learning Approach for Hierarchical Segmentation of COVID-19 via CT Images. Neural Process Lett. 2022;54(5):3771-3792. https://doi.org/10.1007/s11063-022-10785-x.Epub 2022 Mar 16. PMID: 35310011
23. Ter-Sarkisov A. Covid-ct-mask-net: Prediction of covid-19 from ct scans using regional features. Applied Intelligence. 2022. Jan 8:1-2. https://doi.org/10.1007/s10489-021-02731-6
24. Saeedizadeh N, Minaee S, Kafieh R, Yazdani S, Sonka M. COVID TV-Unet: Segmenting COVID-19 chest CT images using connectivity imposed Unet. Computer Methods and Programs in Biomedicine Update. 2021. Jan 1;1:100007. https://doi.org/10.1016/j.cmpbup.2021.100007
25. Xu X, Wen Y, Zhao L, Zhang Y, Zhao Y, Tang Z, et al. CARes-UNet: Content-aware residual UNet for lesion segmentation of COVID-19 from chest CT images. Medical Physics. 2021. Nov;48(11):7127-40. https://doi.org/10.1002/mp.15231
26. Yin S, Deng H, Xu Z, Zhu Q, Cheng J. SD-UNet: A Novel Segmentation Framework for CT Images of Lung Infections. Electronics. 2022. Jan 1;11(1):130. https://doi.org/10.3390/electronics11010130
27. Markowetz, F. All models are wrong and yours are useless: making clinical prediction models impactful for patients. npj Precis. Onc. 8, 54 (2024). https://doi.org/10.1038/s41698-024-00553-6

Privacy-Preserving Tabular Data Generation: Application to Sepsis Detection

Eric Macias-Fassio[1,2](✉), Aythami Morales[2], Cristina Pruenza[1], and Julian Fierrez[2]

[1] Instituto de Ingeniería del Conocimiento, Madrid, Spain
{eric.macias,cristina.pruenza}@iic.uam.es
[2] BiDA-Lab, Universidad Autónoma de Madrid, 28049 Madrid, Spain
{aythami.morales,julian.fierrez}@uam.es

Abstract. The biomedical field is among the sectors most impacted by the increasing regulation of Artificial Intelligence (AI) and data protection legislation, given the sensitivity of patient information. However, the rise of synthetic data generation methods offers a promising opportunity for data-driven technologies. In this study, we propose a statistical approach for synthetic data generation applicable in classification problems. We assess the utility and privacy implications of synthetic data generated by Kernel Density Estimator and K-Nearest Neighbors sampling (KDE-KNN) within a real-world context, specifically focusing on its application in sepsis detection. The detection of sepsis is a critical challenge in clinical practice due to its rapid progression and potentially life-threatening consequences. Moreover, we emphasize the benefits of KDE-KNN compared to current synthetic data generation methodologies. Additionally, our study examines the effects of incorporating synthetic data into model training procedures. This investigation provides valuable insights into the effectiveness of synthetic data generation techniques in mitigating regulatory constraints within the biomedical field.

Keywords: Synthetic data · Machine learning · Sepsis detection

1 Introduction

The exponential growth of Artificial Intelligence (AI) has sparked a revolutionary wave across various sectors with its profound impact particularly evident in the biomedical field. AI's ability to analyze vast amounts of data quickly and accurately has transformed medical research, diagnosis, and treatment. In recent years, there has been significant progress in the application of machine learning (ML) and deep learning models for early disease diagnosis [42]. These methodologies have exhibited substantial potential in identifying a diverse range

---

Supported by organization Instituto de Ingeniería del Conocimiento, Hospital Universitario Son Llátzer, the Fundación Instituto de Investigación Sanitaria Illes Balears (Spain), and BBforTAI (PID2021-127641OB-I00 MICINN/FEDER).

of diseases, including cancer [40], cardiovascular disease [29], and Parkinson's disease [15], or conditions such as fatigue [1]. Through sophisticated algorithms and analysis of extensive datasets, these models can potentially identify subtle patterns and markers indicative of these conditions at their early stages.

However, many governments are introducing strict regulations for personal data processing and AI applications such as new European Union AI act[1], CCPA[2] (Unitated States), and LGPD[3] (Brazil), which enforce data protection measures. A significant development in the regulatory landscape of AI has occurred with the enactment of the AI Act within the European Union. This legislative framework is designed to oversee and govern the application of AI models. In the realm of biomedical research, cautious consideration must be exercised when employing patient data for the training of AI models. Patient data, characterized by its sensitive nature [30], is subject to stringent protection under data protection laws, necessitating the preservation of privacy [5].

A solution that can potentially overcome these limitations involves the generation of fully Synthetic Data (SD) as an alternative to real data. SD is artificial data generated by a trained model and built to replicate real data by taking into account its distribution (mean, variance) and structure (e.g. correlation between attributes) [12]. The utilization of SD generation emerges as a versatile methodology in machine learning, extending its applications across two domains: augmenting datasets to enhance model training [4,39] and safeguarding the privacy of sensitive information [35]. Henceforth, this study introduces a straightforward technique for SD generation and conducts a comparative evaluation against state-of-the-art methodologies in terms of both utility and privacy considerations. The evaluation of these methods is performed within the context of a real-world application, specifically the early diagnosis of sepsis.

In more detail, the main contributions of this work are the following:

- We propose KDE-KNN, a statistical method to generate synthetic data for training and evaluating supervised learning algorithms.
- We evaluated the utility and privacy of the generated synthetic data using different supervised algorithms in the context of sepsis detection. Our findings demonstrate that KDE-KNN outperforms existing methods in generating synthetic tabular data for sepsis detection.
- We assessed the generalization capacity of KDE-KNN using two real databases with more than 2000 patients. Our results suggest that KDE-KNN has certain advantages in terms of generalization over other methods.

2 Related Works

We have divided this section in two parts: (i) synthetic tabular data generation approaches in Healthcare, and (ii) machine learning models for predicting sepsis.

---

[1] https://artificialintelligenceact.eu/
[2] https://oag.ca.gov/privacy/ccpa
[3] https://iapp.org/resources/article/brazilian-data-protection-law-lgpd-english-translation/

2.1 Synthetic Tabular Data Generation Approaches in Healthcare

Synthetic data (SD) have been considered for a long time (e.g., Rubin [36] and Little [27]) as records of synthetic values instead of real values for different purposes. Nowadays, the concept of SD has evolved to encompass artificial data generated by trained models, designed to emulate real data by faithfully capturing its distributional (such as mean and variance) and structural attributes (including correlations between attributes) [12]. SD generation stands out as a highly promising yet largely underexploited technology for fulfilling privacy-preserving laws. In the biomedical sector, synthetic data generation has been mainly investigated in medical imaging [19], Electronic Health Records (EHR) free-text content [17] and EHR tabular data [48].

The present study focuses on synthetic EHR tabular data generation, as it is the predominant type of data used to develop ML models to aid healthcare decision-making [21]. Tabular healthcare-related data stored in EHR contain vast and diverse amounts of patient-related data. Typically, each row in a healthcare tabular dataset represents a single data record containing descriptive patient details such as date of birth, gender, and demographic information, along with sensitive attributes primarily consisting of longitudinal data. This longitudinal data comprises a series of medical events occurring at various time points, encompassing diagnoses, laboratory test results, and prescription information [10].

In the healthcare context, numerous approaches to generating synthetic data can be found in the literature. Among these, one widely utilized algorithm is the Synthetic Minority Oversampling Technique (SMOTE) [7], representing a straightforward method for generating synthetic tabular data [44]. This algorithm operates by synthesizing new data through interpolation of the existing samples. Another statistical approach to generate synthetic data involves Kernel Density Estimation (KDE) based models. Our framework for synthetic data generation in the healthcare context relies on KDE, chosen for its non-parametric nature and demonstrated efficacy, particularly in small datasets, which are prevalent in the biomedical field [34].

Additional methodologies used for SD generation involves the utilization of generative models, which include Generative Adversarial Networks (GANs) and Diffusion Models (DMs).

Since their inception in 2014 [16], GANs have demonstrated exceptional capability in the production of synthetic image data [32]. For this reason, the application of GANs to other data types, such as tabular data, is a popular topic in the AI research community [20]. Some GAN-based synthetic tabular data generation approaches in Healthcare are ehrGAN [8], medGAN [9], GcGAN [49]. However, owing to the difficulties associated with training these models, as well as constraints related to sample size, we opt not to evaluate such models in this study.

On the other hand, we have Diffusion Models (DMs). DMs represent another class of generative models which have been widely used in the computer vision field. Notably, recent advancements have led to the development of architectures tailored to exploit diffusion models for tabular data, such as TabDDPM

[26], which has demonstrated significant potential and promising outcomes in this regard. For these reasons, in this study we have evaluated the performance between SMOTE, TabDDPM and KDE-based generative models.

2.2 Machine Learning Models for Predicting Sepsis

Sepsis is defined as life-threatening organ dysfunction caused by a dysregulated host response to infection [43]. In 2017, approximately 20% of all global deaths were attributed to sepsis [37]. Early diagnosis of sepsis is crucial in the clinical setting, as it could help to significantly improve patient outcomes [47], but early and accurate sepsis detection is still a challenging clinical problem [3]. For this reason, several ML algorithms have been designed to predict sepsis using retrospective data [6,14,22,23,25,31]. To our current knowledge, existing algorithms for sepsis prediction operate within a defined temporal window, typically forecasting the likelihood of sepsis onset within a specific time lapse, such as the next 24 hours. In our study, we want to overcome temporal constraints by seeking to predict the presence or absence of sepsis without temporal limitations. Therefore, we frame the task of sepsis detection as a classification problem, with the aim of addressing the question: Will patient A develop sepsis in the future? Furthermore, we substantiate our findings through validation in an external cohort for robustness and generalizability.

3 Materials and Methods

In this study we have used 2 databases: i) Mannheim database (MaDB) used for training our models and building the synthetic datasets; ii) Son Llàtzer hospital database (SLDB) used as external validating dataset to evaluate the trained models.

3.1 Mannheim Database

We used the University Medical Centre Mannheim database of patients admitted to Intensive Care Unit (ICU) [38]. This database contains a total of 1275 patients, 979 with non-sepsis and 296 with sepsis. Initially, the MaDB comprised 42 timelines of features and the diagnosis of sepsis at each time step. However, for comparative analysis with the SLDB, it was necessary to align the feature sets. Consequently, only 27 features were found to be common between both databases. Among these features we have the age of the patient and lab results (Table 1).

For our study we did not use temporal data, instead we set a cut-off value at 9 hours as we estimated that in this time period all clinical tests could be performed and laboratory results could be collected. If a test has been performed several times during this period, the last value is used. In this way we constructed a dataset where our predictor variables were collected in that time interval and

Table 1. Description of the 27 variables present in the databases.

ID	Feature	Description
1	heart_rate	Number of heartbeats per minute
2	leukocytes	Cells of the immune system
3	temperature	Body temperature
4	respiratory_rate	Number of breaths a person takes per minute
5	bilirubin	Compound originating from heme catabolism [46]
6	blood_urea_nitrogen	Amount of urea nitrogen in the blood
7	creatinine	The end product of creatine phosphate metabolism [24]
8	diastolic_bp	Blood pressure measurement
9	fraction_of_inspired_o2	Fraction of oxygen present in the air that a person inhales
10	systolic_bp	Blood pressure measurement
11	thrombocytes	Blood cells
12	lactate	Metabolite of glucose
13	bicarbonate	Electrolyte [41]
14	c-reactive_protein	Molecule secreted in response to inflammatory cytokines [11]
15	hemoglobin	Protein found in red blood cells
16	lymphocytes	Cells of the immune system
17	sodium	Electrolyte [41]
18	pancreatic_lipase	Enzyme [28]
19	procalcitonin	Peptide
20	oxygen_saturation	Percentage of hemoglobin bound to oxygen [18]
21	blood_glucose	Concentration of glucose
22	chloride	Electrolyte [41]
23	calcium	Electrolyte [41]
24	potassium	Electrolyte [41]
25	alanine_transaminase	Enzyme [45]
26	aspartate_transaminase	Enzyme [45]
27	age	Years

the objective was to predict whether or not a patient will develop sepsis in the future (classification problem).

On the other hand, the Mannheim database (MaDB), contains temporal data that allow precise tracking of sepsis onset times for patients, as evidenced in Table 2. The notable variability in the timing of sepsis manifestation within this dataset underscores its inherent heterogeneity. However, we do not use this temporal information, because we treat the detection of sepsis as a binary classification problem. The MaDB has been used to train and test models and generate synthetic data.

3.2 Son Llàtzer Hospital Database

We used a database from Son Llàtzer Hospital of patients admitted to emergency and ICU. The Son Llàtzer DataBase (SLDB) contains 2028 patients in total, 1014 with non-sepsis and 1014 with sepsis. In this database, we also selected the 27 common features with the MaDB. However, within the SLDB, the precise

Table 2. Main characteristics of the databases, including the number of features and patients, as well as the mean, minimum and maximum time of sepsis onset (in hours) and the service where the data were collected.

DB	Patients	Features	Mean(t)	Min(t)	Max(t)	Hospital service
MaDB	979 non-sepsis/ 296 sepsis	27	208.7	39.5	1385	ICU
SLDB	1014 non-sepsis/ 1014 sepsis	27	36	24	48	ICU/emergency

mean sepsis onset time remains unknown. According to insights from the medical team, the mean sepsis onset time is estimated to range between 24 to 48 hours. We employed this database for external validation, acknowledging significant disparities in sepsis onset times compared to our primary dataset. Notably, there are substantial variations in data distribution between the two databases. Thus, we perceived this as an opportunity to assess the generalization capacity of our models across diverse demographic populations.

3.3 Sepsis Prediction Models

Predicting sepsis onset remains a critical challenge in clinical practice due to its rapid progression and potentially life-threatening consequences [43]. Early detection and intervention are paramount for improving patient outcomes and reducing mortality rates associated with this severe condition. Our study evaluates three distinct ML models and assesses their performance based on the Area Under the Curve (AUC) score.

- Random Forest (RF). It is a widely used machine learning algorithm that belongs to the ensemble learning family, characterized by the construction of multiple decision trees during training. For classification tasks, RF outputs the predicted class, which in the context of sepsis prediction signifies whether a patient is likely to develop sepsis or not.
- Support Vector Machine (SVM). Unlike traditional classifiers that aim to find a decision boundary that separates classes, SVM seeks to find the hyperplane that best divides the classes while maximizing the margin between them. In our experiments we have used two SVM changing the type of kernel: *i)* SVM with a linear kernel; *ii)* SVM with a radial basis function (rbf) kernel.

The hyperparameters of the models were tuned using the Optuna library [2]. Specifically, we employed a TPE (Tree-structured Parzen Estimator) sampler with 40 trials to maximize AUC. For the Random Forest (RF) model, the optimal hyperparameters were determined as follows: bootstrap was set to False, max_depth to 20, max_features to 5, min_samples_leaf to 5, min_samples_split to 12, and n_estimators to 500. As for the SVM models, both of them had C set to 1. Other parameters that were not mentioned stay by default according to the scikit-learn library [33].

3.4 Statistical Data Modelling Approaches

In this paper we have analysed 3 popular statistical data modelling methods:

SMOTE [7]: SMOTE is an oversampling methodology that was initially employed to generate synthetic observations exclusively from the minority class. We expanded this approach to incorporate the majority class as well [26], resulting in the creation of a fully synthetic dataset.

TabDDPM [26]: TabDDPM is a design of denoising diffusion probabilistic models for tabular data. To tackle mixed-type characteristics of tabular data, this architecture integrates gaussian diffusion for capturing the characteristics of continuous features and multinomial diffusion for effectively modeling categorical attributes.

KDE: KDE is a method used to estimate probability density functions. By constructing this distribution, we gain the ability to generate synthetic data samples through sampling. This capability allows for the creation of synthetic datasets representative of the underlying probability distribution. We conducted experiments using multivariate KDE, taking into account the interdependencies between features. This allows us to capture complex relationships and dependencies across multiple variables simultaneously.

3.5 KDE-KNN: Privacy-Preserving Synthetic Clinical Data

Our proposed methodology is founded on the integration of KDE and K-Nearest Neighbors (KNN). The idea is to use a multivariate gaussian KDE to approximate the probability density function of the original dataset features and then sample it to generate synthetic datasets. However, as the feature space can be very large, we train a KNN to validate the synthetic samples. The procedural steps to construct our synthetic dataset are the following:

1. Training a KNN model: A KNN model is trained using the provided training dataset.
2. Data preparation for KDE: The training dataset is partitioned into two distinct groups, patients with sepsis (18.55%) and without sepsis (81.45%).
3. Multivariate KDE construction: Statistical independent multivariate KDE distributions are trained for each subgroup.
4. Sampling synthetic data: Sampling is performed from each multivariate KDE model, generating 540 synthetic patients with sepsis and 540 synthetic patients without sepsis.
5. Validation using KNN model: Validation of the synthetic samples is conducted by the trained KNN model. Synthetic data originated from the KDE model built with non-sepsis data should be classified as non-sepsis by the KNN model. Any discrepancies lead to discarding the data point.

Steps 4 and 5 are iteratively executed until we attain a total of 540 synthetic data points for patients with sepsis and 540 synthetic data points for patients without sepsis. This process ensures the creation of a balanced synthetic dataset representative of both septic and non-septic patient populations.

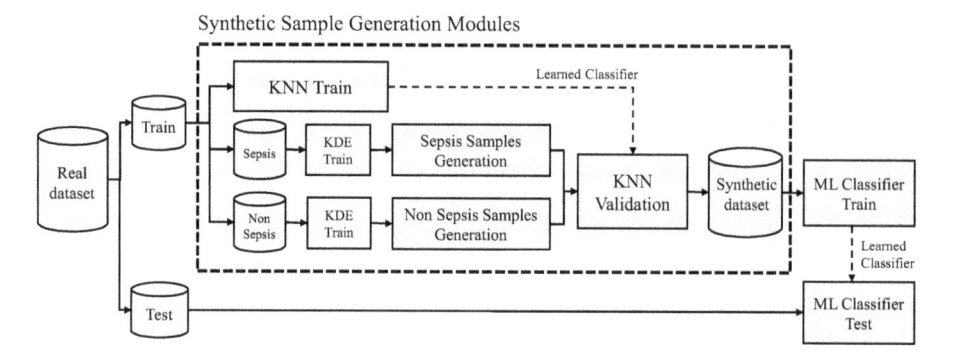

Fig. 1. Block diagram of our proposed KNN-KDE method for synthetic data generation including the generation modules based on two Kernel Density Estimators (Sepsis and Non-Sepsis) and K-NN sampling.

For clarification, we close this section by visualizing our proposed synthetic method as a flowchart, illustrated in Fig. 1.

4 Experiments and Results

In this section, we assess the influence of synthetic data on the performance of sepsis detection models.

4.1 Experimental Protocol

Initially, our study is based on two distinct sepsis databases: the MaDB and the SLDB. The MaDB served as the primary dataset for model training/testing and synthetic data generation, while the SLDB was exclusively utilized for external validation purposes. Regarding the preprocessing of the datasets, missing values were imputed with the median and the data were standardized.

Our first experimental phase involved evaluating model performance using real data exclusively. To accomplish this, we employed the MaDB and we partitioned the data into training (85%) and testing (15%) sets, repeating the experiment three times while changing the seed. Additionally, each partition underwent an external validation using the SLDB. Notably, our analysis revealed that the performance of the most effective models remained consistent across different partitions, suggesting minimal impact of partitioning on model performance.

The second phase of experimentation was dedicated to assessing the utility of synthetic data. Our focus was on evaluating model performance exclusively using synthetic data. To achieve this, we generated a fully synthetic balanced dataset comprising 540 samples with sepsis and an equal number of 540 samples without sepsis. This balanced dataset mirrored the size of our original imbalanced training set. Building upon the stability observed in model performance across the three partitions in Experiment 1, one of the partitions was randomly

Table 3. Results of Experiment 1 using real data. The result is shown in terms of AUC ± variance as each model was trained 3 times.

Model	MaDB	SLDB
RF	**0.6708 ± 0.0169**	0.6469 ± 0.0313
SVM lineal kernel	0.5426 ± 0.0581	0.6120 ± 0.0701
SVM rbf kernel	0.6194 ± 0.0119	**0.6952 ± 0.0282**

selected for this subsequent analysis. This selection yielded both a training set and a test set sourced from the MaDB. The training set acted as a seed for generating synthetic data, employing the methodologies outlined in Section 3.4. For each method, three distinct synthetic balanced datasets were generated. Following this, the quality of the synthetic data was evaluated using the test set. Additionally, an external validation was conducted using the SLDB to ensure the reliability and validity of our findings (Experiment 2). The results of this experiment showed that the best method to generate synthetic data in the context of sepsis prediction is KDE-KNN. Finally, in the Experiment 3, we examined how the incorporation of both real and synthetic data in the training set influenced model performance.

The third phase of our experimentation aimed to assess data privacy preservation using our KDE-KNN method. In the Experiment 4, we investigated the proximity of synthetic data to real data using the Mean Distance to Closest Record (DCR) metric [50]. Mean DCR calculates the average distance between synthetic samples and their closest real data points.

4.2 Real Data and Synthetic Data Utility

The findings from Experiment 1 are presented in Table 3. This experiment involves evaluating model performance in terms of AUC using exclusively real data. The results indicates that the RF model demonstrates superior performance in the test set. Nevertheless, concerning generalization, the results indicates a lower performance of the RF model, while the SVM with the rbf kernel demonstrates better generalization capabilities.

Experiment 2 aimes to assess the utility of synthetic data. The outcomes of Experiment 2 are detailed in Table 4. This experiment involved evaluating model performance using balanced synthetic datasets for training.

The findings indicates a notable enhancement in model performance when employing balanced synthetic datasets. Remarkably, balanced synthetic data appear to outperform real imbalanced data. Specifically, the SVM model with the rbf kernel demonstrates superior performance when trained on synthetic data generated using the KDE-KNN method. Furthermore, enhanced model performance is evident in the external validation database. These results may be due to the reduced heterogeneity of the external database and the earlier onset of sepsis in patients, suggesting that our models perform better when sepsis occurs within the 24-48 hour timeframe. Additionally, we emphasize the minimal variance observed in synthetic datasets generated through our method.

Table 4. Results of Experiment 2 using synthetic data. The result is shown in terms of AUC ± variance as each model was trained with 3 synthetic datasets. The ML classifier was trained with synthetic samples generated from a subset of samples obtained from MaDB.

Data Generation Method	ML Classifier	MaDB	SLDB
SMOTE	RF	0.6721 ± 0.0144	0.4560 ± 0.0491
	SVM (lineal)	0.6309 ± 0.0346	0.6583 ± 0.0453
	SVM (rbf)	0.6771 ± 0.0212	0.4437 ± 0.0596
TabDDPM [26]	RF	0.6942 ± 0.0102	0.5187 ± 0.0804
	SVM (lineal)	0.6697 ± 0.0207	0.6446 ± 0.046
	SVM (rbf)	0.7020 ± 0.0095	0.6949 ± 0.0246
KDE	RF	0.6495 ± 0.0051	0.6261 ± 0.00255
	SVM (lineal)	0.6449 ± 0.0017	0.7202 ± 0.0215
	SVM (rbf)	0.6748 ± 0.0072	0.7114 ± 0.0019
KDE-KNN [ours]	RF	0.6914 ± 0.0097	0.7650 ± 0.0049
	SVM (lineal)	0.7092 ± 0.0064	0.7541 ± 0.0040
	SVM (rbf)	**0.7129 ± 0.0062**	**0.7682 ± 0.0016**

In Experiment 3, our goal was to examine how the combination of real and synthetic data during training affects model performance. The findings from Experiment 3 are presented in Table 5. For this analysis, we selected the best model and the best synthetic method from Experiment 2, which were identified as the SVM model with an rbf kernel and KDE-KNN as the synthetic method. We proceeded to train the SVM model using varying proportions of real and synthetic data generated by KDE-KNN, as illustrated in Table 5. Experiments combining real and synthetic data were performed 3 times using different seeds to sample the data. The findings indicate that augmenting the percentage of synthetic data generated with KDE-KNN in the training set leads to an improvement in the model performance, attributable to the enhanced balance of the dataset. The Fig. 2 shows the normalized distributions of 4 features in the real and synthetic databases. Note that the differences between both real databases are significant, and how the distribution of the synthetic samples tend to be realistic.

Table 5. Results of Experiment 3, combining real and synthetic data in the training set using the SVM model with rbf kernel. The results are shown in terms of AUC ± variance.

% Real	% Synthetic	MaDB	SLDB
100	0	0.6267 ± 0	0.6806 ± 0
80	20	0.6828 ± 0.0177	0.7329 ± 0.0121
60	40	0.6874 ± 0.0047	0.7319 ± 0.0195
40	60	0.7033 ± 0.0066	0.7515 ± 0.0090
20	80	0.7160 ± 0.0099	0.7589 ± 0.0079
0	100	0.7129 ± 0	0.7682 ± 0

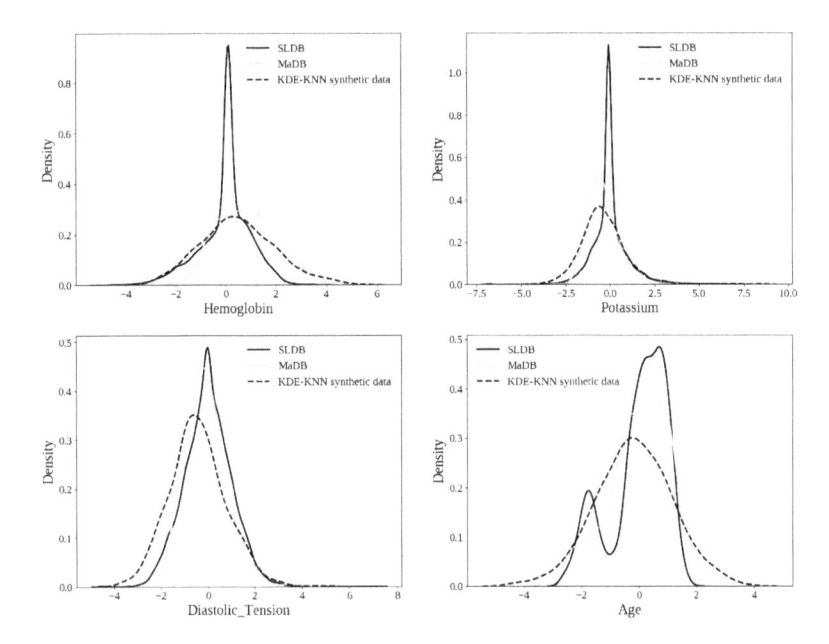

Fig. 2. Distribution of 4 features from the two real datasets and the synthetic dataset. The solid black line represents the data distribution from the SLDB, the grey line represents the distribution from MaDB, and the dashed line represents the distribution of synthetic data generated by KDE-KNN. All features were normalized using a z-score normalization technique [13].

4.3 Privacy Preservation Result

In Experiment 4, we conducted an analysis of the mean Distance to Closest Record (DCR) [50] between synthetic samples and their nearest real data points. The DCR is calculated as the Euclidean distance between a real sample and the closest synthetic sample. Low DCR values suggest that synthetic samples closely resemble real data points, potentially compromising privacy requirements. Conversely, higher DCR values indicate that the generative model can produce novel records rather than mere replicas of existing data. It is important to note that out-of-distribution data, such as random noise, can also yield high DCR values. Therefore, DCR must be evaluated alongside machine learning efficiency considerations [26]. The Fig. 3 illustrates the compromise between privacy-preserving generation and realism of the synthetic samples.

The Fig. 4 presents the probability distributions of DCR for the real samples (d_{R-R}) and the 3 generation approaches evaluated in previous experiments (d_{R-S}). For SMOTE, the mean DCR value is 0.989, while for KDE-KNN and TabDDPM, the values are 4.971 and 7.463 respectively. Comparing these results with the mean distance between real data, which is 2.715, we observe that both TabDDPM and KDE-KNN demonstrate efficacy in generating synthetic data that preserves privacy, exhibiting superior performance compared to SMOTE.

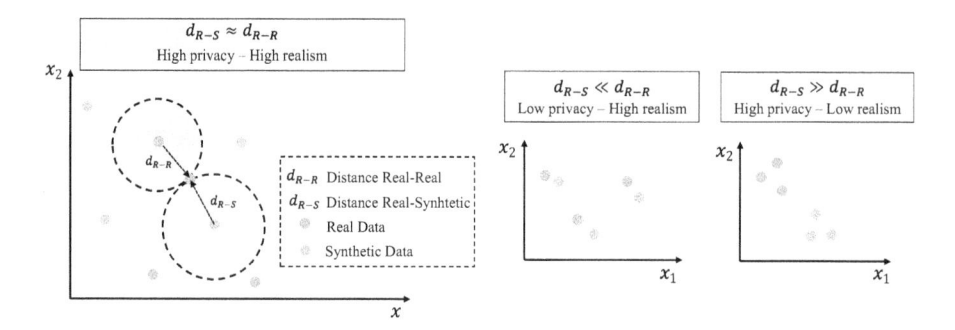

Fig. 3. Compromise between privacy and realism of synthetic samples. The graphs represent the distance between real and synthetic samples in a conceptual 2-dimensional space.

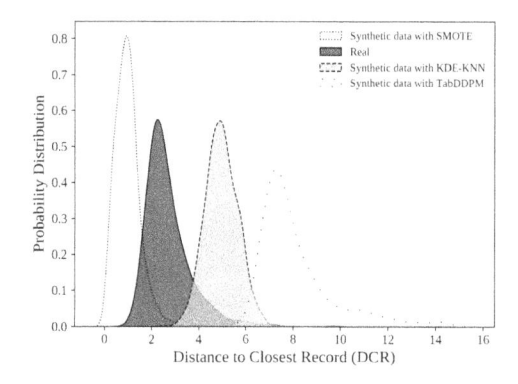

Fig. 4. Probability distribution of the Distance to Closest Record (DCR) for real samples and synthetic samples generated with the 3 generation approaches evaluated in our experiments.

5 Conclusions

Motivated by the imperative of adhering to data privacy regulations, we introduce KDE-KNN, a statistical method for generating tabular synthetic data. Through an extensive evaluation within the context of sepsis detection, we assessed this method in terms of both utility and privacy. Remarkably, our findings suggested that synthetic data outperformed real data in sepsis detection. We attributed this phenomenon to the fact that real dataset was quite imbalanced while synthetic dataset was balanced. For this reason, KDE-KNN, also would be a good method to balance datasets. Moreover, our findings have been corroborated through validation in an external database, reinforcing the generalizability potential of our synthesis approach. Additionally, our results affirmed the efficacy of KDE-KNN in preserving privacy, as evidenced by the distance observed between synthetic and real data points. In conclusion, KDE-KNN emerges as a promising method for not only enhancing dataset utility but also safeguarding data privacy, making it a valuable tool in various data-driven applications.

Acknowledgments. This research work is partly supported by National Natural Science Foundation of China No.62472177, No.62172001, and Science and Technology Commission of Shanghai Municipality (Grant No. 22DZ2229004).

References

1. Acien, A., Morales, A., Vera-Rodriguez, R., Fierrez, J., Mondesire-Crump, I., Arroyo-Gallego, T., et al.: Detection of mental fatigue in the general population: Feasibility study of keystroke dynamics as a real-world biomarker. JMIR Biomedical Engineering **7**(2), e41003 (2022)
2. Akiba, T., Sano, S., Yanase, T., Ohta, T., Koyama, M.: Optuna: A next-generation hyperparameter optimization framework. In: International Conference on Knowledge Discovery and Data Mining (2019)
3. Alanazi, A., Aldakhil, L., Aldhoayan, M., Aldosari, B.: Machine learning for early prediction of sepsis in Intensive Care Unit (ICU) patients. Medicina **59**(7), 1276 (2023)
4. Boutros, F., Struc, V., Fierrez, J., Damer, N.: Synthetic data for face recognition: Current state and future prospects. Image Vis. Comput. **135**, 104688 (2023)
5. Busch, C., et al.: Privacy and Security Matters in Biometric Technologies. Springer (2024)
6. Camacho-Cogollo, J.E., Bonet, I., Gil, B., Iadanza, E.: Machine learning models for early prediction of sepsis on large healthcare datasets. Electronics **11**(9) (2022)
7. Chawla, N.V., Bowyer, K.W., Hall, L.O., Kegelmeyer, W.P.: SMOTE: synthetic minority over-sampling technique. Journal of Artificial Intelligence Research **16**, 321–357 (2002)
8. Che, Z., Cheng, Y., Zhai, S., Sun, Z., Liu, Y.: Boosting deep learning risk prediction with generative adversarial networks for electronic health records. In: IEEE International Conference on Data Mining (ICDM). pp. 787–792 (2017)
9. Choi, E., Biswal, S., Malin, B., Duke, J., Stewart, W.F., Sun, J.: Generating multi-label discrete patient records using generative adversarial networks. In: Machine learning for healthcare conference. pp. 286–305 (2017)
10. Chong, K.M.: Privacy-preserving healthcare informatics: A review. In: Web of Conferences. vol. 36, p. 04005 (2021)
11. Du Clos, T.W.: Function of c-reactive protein. Ann. Med. **32**(4), 274–278 (2000)
12. El Emam, K., Hoptroff, R.: The synthetic data paradigm for using and sharing data. Cutter Executive Update **19**(6), 1–12 (2019)
13. Fierrez-Aguilar, J., Ortega-Garcia, J., Gonzalez-Rodriguez, J.: Target dependent score normalization techniques and their application to signature verification. IEEE Transactions on Systems, Man, and Cybernetics, Part C (Applications and Reviews) **35**(3), 418–425 (2005)
14. Giannini, H.M., Ginestra, J.C., Chivers, C., Draugelis, M., Hanish, A., Schweickert, W.D., Fuchs, B.D., Meadows, L., Lynch, M., Donnelly, P.J., et al.: A machine learning algorithm to predict severe sepsis and septic shock: Development, implementation and impact on clinical practice. Crit. Care Med. **47**(11), 1485 (2019)
15. Gomez, L.F., Morales, A., Fierrez, J., Orozco-Arroyave, J.R.: Exploring facial expressions and action unit domains for Parkinson detection. PLoS ONE **18**(2), e0281248 (2023)

16. Goodfellow, I., Pouget-Abadie, J., Mirza, M., Xu, B., Warde-Farley, D., Ozair, S., Courville, A., Bengio, Y.: Generative Adversarial Nets. Advances in Neural Information Processing Systems **27** (2014)
17. Guan, J., Li, R., Yu, S., Zhang, X.: Generation of synthetic electronic medical record text. In: IEEE International Conference on Bioinformatics and Biomedicine. pp. 374–380 (2018)
18. Hafen, B.B., Sharma, S.: Oxygen saturation. StatPearls Publishing (2018)
19. Han, C., Hayashi, H., Rundo, L., Araki, R., Shimoda, W., Muramatsu, S., Furukawa, Y., Mauri, G., Nakayama, H.: GAN-based synthetic brain MR image generation. In: IEEE International Symposium on Biomedical Imaging. pp. 734–738 (2018)
20. Hazra, D., Byun, Y.C.: SynSigGAN: Generative Adversarial Networks for synthetic biomedical signal generation. Biology **9**(12), 441 (2020)
21. Hernadez, M., Epelde, G., Alberdi, A., Cilla, R., Rankin, D.: Synthetic tabular data evaluation in the health domain covering resemblance, utility, and privacy dimensions. Methods Inf. Med. **62**, 19–38 (2023)
22. Horng, S., Sontag, D.A., Halpern, Y., Jernite, Y., Shapiro, N.I., Nathanson, L.A.: Creating an automated trigger for sepsis clinical decision support at emergency department triage using machine learning. PLoS ONE **12**(4), e0174708 (2017)
23. Islam, M.M., Nasrin, T., Walther, B.A., Wu, C.C., Yang, H.C., Li, Y.C.: Prediction of sepsis patients using machine learning approach: a meta-analysis. Comput. Methods Programs Biomed. **170**, 1–9 (2019)
24. Kashani, K., Rosner, M.H., Ostermann, M.: Creatinine: from physiology to clinical application. Eur. J. Intern. Med. **72**, 9–14 (2020)
25. Kausch, S.L., Moorman, J.R., Lake, D.E., Keim-Malpass, J.: Physiological machine learning models for prediction of sepsis in hospitalized adults: An integrative review. Intensive Crit. Care Nurs. **65**, 103035 (2021)
26. Kotelnikov, A., Baranchuk, D., Rubachev, I., Babenko, A.: TabDDPM: Modelling tabular data with diffusion models. In: International Conference on Machine Learning. pp. 17564–17579 (2023)
27. Little, R.J., et al.: Statistical analysis of masked data. Journal of Official Statistics-stockholm- **9**, 407–407 (1993)
28. Lowe, M.E.: Structure and function of pancreatic lipase and colipase. Annu. Rev. Nutr. **17**(1), 141–158 (1997)
29. Miao, L., Guo, X., Abbas, H.T., Qaraqe, K.A., Abbasi, Q.H.: Using machine learning to predict the future development of disease. In: International conference on UK-China emerging technologies (UCET). pp. 1–4 (2020)
30. Morales, A., Fierrez, J., Vera-Rodriguez, R., Tolosana, R.: SensitiveNets: Learning agnostic representations with application to face images. IEEE Trans. Pattern Anal. Mach. Intell. **43**(6), 2158–2164 (2020)
31. Nemati, S., Holder, A., Razmi, F., Stanley, M.D., Clifford, G.D., Buchman, T.G.: An interpretable machine learning model for accurate prediction of sepsis in the icu. Crit. Care Med. **46**(4), 547–553 (2018)
32. Neves, J.C., Tolosana, R., Vera-Rodriguez, R., Lopes, V., Proenca, H., Fierrez, J.: Gan fingerprints in face image synthesis. In: H. T. Sencar, L. Verdoliva, N.M. (ed.) Multimedia Forensics. pp. 175–204. ACVPR (April 2022)
33. Pedregosa, F., Varoquaux, G., Gramfort, A., Michel, V., Thirion, B., Grisel, O., Blondel, M., Prettenhofer, P., Weiss, R., Dubourg, V., Vanderplas, J., Passos, A., Cournapeau, D., Brucher, M., Perrot, M., Duchesnay, E.: Scikit-learn: Machine learning in Python. J. Mach. Learn. Res. **12**, 2825–2830 (2011)

34. Plesovskaya, E., Ivanov, S.: An empirical analysis of KDE-based generative models on small datasets. Procedia Computer Science **193**, 442–452 (2021)
35. Rankin, D., Black, M., Bond, R., Wallace, J., Mulvenna, M., Epelde, G., et al.: Reliability of supervised machine learning using synthetic data in health care: Model to preserve privacy for data sharing. Med. Inform. **8**(7), e18910 (2020)
36. Rubin, D.B.: Statistical disclosure limitation. Journal of Official Statistics **9**(2), 461–468 (1993)
37. Rudd, K.E., Johnson, S.C., Agesa, K.M., Shackelford, K.A., Tsoi, D., Kievlan, D.R., Colombara, D.V., Ikuta, K.S., Kissoon, N., Finfer, S., et al.: Global, regional, and national sepsis incidence and mortality, 1990–2017: analysis for the global burden of disease study. The Lancet **395**(10219), 200–211 (2020)
38. Schamoni, S., Hagmann, M., Riezler, S.: Ensembling neural networks for improved prediction and privacy in early diagnosis of sepsis. In: Machine Learning for Healthcare Conference. pp. 123–145 (2022)
39. Shafique, R., Rustam, F., Choi, G.S., Díez, I.d.l.T., Mahmood, A., Lipari, V., Velasco, C.L.R., Ashraf, I.: Breast cancer prediction using fine needle aspiration features and upsampling with supervised machine learning. Cancers **15**(3), 681 (2023)
40. Sharma, A., Rani, R.: A systematic review of applications of machine learning in cancer prediction and diagnosis. Archives of Computational Methods in Engineering **28**(7), 4875–4896 (2021)
41. Shrimanker, I., Bhattarai, S.: Electrolytes. StatPearls Publishing (2019)
42. Siddiq, M.: Use of machine learning to predict patient developing a disease or condition for early diagnose. International Journal of Multidisciplinary Sciences and Arts **1**(1) (2022)
43. Singer, M., Deutschman, C.S., Seymour, C.W., Shankar-Hari, M., Annane, D., Bauer, M., Bellomo, R., Bernard, G.R., Chiche, J.D., Coopersmith, C.M., et al.: The third international consensus definitions for sepsis and septic shock (sepsis-3). JAMA **315**(8), 801–810 (2016)
44. Sinha, N., Kumar, M.G., Joshi, A.M., Cenkeramaddi, L.R.: DASMcC: Data augmented SMOTE multi-class classifier for prediction of cardiovascular diseases using time series features. IEEE Access **11**, 117643–117655 (2023)
45. Sookoian, S., Pirola, C.J.: Alanine and aspartate aminotransferase and glutamine-cycling pathway: their roles in pathogenesis of metabolic syndrome. World J. Gastroenterol. **18**(29), 3775 (2012)
46. Vítek, L., Tiribelli, C.: Bilirubin: The yellow hormone? J. Hepatol. **75**(6), 1485–1490 (2021)
47. Weber, B., Henrich, D., Hildebrand, F., Marzi, I., Leppik, L.: The roles of extracellular vesicles in sepsis and systemic inflammatory response syndrome. Shock **59**(2), 161 (2023)
48. Yale, A., Dash, S., Dutta, R., Guyon, I., Pavao, A., Bennett, K.P.: Generation and evaluation of privacy preserving synthetic health data. Neurocomputing **416**, 244–255 (2020)
49. Yang, F., Yu, Z., Liang, Y., Gan, X., Lin, K., Zou, Q., Zeng, Y.: Grouped correlational Generative Adversarial Networks for discrete electronic health records. In: IEEE International Conference on Bioinformatics and Biomedicine. pp. 906–913 (2019)
50. Zhao, Z., Kunar, A., Birke, R., Chen, L.Y.: CTAB-GAN: Effective table data synthesizing. In: Asian Conference on Machine Learning. pp. 97–112 (2021)

SVD-Grad-CAM: Singular Value Decomposition filtered Gradient Weighted Class Activation Map

Gokaramaiah Thota[✉], K. Nagaraju, and Sathya Babu Korra

CSE Department, Indian Institute of Information Technology Design and Manufacturing Kurnool, Jagannathagattu Hill, Kurnool 518 008, Andhra Pradesh, India
{322cs0001,knagaraju,ksb}@iiitk.ac.in
https://iiitk.ac.in/

Abstract. The class activation map (CAM) is useful in identifying significant image features that the convolutional neural network (CNN) model is considering while making the prediction. This is critical especially in medical diagnosis like scenarios. However, existing gradient-based methods like Grad-CAM often produce low-quality visualization results due to gradient errors despite their computational efficiency. On the other hand, non-gradient methods like Score-CAM produce quality visualization that comes with high computational costs. The proposed method SVD filters Grad-CAM (SVD-Grad-CAM), which leverages singular value decomposition (SVD) to overcome the limitations of Grad-CAM. SVD-Grad-CAM filters gradients within the gradient matrix to compute the weight of the feature map for a specific class. This filtering process is achieved by selecting the top k principal components from the SVD decomposition, which discards less important patterns and potential error data. Consequently, SVD-Grad-CAM enhances the quality of Grad-CAM by reducing the clutter of multiple region highlights. The MURA dataset, focusing on elbow study type, is utilized to assess CAM visualization quality, with a DenseNet-169 CNN model fine-tuned via transfer learning. A total of 564 validation radiographs are used in empirical comparison, showing that SVD-Grad-CAM improves average drop, average increase, maximum coherency, and Average DCC by 30%, 21.67%, 19.91% and 22.56% respectively, in comparison to Grad-CAM.
  Code::https://github.com/ramaiahthota02/SVD-Grad-CAM-v1.git

Keywords: Class Activation Map · Grad-CAM · Musculoskeletal Disorders · Singular Value Decomposition · Visual Interpretation

1 Introduction

Deep learning models such as CNN are successfully used in computer vision tasks like prediction, object segmentation, medical diagnosis, etc. [1]. However, their

---

Supplementary Information The online version contains supplementary material available at https://doi.org/10.1007/978-3-031-78198-8_7.

© The Author(s), under exclusive license to Springer Nature Switzerland AG 2025
A. Antonacopoulos et al. (Eds.): ICPR 2024, LNCS 15312, pp. 90–105, 2025.
https://doi.org/10.1007/978-3-031-78198-8_7

black-box nature raises concerns about accountability and transparency in critical applications like medical diagnosis [5]. This lack of explainability can hinder trust in the model's predictions, potentially leading to hesitation in adopting them for real-world medical diagnosis [7]. The common methods of interpreting CNN decisions are visualizing learned weights, activation maximization, and CAM.

The learned weights are visualized in CNNs to detect higher-level features in the training process. However, these visualizations didn't show how CNN makes its ultimate decision [21]. Gradient-descending techniques like activation maximization were introduced to enhance specific neurons and understand which input they respond. But, these techniques still don't explain how CNNs make decisions for a particular image [11].

CAM is a widely used technique in CNNs that highlights important regions for specific class predictions [24]. It is used in applications like semantic segmentation [22], report generation [20], medical imaging [12], etc. CAM utilizes the feature maps from the final layer of a CNN alongside the learned weights from the fully connected classification layer. Its effectiveness relies on these learned weights, which are influenced by the network architecture. Consequently, Selvaraju et al. [17] introduced Grad-CAM as an alternative approach. This method calculates weights by taking the average of the gradient matrix. The gradient matrix comprises the partial derivatives that flow backwards through the layers of the neural network during backpropagation, starting from the output layer. However, there is an issue with both CAM and Grad-CAM methodologies, as they tend to generate low-quality visualization. Despite attempts to enhance CAM visualization quality using methods such as Smooth-Grad [18] and Extended-Grad-CAM, they have not achieved satisfactory results [8].

To improve visualization of CAM quality, researchers developed non-gradient methods that utilize changes in class scores. Examples of these methods include Score-CAM [19], Ablation-CAM [16], and Eigen-CAM [9], which have been effective in producing high-quality visualization. However, these methods require significant computational resources. Grad-CAM is popularly used in medical imaging to interpret the CNN decision [13]. Despite its popularity, Grad-CAM's visualization quality is still compromised due to gradient errors during backpropagation and multiple region highlights, as shown in Figure 1.

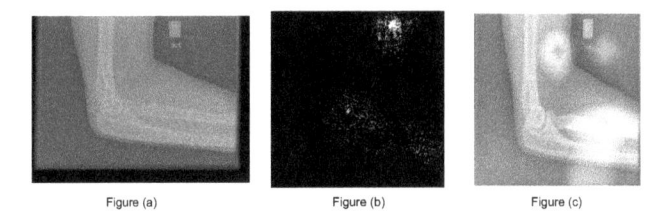

Figure (a)　　　　Figure (b)　　　　Figure (c)

Fig. 1. Figure 1(a) illustrates the elbow study type, while Figure 1(b) describes a saliency map of Figure 1(a) highlighting distortions in gradients. Figure 1(c) showcases Grad-CAM visualization, emphasizing multiple region highlights.

In this paper, a new method called SVD-Grad-CAM has been proposed to address the above-mentioned challenges. This method utilizes SVD to filter gradients from the gradient matrix during weight computation. Empirical evaluations of this method have been conducted on the musculoskeletal radiograph (MURA) dataset [10] with the DenseNet-169 CNN model [3], using metrics including average drop, average increase, maximum coherency [23] complexity and average DCC [14]. The study proposes several strategies to improve the quality of visualization of Grad-CAM by utilizing the SVD.

- The SVD-Grad-CAM method is proposed to filter the gradient matrix to compute the weights such that it selects the best patterns or neglects the error data in the feature maps to improve the quality visualization and reduce the multiple region highlights.
- The study quantitatively assessed the visualization quality of CAMs using the following measures: average drop 12, average increase 13, maximum coherency 15, complexity 16, and Average DCC 17.
- The DenseNet-169 CNN model was trained on a musculoskeletal radiograph dataset with elbow study type for quantitative evaluation of SVD-Grad-CAM and existing CAMs.

The paper presents its research findings in a structured sequence. It begins by reviewing relevant literature on CAM in Section 2. Then, it outlines the proposed methodology in Section 3, discusses the dataset, evaluation metrics, experiments, and the results in Section 4, and concludes with future works in Section 5.

2 Related Work

CAMs are essential tools for interpreting CNNs. They enable us to visualize the parts of an input image that contribute the most to the network's decision-making process. There are two main types of CAMs: gradient-based and non-gradient-based.

Gradient-based CAMs use the gradients that flow through the CNN during the backpropagation process. They analyze these gradients to produce class activation maps. Some examples of gradient-based CAMs are CAM, Grad-CAM (Gradient-weighted Class Activation Mapping), Grad-CAM++, Smooth-Grad-CAM, and Extended-Grad-CAM.

Non-gradient-based CAMs derive weights from changes in class scores instead of gradients. Examples of non-gradient-based CAMs include Score-CAM, Ablation-CAM, and Eigen-CAM. While gradient-based CAMs are computationally efficient, they may produce low-quality CAM visualizations due to gradient errors during the backpropagation process. This can result in inaccuracies in highlighting the relevant parts of the input image. Non-gradient-based CAMs offer an alternative approach that avoids these gradient-related issues, potentially producing more reliable visualizations.

Zhou *et al.* used global average pooling (GAP) [6] to implement the CAM technique, which establishes the relationship between feature maps and weights in fully connected layers in classification [24]. The CAM formula utilizes w_k^c learned weights in a fully connected layer and the feature map $M_{i,j}^k$ originating from the final convolutional layer, with a specific focus on class c. CAM visualization is defined as:

$$L_{CAM}^c = \sum_{i,j} w_k^c M_{i,j}^k \tag{1}$$

However, CAM has some limitations such as requiring retraining for pre-trained models for weights. Grad-CAM introduces a gradient-based approach to visualize deep learning model decisions without altering the model's architecture or necessitating retraining. In the case of Grad-CAM, the values of the weights are established through the computation of gradients available in backpropagation. The CAM formula utilizes w_k^c learned weights in a fully connected layer given in Equation 1. In the case of Grad-CAM, the values of the weights are established through the computation of gradients related to the class score Y^c concerning the feature maps $M_{i,j}^k$:

$$w_k^c = \frac{1}{Z} \sum_i \sum_j \frac{\partial Y^c}{\partial M_{i,j}^k} \tag{2}$$

Despite its utility, Grad-CAM has limitations, including issues with visualization quality and localization accuracy and difficulty detecting objects with minimal spatial footprint. Grad-CAM++ [2] builds upon Grad-CAM, providing more detailed explanations using a technique that calculates pixel-wise weights using higher-order gradients, offering a refined heatmap. However, Grad-CAM++ is resource-intensive due to the computation of higher-order gradients and may not handle noise effectively.

Extended-Grad-CAM [4], a variant of Grad-CAM, applies a Gaussian filter to activation maps before creating the class activation map. This technique aims to enhance map accuracy, reliability, and robustness to input noise, offering a smooth visualization of important features. The implementation of Extended-CAM involves computing the Gaussian filter $G(x,y)$ is defined as:

$$G(x, y; \mu, \sigma) = \frac{1}{2\pi\sigma^2} \exp\left(-\frac{(x-\mu_x)^2 + (y-\mu_y)^2}{2\sigma^2}\right) \tag{3}$$

The Gaussian filter $G(x,y)$ is applied to the activation map A^k of the last convolutional layer.

Smooth-Grad-CAM [18] introduces a novel approach by adding Gaussian noise (σ) to images and applying Grad-CAM over them. This technique reduces noise, resulting in improved feature visualization. Smooth-Grad-CAM combines the strengths of smoothing and gradient-based methods to enhance model interpretability, but satisfactory visualization quality is not achieved. x is the input image and noisy is defined as;

$$x_i = x + N(0, \sigma^2) \tag{4}$$

Score-CAM extends CAM by providing visual explanations without relying on gradients. It calculates the importance of each activation map based on its score for the target class, outperforming previous CAM methods in recognition and localization tasks [19]. Nevertheless, Score-CAM is computationally expensive and sensitive to model architecture. Eigen-CAM arranges each feature map into a column vector and applies the SVD [9]. It utilizes the right singular matrix V^T to compute the weights and employs the first eigenvector, providing insightful visualizations of model decision-making. However, it's worth noting that Eigen-CAM may not inherently support multi-class discrimination. Ablation-CAM is a method used to determine the significance of individual feature map units for a specific class [16]. It involves zeroing out various regions of an input image and monitoring the changes in the model's output. Compared to other approaches such as Grad-CAM and Score-CAM, Ablation-CAM is less prone to noise and artifacts. However, careful consideration of the regions must be ablated, and multiple evaluations of the model demand high resources and time consuming.

Grad-CAM overcomes CNN architecture dependency limitations using a gradient-based approach, which provides localization maps without altering the model's architecture. However, it still faces challenges with low-quality visualization and detecting small objects. Grad-CAM++ improves upon Grad-CAM by pinpointing exact object locations. However, it is resource-intensive. Score-CAM extends CAM by providing explanations without relying on gradients, but it is computationally expensive. Extended-Grad-CAM enhances accuracy and robustness to noise by applying a Gaussian filter to gradients. Smooth-Grad-CAM combines smoothing and gradient-based methods to reduce noise and improve feature visualization but has not achieved satisfactory visualization quality.

3 Methodology

The CNN layer l produces k feature maps denoted as $M^k \in \mathbb{R}^{P \times Q \times R}$. These maps are then used alongside class scores Y^c obtained from fully connected layers to compute the gradient matrix G, representing the partial derivatives $\frac{\partial Y^c}{\partial M^k_{i,j}}$. This gradient matrix captures errors in gradient propagation during back-propagation.

Directly deriving weights from the gradient matrix G can potentially compromise the accuracy of CAM visualizations. The objective is to filter out erroneous data and undesirable patterns to ensure precise weight computation for each feature map. To achieve this, Principal Component Analysis (PCA) is applied to G using SVD. The top n principal components are retained to eliminate noise and subtle patterns, enabling the computation of precise weights.

Subsequently, these computed weights and the feature maps are used to generate the visualization of SVD-Grad-CAM, as depicted in Figure 2. This technique enhances the accuracy of CAM visualization by effectively filtering out noise and unwanted patterns, thus providing a clearer local discrimination.

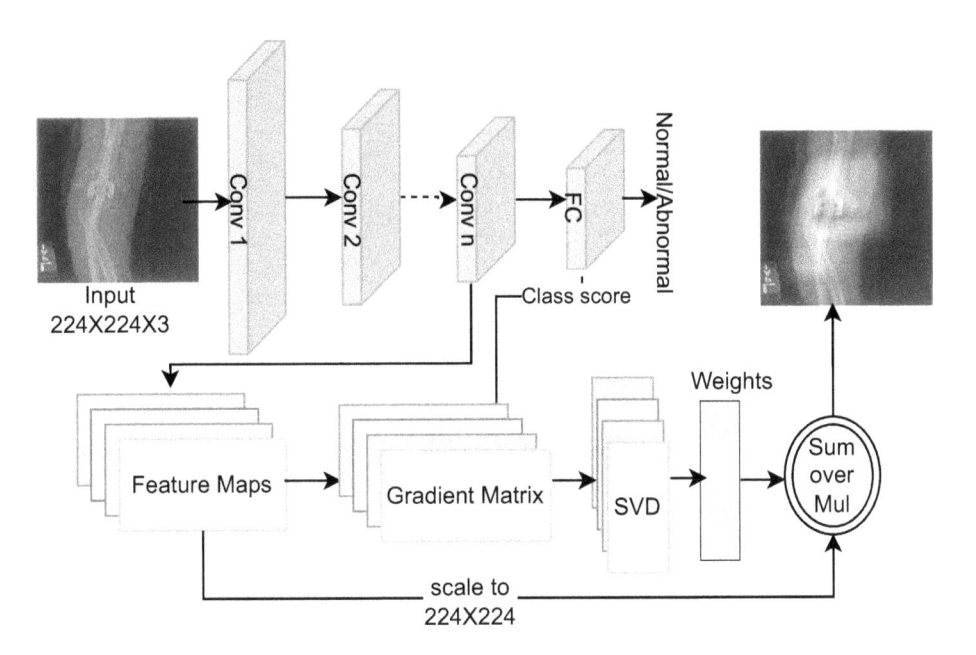

Fig. 2. The system diagram of SVD-filtered Grad-CAM that computes weights from the singular matrix.

The SVD is applied to the gradient matrix G, its partial derivative $\frac{\partial Y^c}{\partial M^k_{i,j}}$ is represented as:

$$G = \begin{bmatrix} \frac{\partial Y^c}{\partial M^k_{1,1}} & \frac{\partial Y^c}{\partial M^k_{1,2}} & \cdots & \frac{\partial Y^c}{\partial M^k_{1,j}} \\ \frac{\partial Y^c}{\partial M^k_{2,1}} & \frac{\partial Y^c}{\partial M^k_{2,2}} & \cdots & \frac{\partial Y^c}{\partial M^k_{2,j}} \\ \vdots & \vdots & \ddots & \vdots \\ \frac{\partial Y^c}{\partial M^k_{i,1}} & \frac{\partial Y^c}{\partial M^k_{i,2}} & \cdots & \frac{\partial Y^c}{\partial M^k_{i,j}} \end{bmatrix} \tag{5}$$

Let the i indices ranging from 1 to P, and j be indices ranging from 1 to Q. k represent the kth feature map its value ranges from 1 to R. The principal component analysis processes the gradient matrix G employing SVD. This serves the purpose of minimizing gradient errors and singular values obtained through SVD signify the directions of maximum data variability, thereby forming the foundation for determining the weights of the feature map. The SVD of the gradient matrix, denoted as $\frac{\partial Y^c}{\partial M^k_{i,j}}$, is defined as follows:

$$G \approx U_k \Sigma_k V_k^T \tag{6}$$

In SVD, Σ_k represents the diagonal matrix of singular values, U_k contains the left singular vectors, and V_k^T contains the right singular vectors. By selecting the top n components of singular values, attention is focused on the most significant directions, which are crucial for determining the weights of the feature map M^k.

Moreover, there are four ways to compute the weights w_k^c, each utilizing different combinations of these SVD components:

a) **Using left singular matrix U_k:**

$$\mathbf{w}_k^c = \frac{1}{n \cdot P} \sum_{j=1}^{n} \left(\sum_{i=1}^{P} \mathbf{U}_{k i,j} \right) \tag{7}$$

Here, it calculates w_k^c of the feature map M^K by averaging all rows of the left singular matrix U_k within the top n principal components.

b) **Using right singular matrix V_k^T:**

$$\mathbf{w}_k^c = \frac{1}{n \cdot Q} \sum_{j=1}^{n} \left(\sum_{i=1}^{Q} \mathbf{V}_{k j,i}^{\mathbf{T}} \right) \tag{8}$$

Here, it calculates w_k^c of the feature map M^K by averaging all columns of the left singular matrix V_k^T within the top n principal components

c) **Using the product of $U_k V_k^T$:**

$$\mathbf{w}_k^c = \frac{1}{n \cdot P} \sum_{j=1}^{n} \left(\sum_{i=1}^{P} (\mathbf{U}_k \mathbf{V}_k^T)_{i,j} \right) \tag{9}$$

This equation computes w_k^c of the feature map M^K as the mean of all elements in the product of the matrices U_k and V_k^T, considering top n principal components and P columns.

d) **Using reconstructed gradient matrix:**

$$\mathbf{w}_k^c = \frac{1}{n \cdot P} \sum_{j=1}^{n} \left(\sum_{i=1}^{P} (\mathbf{U}_k \Sigma_k \mathbf{V}_k^T)_{i,j} \right) \tag{10}$$

In this equation, w_k^c for the feature map M^K is calculated as the average of all elements in the product of the matrices U_k, Σ_k, and V_k^T, considering the top n principal components and P columns.

There are four ways to compute the w_k^c for the feature map M^k, and one way is employed among them. To define the SVD-Grad-CAM visualization, the sum is taken over the product of w_k and M^k for each feature map k. This can be expressed as:

$$L_{\text{SVD-Grad-CAM}}^c = \sum_{i,j} w_k^c M_{i,j}^k \tag{11}$$

Figure 3 displays the four outputs from SVD-Grad-CAM, which result from four ways to compute the weights as outlined in equations 7, 8, 9, and 10. The first two methods mentioned in equations 7 and 8 cannot produce the correct visualization. Therefore, they will not be utilized in future experiments to generate SVD-Grad-CAM. The remaining two methods, using the product of

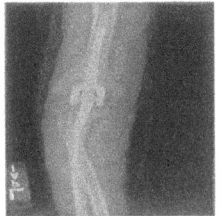

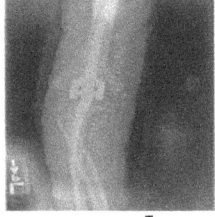

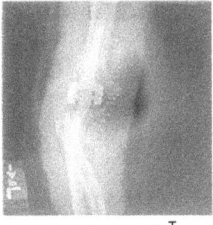

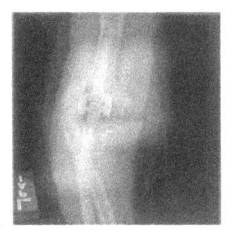

| Weights from U_K | Weights from V_K^T | Weights from $U_K.V_K^T$ | Weights from reconstruted G |

Fig. 3. Each sub-figure represents the SVD-Grad-CAM visualization from the four ways of computing weights. The product of $U_k.V_k^T$ and reconstructed G demonstrate satisfactory performance.

$U_k.V_k^T$ and reconstructed gradient matrix, have produced satisfactory results, as demonstrated in Figure 3. Both methods will continue to be tested, and the superior one will be selected for generating SVD-Grad-CAM visualization. The reconstructed gradient matrix or the combination of $U_k V_k^T$ effectively captures significant data variations crucial for class discrimination. Algorithm 1 delineates the overall process of SVD-Grad-CAM. The SVD also exhibits a noise reduction property, which diminishes noise and enhances the quality of SVD-Grad-CAM visualization by preserving top singular vectors.

Algorithm 1. SVD-filtered-Grad-CAM Algorithm

1: **Input:** X (Input image), $f(X;\theta)$ (The CNN Model)
2: **Output:** $L^c_{\text{SVD-Grad-CAM}}$
3: **Call CNN Model:**
4:     Pass X through $f(X;\theta)$ CNN model for prediction.
5: **Collect Predicted Class Score and Convolutional Layer Output:**
6:     Collect predicted class score Y^c and last convolutional layer output M^k.
7: **Gradient Matrix Computation:**
8:     Compute gradient matrix $G = \frac{\partial Y^c}{\partial M^k_{i,j}}$.
9: **Apply SVD to Gradient Matrix**
10:     Apply SVD to G to obtain U_k, Σ_k, V_k^T.
11: **Weight Calculation:**
12:     Retrieve top n components from $U_k V_k^T$ or $U_k \Sigma_k V_k^T$.
13: **Scaling the feature maps:**
14:     Scale the feature maps M_k to size of the input image X.
15: **SVD-Grad-CAM computation:**
16:     Calculate $L^c_{\text{SVD-Grad-CAM}} = \sum_{i,j} w^c_k M^k_{i,j}$.
17: **Visualization:**
18:     Visualize SVD-Grad-CAM output for interpretability.

4 Experiments and Results

4.1 Dataset and Pre-processing

The MURA dataset encompasses 14,656 study cases across seven distinct study types, such as finger, shoulder, elbow, forearm, humerus, hand, and wrist, each meticulously annotated and categorized [15]. To investigate the efficacy of the SVD-Grad-CAM approach on the MURA dataset, elbow study type radiograph images were used. Regarding elbow studies within the MURA dataset, the training set encompasses 4930 radiograph images, with 2924 representing normal cases and 2006 indicating abnormalities and it is available for download at [10]. The validation set is specifically tailored for elbow studies, comprising 564 radiograph images, of which 234 are categorized as normal and 230 denote abnormalities.

4.2 Setup

The PARAM SHAVAK system at the Institute comprises two elements, each with dual-socket Intel Xeon Gold 6226 processors. It has 96GB of DDR4 RAM, two high-speed network ports, two 16 PCI-E Gen3 slots, and four 4TB SATA or NL-OSAS disks. It supports up to two Nvidia RTX P5000 GPUs for compute-heavy activities, such as deep learning, scientific simulations, and visualizations.

4.3 Evaluation Metrics

The article [14] discusses assessment metrics that offer significant information on the performance of models that use explanation maps.

Average Drop: This measure evaluates the average decrease in confidence for the specific target class c when the model depends solely on the CAM visualization rather than the complete image.

$$\text{Average Drop:} = \left(\max\left(0, \frac{y_c - y_c^{CAM}}{y_c}\right) \right) \cdot 100, \tag{12}$$

where y_c is the output score for class c when using the full image, and y_c^{CAM} is the output score when using the CAM output.

Average Increase: This metric computes the proportion of cases where the model exhibits greater confidence when utilizing the CAM output as opposed to the entire image.

$$\text{Average Increase} = \frac{\mathbb{1}(y_c < y_c^{CAM}) \cdot 100}{l}, \tag{13}$$

where $\mathbb{1}$ represents the indicator function and l is the total number of instances.

Maximum Coherency: In CAM visualization, coherency refers to how well the explanation map matches the input image. The CAM should highlight important features that explain a prediction, and ignore unimportant features. This is represented mathematically as the alignment between the CAM and the image.

$$CAM_c(x \odot CAM_c(x)) = CAM_c(x). \tag{14}$$

The CAM should accurately show the important parts of an image and ignore the rest. The coherency score is then normalized to a range of 0 to 1. The coherency score can be calculated using this formula:

$$\text{Maximum Coherency}(x) = \frac{1}{2}\left(1 + \frac{\text{Cov}(CAM_c(x \circledast CAM_c(x)), CAM_c(x))}{\sigma_{CAM_c}(x \circledast CAM_c(x))\sigma_{CAM_c}(x)}\right) \tag{15}$$

In the equation, Cov signifies the covariance between the convolved CAM and the original CAM, while σ denotes the standard deviation of the CAM at a given position. These statistical measures evaluate the coherence of the CAM in emphasizing significant image regions.

4.4 Minimum Complexity (lower value best):

Complexity measures how good a CAM explanation is. It's quantified using the L1 norm of the CAM, represented as

$$Complexity(x) = ||CAM_c(x)||. \tag{16}$$

The CAM technique is utilized to emphasize the important sections of an image while disregarding the rest.

4.5 Average Drop Complexity Coherency (higher value best):

This metric combines the average drop, complexity, and coherency (ADCC) to give a comparable score for all models. It is defined as

$$ADCC(x) = 3\left(\frac{1}{\text{Coherency}(x)} + \frac{1}{1 - \text{Complexity}(x)} + \frac{1}{1 - \text{AverageDrop}(x)}\right)^{-1}. \tag{17}$$

4.6 Hyperparameters

The model was trained using a learning rate of 0.01, with batches of 8 and early stopping implemented with the patience of 10 throughout 65 epochs. The activation function used was leakyreLU, and the optimizer used was Adam with default values. The model's architecture adopts focal loss as its loss functions with a balance coefficient of β set to 0.25 and a γ value of 2.

4.7 Experiment-1: To determine optimal n components

The study evaluates the performance of the proposed SVD-Grad-CAM with deep neural network architectures, namely, DenseNet-169 trained on the MURA

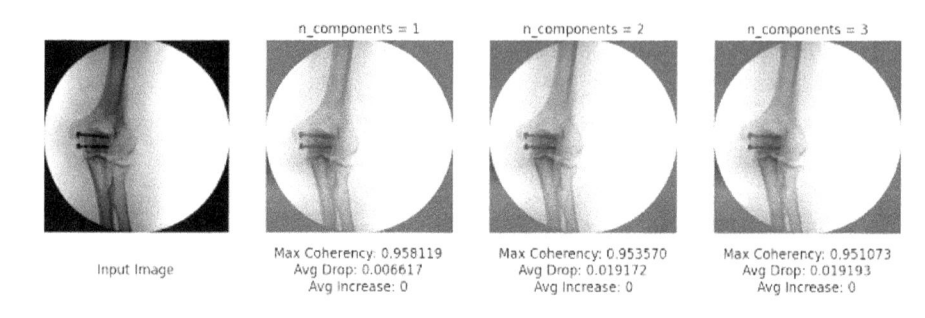

Fig. 4. Illustrates that maximum coherency decreases and average drop increases as the number of principal components (n value) increases.

dataset. The transfer learning technique is used to initialize their weights, and pre-trained weights sourced from the ImageNet dataset are employed. These models utilize sigmoid activation for binary classification tasks at the final output layer. The model's architecture adds a global average pooling layer after the last convolutional layer. This converts the extracted characteristics into a vectorized shape, enabling the model to be interpretable with the help of class activation maps.

The two procedures for computing the weights from SVD are specified in Equation 9 using the combination of $U_k.V_k^T$, and Equation 10 using the reconstructed gradient matrix. These procedures require selecting a value n to determine their principal components. To determine the appropriate value of n, several experiments were conducted. Figure 4 shows a representative image and its corresponding metrics for the procedure using a reconstructed gradient matrix. It demonstrates that the maximum coherency decreases with an increase in the value of n, while the average drop increases. This suggests that $n = 1$ yields the best results for computing the weights from SVD. Similar experiments were conducted for the remaining procedure, concluding that $n = 1$ is the optimal choice. Next, experiments were conducted to determine which procedure best compute the weights.

Table 1. Comparison of metric values to determine the optimal method for computing weights for SVD-Grad-CAM.

Procedure	Avg Drop ↓	Avg Increase ↑	Coherency ↑	Complexity ↓	ADCC ↑
$U_k V_k^T$	6.80	40.12	86.05	18.89	88.36
$U_k \Sigma_k V_k^T$	**3.47**	**43.33**	**88.82**	15.34	90.714

4.8 Experiment-2: To determine the best procedure of SVD-Grad-CAM:

To determine the most effective procedure for computing the weights from SVD, consider two methods outlined in 9 and 10: The combination of $U_k.V_k^T$ and the reconstructed gradient matrix. To determine the best procedure, 154 random radiograph images from the MURA dataset were chosen from the elbow study type. These radiograph images were then analyzed to calculate the average increase, average drop, maximum coherency, complexity and Average DCC. Table 1 provides representative experimental values for each procedure. Notably, the procedure involving the reconstructed gradient matrix has the lowest average drop, highest average increase, and average DCC compared to the other procedure. Therefore, the reconstructed gradient matrix procedure is chosen to compute the weights in all experiments for the SVD-Grad-CAM.

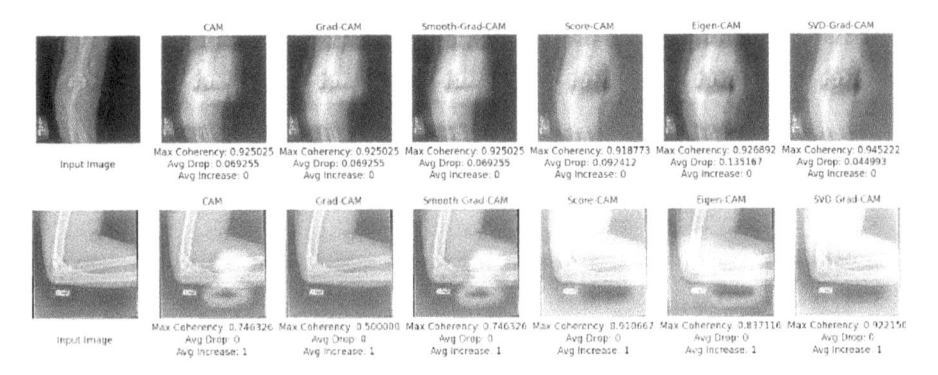

Fig. 5. Shows various CAMs highlighting features and their metric values for the input image, where DenseNet-169 predicted abnormality in the first-row input image, whereas normality in the second-row input image.

4.9 Experiment-3: Comparing the existing CAM methods with proposed SVD-Grad-CAM

The study compares the effectiveness of different methods such as CAM [24], Grad-CAM [17], Smooth-Grad-CAM [18], Score-CAM [19], Ablation-CAM [16] and Eigen-CAM [9] with the proposed SVD-Grad-CAM. The comparison is based on the metrics specified in equations 12, 13, and 15. The DenseNet-169 CNN model was trained using the elbow study type to do this. An entire validation set of 526 radiographs was utilized to calculate the average drop, average increase, and maximum coherency. Two random validation radiograph images were selected, and their average drop, average increase, and maximum coherency were displayed in Figure 5. The experimental values are presented in Table 2. The experimental results are discussed in the following paragraphs.

Table 2. Comparison of different CAMs using the MURA dataset's validation set of elbow study comprising 564 radiograph images.

Method	Avg Drop ↓	Avg Increase ↑	Coherency ↑	Complexity ↓	ADCC ↑
CAM [24]	10.43	**48**	**73.19**	43.81	**70.38**
Grad-CAM [17]	10.43	**48**	**73.19**	43.81	**70.38**
Smooth-Grad [18]	**11.5**	49.50	74.56	25.13	78.80
Ablation-CAM [16]	8.20	52.6	84.05	36.83	77.66
Score-CAM [19]	7.60	53.6	86.05	30.46	81.46
Eigen-CAM [9]	8.95	**54.8**	81.77	27.63	80.73
SVD-Grad-CAM	**7.3**	52.4	**87.77**	**20.62**	**86.26**

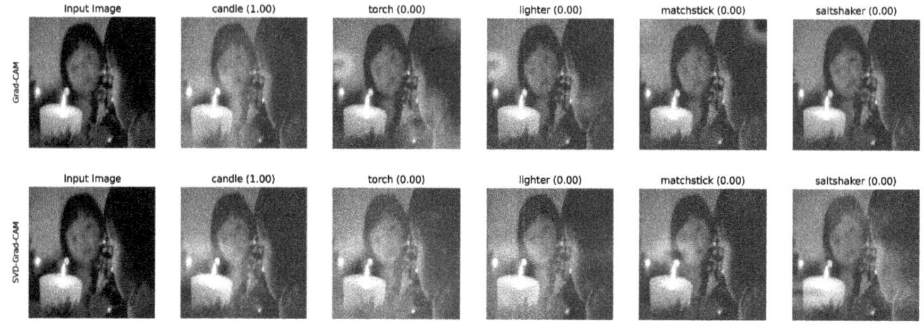

Fig. 6. Qualitative comparison of DenseNet-169 prediction visualization maps using Grad-CAM and SVD-Grad-CAM on ImageNet sample. The proposed SVD-Grad-CAM method uses two principal components to generate the visualization map.

SVD-Grad-CAM has the lowest average drop (7.30), followed closely by SCore-CAM (7.6), indicating that these methods are better at retaining model confidence than others. Eigen-CAM exhibits the highest average increase (54.8), followed by Score-CAM (53.6), and SVD-Grad-CAM (52.4), indicating that these methods are more effective at improving model confidence. SVD-Grad-CAM has the highest maximum coherency and ADCC, followed by Score-CAM and Eigen-CAM, suggesting that these methods maintain good consistency with the input image.

After comparing different CAMs, it was found that SVD-Grad-CAM is the most effective in improving model confidence, as it has the lowest average drop, the best average increase, and the highest coherency. An empirical comparison showed that SVD-Grad-CAM improves average drop, average increase, maximum coherency and Average DCC by 30%, 21.67%, 19.91%, and 22.56% respectively, when compared to Grad-CAM. It's worth noting that Score-CAM and Eigen-CAM also demonstrate strong performance, positioning them as strong contenders.

Fig. 7. Qualitative comparison of various CAM visualization methods on ImageNet sample of DenseNet-169 predictions.

4.10 Experiment 4: Qualitative comparison of existing CAM methods with the proposed SVD-Grad-CAM on ImageNet samples

The DenseNet-169 model was created using transfer learning from ImageNet. A qualitative comparison of the model's prediction visualization maps is illustrated in Figures 7 and 6. The top five predictions made by DenseNet-169 are displayed with their visualization map. SVD-Grad-CAM leverages two principal components in the generation of visualization maps.

5 Conclusion and Future work

Singular value decomposition is applied to the gradient matrix to select optimal patterns or exclude erroneous data. The reconstructed gradient matrix performs the best among the four possible ways to compute weights from the gradient matrix. SVD-Grad-CAM was evaluated using the MURA dataset's elbow study type, and the DenseNet-169 CNN model was trained on the same dataset using transfer learning for evaluation. Evaluation metrics such as average drop, average increase, and maximum coherency are used to assess the effectiveness of SVD-Grad-CAM. Results indicate that SVD-Grad-CAM performs better than CAM, Smooth-Grad, Extended-Grad-CAM, Score-CAM, and Eigen-CAM in terms of average increase and maximum coherency. Notably, Eigen-CAM performs well following SVD-Grad-CAM in terms of average increase.

The $U_k V_k^T$ method is ideal for computing weights, requiring less computation than the reconstructed gradient matrix. Therefore, this technique is being explored in question-answering and object-tracking videos, particularly in scenarios where system resources are crucial.

References

1. Bzdok, D., Krzywinski, M., Altman, N.: Machine learning: a primer. Nat. Methods **14**(12), 1119 (2017)
2. Chattopadhay, A., Sarkar, A., Howlader, P., Balasubramanian, V.N.: Grad-cam++: Generalized gradient-based visual explanations for deep convolutional networks. In: 2018 IEEE winter conference on applications of computer vision (WACV). pp. 839–847. IEEE (2018)
3. Huang, G., Liu, Z., Van Der Maaten, L., Weinberger, K.Q.: Densely connected convolutional networks. In: Proceedings of the IEEE conference on computer vision and pattern recognition. pp. 4700–4708 (2017)
4. Kim, B.J., Koo, G., Choi, H., Kim, S.W.: Extending class activation mapping using gaussian receptive field. Comput. Vis. Image Underst. **231**, 103663 (2023)
5. Kim, C., Lee, S., Lee, K.: Tremor feature extraction for enhanced interpretability of vocal disease classification. In: International Conference on Pattern Recognition. pp. 618–633. Springer (2022)
6. Lin, M., Chen, Q., Yan, S.: Network in network. arXiv preprint arXiv:1312.4400 (2013)
7. Matthew Zeiler, D., Rob, F.: Visualizing and understanding convolutional neural networks. In: ECCV (2014)
8. Mi, J.X., Jiang, X., Luo, L., Gao, Y.: Toward explainable artificial intelligence: A survey and overview on their intrinsic properties. Neurocomputing **563**, 126919 (2024)

9. Muhammad, M.B., Yeasin, M.: Eigen-cam: Class activation map using principal components. In: 2020 international joint conference on neural networks (IJCNN). pp. 1–7. IEEE (2020)
10. Mura. https://stanfordmlgroup.github.io/competitions/mura/ (2018), accessed on 8-April-2023
11. Olah, C., Mordvintsev, A., Schubert, L.: Feature visualization. Distill **2**(11), e7 (2017)
12. Paciorek, A.M., von Schacky, C.E., Foreman, S.C., Gassert, F.G., Gassert, F.T., Kirschke, J.S., Laugwitz, K.L., Geith, T., Hadamitzky, M., Nadjiri, J.: Automated assessment of cardiac pathologies on cardiac mri using t1-mapping and late gadolinium phase sensitive inversion recovery sequences with deep learning. BMC Med. Imaging **24**(1), 43 (2024)
13. Patra, G., Datta, S.: Xai for society 5.0: Requirements, opportunities, and challenges in the current context. XAI Based Intelligent Systems for Society 5.0 pp. 269–293 (2024)
14. Poppi, S., Cornia, M., Baraldi, L., Cucchiara, R.: Revisiting the evaluation of class activation mapping for explainability: A novel metric and experimental analysis. In: Proceedings of the IEEE/CVF Conference on Computer Vision and Pattern Recognition. pp. 2299–2304 (2021)
15. Rajpurkar, P., Irvin, J., Bagul, A., Ding, D., Duan, T., Mehta, H., Yang, B., Zhu, K., Laird, D., Ball, R.L., et al.: Large dataset for abnormality detection in musculoskeletal radiographs. In: 1st Conference on Medical Imaging with Deep Learning (MIDL 2018). Amsterdam, The Netherlands (2018)
16. Ramaswamy, H.G.: Ablation-cam: Visual explanations for deep convolutional network via gradient-free localization. In: proceedings of the IEEE/CVF winter conference on applications of computer vision. pp. 983–991 (2020)
17. Selvaraju, R.R., Das, A., Vedantam, R., Cogswell, M., Parikh, D., Batra, D.: Gradcam: Why did you say that? arXiv preprint arXiv:1611.07450 (2016)
18. Smilkov, D., Thorat, N., Kim, B., Viégas, F., Wattenberg, M.: Smoothgrad: removing noise by adding noise. arXiv preprint arXiv:1706.03825 (2017)
19. Wang, H., Wang, Z., Du, M., Yang, F., Zhang, Z., Ding, S., Mardziel, P., Hu, X.: Score-cam: Score-weighted visual explanations for convolutional neural networks. In: Proceedings of the IEEE/CVF conference on computer vision and pattern recognition workshops. pp. 24–25 (2020)
20. Wang, J., Bhalerao, A., Yin, T., See, S., He, Y.: Camanet: class activation map guided attention network for radiology report generation. IEEE Journal of Biomedical and Health Informatics (2024)
21. Yosinski, J., Clune, J., Nguyen, A., Fuchs, T., Lipson, H.: Understanding neural networks through deep visualization. arXiv preprint arXiv:1506.06579 (2015)
22. Zhai, W., Wu, P., Zhu, K., Cao, Y., Wu, F., Zha, Z.J.: Background activation suppression for weakly supervised object localization and semantic segmentation. Int. J. Comput. Vision **132**(3), 750–775 (2024)
23. Zhang, H., Torres, F., Sicre, R., Avrithis, Y., Ayache, S.: Opti-cam: Optimizing saliency maps for interpretability. arXiv preprint arXiv:2301.07002 (2023)
24. Zhou, B., Khosla, A., Lapedriza, A., Oliva, A., Torralba, A.: Learning deep features for discriminative localization. In: Proceedings of the IEEE conference on computer vision and pattern recognition. pp. 2921–2929 (2016)

Applying Layer-Wise Relevance Propagation on U-Net Architectures

Patrick Weinberger[1,3](✉)[iD], Bernhard Fröhler[1][iD], Anja Heim[1,4,5][iD], Alexander Gall[1,4][iD], Ulrich Bodenhofer[2][iD], and Sascha Senck[1][iD]

[1] Research Group Computed Tomography, University of Applied Sciences Upper Austria, Wels, Austria
patrick.weinberger@fh-wels.at
[2] School of Informatics, Communication, and Media, University of Applied Sciences Upper Austria, Wels, Austria
[3] Johannes Kepler University Linz, Linz, Austria
[4] Faculty of Computer Science and Mathematics, University of Passau, Passau, Germany
[5] Division Development Center X-ray Technology, Fraunhofer Institute for Integrated Circuits IIS, Erlangen, Germany
https://JKU.at, https://www.iis.fraunhofer.de/en/ff/zfp.html, https://www.uni-passau.de

Abstract. For safety critical applications, it is still a challenge to use AI and fulfill all regulatory requirements. Medicine/healthcare and transportation are two fields where regulatory requirements are of fundamental importance. A wrong decision can lead to serious hazards or even deaths. In these fields, semantic segmentation is often utilized to extract features. Especially U-Net architectures are used. This paper shows how to apply layer-wise relevance propagation (LRP) to a trained U-Net architecture. We achieve an efficient explanation of a segmentation by back-propagating the whole resulting image. To tackle the non-linear results of the LRP, we introduce a threshold mechanism in combination with a logarithmic transfer function to preprocess the data for visualization. We demonstrate our method on three use cases: the segmentation of a fiber-reinforced polymer in the field of non-destructive testing, the segmentation of pedestrians in an automotive application, and a lung segmentation example from the medical domain.

Keywords: Layerwise Relevance Propagation · U-Net · Segmentation · Explainable AI

1 Introduction

Convolutional Neuronal Networks (CNNs) are currently used for many tasks like image classification, segmentation, image improvement, and many more [2]. They are starting to be superseded by Transformer-based architectures [36] adopted to computer vision tasks [11,19]. These Transformer-based architectures are outperforming CNNs in most classification tasks [11,19]. However, a key limitation of Transformer-based architectures lies in their data dependency. They often require vast amounts of training data for optimal performance, making them less suitable for domains with limited datasets, such as medical imaging and non-destructive testing (NDT) [1]. CNNs are more data-efficient, and therefore research continues to refine CNNs to bridge the gap with Transformer-based approaches. One such development is ConVNext, which demonstrates promising results [20,38].

In safety-critical fields such as medicine and NDT, CNNs remain the dominant force due to the typically smaller size and availability of datasets [17,22,37]. Reliability and robustness are paramount, as these models often influence decisions with potentially significant consequences. Most currently available techniques for interpreting results of CNNs are focused on detection (classification). However, tasks in these domains extend beyond classification, with semantic segmentation playing a crucial role. Medical applications, in particular, require highly interpretable models for building trust and understanding the reasoning behind the model's predictions. Similar challenges and requirements are present in NDT, where various scanning techniques like X-ray, ultrasound, and thermography are employed for inspecting components during production processes. For both fields, segmentation is an important task to determine the size and shape of objects. Therefore it is crucial to explain how much specific parts of the input impact the segmentation results.

The contributions of this paper are the following: methods for the calculation of LRP for explaining the segmentation results from U-Net architectures; a visualization based on logarithmic scaling for these relevance values; and a demonstration of the effectiveness of these techniques in explaining the segmentation results of these U-Net architectures, based on three use cases.

2 Related Work

The development of explainable AI methods for vision models has so far mainly concentrated on image classification [28,31,33,34]. Therefore, we provide an overview of the methods available for this purpose. Then we introduce the U-Net architecture we use for segmentation. Finally we give an overview of how explainable AI methods can be applied to segmentation.

2.1 Overview Explainable AI in Classification

For image classification, a lot of techniques have been presented, like Gradient-weighted Class Activation Mapping (Grad-CAM) [32], Layer-wise Relevance Propagation (LRP) [3], or Deep Learning Important FeaTures (DeepLIFT) [7]. In the following, we provide a short introduction for the techniques most relevant to the field of segmentation.

Feature Visualization in Neural Networks Feature visualization has been extensively studied to interpret the inner mechanisms of convolutional neural networks. This approach deciphers the representations learned by the neurons at a particular layer. Specifically, it uses optimization to find synthetic inputs that cause neurons to fire, thereby giving insights about the kind of shapes, colors, or textures these neurons recognize. Deep learning models are multi-layered and highly non-linear. Understanding what complex patterns models have learned, and how these learned representations evolve deeper in the network, therefore is still an open problem [41].

Activation Maps in CNNs Several works have proposed to visualize and interpret CNNs through activation maps. Techniques such as Grad-CAM [32] and its refined version Grad-CAM++ [6] have been widely used. This technique uses the gradient information flowing into the last convolutional layer of the model to understand which parts of the image have contributed to the decision-making by creating a heatmap of the image [32].

Shapley Additive Explanations (SHAP) SHAP provide a measure of the contribution of each feature to an overall prediction for a specific sample. Recent work has adapted SHAP for use with deep learning models to derive pixel-level contributions, illuminating regions in the image that the model found most influential for its decision [21].

Local Interpretable Model-agnostic Explanations (LIME) In contrast to the other methods described here, LIME is a technique for explaining blackbox models. It works by fitting a simple model, such as linear regression, locally around the prediction and using this simpler model to explain the original complex one. Recently, researchers have demonstrated its use in image classification tasks, providing a pixel-level explanation for the prediction of the model [25].

Layer-wise Relevance Propagation (LRP) LRP is a method for explaining predictions of neural networks by assigning relevance scores to the input features, thus providing a means to interpret the decision-making process of these models [3]. The principle of conservation of relevance ensures a fair distribution of relevance across all features [27].

2.2 U-Net Architecture

U-Net is a convolutional neural network specifically designed for image segmentation. It features a U-shaped architecture that includes an encoding (contracting) and decoding (expanding) path (see Figure 1). A unique aspect of U-Net is the use of skip connections, where feature maps from the encoding path are concatenated with those in the decoding path. This allows the network to retain high-resolution features, enabling precise localization. The model can work with a low number of training images and accepts images of any size due to the absence of fully connected output layers compared to classification networks [26]. To further improve the performance of U-Net, modifications have been applied to the architecture, like the addition of residual connections [39], nested U-Net architecture [40] and many more [24].

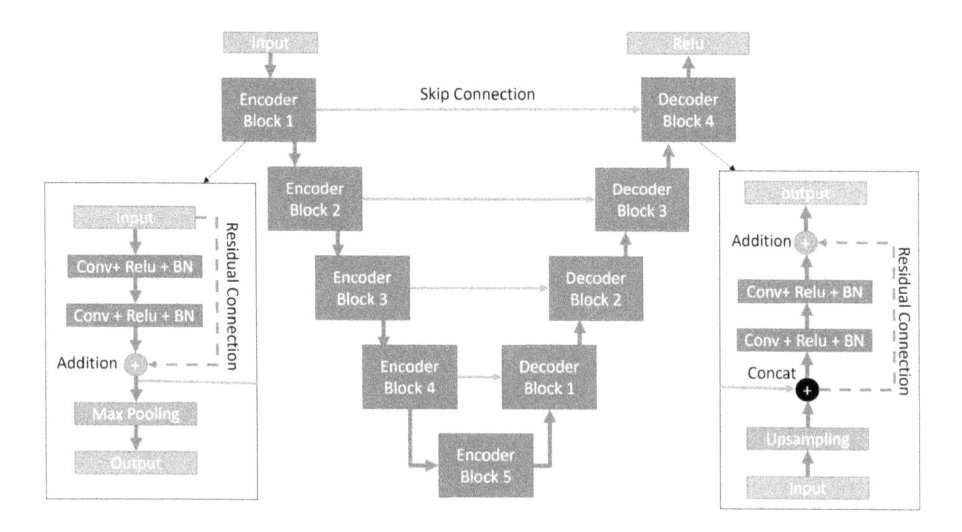

Fig. 1. U-Net Architecture including residual connections

2.3 Interpretable Segmentation

The techniques for explainable classification cannot directly be applied to segmentation tasks. However, many of the methods used to explain segmentation results are adaptions from methods that have been previously developed for classification.

The Grad-CAM method can be used to explain what is important for a semantic segmentation network, specifically, the spatial locations relevant for the decision of a U-Net network. For this purpose, the information from the bottleneck layer is extracted and used [30]. This has the downside that information on the influence of pixels surrounding the segmentation result is not available in the bottleneck layer.

SHAP, in contrast, can also give information about the location of the surrounding pixels which are not segmented. To achieve this, input pixels are grouped, and their contribution to the segmentation is calculated [10]. But with the grouping, it is no longer possible to obtain information at the feature level.

LRP is able to combine the feature level information and also is not only limited to the segmentation. This also allows for determining the influence of surrounding pixels. There are already publications which have shown the use of LRP on U-Net architectures in the field of medical imaging [8,29,35]. In the publications of Chlebus et al. [8] and Schnurr et al. [29], LRP was used for the sensitivity analysis, to determine which MRI channel has the highest impact on the segmentation results.

3 Implementation

In our implementation we decided to use LRP, because it can be applied without modifying the network to be explained. For techniques like Grad-Cam it is necessary to extract certain layers from the bottleneck to explain the segmentation. Besides that, LRP also gives information about the relevance of surrounding pixels of the segmented area. We compute the LRP and apply a logarithmic scaling to improve the visibility of the relevance. To improve the stability we use the ϵ-rule. We further show a method for handling residual connections.

3.1 Calculation of LRP

For all neurons j in a layer, the relevance R_j is calculated as shown in Equation 1 as the sum of the relevances of the following layer weighted by their contribution. For each neuron k in the following layer, the contribution of its relevance R_k is computed by multiplying the activation X_j of the neuron by the weight ω_{jk} and normalizing over all neurons j. In the case of a CNN, j and k are not neurons in the classical sense, they are pixel positions where j represents the input pixels of the CNN layer and k the output pixels. The weights ω_{jk} are the convolutional weights connecting j and k. As in the work of Samek et al. [27], the relevance is redistributed to the input pixels in the image in the backward pass. To do this, the relevance R_k of the previous layer is multiplied as shown in Equation 1. At the first step, the relevance R_k is initialized with the output (result classification) of the model after the forward pass.

$$R_j = \sum_k \frac{x_j \omega_{jk}}{\sum_j x_j \omega_{jk}} R_k \quad (1)$$

To calculate the LRP for a segmentation task, every output pixel is interpreted as an individual classification. The relevance is calculated for every output pixel of the segmentation, and the resulting relevance maps are summed up. Applying the distributive rule allows the optimization of calculating the relevance for multiple output pixels in one pass. This is possible because the weights ω and activations x (Equation 1) are the same for all output pixels.

For computing LRP we employ the Captum framework [18] which allows the interpretation of PyTorch based networks. To enable this framework to work with U-Net architectures, it has been extended by two additional functionalities: The first one is the extension to handle non-linear activation functions for segmentation tasks like Sigmoid or Softmax. This was done by introducing a masking mechanism where every output pixel below a certain threshold is set to zero. As threshold value the same value as for the segmentation (binarization) was used. The masking also allows selecting objects or regions where an explanation is necessary. The second extension is an additional sum layer – this was inspired by the implementation of Montavon et al. [23], where the autograd engine of PyTorch is used for the calculation of LRP. The summation of the output allows to get LRP for the segmentation in one pass.

3.2 Logarithmic Relevance Scaling

LRP was originally developed for classification tasks, where networks like VGG16 and ResNet are popular [3,16,27]. Segmentation can be seen as a pixelwise classification, where every pixel gets assigned a class. The relevance values will be calculated for every output pixel of the U-Net and the input relevances are summed up. Unlike a VGG16 [15] architecture, a U-Net [26] architecture has skip connections to improve the reconstruction ability of the network. These, in combination with the summation of the relevance values from multiple output pixels (segmentation), lead to higher relevance values in comparison to classification tasks. The result of this is shown in Figure 2. Due to the high values, the relevance can no longer be visualized directly in a meaningful way. A few pixels will end up in high peak values for the relevance, which allows no longer a proper visualization.

To compensate these effects of the U-Net architecture, we introduced a new form of threshold in combination with a logarithmic scaling of the LRP results. To keep the information of negative relevance values, we use a special treatment of these values shown in Equation 2b. Negative relevance values can also be important for explanation. Negative relevance values are significant for multi class segmentation, because features are shared in the network. The negative values show which features have not contributed to the decision for a class.

For the calculation we split the relevance values into positive and negative parts as shown in Equation 2a and Equation 2b. This is done by a threshold. Also to cover small values (smaller than 1) without getting negative values after logarithm. To perform this after the threshold the values are divided by the threshold. For negative values the absolute values is taken for the logarithm. Both rescaled parts of the relevance are combined as shown in Equation 2c.

$$f_{\text{pos}}(x) = \log\left(\max(\text{thresh}, x) \cdot \frac{1}{\text{thresh}}\right) \tag{2a}$$

$$f_{\text{neg}}(x) = -\log\left(|\min(-\text{thresh}, x)| \cdot \frac{1}{\text{thresh}}\right) \tag{2b}$$

$$R_{\text{scaled}} = f_{\text{pos}}(x) + f_{\text{neg}}(x) \tag{2c}$$

The result of the logarithmic scaling can be seen in Figure 2. The scaled LRP results enable the identification of details of the dataset in the relevance map.

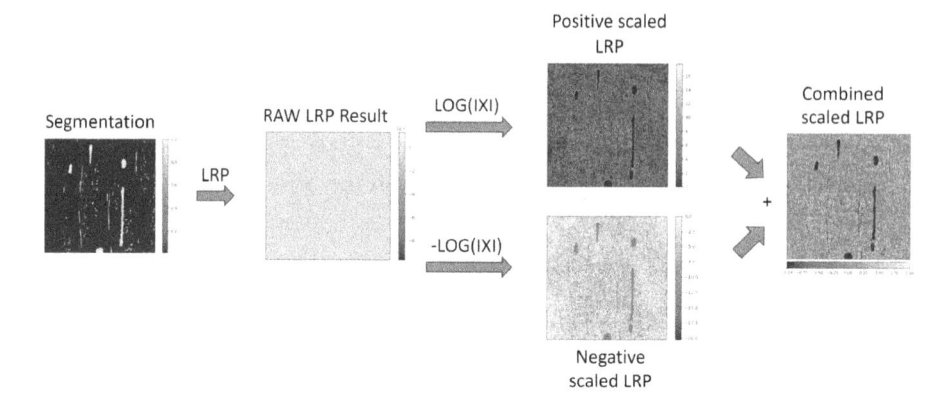

Fig. 2. Application of the logarithmic scaling after the LRP. The first column shows the segmentation result. The second shows the result of calculating the LRP with ϵ rule for all layers. The third column contains the result after logarithmic scaling. The last one shows the final result of adding both parts.

3.3 LRP Epsilon Rule

To avoid division by zero, the ϵ-rule was used [27]. In this rule an epsilon is added as shown in Equation 3.

$$R_j = \sum_k \frac{x_j w_{jk}}{\epsilon + \sum_j x_j w_{jk}} R_k \quad (3)$$

3.4 Residual Connections

A U-Net with residual connections in the encoder and decoder is shown in Figure 1. The residual connections bypass the convolutional layers and allow to forward low level features deeper into the network [13]. LRP residual connections in CNNs need a special treatment [18]. In the Captum implementation, the autograd engine of PyTorch is used to calculate the relevance values. Because of that the derivation of the addition will cause significant increases in LRP values, as shown in Equation 4. To prevent this problem, we handle the addition like a layer, and also use it in the LRP calculation as shown in Equation 1 to distribute the relevance values. The difference of the results are shown in Figure 3, where on the right side high values of up to 2e15 can be observed, whereas the left side with the modification only shows values up to 4000.

$$\frac{d}{dx}(f(x) + x) = f'(x) + 1 \quad (4)$$

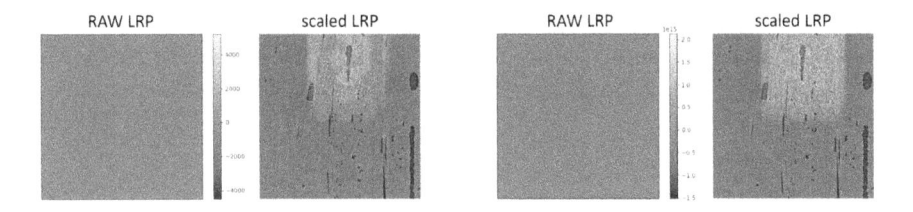

Fig. 3. Comparison of results with modification (left) and without modification (right) for residual connections.

4 Experiment

To test the proposed method, we trained three different architectures based on U-Net on three datasets from different domains. The first dataset is from the field of non-destructive testing (NDT), where the goal was to extract pores and damages from an image scanned by industrial computed tomography. Additional tests were performed on the Cityscapes dataset [9], here the task was to segment pedestrians. Also, tests were conducted using a medical dataset of chest X-ray for the segmentation of lungs. For better visualization, in the following experiments the LRP computations were done on a single selected object (pore, a group of pedestrians or lung).

4.1 Network Architecture

The method was tested on the following U-Net based architectures, implemented in the PyTorch-based framework MONAI [5]:

U-Net [26] Basic U-Net with a feature size for the layers of [32, 64, 128, 256, 512], a stride of one and padding one.
Residual U-Net [39] U-Net with residual connections, a feature size for the layers of [32, 64, 128, 256, 512], a stride of two and padding one.
U-Net++ [40] U-Net where in the skip-connections additional convectional blocks are used, with a feature size for the layers of [32, 32, 64, 128, 256], a stride of one and padding one.

Table 1. Training parameters for the use cases.

	NDT	Cityscapes	Medical
Epochs	30	100	100
Trainings Samples	782	2380	112
Learn Rate	2e-4	2e-4	8e-4
Input Channels	1	3	1

In Table 1 the parameters for the training of the networks are shown. For all use cases, the Generalized Dice Focal Loss was used, as well as the Adam optimizer. All networks were trained both with instance and batch normalization for comparison.

4.2 Selection of Region

To improve the visualization of the segmentation results, only one pore or pedestrian group was selected for the LRP by applying a mask on the output of the networks. If the mask is not applied, it is harder to determine the location of data which is relevant for segmentation. Afterward, the masked regions were used for the calculation of the relevance. This allows to show the relevance of input pixels to a segmented object.

4.3 NDT Use Case

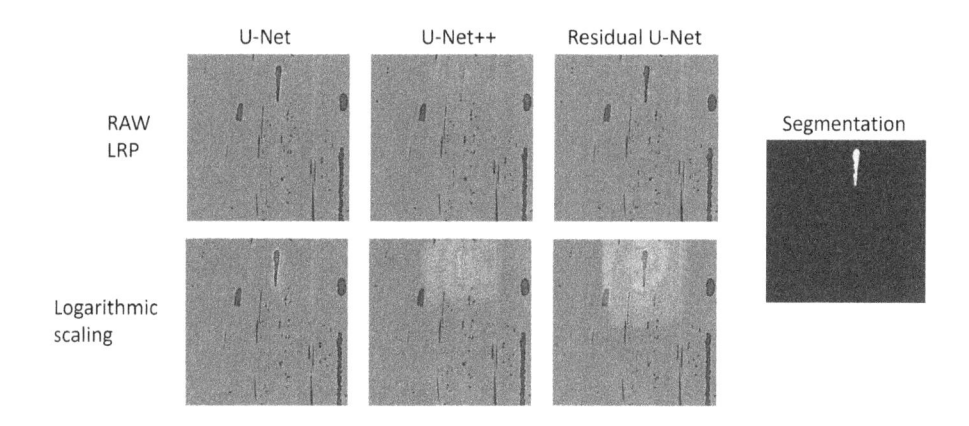

Fig. 4. Comparison of LRP raw values and with logarithmic scaling.

Figure 4 and Figure 5 show the segmentation of a needle-shaped pore. A 2D U-Net was trained on the segmentation of pores in X-ray computed tomography data of a carbon fiber-reinforced polymer specimen. Based on this training, the layer-wise relevance propagation was performed on one pore. We only used the positive relevance information in our examples, because we performed a binary segmentation. Therefore, negative values do not provide additional explanation of the results. In Figure 4, a comparison is shown between the raw LRP values and the logarithmically scaled values with a threshold of 0.1. The logarithmic scaling improves the visibility of the relevant input pixel. The example in Figure 4 show that a few single pixel with high values no longer dominant the results of the LRP. For U-Net++ and residual U-Net models our method reveals that these

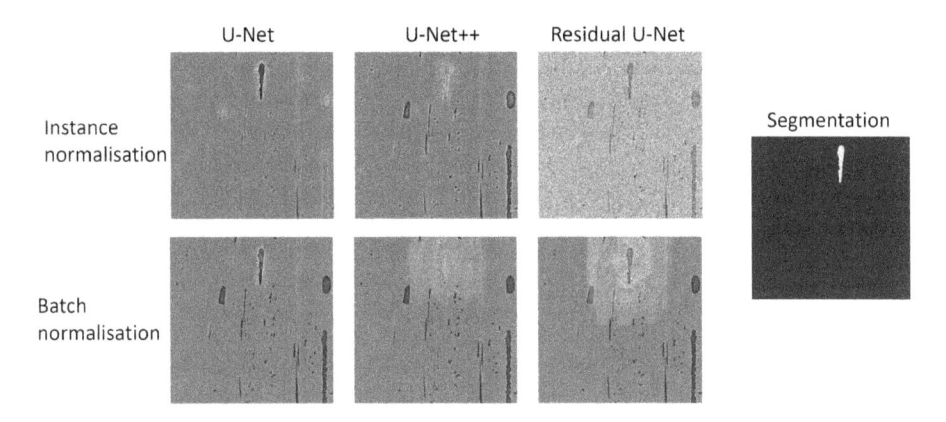

Fig. 5. Comparison of instance and batch normalization using logarithmically scaled LRP values.

focus not only on the area and border of the pore, but also on the surrounding texture. The comparison of instance and batch normalization in Figure 5 shows that a network with batch normalization focuses more on the segmented object. We also see some limitations in the result of the residual U-Net, where batch normalization results in a better focus. For instance normalization it is not really possible to determine the important regions of the input image.

4.4 Cityscape Use Case

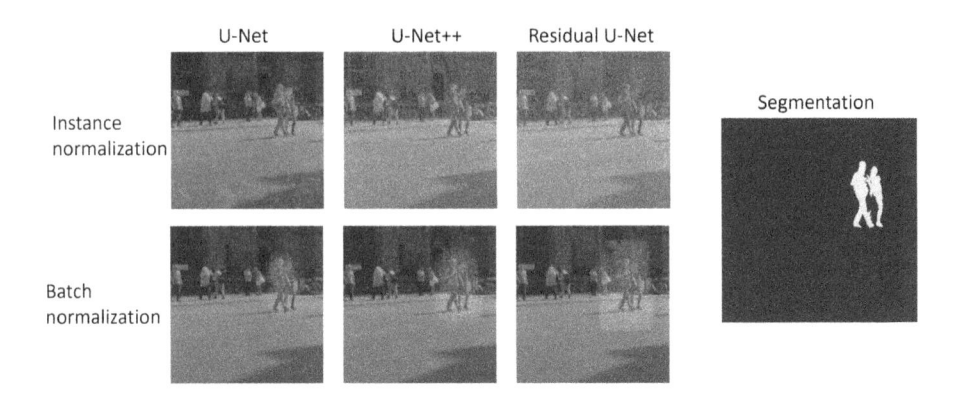

Fig. 6. Explanation of the segmentation of pedestrians with logarithmic scaling applied to the LRP results.

The networks were also trained on the Cityscapes dataset, consisting of RGB images, where the task is to segment pedestrians. As shown in Figure 6 the LRP technique is still able to determine the position of the pixels which are important for the segmentation. It is no longer possible to see detailed features as with the segmentation of the pore in subsection 4.3. When comparing the different normalization types, it turns out that also here, same as with the pore example, the batch normalization helps to improve focus.

4.5 Medical Use Case

In the medical use case the networks have been trained to segment lungs. The dataset used here is provided by the National Library of Medicine, the National Institutes of Health, Bethesda, MD, USA and the Shenzhen No.3 People's Hospital, Guangdong Medical College, Shenzhen, China [4,14]. The dataset consists of chest X-ray images with segmentation masks for the lungs. The networks focus on the shape of the lung as shown in Figure 7. There is a significant difference between both normalization types. As already shown in the other use cases, instance normalization leads to less focus of the relevance on the lung.

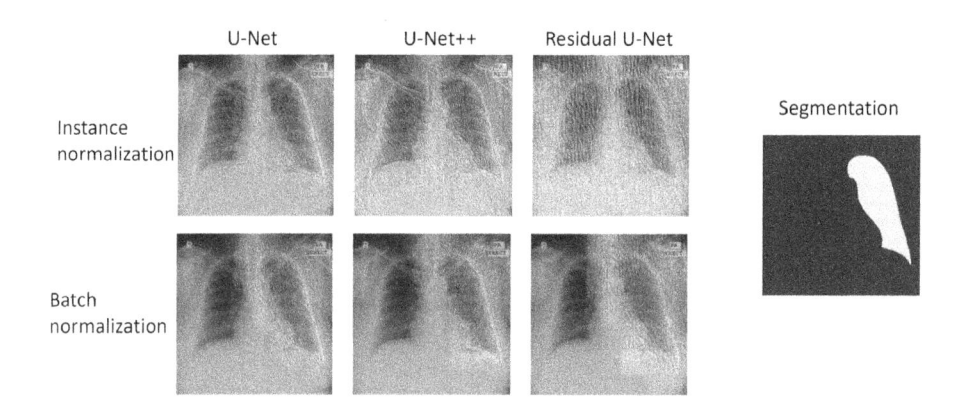

Fig. 7. Explanation of the segmentation of a lung with logarithmic scaling applied to the LRP results.

5 Discussion

Both examples have shown that LRP works well for a standard U-Net architecture and also for a U-Net++. The treatment of residual connections improves the visualization of the relevance in the network, but for instance normalization it is hard to determine the most relevant regions in the input image. This reduces the usefulness of LRP to explain networks with instance normalization. A similar behavior was shown in other publications [16] where LRP was applied to architectures with residual connections. In that work, instead of a VGG16 architecture (original LRP paper), a ResNet architecture was used to perform the LRP, the results are similar to our observations. The relevance map was more blurred and also the retrieval of features and structures was harder. This effect can be observed very well in Figure 5, Figure 6 and Figure 7. In U-Net there are also connections which bypass convolutional layers. These connections are called skip connections (see Figure 1). The skip connections improve spatial accuracy [12]. In the analysis of a U-Net architecture, this improved spatial information is not so relevant to interpret the segmentation results. More important are the high level features which came from the bottleneck after all encoder blocks.

6 Conclusion

We show in our work that layer-wise relevance propagation can be applied to interpret results from multiple U-Net architectures, and demonstrate this with datasets from 3 different domains. This allows explaining segmentation results with a relevance map on the input images. Depending on the task, it can also be shown on which features the network is focused to make the prediction. The example with the cityscape data set showed that for complex tasks it is no longer possible to determine accurate features in the relevance map. The relevance map still gives information about the regions the network is focused on. This limitation results from the structure of a U-Net, specifically from the skip connections in the network, which give additional information to improve the segmentation results. In our research, we found that the maximum values in the raw relevance image for a standard U-Net are higher than for a classification task. For a residual U-Net the values are even higher. We provide a way for handling residual connections, but this handling could potentially benefit from further improvements.

7 Future Work

In our future work, we will focus on the feature retrieval for complex segmentation tasks. We also want to evaluate the impact of architectural aspects like normalization, stride of convolution and residual connections in more detail. A special focus should be on the skip connections. Their contribution to the relevance could be reduced by introducing a specialized rule. In this work, we have shown the impact of batch and instance normalization on the LRP results. We plan to generate more insights on the impact of other normalization methods.

8 Code

The code to reproduce our results can be found at https://github.com/3dct/LRP_UNET

Acknowledgment. The research leading to these results has received funding by research subsidies granted by the government of Upper Austria within the projects "X-PRO", as well as "XPlain", grant no. 895981. The research has also been supported by the European Regional Development Fund in frame of the project Pemowe (BA0100107) in the INTERREG Programm Bayern-Österreich 2021-2027.

References

1. Al-hammuri, K., Gebali, F., Kanan, A., Chelvan, I.T.: Vision transformer architecture and applications in digital health: a tutorial and survey. Visual Computing for Industry, Biomedicine, and Art **6**(1) (2023). https://doi.org/10.1186/s42492-023-00140-9
2. Alzubaidi, L., Zhang, J., Humaidi, A.J., Al-Dujaili, A., Duan, Y., Al-Shamma, O., Santamaría, J., Fadhel, M.A., Al-Amidie, M., Farhan, L.: Review of deep learning: concepts, CNN architectures, challenges, applications, future directions. Journal of Big Data **8**(1) (2021). https://doi.org/10.1186/s40537-021-00444-8
3. Bach, S., Binder, A., Montavon, G., Klauschen, F., Müller, K.R., Samek, W.: On pixel-wise explanations for non-linear classifier decisions by layer-wise relevance propagation. PLoS ONE **10**(7) (2015). https://doi.org/10.1371/journal.pone.0130140
4. Candemir, S., Jaeger, S., Palaniappan, K., Musco, J.P., Singh, R.K., Xue, Z., Karargyris, A., Antani, S., Thoma, G., McDonald, C.J.: Lung segmentation in chest radiographs using anatomical atlases with nonrigid registration. IEEE Trans. Med. Imaging **33**(2), 577–590 (2014). https://doi.org/10.1109/TMI.2013.2290491
5. Cardoso, M.J., Li, W., Brown, R., Ma, N., Kerfoot, E., Wang, Y., Murray, B., Myronenko, A., Zhao, C., Yang, D., Nath, V., He, Y., Xu, Z., Hatamizadeh, A., Zhu, W., Liu, Y., Zheng, M., Tang, Y., Yang, I., Zephyr, M., Hashemian, B., Alle, S., Zalbagi Darestani, M., Budd, C., Modat, M., Vercauteren, T., Wang, G., Li, Y., Hu, Y., Fu, Y., Gorman, B., Johnson, H., Genereaux, B., Erdal, B.S., Gupta, V., Diaz-Pinto, A., Dourson, A., Maier-Hein, L., Jaeger, P.F., Baumgartner, M., Kalpathy-Cramer, J., Flores, M., Kirby, J., Cooper, L.A., Roth, H.R., Xu, D., Bericat, D., Floca, R., Zhou, S.K., Shuaib, H., Farahani, K., Maier-Hein, K.H., Aylward, S., Dogra, P., Ourselin, S., Feng, A.: MONAI: An open-source framework for deep learning in healthcare (2022). https://doi.org/10.48550/arXiv.2211.02701
6. Chattopadhyay, A., Sarkar, A., Howlader, P., Balasubramanian, V.N.: Grad-CAM++: Improved visual explanations for deep convolutional networks. In: IEEE Winter Conference on Applications of Computer Vision (WACV). pp. 839–847 (2018). https://doi.org/10.1109/WACV.2018.00097
7. Chen, H., Lundberg, S., Lee, S.I.: Explaining models by propagating shapley values of local components. Studies in Computational Intelligence **914**, 261–270 (2021). https://doi.org/10.1007/978-3-030-53352-6_24
8. Chlebus, G., Abolmaali, N., Schenk, A., Meine, H.: Relevance analysis of MRI sequences for automatic liver tumor segmentation (2019), http://arxiv.org/abs/1907.11773
9. Cordts, M., Omran, M., Ramos, S., Rehfeld, T., Enzweiler, M., Benenson, R., Franke, U., Roth, S., Schiele, B.: The cityscapes dataset for semantic urban scene understanding. In: IEEE Computer Society Conference on Computer Vision and Pattern Recognition. pp. 3213–3223 (2016). https://doi.org/10.1109/CVPR.2016.350

10. Dardouillet, P., Benoit, A., Amri, E., Bolon, P., Dubucq, D., Credoz, A.: Explainability of image semantic segmentation through shap values. In: Rousseau, J.J., Kapralos, B. (eds.) Pattern Recognition, Computer Vision, and Image Processing. ICPR 2022 International Workshops and Challenges. pp. 188–202. Springer Nature Switzerland, Cham (2023)
11. Dosovitskiy, A., Beyer, L., Kolesnikov, A., Weissenborn, D., Zhai, X., Unterthiner, T., Dehghani, M., Minderer, M., Heigold, G., Gelly, S., Uszkoreit, J., Houlsby, N.: An image is worth 16x16 words: Transformers for image recognition at scale. In: 9th International Conference on Learning Representations, ICLR 2021 (2021), https://openreview.net/forum?id=YicbFdNTTy
12. Drozdzal, M., Vorontsov, E., Chartrand, G., Kadoury, S., Pal, C.: The importance of skip connections in biomedical image segmentation. Lect. Notes Comput. Sci. **10008**, 179–187 (2016). https://doi.org/10.1007/978-3-319-46976-8_19
13. He, K., Zhang, X., Ren, S., Sun, J.: Deep residual learning for image recognition. In: 2016 IEEE Conference on Computer Vision and Pattern Recognition (CVPR). pp. 770–778 (2016). https://doi.org/10.1109/CVPR.2016.90
14. Jaeger, S., Karargyris, A., Candemir, S., Folio, L., Siegelman, J., Callaghan, F., Xue, Z., Palaniappan, K., Singh, R.K., Antani, S., Thoma, G., Wang, Y.X., Lu, P.X., McDonald, C.J.: Automatic tuberculosis screening using chest radiographs. IEEE Trans. Med. Imaging **33**(2), 233–245 (2014). https://doi.org/10.1109/TMI.2013.2284099
15. Karen, S., Andrew, Z.: Very deep convolutional networks for large-scale image recognition. In: 3rd International Conference on Learning Representations (ICLR) Conference Track Proceedings. p. 14 (2015), http://arxiv.org/abs/1409.1556
16. Kauffmann, J., Esders, M., Ruff, L., Montavon, G., Samek, W., Muller, K.R.: From clustering to cluster explanations via neural networks. IEEE Transactions on Neural Networks and Learning Systems (2022). https://doi.org/10.1109/TNNLS.2022.3185901
17. Ker, J., Wang, L., Rao, J., Lim, T.: Deep learning applications in medical image analysis. IEEE Access **6**, 9375–9389 (2017). https://doi.org/10.1109/ACCESS.2017.2788044
18. Kokhlikyan, N., Miglani, V., Martin, M., Wang, E., Alsallakh, B., Reynolds, J., Melnikov, A., Kliushkina, N., Araya, C., Yan, S., Reblitz-Richardson, O.: Captum: A unified and generic model interpretability library for PyTorch (2020)
19. Lin, Y., Cao, Y., Hu, H., Wei, Y., Zhang, Z., Lin, S., Guo, B.: Swin transformer : hierarchical vision transformer using shifted windows. Proceedings of the IEEE International Conference on Computer Vision pp. 9992–10002 (2021)
20. Liu, Z., Mao, H., Wu, C.Y., Feichtenhofer, C., Darrell, T., Xie, S.: A ConvNet for the 2020s (2022), https://arxiv.org/abs/2201.03545
21. Lundberg, S.M., Lee, S.I.: A unified approach to interpreting model predictions. In: Proceedings of the 31st International Conference on Neural Information Processing Systems. NIPS'17, vol. 31, p. 4768-4777. Curran Associates Inc., Red Hook, NY, USA (2017)
22. Mall, P.K., Singh, P.K., Srivastav, S., Narayan, V., Paprzycki, M., Jaworska, T., Ganzha, M.: A comprehensive review of deep neural networks for medical image processing: Recent developments and future opportunities. Healthcare Analytics **4** (2023). https://doi.org/10.1016/j.health.2023.100216
23. Montavon, G., Binder, A., Lapuschkin, S., Samek, W., Müller, K.R.: Layer-Wise Relevance Propagation: An Overview, pp. 193–209. Springer International Publishing, Cham (2019). https://doi.org/10.1007/978-3-030-28954-6_10

24. Mubashar, M., Ali, H., Grönlund, C., Azmat, S.: R2u++: a multiscale recurrent residual u-net with dense skip connections for medical image segmentation. Neural Comput. Appl. **34**(20), 17723–17739 (2022). https://doi.org/10.1007/s00521-022-07419-7
25. Ribeiro, M.T., Singh, S., Guestrin, C.: "why should i trust you?" explaining the predictions of any classifier. NAACL-HLT 2016 - 2016 Conference of the North American Chapter of the Association for Computational Linguistics: Human Language Technologies, Proceedings of the Demonstrations Session pp. 97–101 (2016). https://doi.org/10.18653/v1/n16-3020
26. Ronneberger, O., Fischer, P., Brox, T.: U-Net: Convolutional networks for biomedical image segmentation. In: Medical Image Computing and Computer-Assisted Intervention (MICCAI). pp. 234–241. Springer International Publishing, Cham (2015). https://doi.org/10.1007/978-3-319-24574-4_28
27. Samek, W., Wiegand, T., Müller, K.R.: Explainable artificial intelligence: Understanding, visualizing and interpreting deep learning models (2017), https://arxiv.org/abs/1708.08296
28. Saranya, A., Subhashini, R.: A systematic review of explainable artificial intelligence models and applications: Recent developments and future trends. Decision Analytics Journal **7** (2023). https://doi.org/10.1016/j.dajour.2023.100230
29. Schnurr, A.K., Schoeben, M., Hermann, I., Schmidt, R., Chlebus, G., Schad, L.R., Gass, A., Zoellner, F.G.: Relevance analysis of MRI sequences for MS lesion detection. In: European Society of Magnetic Resonance in Medicine and Biology. vol. 20 (2019)
30. Schorr, C., Goodarzi, P., Chen, F., Dahmen, T.: Neuroscope: An explainable AI toolbox for semantic segmentation and image classification of convolutional neural nets. Applied Sciences (Switzerland) **11**(5), 1–16 (2021). https://doi.org/10.3390/app11052199
31. Schwalbe, G., Finzel, B.: A comprehensive taxonomy for explainable artificial intelligence: a systematic survey of surveys on methods and concepts. Data Min. Knowl. Disc. (2023). https://doi.org/10.1007/s10618-022-00867-8
32. Selvaraju, R.R., Cogswell, M., Das, A., Vedantam, R., Parikh, D., Batra, D.: Grad-CAM: Visual explanations from deep networks via gradient-based localization. In: IEEE International Conference on Computer Vision. pp. 618–626 (2017). https://doi.org/10.1109/ICCV.2017.74
33. Sheu, R.K., Pardeshi, M.S.: A survey on medical explainable ai (xai): Recent progress, explainability approach, human interaction and scoring system. Sensors **22**(20) (2022). https://doi.org/10.3390/s22208068
34. Tjoa, E., Guan, C.: A survey on explainable artificial intelligence (xai): Toward medical xai. IEEE Transactions on Neural Networks and Learning Systems **32**(11), 4793–4813 (2021). https://doi.org/10.1109/tnnls.2020.3027314
35. Tjoa, E., Heng, G., Yuhao, L., Guan, C.: Enhancing the extraction of interpretable information for ischemic stroke imaging from deep neural networks (2019), https://arxiv.org/abs/1911.08136
36. Vaswani, A., Shazeer, N., Parmar, N., Uszkoreit, J., Jones, L., Gomez, A.N., Kaiser, L.u., Polosukhin, I.: Attention is all you need. In: Guyon, I., Luxburg, U.V., Bengio, S., Wallach, H., Fergus, R., Vishwanathan, S., Garnett, R. (eds.) Advances in Neural Information Processing Systems. vol. 30. Curran Associates, Inc. (2017). https://doi.org/10.48550/arXiv.1706.03762
37. Yang, J., Li, S., Wang, Z., Dong, H., Wang, J., Tang, S.: Using deep learning to detect defects in manufacturing: A comprehensive survey and current challenges. Materials **13**, 5755 (2020). https://doi.org/10.3390/ma13245755

38. Zhang, H., Zhong, X., Li, G., Liu, W., Liu, J., Ji, D., Li, X., Wu, J.: BCU-Net: Bridging ConvNeXt and U-Net for medical image segmentation. Comput. Biol. Med. **159**, 106960 (2023). https://doi.org/10.1016/j.compbiomed.2023.106960
39. Zhang, Z., Liu, Q., Wang, Y.: Road extraction by deep residual u-net. IEEE Geosci. Remote Sens. Lett. **15**(5), 749–753 (2018). https://doi.org/10.1109/lgrs.2018.2802944
40. Zhou, Z., Rahman Siddiquee, M.M., Tajbakhsh, N., Liang, J.: Unet++: A nested u-net architecture for medical image segmentation. Lect. Notes Comput. Sci. **11045**, 3–11 (2018). https://doi.org/10.1007/978-3-030-00889-5_1
41. Zimmermann, R.S., Borowski, J., Geirhos, R., Bethge, M., Wallis, T.S., Brendel, W.: How well do feature visualizations support causal understanding of CNN activations? In: Ranzato, M., Beygelzimer, A., Dauphin, Y., Liang, P., Vaughan, J.W. (eds.) Advances in Neural Information Processing Systems. vol. 34, pp. 11730–11744. Curran Associates, Inc. (2021), https://neurips.cc/virtual/2021/poster/27775

Visualizing Dynamics of Federated Medical Models via Conversational Memory Elements

Sanidhya Kumar, Varun Vilvadrinath, Jignesh S. Bhatt[✉], and Ashish Phophalia

Indian Institute of Information Technology Vadodara, Gandhinagar, India
{202151138,202151195,jignesh.bhatt,ashish_p}@iiitvadodara.ac.in

Abstract. The recent shift towards distributed learning in medical AI has prompted investigations into crucial operational dynamics. This includes determination of the point when a federated model attains training saturation and examination of the influence of model(s) within a federated set-up. To this end, recently coined conversational memory elements (CMEs)-based cognitive analysis is explored in a federated environment. The key fact is to establish conversation with federated models by associating the learning with changes in weights, which are archived in CMEs. The CMEs for a federated learning setup are defined as Jacobian matrix structures derived from contiguous changes among weights over training epochs. Covariance map and auto-correlation function (ACF) are then derived from this extracted CMEs for each federated client. Such analysis yield interesting qualitative insights for visualizing operational dynamics of federated medical models. For experiments, three prototype federated clients with different datasets are used. The federated server is based on ResNet50, while clients use ResNet50, DenseNet121, and AlexNet with skip connections. Training is done on the Brain Tumor Classification dataset across four practically observed scenarios. Two significant outcomes are obtained: First, the integration of CMEs helps indicating training saturation points by analyzing evolving patterns in estimated covariance maps and thus offering valuable pointers for resource optimization. Second, ACF plots effectively distinguish dominant and weaker clients to facilitate informed decisions for targeted strategies.

Keywords: Cognitive AI · Conversational Memory Elements · Federated Learning · Medical AI

1 Introduction

In recent years, federated learning (FL) has emerged as a promising decentralized learning paradigm for neural model training. It offers solutions to pressing concerns surrounding data privacy while showcasing its efficacy across diverse domains [2,7,9]. The federated medical architectures [1,2] entail individual training of models by different hospitals, with only the learned weights communicated to a federated (shared) server for a common aggregation policy [17]. Subsequently, thus aggregated weights are considered for further training and validation undergo a reciprocal cycle of dissemination to individual hospitals. Fig. 1 represents an illustration of federated medical model configuration with N hospitals connected as federated clients with a federated server. Each of these clients may have has its own local neural network owning private data. Upon completion of training of individual clients, the weights are then aggregated at the federated server and disseminated back to every client for further training until the convergence.

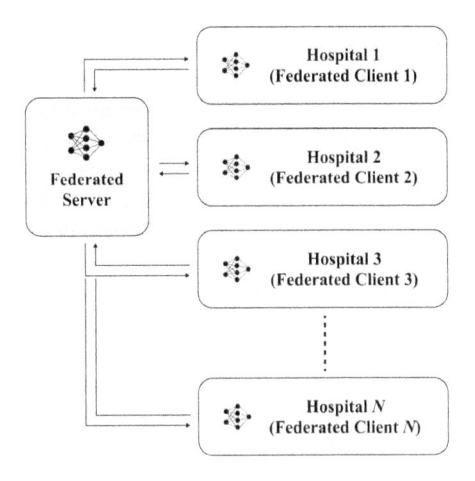

Fig. 1. An illustration of medical federated learning configuration.

Despite its promising prospects, medical FL poses several practical challenges that necessitate investigation [11,13]. One such challenge lies in determining an optimal point of training of client models, which could help preserve valuable time and computational resources. A recent research on conversational memory elements (CME) [19] has addressed the issue within a lumped (non-federated) model setup where they propose to establish conversation with medical AI networks [19]. In this paper, we seek to scale up these interesting findings to distributed learning especially for medical FL which remains unexplored.

What makes this approach novel is its application of CMEs in the decentralized FL context, addressing FL-specific challenges like data heterogeneity and asynchronous updates. By monitoring training dynamics at the client level and tailoring optimization strategies based on CME data, the methodology enhances FL training. Additionally, the use of covariance maps and autocorrelation function plots for visualizing FL dynamics and detecting redundancies contribute to a more robust and nuanced understanding of the FL process. This tailored approach leads to early convergence indicators and help better resource allocation, addressing the current demands of federated learning.

1.1 Literature Review

The exploration of medical AI models [18] and the development of analytical neural networks [16] are actively studied, nevertheless still posing open questions in the research. The term federated learning (FL), originally coined by Google [10,21], has captured significant attention within the deep learning community. Owing to its versatile applications in real-world scenarios [2,7,9,22], its impact is particularly pronounced in domains like healthcare; especially in oncology [5], where collaborative efforts among hospitals to train models for disease diagnosis have exemplified its efficacy. In this context, each hospital typically possesses a dataset characterized by variations in type and volume, reflecting the hospital's specialization and size. Moreover, real-world deployments have illustrated both the potential and the challenges of FL. For example, a work shown in [8] described the deployment of FL in training language models for the Google Keyboard, showcasing practical challenges and solution in a real-world application.

Researchers have extensively experimented with federated learning algorithms, such as FedAvg [14], which showed effectiveness in federated settings but noted slow convergence in heterogeneous environments like diverse hospital datasets in healthcare. Another approach focuses on lightweight client selection to identify dominant and weak clients, reducing federated rounds by 74%, though without mechanisms to detect diminishing returns [6]. In medical image computing, federated learning architectures face challenges with non-IID data and communication constraints, addressed by methods like FedProx [12] and clustered FL [20]. FedSeq [4] introduces sequentially trained super-clients to simulate centralized training in heterogeneous environments while preserving privacy. Optimization techniques like momentum-based methods and personalized FL frameworks are employed to improve convergence amidst system heterogeneity and varied client computational resources.

It is interesting to note that, all these studies lacked a systematic approach to determine optimal training cessation points and assessment of individual client contributions effectively. One can find literature on visualisation of neural networks in general but to the best of our knowledge there are no literature on visualization of convergence dynamics in a federated learning setup. Earlier approaches primarily focused on global model accuracy and loss metrics to track convergence, often plotted against communication rounds. While these metrics are useful, they fail to provide insights into operational dynamics and individual client contributions, particularly in heterogeneous environments. Understanding these internal dynamics is crucial for improving convergence and optimizing FL systems. As proposed in [19] that changes in the weights is indicative of learning and it is conveniently archived in CMEs. It is further suggested to derive two gross statistical measures: covariance maps and ACF plots from these CMEs. In this paper, we adapt the CME for medical FL and provide useful qualitative insights on operational dynamics.

1.2 Our contribution

The key contribution of this paper are as follows:

- **Visualization of Convergence Dynamics:** We examine the interaction between federated client devices and the server in distributed learning environments using CMEs [19] for federated learning environment. These visualizations provide valuable insights and represent a significant step towards the explainability of federated learning.

- **Optimizing Iterative Model Training:** By deriving covariance maps from the CMEs, we aim to identify the point of diminishing returns in medical FL client performance. This helps minimize computational overheads and enhances resource efficiency in distributed environments, ensuring that additional iterations yield meaningful improvements.

- **Client Characterization for Enhanced Performance:** Autocorrelation function (ACF) is defined using the CME, which helps distinguish between weak and dominant clients, enabling hospitals to make informed decisions and optimize model performance effectively in FL setups. This strategic identification supports better resource allocation and enhances overall system performance in resource-constrained environments like healthcare.

2 Problem Definition

Given different datasets and diverse learning approaches within federated medical setups, our objective is to generate visualization of convergence dynamics for the medical FL clients. In particular, 1) to analyze the integration of CMEs, help identifying training saturation points within federated setups by analyzing evolving patterns in estimated covariance maps derived from CMEs, and 2) to derive and utilize ACF plots to distinguish dominant and weaker clients by looking at the stability of the ACF plots over rounds of training. Specifically, we adapt the concepts of CMEs [19] for a federated setting, where the weights are influenced not only by the local model but also by other clients. Consequently, this impacts the covariance map and ACF plots. For this purpose, we simulate various operational pipelines as detailed in the following section. The primary motivation behind this endeavor is to provide visual insights for resource optimization and to facilitate informed decisions for targeted strategies in medical FL setup.

3 Proposed Approach

In this section, we first introduce the integration of CMEs into the federated learning setup, and then the medical FL architecture equipped with the CMEs shall be discussed.

3.1 Adapting CME for Federated Learning

In this subsection, we present theoretical details on how the concept of CMEs are adapted for medical FL. The CMEs are originally defined in [19] for gaining technical understanding for medical AI models. The idea is coined in [19] to establish conversation with a neural network to better understand the learning process. The CMEs are represented by a Jacobian matrix structure denoted as C. Here, we add two key terminologies: training rounds called as t-rounds at federated server and epoch at federated client setup. It is evident that each t-round consists of a set of epochs. Following each t-round, weights and biases are aggregated and updated at the federated server. Subsequently, the federated server communicates these updated weights and biases to the federated clients, enabling them to commence their next training round. Here, we define CME in a federated learning setup for a single federated client as:

$$C_i = \begin{pmatrix} \Delta w_1^i \\ \Delta w_2^i \\ \Delta w_3^i \\ \vdots \\ \Delta w_n^i \end{pmatrix} \quad \forall i \in [2, epochs]. \tag{1}$$

Here, Δw_k^i represents the weight differences between the current epoch i and the previous epoch $i-1$ for a layer as shown in equation (2) and n represents the total number of neuron-neuron connections in that layer. Each Δw_k^i can be defined as,

$$\Delta w_k^i = \frac{w_k^i - w_k^{i-1}}{w_k^i}; \quad \forall k \in [0, n] \quad \forall i \in [2, epochs]. \tag{2}$$

In order to implement this within Fig. 1, the matrix C_i is initialized to zero and then inserted between every layer of each federated client in every epoch. We archive the changes in each CME, denoted as C_i (equation (1)).

In general, covariance map of an epoch i, denoted as $Cov\_Map_{C_i}$, is defined for federated learning setup as,

$$Cov\_Map_{C_i} = E\{[C_i - \mu_{C_i}][C_i - \mu_{C_i}]^T\}, \tag{3}$$

where $E(.)$ denotes expectation operator, μ_C is the mean of a CME, and $(.)^T$ represents the transpose of the matrix. This map measures correlations among the CMEs, allowing us to visualize the learning of a FL network by observing the variance-covariance patterns derived from CMEs.

Finally, for federated learning setup the auto-correlation function ACF_C is given by,

$$ACF_C = E\{C_{\Delta w}(i) C_{\Delta w}(i+1)\}, \tag{4}$$

where $C_{\Delta w}(i)$ is a random process using CME vector values for all epochs. The function in equation (4) traces the changing gradients over epochs, providing visualization into the varying correlations until the convergence. Plotting the ACFs enables visualizing network stability as they reach steady-state over consequent t-rounds and would help comment on the convergence of the learning process.

3.2 Medical FL architecture equipped with CMEs

The proposed methodology leverages conversational memory elements (CMEs) to provide granular insights into federated learning (FL). By integrating CMEs (see section 3.1) within federated clients, particularly between critical layers, detailed weight differences across epochs are captured, allowing for visualization of operational dynamics. Covariance maps (Cov_Map) and autocorrelation function (ACF) plots derived from the archived CME values help in understanding weight stabilization, training stability, and redundancy among clients. This client-specific analysis enables the identification of effective learners, informed client participation, and optimization of resources, ultimately improving training efficiency.

In this section, we propose methodology and architectural details to showcase the utilization of CMEs in medical FL setup. Fig. 2 shows a prototype medical FL setup equipped with CMEs. We consider a typical setup with three clients, each representing a distinct entity possessing data and participating in the FL process. While it is possible for these clients to share identical datasets, in practical scenarios, each client typically holds different data. This variation in data can stem from differences in dataset sizes, richness, and diversity among the federated clients (hospitals). Therefore, we safely assume that each federated client (hospital) possesses a separate dataset, with varying sizes as depicted in the figure.

Architecture: As shown in Fig. 2, our federated server incorporates a ResNet50, whereas each federated client possesses different models: client 1 utilizes ResNet50, client 2 employs DenseNet121, and client 3 uses AlexNet with custom skip connections. To establish a standardized aggregation protocol, we ensure that each federated client's final three layers are extended with custom layers (refer Fig. 2): the antepenultimate layer (l-2) consists of 128 neurons, the penultimate layer (l-1) consists of 128 neurons and the final layer comprises 4 neurons to accommodate the classification into four distinct classes. The aggregation method employed here is equal-weighted averaging [15].

Inserting CMEs: Now, to visualize the operational dynamics of this typical medical FL setup shown in Fig. 2, we derive covariance map and ACF plots by inserting CMEs within federated clients. Essentially, the CMEs are blank matrix

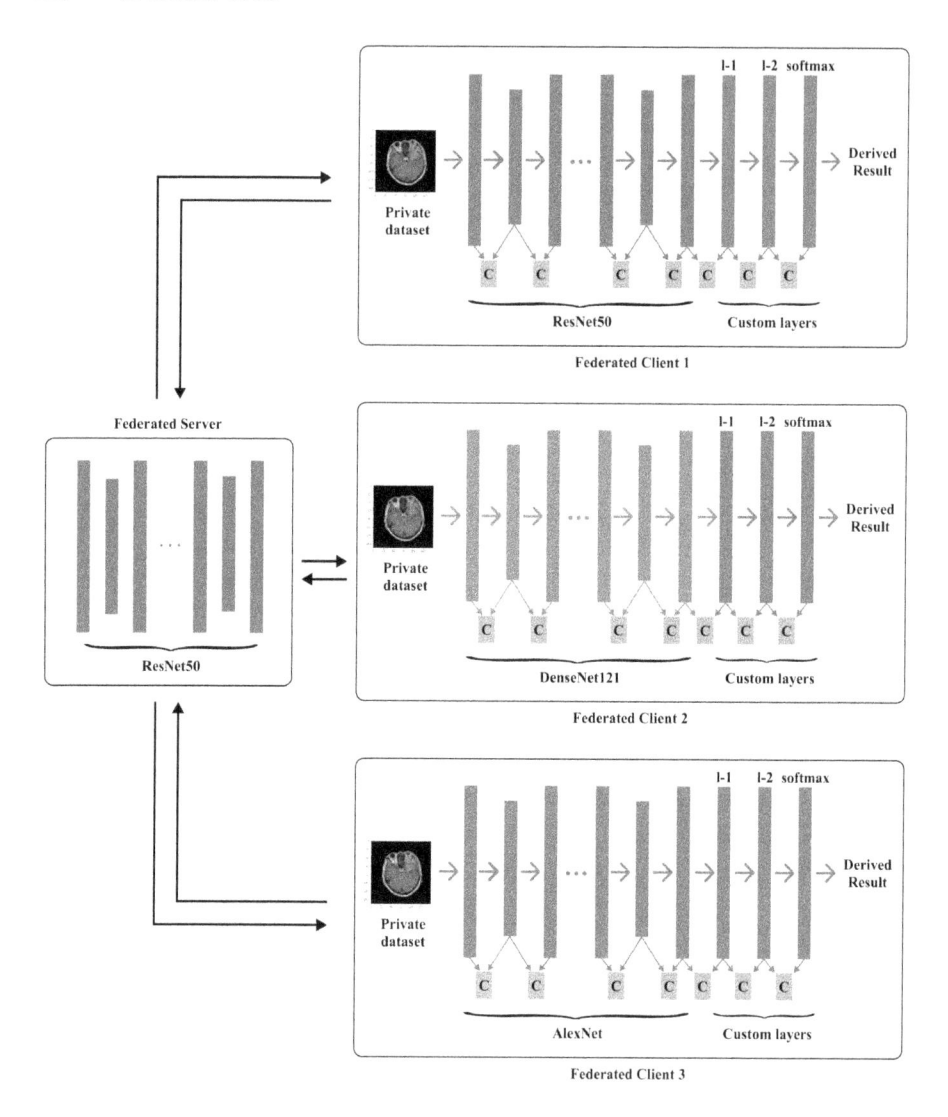

Fig. 2. Proposed medical FL framework with CMEs for visualizing operational dynamics. The symbol C represents CME.

structure individually inserted between layers of each federated client as shown in Fig. 2. Note that CMEs are not inserted between federated clients; instead, they are at each federated client within the client's environment. These are designed to capture the weight differences between epoch i and epoch $i-1$ for each federated connection within the setup.

Inference: The values of CMEs which are found between the antepenultimate and penultimate layers are archived and then analyzed to discern patterns learned through covariance maps. When the variance of these patterns stabilizes and the mean remains consistent, a model is considered to be trained to its capacity. The ACF plots are calculated from archived CME values to determine when to halt a training as well as identify redundant clients in a setup, as they are likely to exhibit minimal contribution to the overall learning.

4 Experimental Results

In this section, experimental results pertaining to the medical dataset are presented, including the machine configurations, the methodology employed in conducting the experiment, and the resultant findings across various experimental scenarios. Table 1 lists the four practically observed scenarios which are considered in our experiments.

Table 1. Practically observed scenarios and prototype federated clients.

	Dataset at clients	Model at clients	Federated client 1	Federated client 2	Federated client 3
Scenario 1	Different	Different	ResNet50	DenseNet121	AlexNet with skip connections
Scenario 2	Different	Same	ResNet50	ResNet50	ResNet50
Scenario 3	Same	Different	ResNet50	DenseNet121	AlexNet with skip connections
Scenario 4	Same	Same	ResNet50	ResNet50	ResNet50

4.1 Medical Dataset

For the experiment, the well-known benchmark brain tumor classification dataset from Kaggle [3] is used. It comprises of images depicting various types of brain tumors, including glioma, meningioma, pituitary tumors, as well as images indicating the absence of tumors. A few images are displayed in Fig 3 for illustration purpose. To enhance the diversity and comprehensiveness for the federated setup, the training and test files are merged, resulting in a combined dataset containing 3264 images. In scenarios where the data is assumed to be different

(Scenario 1 and 2 from Table 1), the dataset is shuffled and equipartitioned before being shared to all the federated clients. In cases where the data is assumed to be identical (Scenario 3 and 4 from Table 1), the dataset is loaded directly onto the clients without modification.

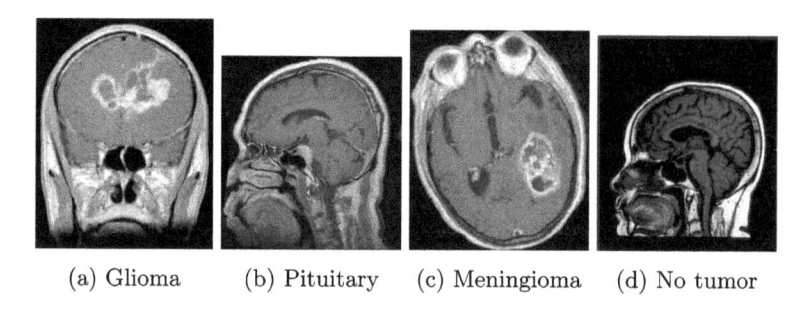

(a) Glioma   (b) Pituitary   (c) Meningioma   (d) No tumor

Fig. 3. A few images from the medical dataset [3] with their class labels.

4.2 Machine Specifications

The experimentation is conducted on Paramshavak HPC supercomputer. The system configuration consists of an x86_64 architecture with an Intel (R) Xeon (R) Gold 6139 CPU clocked at 2.30GHz, with a total of 72 CPU cores. Hyperthreading technology enabled two threads per core, resulting in a total of 144 threads available for parallel processing. Virtualization support is provided using VT-x. Additionally, it uses NVIDIA GPUs for accelerated computations, specifically a Quadro P400 and a Quadro GP100, operating with CUDA version 11.0. Lastly, the system is equipped with 92 gigabytes of RAM and 12 terabytes of HDD storage. These hardware resources facilitated efficient execution of the federated learning experiments with conversational memory elements.

4.3 Experimental Setup and Result Analysis

We consider training for multiclass classification using a federated model setup as depicted in Fig. 2. We carry out a threaded environment for simulating the FL setup. The deliberate selection of this methodological framework (Fig 2) aimed to enhance computational efficiency while mitigating the limitations imposed by finite resources. Three clients are involved in this experiment to simulate three different hospitals. Three parallel threads are executed with an average aggregation policy to maintain simplicity. Three models are selected: ResNet 50, DenseNet121, and an adapted version of AlexNet with skip connections.

Additional layers are introduced into each model. The defined CMEs are then inserted in between every layers, allowing for a detailed study of operational dynamics. Four distinct scenarios are implemented to cover a diverse range of real-life situations as described in the Table 1. For same model scenarios we have considered ResNet50 as it gives best results with 95.3% accuracy with 25 epochs. In our analysis of the results, we are looking for patterns in covariance maps of CMEs situated between the antepenultimate and penultimate layers along with stability in the ACF plots. These metrics will provide visual insights into the convergence behavior and performance of the federated learning process.

It is essential to understand that each t-round represents a round of training, at the end of which weights are transmitted to the federated server for aggregation. Before the commencement of the subsequent t-round, the updated aggregated weights are disseminated back to all federated clients. Each training round encompasses 10 epochs, and a total of 10 such t-rounds are executed in all four scenarios. Note that ResNet50 is considered as federated server in all scenarios.

Upon completion of the training phase, we archived CMEs after each t-round. Subsequently, the covariance map is plotted at each training round. Along with this, the overall ACF plot is also derived, which showcased layers and their corresponding effective ACF value at the conclusion of each training round. Both of these parameters are calculated based on the formulas and mathematical understanding as described in section 1.3. Once these information are extracted, various insights can be drawn for each scenarios as mentioned in Table 1, addressing the problem definition outlined in this paper.

4.3.1 Scenario 1: Different Dataset and Different Model

This scenario covers the case when all the federated clients utilize different datasets and all possess different local models: ResNet50 for client 1, DenseNet121 for client 2, and AlexNet with skip connections for client 3.

From Fig. 4(a) it is evident that for federated client 1, after t-round 1, initially, the covariance map resembled a noisy pattern, gradually evolving into a certain pattern by t-round 4 (Fig. 4(b)), which stabilizes by t-round 7 as seen in Fig. 4(c). It suggests that beyond t-round 7, further training may not yield significant benefits, as the model for that client has reached saturation. Stabilization of covariance pattern is indicative of optimal learning by a client. Similar observations can be made by looking at Fig. 4(d, e,f) for drawn clients 2 and Fig. 4(g, h, i) for client 3, with their covariance patterns becoming apparent after t-round 6 (Fig. 4(f)) and t-round 10 (Fig. 4(i)), respectively.

The ACF plots Fig. 4(j, k, l) indicate that client 1 (Fig. 4(j)) and client 3 (Fig. 4(l)) are in a state of continuous learning, as evidenced by the variations in their plots across t-rounds. In contrast, the ACF plot for client 2 (Fig. 4(k))

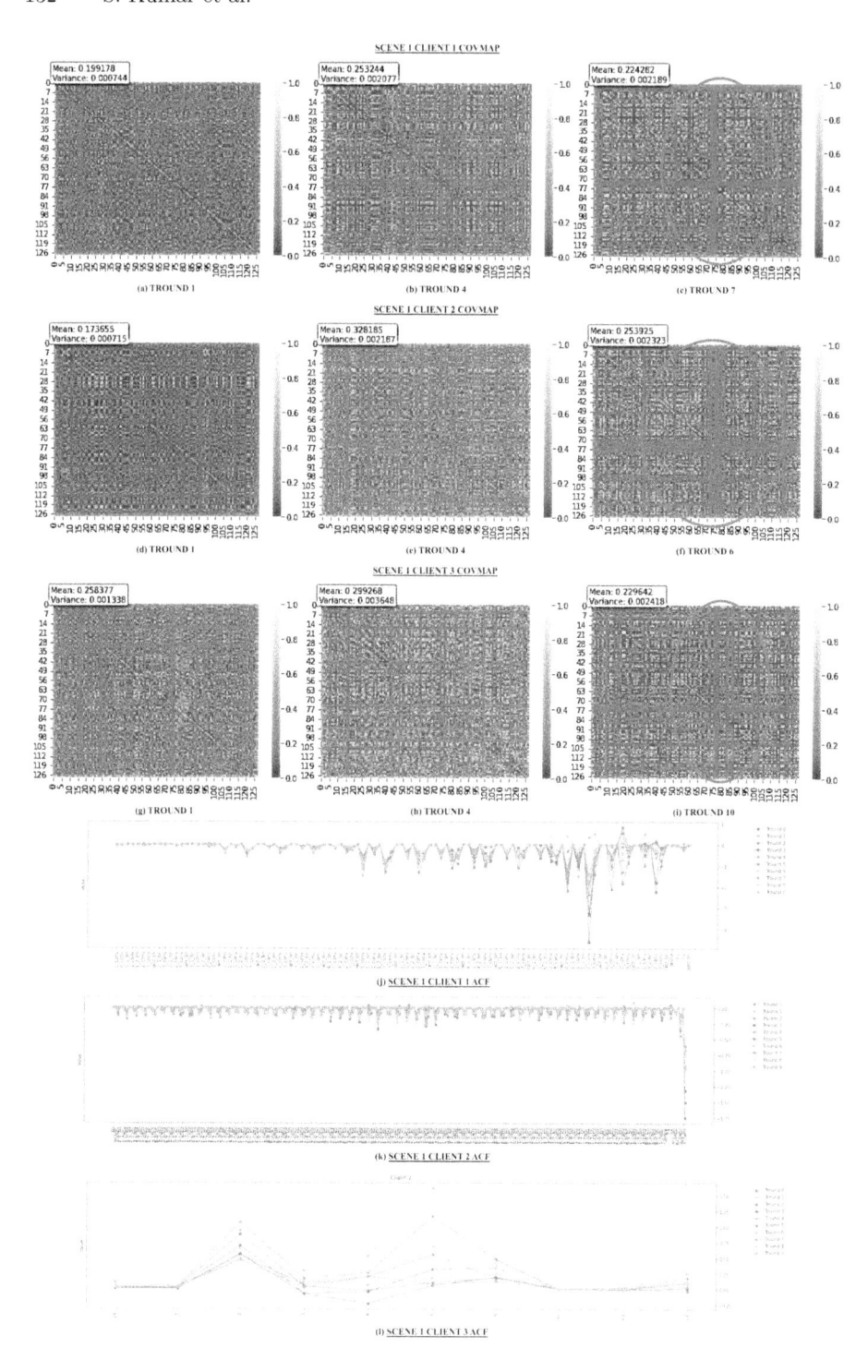

Fig. 4. Scenario 1 - Covariance Maps and ACF Plots

remains relatively stable, implying minimal learning, which suggests very few changes in weights. Client 2 results in a consistently stable ACF plot compared to clients 1 and 3 which may indicate that it requires little learning effort from the start. The ACF results are inline with the results of covariance maps as the emergence of stable clients is accompanied by a more distinct pattern in the covariance map, as illustrated in the Fig. 4.

4.3.2 Scenario 2: Different Datasets and Same Model
This scenario covers the case when all the federated clients utilize different datasets and all possess same local model that is ResNet50.

Here, Fig. 5(a) makes it clear that, for federated client 1, the covariance map first resembled a noisy pattern. By t-round 6, however, this pattern had evolved into a specific one (Fig. 5(b)), and by t-round 7, it had stabilized (Fig. 5(c)). It implies that after t-round 7, there would be no more substantial gains from training since the client's model has achieved saturation. Fig. 5(d, e,f) for client 2 and Fig. 5(g, h, i) for client 3 show similar trends in their covariance patterns, which stabilize at t-rounds 8 (Fig. 5(f)) and t-rounds 8 (Fig. 5(i)), respectively. Based on the differences in their plots across t-rounds, the ACF plots Fig. 5(j, k, l) show that client 1 (Fig. 5(j)) is in a continuous learning state. On the other hand, the ACF plot for client 2 (Fig. 5(k)) and client 3 (Fig. 5(l)) shows limited learning as it stays rather steady. In contrast to client 1, client 2 and 3 dominates the process and may require minimal learning effort from the beginning, producing an ACF plot that is constantly stable.

4.3.3 Scenario 3: Same Dataset and Different Models
In this scenario, all federated clients utilize identical datasets, each employing distinct local models: ResNet50 for client 1, DenseNet121 for client 2, and AlexNet with skip connections for client 3. Drawing parallels with the observations made in Sections 4.3.1 and 4.3.2, similar insights can be drawn by analysis.

4.3.4 Scenario 4: Same Dataset and Same Model
In this scenario, all federated clients use the same dataset and have same local model that is ResNet50. Similar conclusions can be drawn by analysis based on the discussions in sections 4.3.1 and 4.3.2.

4.3.5 Quantitative Analysis
The accuracy results of the classification problem are displayed in Table 2 for all scenarios. These results are consistent with the visual results.

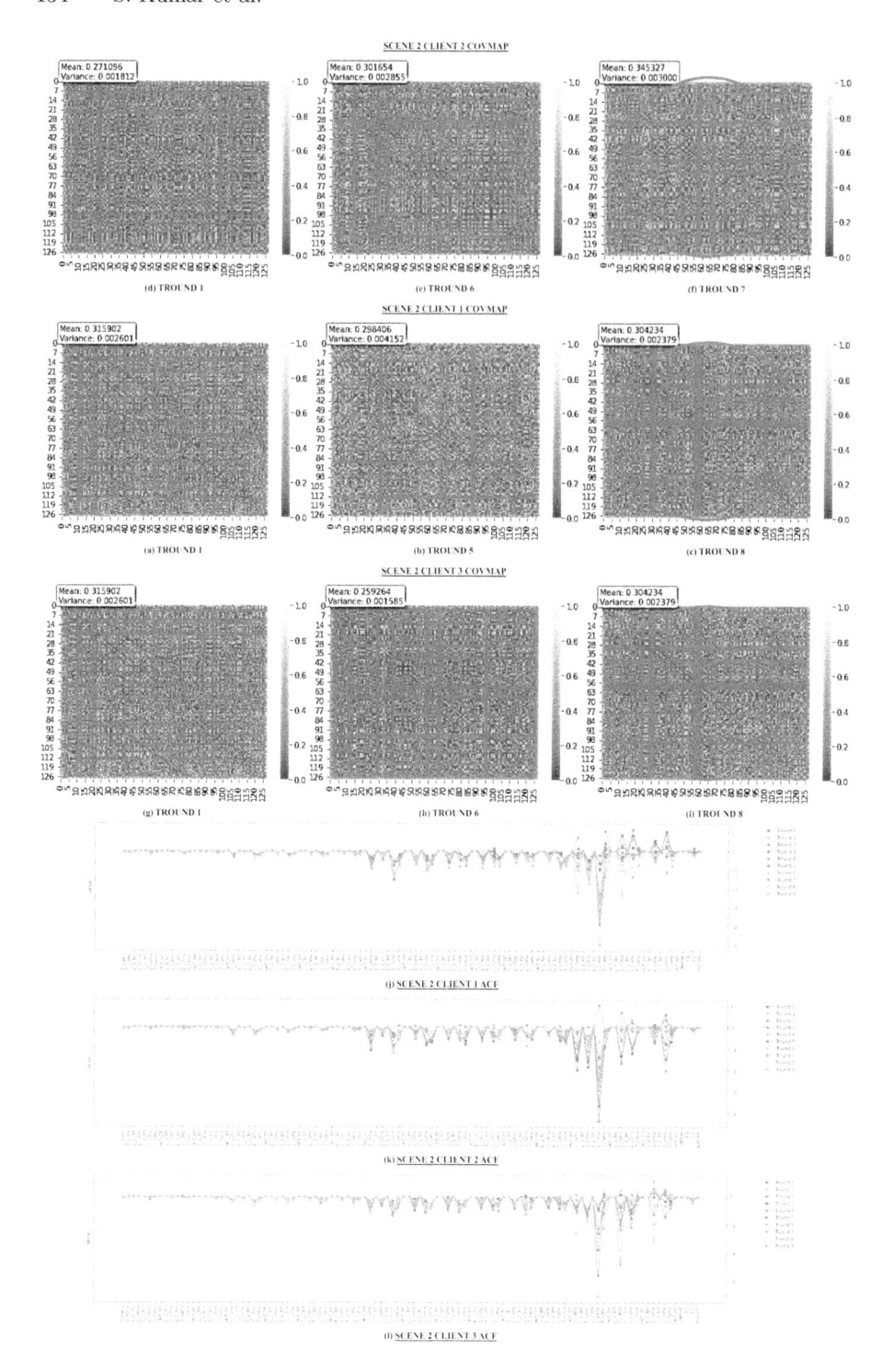

Fig. 5. Scenario 2 - Covariance Maps and ACF Plots

Table 2. Accuracies achieved with different federated scenarios

	Dataset at clients	Model at clients	Federated client 1	Federated client 2	Federated client 3
Scenario 1	Different	Different	91.3%	92.4%	85.62%
Scenario 2	Different	Same	90.5%	94.6%	94.4%
Scenario 3	Same	Different	95.4%	95.6%	93.75%
Scenario 4	Same	Same	96.3%	95.89%	95.83%

5 Conclusion

In this work, we have visualized the operational dynamics of federated medical AI models by utilizing CMEs within a federated learning framework. Using this cognitive approach, we obtain qualitative insights crucial for resource optimization and well-informed decision-making in medical AI development by examining evolving patterns in covariance maps and ACF plots generated from CMEs respectively. Following major observations are drawn from this research endeavour:

1. The integration of conversational memory elements helps in visualizing saturation point during training within a federated setup. This is accomplished by monitoring the evolving covariance patterns.
2. ACF plots offer a means to discern the dominant clients and identify the relatively weaker clients via stability of learning.
3. The insights obtained help understand the point of diminishing returns which can be leveraged for optimizing training time and effective utilization of available resources.

Future research includes extracting quantitative insights from the CMEs, utilizing them to comprehend weak and dominant layers within an architecture, and ultimately proposing methods for designing an optimal federated network.

Acknowledgements. This work is supported by the IIIT Vadodara grants to Dr. Jignesh S. Bhatt and Dr. Ashish Phophalia. Thanks to Gujarat Council on Science and Technology (GUJCOST) for providing the Paramshavak supercomputer for conducting exhaustive experiments in this research work. Special thanks to Ms. Swati Rai, PhD scholar at IIIT Vadodara, for helpful discussion.

References

1. Aledhari, M., Razzak, R., Parizi, R.M., Saeed, F.: Federated learning: A survey on enabling technologies, protocols, and applications. IEEE Access **8**, 140699–140725 (2020). https://doi.org/10.1109/ACCESS.2020.3013541
2. Antunes, R.S., André da Costa, C., Küderle, A., Yari, I.A., Eskofier, B.: Federated learning for healthcare: Systematic review and architecture proposal. ACM Trans. Intell. Syst. Technol. **13**(4) (may 2022). https://doi.org/10.1145/3501813

3. Bhuvaji, S., Kadam, A., Bhumkar, P., Dedge, S., Kanchan, S.: Brain tumor classification (mri) (2020). https://doi.org/10.34740/KAGGLE/DSV/1183165, https://www.kaggle.com/dsv/1183165
4. Chen, Z., Li, D., Ni, R., Zhu, J., Zhang, S.: Fedseq: A hybrid federated learning framework based on sequential in-cluster training. IEEE Syst. J. **17**(3), 4038–4049 (2023). https://doi.org/10.1109/JSYST.2023.3243694
5. Chowdhury, A., Kassem, H., Padoy, N., Umeton, R., Karargyris, A.: A review of medical federated learning: Applications in oncology and cancer research. In: International MICCAI Brainlesion Workshop. pp. 3–24. Springer (2021)
6. Constantinou, G., You, S., Shahabi, C.: Towards scalable and efficient client selection for federated object detection. In: 2022 26th International Conference on Pattern Recognition (ICPR). pp. 5140–5146 (2022). https://doi.org/10.1109/ICPR56361.2022.9956464
7. Ghimire, B., Rawat, D.B.: Recent advances on federated learning for cybersecurity and cybersecurity for federated learning for internet of things. IEEE Internet Things J. **9**(11), 8229–8249 (2022). https://doi.org/10.1109/JIOT.2022.3150363
8. Hard, A., Rao, K., Mathews, R., Ramaswamy, S., Beaufays, F., Augenstein, S., Eichner, H., Kiddon, C., Ramage, D.: Federated learning for mobile keyboard prediction (2019), https://arxiv.org/abs/1811.03604
9. Khan, L.U., Saad, W., Han, Z., Hossain, E., Hong, C.S.: Federated learning for internet of things: Recent advances, taxonomy, and open challenges. IEEE Communications Surveys & Tutorials **23**(3), 1759–1799 (2021)
10. Konečný, J., McMahan, H.B., Yu, F.X., Richtarik, P., Suresh, A.T., Bacon, D.: Federated learning: Strategies for improving communication efficiency. In: NIPS Workshop on Private Multi-Party Machine Learning (2016), https://arxiv.org/abs/1610.05492
11. Li, T., Sahu, A.K., Talwalkar, A., Smith, V.: Federated learning: Challenges, methods, and future directions. IEEE Signal Process. Mag. **37**(3), 50–60 (2020). https://doi.org/10.1109/MSP.2020.2975749
12. Li, T., Sahu, A.K., Zaheer, M., Sanjabi, M., Talwalkar, A., Smith, V.: Federated optimization in heterogeneous networks (2020), https://arxiv.org/abs/1812.06127
13. Mammen, P.M.: Federated learning: Opportunities and challenges. CoRR **abs/2101.05428** (2021), https://arxiv.org/abs/2101.05428
14. McMahan, H.B., Moore, E., Ramage, D., Hampson, S., y Arcas, B.A.: Communication-efficient learning of deep networks from decentralized data (2023), https://arxiv.org/abs/1602.05629
15. Moshawrab, M., Adda, M., Bouzouane, A., Ibrahim, H., Raad, A.: Reviewing federated learning aggregation algorithms; strategies, contributions, limitations and future perspectives. Electronics **12**(10) (2023). https://doi.org/10.3390/electronics12102287, https://www.mdpi.com/2079-9292/12/10/2287
16. Piduguralla, M., Bhatt, J.S.: An analytical cnn: Use of wavelets for learning image structures in cross-domain generalization. In: 2024 National Conference on Communications (NCC). pp. 1–6 (2024). https://doi.org/10.1109/NCC60321.2024.10485918
17. Pillutla, K., Kakade, S.M., Harchaoui, Z.: Robust aggregation for federated learning. IEEE Trans. Signal Process. **70**, 1142–1154 (2022). https://doi.org/10.1109/TSP.2022.3153135
18. Rai, S., Bhatt, J.S., Patra, S.K.: Deep learning in medical image analysis: Recent models and explainability. In: Explainable AI in Healthcare, pp. 23–49. Chapman and Hall/CRC (2024)

19. Rai, S., Bhatt, J.S., Patra, S.K., Ambadkar, T.: A cognitive behavioral AI: novel conversational memory elements for technical understanding of medical deep denoisers. In: 5th IEEE International Conference on Cognitive Machine Intelligence, CogMI 2023, Atlanta, GA, USA, November 1-4, 2023. pp. 41–48. IEEE (2023). https://doi.org/10.1109/COGMI58952.2023.00016
20. Sattler, F., Müller, K.R., Samek, W.: Clustered federated learning: Model-agnostic distributed multi-task optimization under privacy constraints (2019), https://arxiv.org/abs/1910.01991
21. Yang, Q., Fan, L., Tong, R., Lv, A.: Ieee federated machine learning. IEEE Federated Machine Learning - White Paper pp. 1–18 (2021)
22. Yang, Q., Liu, Y., Chen, T., Tong, Y.: Federated machine learning: Concept and applications. ACM Trans. Intell. Syst. Technol. **10**(2) (jan 2019). https://doi.org/10.1145/3298981

Generating Counterfactual Trajectories with Latent Diffusion Models for Concept Discovery

Payal Varshney[1,2]( ), Adriano Lucieri[1,2], Christoph Balada[1,2], Andreas Dengel[1,2], and Sheraz Ahmed[2]

[1] Rheinland-Pfälzische Technische Universität Kaiserslautern-Landau, Kaiserslautern, Germany
[2] German Research Center for Artificial Intelligence GmbH (DFKI), Kaiserslautern, Germany
{payal.varshney,adriano.lucieri,christoph.balada, andreas.dengel,sheraz.ahmed}@dfki.de

Abstract. Trustworthiness is a major prerequisite for the safe application of opaque deep learning models in high-stakes domains like medicine. Understanding the decision-making process not only contributes to fostering trust but might also reveal previously unknown decision criteria of complex models that could advance the state of medical research. The discovery of decision-relevant concepts from black box models is a particularly challenging task. This study proposes *Concept Discovery through Latent Diffusion-based Counterfactual Trajectories* (CDCT), a novel three-step framework for concept discovery leveraging the superior image synthesis capabilities of diffusion models. In the first step, CDCT uses a Latent Diffusion Model (LDM) to generate a counterfactual trajectory dataset. This dataset is used to derive a disentangled representation of classification-relevant concepts using a Variational Autoencoder (VAE). Finally, a search algorithm is applied to identify relevant concepts in the disentangled latent space. The application of CDCT to a classifier trained on the largest public skin lesion dataset revealed not only the presence of several biases but also meaningful biomarkers. Moreover, the counterfactuals generated within CDCT show better FID scores than those produced by a previously established state-of-the-art method, while being 12 times more resource-efficient. Unsupervised concept discovery holds great potential for the application of trustworthy AI and the further development of human knowledge in various domains. CDCT represents a further step in this direction.

Keywords: Explainablility · Counterfactuals · Concept Based Explanations · Latent Diffusion Models · Dermoscopy · Concept Discovery

---
P. Varshney and A. Lucieri—These authors contributed equally to this work.

Supplementary Information The online version contains supplementary material available at https://doi.org/10.1007/978-3-031-78198-8_10.

© The Author(s), under exclusive license to Springer Nature Switzerland AG 2025
A. Antonacopoulos et al. (Eds.): ICPR 2024, LNCS 15312, pp. 138–153, 2025.
https://doi.org/10.1007/978-3-031-78198-8_10

1 Introduction

Deep learning (DL) algorithms have gained immense popularity for their outstanding performance in the past decade [25,26,34]. The inherent non-linearity and over-parametrization of these algorithms make it particularly challenging to comprehend their reasoning processes. This, among other factors, contributes to a general lack of trust in such black-box models, constituting a major impediment to their application in safety-critical domains such as medicine. With the General Data Protection Regulation [38] and the upcoming European Artificial Intelligence (AI) Act [36], DL-based AI systems must now also comply with regulatory requirements directly concerning the topic of trust. Research on eXplainable AI (XAI) aims at increasing trust by providing more profound insights into the decision-making of opaque black-box models.

The correct interpretation of an explanation by the explainee is fundamental for yielding an understanding of the opaque decision-making process, which is a crucial prerequisite for building trust [23]. Many popular XAI methods only provide information about the relevance of individual input features [27,31,43]. The interpretation of these methods by explainees often suffers from a lack of context about the informative value of feature relevance. Concept-based explanation methods promise to remedy this by allowing to localize [21] human-aligned concepts and quantify [16] their relevance for decision-making. However, the acquisition of fine-grained concept annotations is expensive and inflexible. First efforts have been made towards unsupervised concept discovery [3,11,19], which could reduce the burden of collecting labels for concepts but also comes with the prospect of discovering new, previously unknown concepts. Existing methods are either based on strong structural assumptions [11] or make use of Generative Adversarial Networks (GANs) [3,19] that are usually challenging to train.

This work proposes *Concept Discovery through Latent Diffusion-based Counterfactual Trajectories* (CDCT), the first concept discovery framework based on counterfactual trajectories generated by state-of-the-art latent diffusion models (LDMs) [28]. The first step encompasses the generation of a counterfactual trajectory dataset using a text-conditioned LDM guided by the target classifier. This step aims to extract relevant semantic changes that describe the classifier's decision boundaries. The counterfactual trajectory dataset is used in the subsequent step to compute a disentangled representation of the classifier-relevant features using a Variational Autoencoder (VAE). In the final step, the VAE's latent space is exploited for concept discovery through a search algorithm that is built upon [19]. The contribution of our work is as follows:

- we propose CDCT, the first concept discovery framework that leverages latent diffusion-based counterfactual trajectories.
- we demonstrate improved counterfactual explanation capabilities by combining latent diffusion models with classifier guidance.
- we apply CDCT to a skin lesion classifier, demonstrating its ability to reveal classifier-specific concepts including biases and new biomarkers.

2 Related Work

The field of eXplainable AI can be subdivided by the output modality of a given XAI method. Several popular examples in the image domain are based on the quantification of feature relevance through surrogate models [27], activations [43], or gradients [31]. Recently, human-centric XAI methods like counterfactuals [7,35,39] or concept-based explanation methods [16,18] struck particular interest in the community, as they focus primarily on facilitating the interpretation of explanations by stakeholders and thus achieving their trust.

2.1 Generative Counterfactual Explanations

Counterfactuals constitute alternative classification scenarios with different outcomes given a marginal modification of a reference image. Wachter et al. [39] first introduced a method for the generation of counterfactuals through optimization.

The closest related works use diffusion models (DMs) to generate counterfactuals [4,13,14,29,30]. Some works use an unconditional diffusion model for the generation of counterfactuals guided by the target classifier [13,30]. Augustin et al. [4] utilize adaptive parameterization and cone regularization of gradients. Sanchez et al. [29] made use of a conditional diffusion model to generate healthy counterfactual examples of brain images to localize abnormal lesions. In [14], an adversarial pre-explanation is improved by diffusion-based inpainting to generate minimal counterfactuals. All previous works on DM-based counterfactual generation operate in the pixel space. Our proposed work instead leverages the more recent latent diffusion model to make use of improved synthesis capabilities with better computational efficiency, while focusing on semantic changes through guidance in the latent instead of the pixel space.

2.2 Concept-based Explainability

Concept-based XAI methods aim at the quantification and localization of higher-level concepts that are aligned with human cognitive processes. Concept-based explanations in pedestrian detection might for instance highlight a group of pixels as belonging to the concept *face*. Supervised concept-based explanation methods [8,16,18,21] rely on concept annotations, which can be expensive to obtain. Moreover, in complex domains like medicine, often experts are required for such annotations. However, some works approached the unsupervised discovery of novel concepts. One line of work deals with the identification of concepts from a classifier's latent representation [9,11,37,40,41]. Ghorbani et al. [11], for instance, proposed a framework for concept discovery based on the segmentation of images, clustering of their latent representations, and quantification of the conceptual influence of clusters. Fel et al. [9] extend the work of [11] and [41] to recursively decompose concepts across layers and localize them. In another approach, Achtibat et al. [1] introduced Concept Relevance Propagation (CRP), combining local and global perspectives to address both the 'where' and 'what' questions for individual predictions. Additionally, Poeta et al. [24] provided a comprehensive survey on concept-based explainable AI, detailing various methodologies and their applications.

Another relevant line of work regards the classifier as a black box and aims at the discovery of concepts through the generative manipulation of input samples [3,10,19,33]. Lang et al. [19], for instance, modified the StyleGAN architecture to systematically examine the GAN's style space for class-influential concept dimensions. Ghandeharioun et al. [10] approach concept discovery by enforcing the disentanglement of a GAN's latent concepts, encouraging higher distances between dissimilar and proximity between similar concepts. In contrast to existing concept discovery frameworks, CDCT leverages the power of state-of-the-art, conditional latent diffusion models for the generation of counterfactual trajectories to disentangle and identify classifier-relevant information.

3 Methodology

CDCT is a three-step framework for unsupervised concept discovery. In the first step, a latent diffusion model with classifier guidance is used to generate a counterfactual trajectory dataset, capturing decision-relevant concepts of the target classifier. This dataset is used to obtain a disentangled representation of concepts with the help of a Variational Autoencoder. Finally, decision-relevant dimensions are identified within the VAE by analyzing their impact on the classifier's output. An overview of the framework is presented in figure 1.

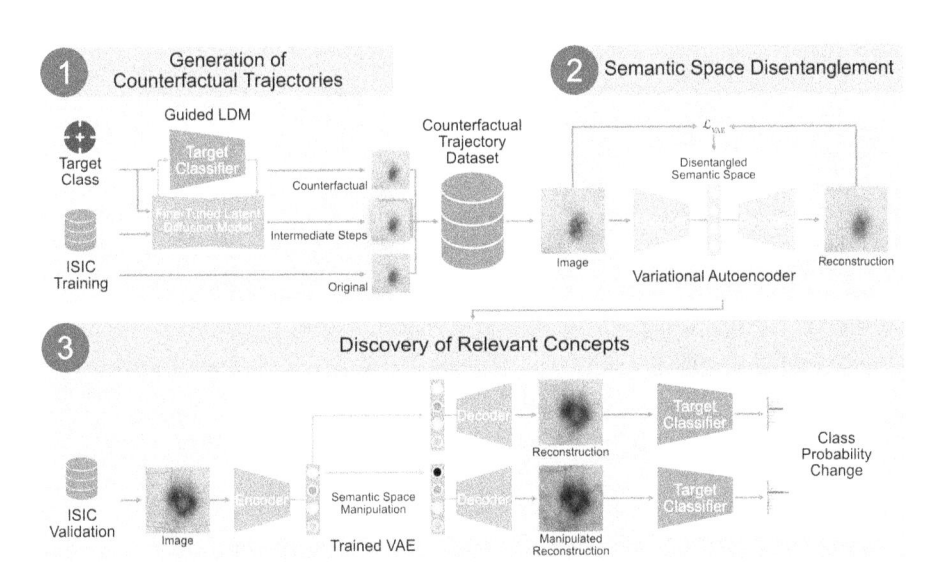

Fig. 1. CDCT is a three-step concept discovery framework. An LDM with classifier guidance is used to generate a counterfactual trajectory dataset. A VAE is trained on this trajectory dataset to disentangle decision-relevant features. Finally, class-relevant dimensions are identified by manipulating the VAE's latent space and observing the target classifier's output.

3.1 Generation of Counterfactual Trajectories

The first step aims to extract decision-relevant variations of the input image from the target classifier into a counterfactual trajectory dataset. In contrast to previous works [4,13,29], our proposed framework utilizes a latent diffusion model for a counterfactual generation to benefit from the improved generation efficiency and quality. Text conditioning is used to structure the latent space in accordance with known concepts. Figure 2 depicts the counterfactual generation process with detailed information about the guided step t.

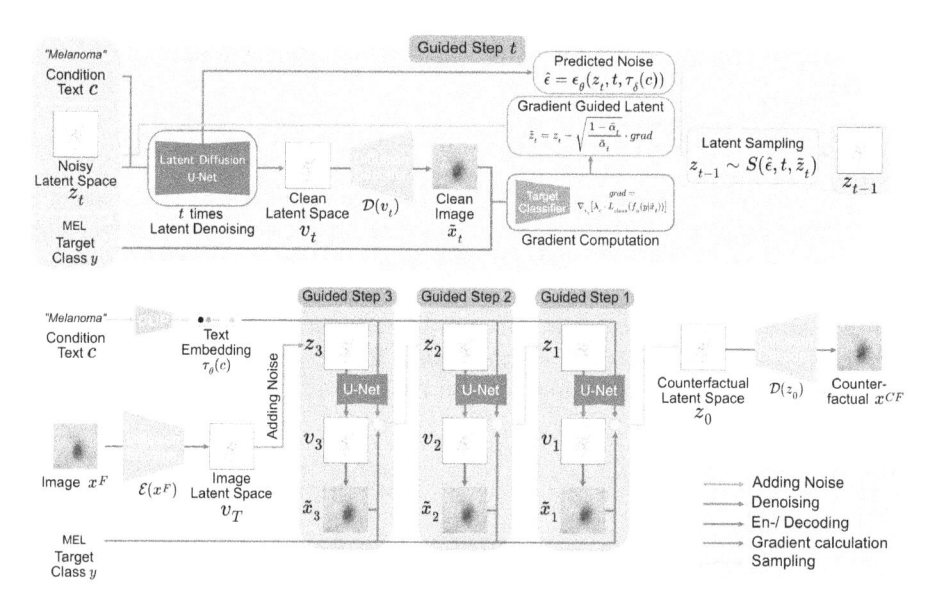

Fig. 2. The counterfactual generation process starts by encoding the image $\mathcal{E}(x^F)$ and perturbing it to obtain z_T (here $T = 3$). The text encoder transforms the condition (c) into an embedding $\tau_\theta(c)$. Both z_T and $\tau_\theta(c)$ are fed to the diffusion model for denoising. At guided step t, we denoise the noisy latent z_t, t times to produce the clean latent v_t, which is decoded into a clean image $\tilde{x}_t$ to calculate the gradient of loss $\mathcal{L}_{\text{class}}$ for updating z_t. We sample the previous less noisy latent z_{t-1} from the estimated noise $\hat{\epsilon}$ and the updated noisy latent $\tilde{z}_t$. In the final guided step, z_0 is decoded to yield the counterfactual image (x^{CF}).

We start the process with an original image x^F and utilize the encoder $\mathcal{E}$ to transform it into a lower-dimensional latent space. In the forward process, we take its latent representation to calculate a noisy version z_t with variance schedule $\beta_t \in (0,1)$, where $\alpha_t = 1 - \beta_t$, $\bar{\alpha}_t = \prod_{k=1}^{t} \alpha_k$ and $1 \leq t \leq T$.

$$z_T = \sqrt{\bar{\alpha}_t}\, \mathcal{E}(x^F) + \epsilon_t \sqrt{1 - \bar{\alpha}_t}, \quad \text{where} \quad \epsilon_t \sim \mathcal{N}(\mathbf{0}, \mathbf{I}) \tag{1}$$

Subsequently, we traverse the reverse Markov chain to generate a counterfactual with adapted classifier guidance introduced by Jeanneret et al. [13] applied to the latent space. In the reverse process, we iteratively go through the following steps until $t = 0$:

1. The LDM produces the cleaned latent vector v_t by denoising z_t, t times.
2. Then, v_t is decoded to obtain its representation in the image space $\tilde{x}_t$.
3. The target classifier f_ϕ is then applied on $\tilde{x}_t$ to calculate the gradients of the classification loss $\mathcal{L}_{\text{class}}$.

$$\nabla_{z_t} L(z_t, y) = \frac{1}{\sqrt{\bar{\alpha}_t}} \nabla_{v_t} \left[\lambda_c \cdot L_{class}(f_\phi(y|\tilde{x}_t)) \right] \tag{2}$$

4. The noisy latent z_t is guided using the gradient to produce $\tilde{z}_t$.
5. z_{t-1} is finally computed from $\tilde{z}_t$ and the noise $\hat{\epsilon}$ predicted at timestamp t.

We utilize the PNDM [20] sampling method represented as $S(\hat{\epsilon}, t, \tilde{z}_t) \rightarrow z_{t-1}$ to get the less noisy sample at each timestamp t. After t repetitions of the aforementioned steps, the counterfactual image is generated by decoding z_0 using the decoder $x^{CF} = \mathcal{D}(z_0)$. A sequence of clean images produced throughout the guiding process, commencing with the factual image and concluding with the counterfactual image, is referred to as a counterfactual trajectory.

3.2 Semantic Space Disentanglement

Counterfactual explanations may simultaneously alter multiple features, complicating the discovery of single, class-relevant concepts. Therefore, the counterfactual trajectory dataset is used in the second step to derive a disentangled representation of decision-relevant biomarkers for the target classifier.

Variational Autoencoders [17] are known for their ability to learn disentangled representations [6]. To enhance the reconstruction fidelity of the VAE, three additional loss functions are integrated besides the default reconstruction loss ($\mathcal{L}_{\text{Rec}}$) and the regularization term ($\mathcal{L}_{\text{KLD}}$). The L1 loss ($\mathcal{L}_{\text{L1}}$) is added to minimize pixel-wise differences, while the Structural Similarity Index Measure (SSIM) [42] ($\mathcal{L}_{\text{SSIM}}$) ensures perceptual image similarity. Lastly, a perceptual loss ($\mathcal{L}_{\text{Perc}}$) based on all the layers of a pre-trained VGG19 is used to capture high-level semantic features. Weighting the KLD loss by w_{kld} balances the trade-off between reconstruction fidelity and disentanglement. The final objective function for training the VAE can be stated as:

$$\mathcal{L}_{\text{VAE}} = \mathcal{L}_{\text{Rec}} + w_{kld} \cdot \mathcal{L}_{\text{KLD}} + \mathcal{L}_{\text{L1}} + \mathcal{L}_{\text{SSIM}} + \mathcal{L}_{\text{Perc}}$$

The VAE is trained on the counterfactual trajectory dataset to disentangle the decision-relevant semantic features.

3.3 Discovery of Relevant Concepts

Finally, our framework modifies the AttFind [19] algorithm to identify decision-relevant concepts in the disentangled latent representation. Algorithm 1 identifies the top latent space dimensions L_y for a class with a given set of images. These dimensions yield the highest average increase in class probability when changing its value during reconstruction through the VAE. In contrast to [19], our approach modifies directions in the range $D_y \in [-3, +3]$ to account for the standard normal distribution, as high magnitudes do not yield artifacts or unrealistic changes but lead to properly emphasized concepts. Moreover, concepts that yield inconsistent change directions were not filtered out, as we argue that conceptual manifestation is not necessarily monotonous on a linear trajectory.

Algorithm 1. Identify Relevant Latent Space Dimensions for a Specific Class

---

Input: Target classifier f_ϕ, encoder $\mathcal{E}$, decoder $\mathcal{D}$,
Set X of images with $\forall x \in X, \quad f_\phi(x) \neq y$,
Output: Set L_y of top latent space dimensions & set D_y of their directions.
for x in X **do**
    $l \leftarrow \mathcal{E}(x), \quad l \in \mathbb{R}^m$ // Encoding
    for $i = 1, \ldots, l$ **do**
        for d in $[-3, -2, -1, 0, 1, 2, 3]$ **do**
            $l[i] = d$
            $\tilde{x} \leftarrow \mathcal{D}(l)$ // Decoding
            Set $\Delta[x, l, d] = f_{\phi_y}(\tilde{x}) - f_{\phi_y}(\mathcal{D}(\mathcal{E}(x)))$ //Diff. softmax probability class y
        end for
    end for
end for
Set $\bar{\Delta}[l, d] = \text{Mean}(\Delta[x, l, d])$ over all $x \in X$
$L_y, D_y \leftarrow \text{argsort}(\bar{\Delta}[l, d], \text{descending order, top-K})$

---

4 Experiments & Results

4.1 Datasets & Classification Models

Data from the International Skin Imaging Collaboration (ISIC) challenges (2016-2020)[1] is used for experimentation. The duplicate removal strategy proposed in Cassidy et al. [5] is followed, resulting in a consolidated dataset of 29,468 samples, further split into train, validation, and test sets. The dataset consists of eight distinct classes, namely *Melanocytic Nevus* (NV), *Squamous Cell Carcinoma* (SCC), *Benign Keratosis* (BKL), *Actinic Keratosis* (AK), *Basal Cell Carcinoma* (BCC), *Melanoma* (MEL), *Dermatofibroma* (DF), and *Vascular Lesions* (VASC). The ISIC 2016, ISIC 2017, PH2 [22], and Derm7pt [15] datasets were

---
[1] The data is available at https://challenge.isic-archive.com/challenges/.

used for the training of concept classifiers to provide missing conditioning labels. The selected dermoscopic concepts are *Streaks, Pigment Networks, Dots & Globules, Blue-Whitish Veils, Regression Structures*, and *Vascular Structures*.

In this study, a ResNet50 [12] is trained on the consolidated ISIC training dataset, serving as the target classifier. The individual concept classifiers are based on the target classifier, fine-tuned on the respective concept dataset.

4.2 Generation of Counterfactual Trajectories

Stable Diffusion (SD) 2.1[2] is used as a baseline for the latent diffusion model. To enhance conditioning information for fine-tuning, captions are generated for training images including diagnostic details and skin lesion concepts. Concept classifiers were used to predict the presence or absence of concepts wherever no annotations were available. Only present concepts were used as conditioning information. The SD model is fine-tuned on the consolidated ISIC training dataset with text conditioning for 15K steps and a learning rate of $1e^{-5}$, generating realistic skin lesion samples.

The methodology described in section 3.1 is applied to generate counterfactuals. Experimentation revealed that a gradient loss scale of $\lambda_c > 10$ introduces unrealistic artifacts on the counterfactual images. A value of $\lambda_c = 4$ has therefore been chosen for all results. High values for the initial noise level t compromised the image quality, while $t = 10$ yielded a good trade-off between generative capacity and image quality. Figure 3 shows exemplary counterfactuals from one source image to all other target classes, showcasing the ability of the LDM to generate class-selective biomarkers. More examples can be found in the supplementary material A.3.

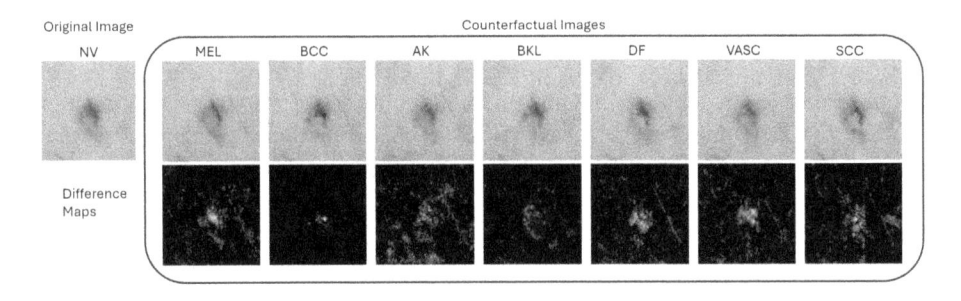

Fig. 3. An image of the *Nevus* class alongside its counterfactual images in all other target classes. The second row shows difference maps, providing an easy way to identify the areas of alteration between the original image and each counterfactual.

Counterfactuals generated by different versions of CDCT are quantitatively compared with the results from DiME [14] in table 1. The flip ratio measures the

---

[2] Available at https://huggingface.co/stabilityai/stable-diffusion-2-1.

frequency at which generated counterfactual images are classified as the target class by the classification model. A qualitative comparison can be found in the supplementary material A.4.

Table 1. L1, L2, FID, and Flip Ratio (FR) scores for counterfactuals generated by different versions of the proposed approach, compared with results from Jeanneret et al. [13]. Subscripts *ft* describe LDM instances fine-tuned on ISIC, while *wo-ft* refers to pre-trained LDMs. All reported results are computed on the consolidated ISIC test dataset for *MEL* and *NV* classes. The best values are highlighted in bold.

Method	L1	L2	FID	FR
DiME [13]	0.02969	0.00176	60.79377	0.86710
Unconditional $CDCT_{wo-ft}$	0.01930	0.00078	17.36892	1.0
Unconditional $CDCT_{ft}$	0.01858	**0.00073**	13.45541	1.0
Conditional $CDCT_{wo-ft}$	0.01930	0.00078	17.31825	1.0
Conditional $CDCT_{ft}$	**0.01857**	**0.00073**	**13.41800**	1.0

In every version, CDCT shows a significant improvement in overall metrics, indicating the superiority of counterfactuals generated by LDMs, over traditional approaches. The fine-tuning of the diffusion model yielded a notable improvement in FID, L1, and L2 measures. This aligns with the discovery, that the pre-trained LDM is completely incapable of synthesizing skin lesion images (see supplementary material A.2). The conditional generation of counterfactuals using the target class yielded further minor improvements. It is notable that in comparison to DiME, CDCT achieves a perfect flip ratio, indicating a stronger generative capacity of the LDM. Due to its superior performance, the conditional, fine-tuned CDCT is chosen for all further experiments.

Counterfactual trajectories capture the progression from a factual image to a counterfactual, as shown in figure 4. The counterfactual trajectory dataset is generated by computing trajectories for each ISIC training image into each alternate target class. Each trajectory consists of intermediate images from the 10 guided steps as well as the final counterfactual for 7 target classes, resulting in 78 images per sample. With 23,868 samples in the ISIC training dataset, the resulting counterfactual trajectory dataset contains 1,861,704 samples.

4.3 Semantic Space Disentanglement

A VAE was trained on the counterfactual trajectory dataset to capture all variations reflected by the data. The encoder comprises six ResNet-based downsampling blocks and final convolution layers, and the decoder architecture mirrors the encoder executing the reverse process of the encoding phase. An extensive hyperparameter search has been conducted, using different loss combinations, weightings, batch sizes, learning rates, and corresponding schedules. The

Counterfactual Trajectories for Concept Discovery   147

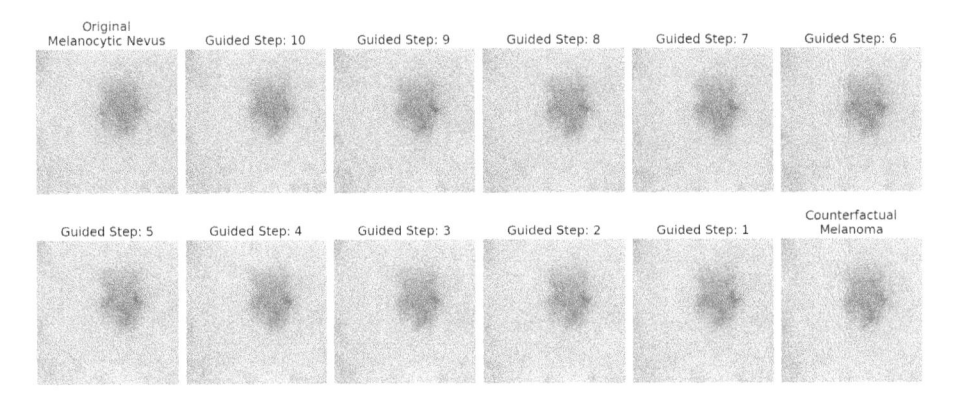

Fig. 4. Counterfactual trajectory for a *Nevus* with target class *Melanoma*. Along the process, the manifestation of darker, atypical pigment structures can be observed.

experiments indicate that there is an inherent trade-off between the disentanglement provided by the KLD loss and the reconstruction quality of the VAE. All reported results are based on a VAE trained for 23 epochs with the Adam optimizer using a batch size of 128, a learning rate of $2.5e^{-5}$, and an exponential decay learning rate scheduler with a decay factor γ of 0.95. ISIC test images along with their reconstructions are shown in supplementary material A.5.

4.4 Discovery of Relevant Concepts

VAE latent dimensions relevant for the target classifier are determined by algorithm 1 using the ISIC validation dataset. Figure 5 shows a small selection of the most relevant dimensions identified by CDCT. The success rate of a dimension is provided in the sub-caption for each concept and describes the fraction of test cases where the target class probability increased when altering said dimension.

For *Melanoma*, it has been found that many dimensions are related to darkening the skin lesion area while brightening the surrounding skin (e.g., figure 5b). Moreover, some dimensions (e.g., figure 5a) added darker spots and texture within the skin lesion to promote the prediction of *Melanoma*. Another dimension adds dark corner artifacts to the images, evident in figure 5c. The inverse of that dimension was found to promote the prediction of *Nevus*.

Similarly, for other classes, mostly concepts related to overall color are identified. Several relevant dimensions for *Nevus* altered the overall hue of the image to a red color (see figure 5d). For *Basal Cell Carcinoma*, the most relevant dimensions were related to the brightening of the whole image (see figure 5e). For *Dermatofibroma*, the most relevant dimension added white structures in the center of the lesions (see figure 5f). The relevance of the concepts is also emphasized by the high success rates of concepts, as shown in the sub-captions.

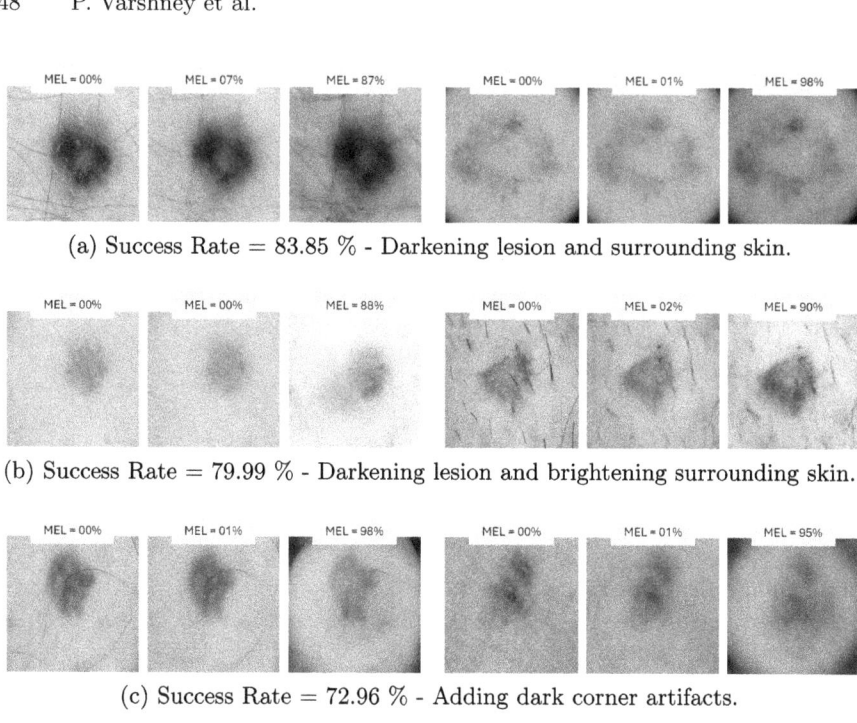

(a) Success Rate = 83.85 % - Darkening lesion and surrounding skin.

(b) Success Rate = 79.99 % - Darkening lesion and brightening surrounding skin.

(c) Success Rate = 72.96 % - Adding dark corner artifacts.

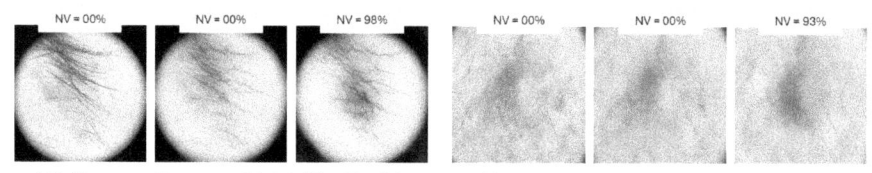

(d) Success Rate = 83.74 % - Reddening of lesion and surrounding skin area.

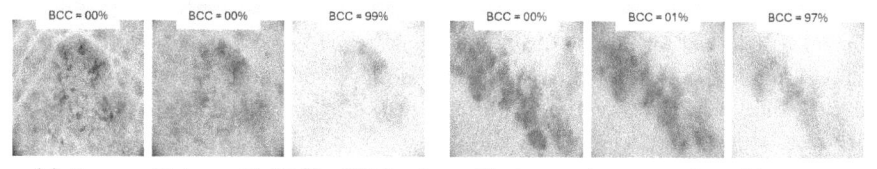

(e) Success Rate = 80.27 % - Whitening of lesion and surrounding skin area.

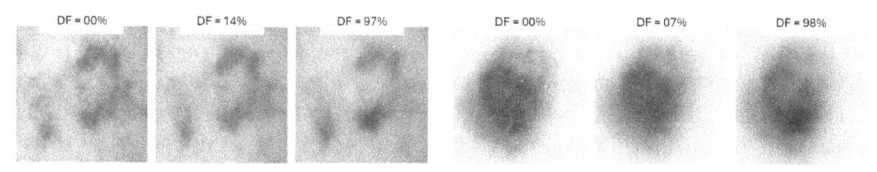

(f) Success Rate = 65.18 % - Adding central white patch on the lesion.

Fig. 5. Examples of discovered concepts using CDCT with the ISIC dataset. Each row shows two examples of a concept, where the original image, the reconstruction, and the manipulated reconstruction are aligned from left to right.

5 Discussion

In the previous section, the proposed three-step framework has been successfully applied for the global explanation of a skin lesion classifier. The results strongly suggest that CDCT successfully outperforms DiME [13] regarding the quality of generated counterfactuals. Using an LDM allowed for a reduction of the required denoising steps leading to a significant performance improvement in the generation of counterfactuals as compared to DiME (at least 12x speed-up). Moreover, the reduced number of denoising steps together with the generation in the latent space allowed the omission of the perceptual and the L1 loss employed in [13] while yielding better counterfactuals. The conditioning of the LDM for the generation of counterfactual trajectories turned out to yield insignificant improvements in the quality of counterfactuals, but might aid the generation of more diverse and realistic counterfactuals.

Through CDCT several disentangled dimensions were identified. The success rates of the presented dimensions confirm their relevance for the target classifier, and therefore their utility for explanation. By manual inspection, the plausibility of some high-performing dimensions could be established. For instance, in the case of *Melanoma*, the discovered concept of dark corner artifacts is a bias inherent to ISIC, previously highlighted in the literature as well [32]. Other dimensions potentially associated with dataset biases include redness, which tends to increase the prediction confidence of *Nevus*, image brightening, which enhances the likelihood of predicting *Basal Cell Carcinoma*, and image darkening, which promotes the probability of *Melanoma* prediction. Upon nearer investigation of the dataset, it can be observed that these phenomena also reflect in the dataset's statistics (see supplementary material A.1).

Apart from biases, CDCT has also yielded concepts that can be clinically proven. One dimension that led to a particularly high increase in *Dermatofibroma* prediction probability resulted in the introduction of white structures to the center of the lesion. This concept has been already documented in medical literature as a white, scar-like structure [2] that is highly indicative of the diagnosis, but not yet proven in DL-based classifiers.

6 Limitations & Future Work

Although CDCT was successfully used to discover previously unknown concepts in this work, some limitations remain. Despite the automated selection of classifier-relevant dimensions, a persisting challenge in concept discovery is the manual interpretation of concepts that often requires the help of domain experts. Future works might address this limitation by providing more sophisticated analysis and visualization methods that segment, aggregate, and cluster changes induced by the manipulation of concept dimensions. Moreover, the exploration of new concepts is drastically hampered by the existence of significant biases in the dataset. The linear process of concept discovery might have to be replaced in practice by an iterative approach that involves a debiasing step, which would facilitate the discovery of meaningful concepts and improve the classifier's validity. Finally, the utilization of the VAE architecture for deriving a disentangled representation of counterfactual trajectories might be further improved in the future. The VAE training presented a crucial trade-off between reconstruction fidelity and disentanglement through the weight of the KLD loss. As skin lesion classification relies on fine-grained features like subtle textures, reconstruction was favored over disentanglement. This, however, led to one entangled dimension that stood out for its comparatively high relevance throughout all classes, introducing inconclusive changes to the images. Future work should address this issue by applying more sophisticated disentanglement techniques with better fine-grained reconstruction capability, such as StyleGAN.

7 Conclusion

In high-stakes scenarios such as healthcare, DL models play a pivotal role in early disease detection, potentially saving lives. It is essential to provide human-aligned explanations which offer insights into the complex decision-making of models to gain the trust of their users. These insights can potentially reveal new biomarkers relevant in clinical practice. This study provides a new automated framework for the discovery of concepts based on the synthesis of counterfactual trajectories using LDMs.

The proposed CDCT framework is the first to use LDMs with classifier guidance to generate counterfactual explanations. Its proposed counterfactual generation step yields better FID scores compared to the previous DiME [13] method while being up to 12 times more resource-efficient. A counterfactual trajectory dataset is constructed, reflecting relevant semantic changes along the decision boundaries of the target classifier. A disentangled representation of these classifier-relevant cues is derived using a VAE. The automatic and unsupervised exploration of this latent representation yielded valuable insights into the decision-making behavior of a skin lesion classifier, revealing not only biases but also previously unknown biomarkers, supported by first evidence in the medical literature. CDCT can be applied to arbitrary application domains such as radiology and histology, providing a combination of local and global explanations, and therefore paving the way to trustworthy AI finding its way into clinical practice.

Acknowledgements. The project was funded by the Federal Ministry for Education and Research (BMBF) with the grant number 03ZU1202JA.

References

1. Achtibat, R., Dreyer, M., Eisenbraun, I., Bosse, S., Wiegand, T., Samek, W., Lapuschkin, S.: From attribution maps to human-understandable explanations through concept relevance propagation. Nature Machine Intelligence **5**(9), 1006–1019 (2023)

2. Agero, A.L.C., Taliercio, S., Dusza, S.W., Salaro, C., Chu, P., Marghoob, A.A.: Conventional and polarized dermoscopy features of dermatofibroma. Arch. Dermatol. **142**(11), 1431–1437 (2006)
3. Atad, M., Dmytrenko, V., Li, Y., Zhang, X., Keicher, M., Kirschke, J., Wiestler, B., Khakzar, A., Navab, N.: Chexplaining in style: Counterfactual explanations for chest x-rays using stylegan. arXiv preprint arXiv:2207.07553 (2022)
4. Augustin, M., Boreiko, V., Croce, F., Hein, M.: Diffusion visual counterfactual explanations. Adv. Neural. Inf. Process. Syst. **35**, 364–377 (2022)
5. Cassidy, B., Kendrick, C., Brodzicki, A., Jaworek-Korjakowska, J., Yap, M.H.: Analysis of the isic image datasets: Usage, benchmarks and recommendations. Med. Image Anal. **75**, 102305 (2022)
6. Chen, R.T., Li, X., Grosse, R.B., Duvenaud, D.K.: Isolating sources of disentanglement in variational autoencoders. Advances in neural information processing systems **31** (2018)
7. Dandl, S., Molnar, C., Binder, M., Bischl, B.: Multi-objective counterfactual explanations. In: International Conference on Parallel Problem Solving from Nature. pp. 448–469. Springer (2020)
8. Espinosa Zarlenga, M., Barbiero, P., Ciravegna, G., Marra, G., Giannini, F., Diligenti, M., Shams, Z., Precioso, F., Melacci, S., Weller, A., et al.: Concept embedding models: Beyond the accuracy-explainability trade-off. Adv. Neural. Inf. Process. Syst. **35**, 21400–21413 (2022)
9. Fel, T., Picard, A., Bethune, L., Boissin, T., Vigouroux, D., Colin, J., Cadène, R., Serre, T.: Craft: Concept recursive activation factorization for explainability. In: Proceedings of the IEEE/CVF Conference on Computer Vision and Pattern Recognition. pp. 2711–2721 (2023)
10. Ghandeharioun, A., Kim, B., Li, C.L., Jou, B., Eoff, B., Picard, R.W.: Dissect: Disentangled simultaneous explanations via concept traversals. arXiv preprint arXiv:2105.15164 (2021)
11. Ghorbani, A., Wexler, J., Zou, J.Y., Kim, B.: Towards automatic concept-based explanations. Advances in neural information processing systems **32** (2019)
12. He, K., Zhang, X., Ren, S., Sun, J.: Deep residual learning for image recognition. In: Proceedings of the IEEE conference on computer vision and pattern recognition. pp. 770–778 (2016)
13. Jeanneret, G., Simon, L., Jurie, F.: Diffusion models for counterfactual explanations. In: Proceedings of the Asian Conference on Computer Vision. pp. 858–876 (2022)
14. Jeanneret, G., Simon, L., Jurie, F.: Adversarial counterfactual visual explanations. In: Proceedings of the IEEE/CVF Conference on Computer Vision and Pattern Recognition. pp. 16425–16435 (2023)
15. Kawahara, J., Daneshvar, S., Argenziano, G., Hamarneh, G.: Seven-point checklist and skin lesion classification using multitask multimodal neural nets. IEEE J. Biomed. Health Inform. **23**(2), 538–546 (2018)
16. Kim, B., Wattenberg, M., Gilmer, J., Cai, C., Wexler, J., Viegas, F., et al.: Interpretability beyond feature attribution: Quantitative testing with concept activation vectors (tcav). In: International conference on machine learning. pp. 2668–2677. PMLR (2018)
17. Kingma, D.P., Welling, M.: Auto-encoding variational bayes. arXiv preprint arXiv:1312.6114 (2013)
18. Koh, P.W., Nguyen, T., Tang, Y.S., Mussmann, S., Pierson, E., Kim, B., Liang, P.: Concept bottleneck models. In: International conference on machine learning. pp. 5338–5348. PMLR (2020)

19. Lang, O., Gandelsman, Y., Yarom, M., Wald, Y., Elidan, G., Hassidim, A., Freeman, W.T., Isola, P., Globerson, A., Irani, M., et al.: Explaining in style: Training a gan to explain a classifier in stylespace. In: Proceedings of the IEEE/CVF International Conference on Computer Vision. pp. 693–702 (2021)
20. Liu, L., Ren, Y., Lin, Z., Zhao, Z.: Pseudo numerical methods for diffusion models on manifolds. arXiv preprint arXiv:2202.09778 (2022)
21. Lucieri, A., Bajwa, M.N., Dengel, A., Ahmed, S.: Explaining ai-based decision support systems using concept localization maps. In: International Conference on Neural Information Processing. pp. 185–193. Springer (2020)
22. Mendonça, T., Ferreira, P.M., Marques, J.S., Marcal, A.R., Rozeira, J.: Ph 2-a dermoscopic image database for research and benchmarking. In: 2013 35th annual international conference of the IEEE engineering in medicine and biology society (EMBC). pp. 5437–5440. IEEE (2013)
23. Palacio, S., Lucieri, A., Munir, M., Ahmed, S., Hees, J., Dengel, A.: Xai handbook: towards a unified framework for explainable ai. In: Proceedings of the IEEE/CVF International Conference on Computer Vision. pp. 3766–3775 (2021)
24. Poeta, E., Ciravegna, G., Pastor, E., Cerquitelli, T., Baralis, E.: Concept-based explainable artificial intelligence: A survey. arXiv preprint arXiv:2312.12936 (2023)
25. Ranjan, R., Sankaranarayanan, S., Bansal, A., Bodla, N., Chen, J.C., Patel, V.M., Castillo, C.D., Chellappa, R.: Deep learning for understanding faces: Machines may be just as good, or better, than humans. IEEE Signal Process. Mag. **35**(1), 66–83 (2018)
26. Rank, N., Pfahringer, B., Kempfert, J., Stamm, C., Kühne, T., Schoenrath, F., Falk, V., Eickhoff, C., Meyer, A.: Deep-learning-based real-time prediction of acute kidney injury outperforms human predictive performance. NPJ digital medicine **3**(1), 139 (2020)
27. Ribeiro, M.T., Singh, S., Guestrin, C.: "why should i trust you?" explaining the predictions of any classifier. In: Proceedings of the 22nd ACM SIGKDD international conference on knowledge discovery and data mining. pp. 1135–1144 (2016)
28. Rombach, R., Blattmann, A., Lorenz, D., Esser, P., Ommer, B.: High-resolution image synthesis with latent diffusion models. In: Proceedings of the IEEE/CVF conference on computer vision and pattern recognition. pp. 10684–10695 (2022)
29. Sanchez, P., Kascenas, A., Liu, X., O'Neil, A.Q., Tsaftaris, S.A.: What is healthy? generative counterfactual diffusion for lesion localization. In: MICCAI Workshop on Deep Generative Models. pp. 34–44. Springer (2022)
30. Sanchez, P., Tsaftaris, S.A.: Diffusion causal models for counterfactual estimation. In: Conference on Causal Learning and Reasoning. pp. 647–668. PMLR (2022)
31. Selvaraju, R.R., Cogswell, M., Das, A., Vedantam, R., Parikh, D., Batra, D.: Gradcam: Visual explanations from deep networks via gradient-based localization. In: Proceedings of the IEEE international conference on computer vision. pp. 618–626 (2017)
32. Sies, K., Winkler, J.K., Fink, C., Bardehle, F., Toberer, F., Kommoss, F.K., Buhl, T., Enk, A., Rosenberger, A., Haenssle, H.A.: Dark corner artefact and diagnostic performance of a market-approved neural network for skin cancer classification. JDDG: Journal der Deutschen Dermatologischen Gesellschaft **19**(6), 842–850 (2021)
33. Song, Y., Shyn, S.K., Kim, K.s.: Img2tab: Automatic class relevant concept discovery from stylegan features for explainable image classification. arXiv preprint arXiv:2301.06324 (2023)

34. Sturman, O., von Ziegler, L., Schläppi, C., Akyol, F., Privitera, M., Slominski, D., Grimm, C., Thieren, L., Zerbi, V., Grewe, B., et al.: Deep learning-based behavioral analysis reaches human accuracy and is capable of outperforming commercial solutions. Neuropsychopharmacology **45**(11), 1942–1952 (2020)
35. Van Looveren, A., Klaise, J.: Interpretable counterfactual explanations guided by prototypes. In: Joint European Conference on Machine Learning and Knowledge Discovery in Databases. pp. 650–665. Springer (2021)
36. Veale, M., Zuiderveen Borgesius, F.: Demystifying the draft eu artificial intelligence act-analysing the good, the bad, and the unclear elements of the proposed approach. Computer Law Review International **22**(4), 97–112 (2021)
37. Vielhaben, J., Bluecher, S., Strodthoff, N.: Multi-dimensional concept discovery (mcd): A unifying framework with completeness guarantees. arXiv preprint arXiv:2301.11911 (2023)
38. Voigt, P., Von dem Bussche, A.: The eu general data protection regulation (gdpr). A Practical Guide, 1st Ed., Cham: Springer International Publishing **10**(3152676), 10–5555 (2017)
39. Wachter, S., Mittelstadt, B., Russell, C.: Counterfactual explanations without opening the black box: Automated decisions and the gdpr. Harv. JL & Tech. **31**, 841 (2017)
40. Wang, B., Li, L., Nakashima, Y., Nagahara, H.: Learning bottleneck concepts in image classification. In: Proceedings of the IEEE/CVF Conference on Computer Vision and Pattern Recognition. pp. 10962–10971 (2023)
41. Zhang, R., Madumal, P., Miller, T., Ehinger, K.A., Rubinstein, B.I.: Invertible concept-based explanations for cnn models with non-negative concept activation vectors. In: Proceedings of the AAAI Conference on Artificial Intelligence. vol. 35, pp. 11682–11690 (2021)
42. Zhao, H., Gallo, O., Frosio, I., Kautz, J.: Loss functions for neural networks for image processing. arXiv preprint arXiv:1511.08861 (2015)
43. Zhou, B., Khosla, A., Lapedriza, A., Oliva, A., Torralba, A.: Learning deep features for discriminative localization. In: Proceedings of the IEEE conference on computer vision and pattern recognition. pp. 2921–2929 (2016)

Fusing Forces: Deep-Human-Guided Refinement of Segmentation Masks

Rafael Sterzinger[✉], Christian Stippel, and Robert Sablatnig

Computer Vision Lab, TU Wien, Vienna, Austria
{rafael.sterzinger,christian.stippel,robert.sablatnig}@tuwien.ac.at

Abstract. Etruscan mirrors constitute a significant category in Etruscan art, characterized by elaborate figurative illustrations featured on their backside. A laborious and costly aspect of their analysis and documentation is the task of manually tracing these illustrations. In previous work, a methodology has been proposed to automate this process, involving photometric-stereo scanning in combination with deep neural networks. While achieving quantitative performance akin to an expert annotator, some results still lack qualitative precision and, thus, require annotators for inspection and potential correction, maintaining resource intensity. In response, we propose a deep neural network trained to interactively refine existing annotations based on human guidance. Our human-in-the-loop approach streamlines annotation, achieving equal quality with up to 75% less manual input required. Moreover, during the refinement process, the relative improvement of our methodology over pure manual labeling reaches peak values of up to 26%, attaining drastically better quality quicker. By being tailored to the complex task of segmenting intricate lines, specifically distinguishing it from previous methods, our approach offers drastic improvements in efficacy, transferable to a broad spectrum of applications beyond Etruscan mirrors.

Keywords: Binarization · Interactive Segmentation · Human-in-the-Loop · Etruscan Art · Cultural Heritage

1 Introduction

With more than 3,000 identified specimens, Etruscan hand mirrors represent one of the biggest categories within Etruscan art. On the front, these ancient artworks feature a highly polished surface, whereas, on the back, they typically depict engraved and/or chased figurative illustrations of Greek mythology [5]. A primary component of their examination involves the labor- and cost-intensive task of manually tracing the artworks; an exemplary mirror is illustrated in Fig. 2 together with the sought-after tracing.

This project has received funding from the Austrian Science Fund/Österreichischer Wissenschaftsfonds (FWF) under grant agreement No. P 33721

© The Author(s), under exclusive license to Springer Nature Switzerland AG 2025
A. Antonacopoulos et al. (Eds.): ICPR 2024, LNCS 15312, pp. 154–169, 2025.
https://doi.org/10.1007/978-3-031-78198-8_11

(a) Initial Prediction **Y**   (b) Human Interaction **Δ**   (c) Refined Prediction **Y'**

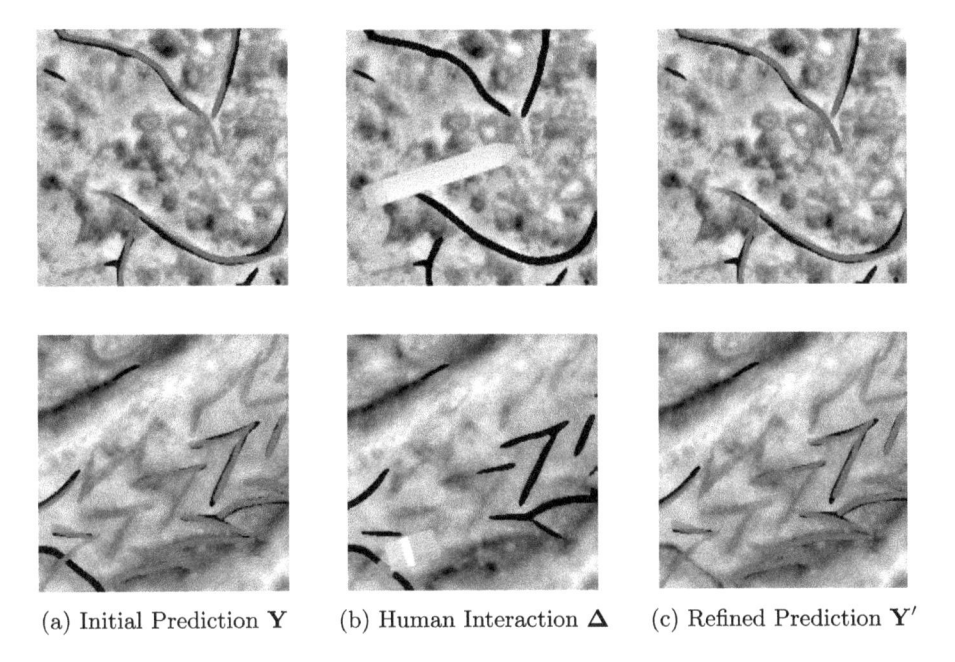

Fig. 1. Illustrating interactive refinement of segmentation masks: Starting from an initial segmentation **Y**, the user can add ($\mathbf{\Delta}^+$) or erase ($\mathbf{\Delta}^-$) parts to bring it closer to the ground truth $\mathbf{Y}^*$ (in blue), creating an updated mask $\mathbf{Y}^\Delta$. Next, using a separate model conditioned on the human input **Δ** and **Y**, we aim that for the refined segmentation **Y'** it holds that $||\mathbf{Y'} - \mathbf{Y}^*||_1 < ||\mathbf{Y}^\Delta - \mathbf{Y}^*||_1$.

In previous works, Sterzinger et al. [13] propose a methodology to automate the segmentation process through photometric-stereo scanning in combination with deep learning; expediting the process of manual tracing and contributing to increased objectivity. Although their segmentation model – trained on depth maps of Etruscan mirrors to recognize intentional lines over scratches – already quantitatively achieves performance on par with an expert annotator, in some instances it lags behind. Based on this, manual inspection and potential refinement by humans are still required, therefore, although alleviated, the tracing remains resource-intensive.

In this paper, we continue their line of work and propose a methodology to simplify the remaining required refinement by adding interactivity to the process: Starting from an initial prediction, we aim to reach qualitatively satisfying results as quickly as possible while keeping necessary labor to a minimum. We achieve this by training a deep neural network to refine the initial segmentation based on a series of hints, i.e., parts being added or erased, illustrated in Fig. 1.

In summary, our contribution entails the development of an interactive refinement network for improved annotation results obtained in less time, requiring less labor. Compared to refining the initial segmentation manually, fusing forces and performing the refinement interactively offers not only a drastic reduction in labor (up to -75%) but also expedites the process by attaining significant relative performance improvements over manual labeling (up to +26%). We differentiate ourselves from prior work by proposing a methodology tailored specifically to the task of segmenting intricate lines scattered across, in our case, Etruscan mirrors versus, e.g., segmenting locally-concentrated hepatic lesions [1]. Additionally, instead of starting from scratch, we start from an initial prediction, a step required due to the non-locality of lines as otherwise labor would be drastically higher.

Finally, we provide public access to both the code and data utilized in this work (see github.com/RafaelSterzinger/etmira-interaction) to promote transparency and reproducibility.

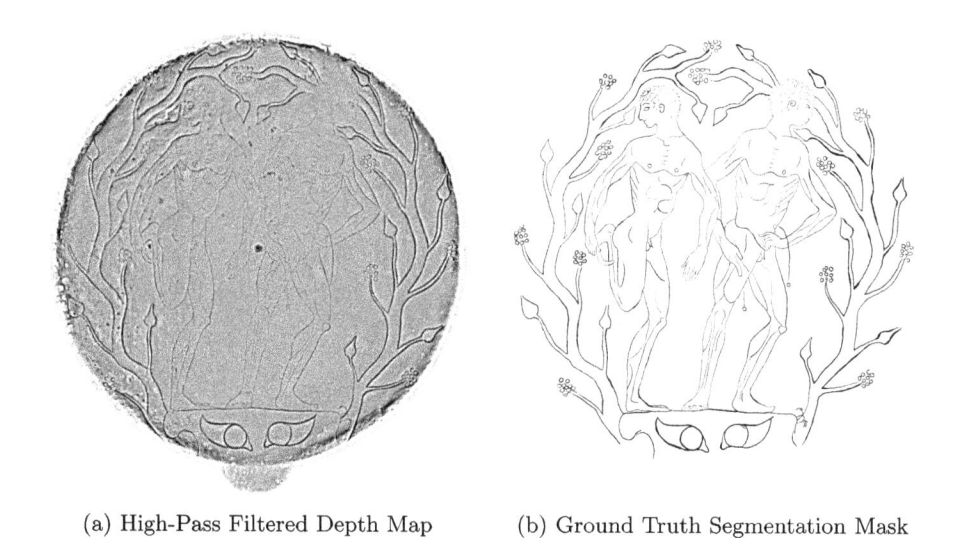

(a) High-Pass Filtered Depth Map   (b) Ground Truth Segmentation Mask

Fig. 2. Etruscan mirrors typically feature scenes from Greek mythology. During their examination, archaeologists seek to extract the drawings for visualization.

2 Related Work

Segmentation: In the field of image segmentation, techniques are led by advanced deep learning architectures such as the UNet [12], DeepLabV3++ [2], Pyramid Attention Network [7], etc. These advancements are particularly propelled by industries where precise segmentation is paramount: For example, in medical imaging, intricate segmentations are crucial for identifying vascular structures within the retina, a crucial aspect for diagnosing retinal diseases [8].

Photometric Stereo: When considering historical artifacts where the content of interest is engraved or chased into the object, as is the case with Etruscan mirrors, instead of RGB, modalities that capture surface details are potentially better suited. Photometric Stereo (PS), a technique introduced by Woodham [17], allows for capturing such details, providing insights into the surface geometry of an object. For instance, McGunnigle and Chantler [10] extract handwriting on paper based on depth profiles. In addition to this, PS is also employed, e.g., to detect cracks in steel surfaces [6], extract leaf venation [20], or detect air voids in concrete [16]. In the context of Etruscan mirrors, Sterzinger et al. [13] resort to a deep-learning-based segmentation approach due to the damage these mirrors have sustained. By integrating PS-scanning with deep segmentation, they learn to recognize intentional lines over scratches.

Interactive Segmentation: Independent of the segmentation methodology employed, resulting masks might not meet performance requirements and, therefore, require correction. With this regard, Li et al. [8], introduce IterNet, a UNet-based iterative approach to enforce connectivity of retinal vessels post segmentation, requiring no-external input. Similarly, interactive methods exist that incorporate human expertise within the process. Xu et al. [18] and Mahadevan et al. [9] focus on object segmentation based on mouse clicks. On the other hand and closest to our work, Amrehn et al. [1], propose an approach that refines the segmentation based on pictorial scribbles for hepatic lesions.

3 Methodology

In the following we will detail our methodology comprised of:

- the dataset; general information, splitting the data into training, validation, and testing, as well as, the preprocessing of depth maps
- the simulation of human interaction; details on the statistics of engravings, acquiring individual line segments, and the procedure for adding and erasing
- the architecture; describing the overall deep neural network used for refining the initial segmentation

3.1 Dataset

Our dataset includes a diverse array of Etruscan mirrors from public collections in Austria. It consists of PS-scans of 59 mirrors, with 53 located at the Kunsthistorischen Museum (KHM) Wien and the remaining 6 scattered throughout Austria. Annotations were acquired for 19 mirrors, encompassing 19 backsides and 10 fronts, resulting in a total of 29 annotated examples. Notably, engravings predominantly adorn the backside to avoid interference with reflectance, however, they are also occasionally found on the front, albeit with less density, near the handle or around the border. For information on the acquisition process, we refer the reader to Sterzinger et al. [13].

Dividing these annotations into training, validation, and test sets is challenging due to three factors: limited sample size, strong variations in the density of engravings, and overall mirror conditions. Mirrors with dense engravings are prioritized for training due to the stronger learning signal they offer. To ensure fair evaluation, we select three mirrors of different conditions and engraving densities for testing: one from Wels and two from the KHM Wien. We create non-overlapping patches of size 512×512 pixels, shuffle, and split them in half to form the validation and test set of similar underlying distributions. One outlier, characterized by a different art style (points instead of lines), is excluded, leaving 25 annotated samples for training.

Preprocessing With regards to preprocessing, we employ the depth modality (which worked best according to [13]) and remove low frequencies. We accomplish this by subtracting a Gaussian-filtered version of the depth map with values capped between $\mu \pm 3\sigma$. In addition, employing the Segment Anything Model (SAM) [4], global segmentation masks are generated to identify the mirror object within a shot. We use these masks to differentiate between mirror and non-mirror parts (see Fig. 4; compare red versus green, top-left), for instance, to calculate per-channel means and standard deviations only on mirror parts which we use to normalize the input.

Addressing the lack of annotations, a per-patch inference approach is adopted. For validation and testing, non-overlapping quadratic patches measuring 512×512 pixels are extracted. Regarding our training data, we pad four pixels, since $6720 \equiv 0 \pmod{2240}$, to the original resolution ($8,964 \times 6,716$ pixels) to extract 25 overlapping tiles of size $2,988 \times 2,240$ pixels using a stride of half the size; tiles, containing no annotation, are discarded. Diversifying the dataset for each epoch, ten patches per tile are extracted, all resized to dimensions of 256×256 pixels to streamline model complexity.

3.2 Simulation of Human Interaction

In order to simulate realistic human interactions, we first look into the statistics of the annotations included in the dataset; necessary to quantitatively capture human-stroke width. Next, to refine initial predictions, we describe the process of filtering and correcting false positives and negatives. Within this, we motivate and denote the algorithm used to extract line segments. Tying all components together, we finally describe simulating human interaction: Starting from either false positives or negatives, we extract the largest error segment and provide a hint in the form of a line with width taken from the acquired statistics of the ground truth annotations.

Statistics With the goal of simulating realistic interaction, one crucial component to consider is the stroke width. For this, we look into the statistics of the annotations included in our dataset by extracting individual thickness, using Algorithm 1: Starting from a binary mask, the ground truth $\mathbf{Y}^*$ in our case, we obtain distance information via the `euclidean_distance_transform` which, for each pixel, returns the Euclidean distance in pixels to the closest non-mask

Algorithm 1: Calculating Stroke Widths for Statistics

Data: ground truth $\mathbf{Y}^*$

def *get_stroke_widths*($\mathbf{Y}^*$):
    distance_map $\leftarrow$ euclidean_distance_transform($\mathbf{Y}^*$)
    gt_skelet $\leftarrow$ skeletonize($\mathbf{Y}^*$)
    return distance_map[gt_skelet]

pixel. Next, employing `skeletonize` [19], we acquire a skeletonized version of the input (essential the center of lines), used to extract the thickness at each section.

From this information, we calculate initial μ and σ of the collected line widths, which we use to remove outliers (long right tail) using the two-sigma rule, keeping values within two standard deviations, to obtain final $\mu = 6.19$ and $\sigma = 1.49$. Based on this filtered set of stroke widths, we fit a Gamma distribution $\mathcal{G}$ from which we can randomly sample realistic widths. Fig. 3 visualizes the distribution of stroke widths as well as the fitted distribution $\mathcal{G}$.

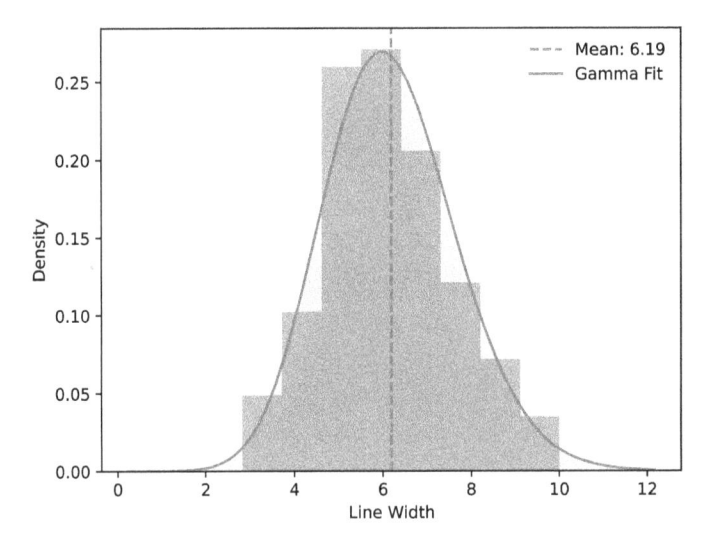

Fig. 3. Illustration of the distribution of stroke widths: After removing outliers from our data, using the two-sigma rule, we fit a Gamma distribution (shape-parameter $a = 49.13$, loc$= -4.28$, scale$= 0.21$).

Operations In general, when an expert annotator is entrusted with the task of refining segmentation masks, one of two operations will be performed: adding missing parts or erasing superfluous ones. Simulating these operations consists of multiple steps: (1) finding an area that requires correction (we assume areas will be selected in decreasing order depending on the magnitude of correction required), (2) deciding on an operation, and (3) performing the operation. In essence, however, steps (1) and (2) go hand in hand, i.e., when deciding on an area based on error, the operation to be performed is already clear.

Let $\mathbf{Y} = f_{init}(\mathbf{X}) \in \mathbb{B}^{H,W}$ be the initial segmentation mask produced by a baseline network f_{init} based on the depth map $\mathbf{X} \in \mathbb{R}^{H,W}$. Given that we generally work with patches, using this initial mask, we find the segment that requires the most correction w.r.t. our ground truth $\mathbf{Y}^*$, considering the pFM (formally introduced in Section 4).

Algorithm 2: False Positive/Negative Detection for $\mathbf{\Delta}^-/\mathbf{\Delta}^+$

Data: ground truth $\mathbf{Y}^*$, prediction $\mathbf{Y}$, line statistics μ and σ

def *get_add*($\mathbf{Y}^*$, $\mathbf{Y}$):
  gt_skelet $\leftarrow$ skeletonize($\mathbf{Y}^*$)
  false_negatives $\leftarrow$ gt_skelet $\wedge \neg \mathbf{Y}$
  return false_negatives

def *get_erase*($\mathbf{Y}^*$, $\mathbf{Y}$, μ, σ):
  // dilate gt for more lenient detection
  expanded_gt $\leftarrow$ dilate(skeletonize($\mathbf{Y}^*$), round($\mu + 2\sigma$))
  pred_skelet $\leftarrow$ skeletonize($\mathbf{Y}$)
  false_positives $\leftarrow \neg$expanded_gt $\wedge$ pred_skelet
  return false_positives

Next, for the remaining steps, we propose Algorithms 2 and 3: For (2), we first employ Algorithm 2 to obtain a binary mask of missing or superfluous skeletonized segments, i.e., false positives or negatives. Note that in order to avoid the correction of minor superfluous parts, in get_erase, we dilate the ground truth to a constant of $\mu + 2\sigma$ s.t. only false positives which drastically diverge from $\mathbf{Y}^*$ will be detected.

After obtaining skeletonized binary masks for false positives and negatives, for step (3), we obtain, for both, the longest line segment utilizing Algorithm 3. Within Algorithm 3, we leverage a key property of skeletonizing: In a pixel-based, skeletonized representation (i.e., one where lines have been reduced to their medial axis, which is 1-pixel wide), a single continuous line, will have exactly two neighbors in its 8-neighborhood, except for endpoints and junctions.

Let $\mathbf{\Delta}^+ \in \{0, +1\}^{H,W}$ and $\mathbf{\Delta}^- \in \{0, -1\}^{H,W}$ denote the missing/superfluous line segment that will be added/erased. Since these operations will be performed interactively, we summarize with $\mathbf{\Delta}$ multiple interactions and thus contains values $\{-1, 0, +1\}$. Finally, we combine previous interactions $\mathbf{\Delta}$ with $\mathbf{\Delta}^+$ or $\mathbf{\Delta}^-$

Algorithm 3: Obtain Edge Segments, Sorted by Length

Data: skeletonized mask **S**

Let $K_{\text{edge}} = \begin{bmatrix} 1 & 1 & 1 \\ 1 & 10 & 1 \\ 1 & 1 & 1 \end{bmatrix}$ and $K_{\text{label}} = \begin{bmatrix} 1 & 1 & 1 \\ 1 & 1 & 1 \\ 1 & 1 & 1 \end{bmatrix}$.

def *get_edges*(**S**):
    conv_skelet ← convolve(**S**, kernel=K_{edge})
    edges ← conv_skelet == 12
    // 8-connectivity
    edge_list ← label_connectivity(edges, kernel=K_{label})
    edge_list ← sort(edge_list, ord='desc')
    return edge_list

---

by leaving previously set values of ±1 fixed, only updating 0-valued values. For simplicity, we introduce $\mathbf{Y}^{\Delta}$, a quantity which denotes the union between the initial prediction $\mathbf{Y}$ and the human interactions $\boldsymbol{\Delta}$, i.e.:

$$\mathbf{Y}^{\Delta}_{i,j} = \begin{cases} 1 & \text{if } \boldsymbol{\Delta}_{i,j} == +1 \\ 0 & \text{if } \boldsymbol{\Delta}_{i,j} == -1 \\ \mathbf{Y}_{i,j} & \text{otherwise.} \end{cases} \quad (1)$$

Interaction After introducing the three necessary steps for the adding/erasing operation, we move on to performing realistic human interactions: We continue from the previously found quantities $\boldsymbol{\Delta}^{+}$ or $\boldsymbol{\Delta}^{-}$ for adding missing/erasing superfluous segments and either pick one of the two at random during training or the longer segment for maximum correction during inference. Within the skeletonized segment, we proceed by randomly sampling a sub-segment of up to eleven pixels (a parameter that we did not vary) and dilating it, based on the statistics of $\mathcal{G}$, with one of the following: (a) a width sampled from the distribution, (b) the mean μ, or (c) a width of $\mu - 2\sigma$.

In Section 4, options (a) and (b) will be evaluated w.r.t. validation performance, and option (c) will be used for the final evaluation on whole mirrors s.t. human interactions are with high probability aligned with $\mathbf{Y}^{*}$, i.e., reduce the risk of strokes being too wide.

Finally, using a separate network f_{iter}, trained to refine $\mathbf{Y}$ conditioned on $\boldsymbol{\Delta}$ and $\mathbf{X}$, we obtain a refined prediction $\mathbf{Y}'$. With this, we motivate the interactivity of our method: Starting over, i.e., $\mathbf{Y} \leftarrow \mathbf{Y}'$, we again find the segment that requires the most correction and update $\boldsymbol{\Delta}$ with newly found $\boldsymbol{\Delta}^{+}/\boldsymbol{\Delta}^{-}$. A general overview of the interactivity is provided by Fig. 4, illustrating inference on a per-patch level, the initial prediction $\mathbf{Y}$ and its refinement over time, based on $\boldsymbol{\Delta}$.

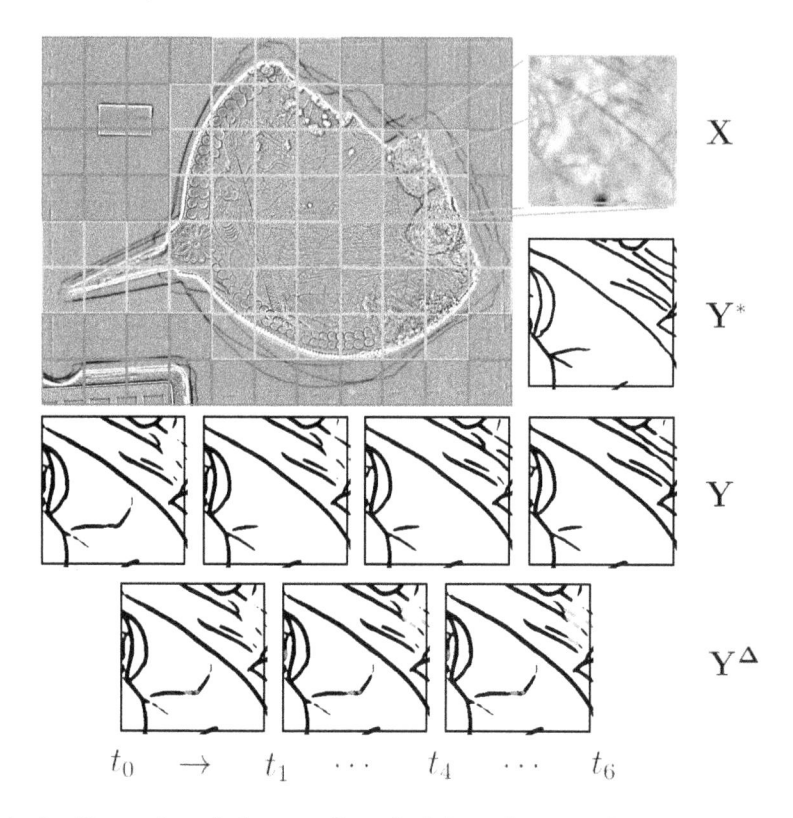

Fig. 4. An illustration of the overall methodology: In general, segmentation is performed on a per-patch level (512 × 512, resized to 256 × 256; red denotes patches that are filtered a priori using SAM [4]). In an interactive paradigm, starting from the initial prediction **Y** at timestep t_0, based on input **X**, a human provides hints in the form of **Δ** (the "union" between **Y** and **Δ** is denoted with $\mathbf{Y^\Delta}$), on which a separately trained network f_{iter} is conditioned on to produce a refined mask at timestep t_1.

3.3 Architecture

With regards to our architecture, we employ a UNet [12] with an EfficientNet-B6 [15] following the proposal by Sterzinger et al. [13] but expand upon the input to condition the network on the (simulated) human input **Δ**. For clarification, the input is now comprised of a $3 \times H \times W$ tensor, including the depth map **X**, the human input **Δ**, as well as the initial prediction **Y** with all three quantities concatenated. Given that our data resources are limited, we train on a per-patch-level employing augmentations among which are rotations, flips, and shifts, optimizing the Dice loss. For the initial prediction **Y**, we employ the exact same methodology as proposed by Sterzinger et al. [13].

4 Evaluation

In this section, we evaluate our design choices: During this process, we report the Intersection-over-Union (IoU) as well as the pseudo-F-Measure (pFM), a metric commonly used for evaluating the binarization quality of handwritten documents. It is thus well-suited for our binarization task, i.e., a task where shifting the mask by a single pixel will have a significant impact on per-pixel metrics. Compared to the standard F-Measure, the pFM relies on the pseudo-Recall (p-Recall) which is calculated based on the skeleton of $\mathbf{Y}^*$ [11]:

$$\text{pFM}(\mathbf{Y}', \mathbf{Y}^*) = \frac{2 \times \text{p-Recall}(\mathbf{Y}', \mathbf{Y}^*) \times \text{Precision}(\mathbf{Y}', \mathbf{Y}^*)}{\text{p-Recall}(\mathbf{Y}', \mathbf{Y}^*) + \text{Precision}(\mathbf{Y}', \mathbf{Y}^*)} \quad (2)$$

Given that we work within an interactive paradigm, we are required to also provide a metric that excludes the human input $\mathbf{\Delta}$ from the evaluation and hence report the relative pFM improvement over $\mathbf{Y}^{\mathbf{\Delta}}$, i.e.:

$$\text{pFM}_{\mathbf{\Delta}}(\mathbf{Y}', \mathbf{Y}^{\mathbf{\Delta}}, \mathbf{Y}^*) = \frac{\text{pFM}(\mathbf{Y}', \mathbf{Y}^*) - \text{pFM}(\mathbf{Y}^{\mathbf{\Delta}}, \mathbf{Y}^*)}{\text{pFM}(\mathbf{Y}^{\mathbf{\Delta}}, \mathbf{Y}^*)} \quad (3)$$

In addition, based on the fact that during training we introduce randomness, i.e., by chance, missing parts can be added ($\mathbf{\Delta}^+$) or superfluous ones erased ($\mathbf{\Delta}^-$), and that sub-segments are sampled and dilated at random, we evaluate on the test/validation set five times and report the average.

4.1 Training

Our model f_{iter} is trained on an NVIDIA RTX A5000 until convergence, i.e., no improvement $\geq 1e-3$ w.r.t. the pFM$_{\mathbf{\Delta}}$ (see Equation 3) for ten consecutive epochs, using a batch size of 32 and a learning rate of $3e-4$. As a loss function, we employ a generalized Dice overlap (Dice loss) that is well suited for highly unbalanced segmentation masks [14] and optimize it using Adam [3]. Additionally, we incorporate a learning rate scheduler that also monitors the pFM$_{\mathbf{\Delta}}$ on our validation set: If there is no improvement for three consecutive training epochs, the learning rate is halved.

4.2 Ablation Study

In the following, we present our ablation study, focusing on input options, different stroke widths (widths kept fixed and sampled randomly), as well as the necessity of our two operations (add and erase).

Input Options: Starting with the evaluation of different input options and their impact on the predictive patch-wise performance of f_{iter} (fixed stroke width, one interaction; results are denoted in Table 1): As expected, simply iterating over the initial prediction $\mathbf{Y}$ (stemming from network f_{init}) results in no improvement, rendering the human an essential part of the refinement process. Moreover, by means of human guidance, i.e., providing $\mathbf{\Delta}$, the network can effectively

Table 1. Evaluating input options and their effect on the per-patch predictive performance (fixed stroke width, one interaction): Iterating over $\mathbf{Y}$ again does not cause improvement whereas providing $\mathbf{\Delta}$ yields ca. +6% over $\mathbf{Y^\Delta}$. Note that, although part of the input, we hide $\mathbf{X}$ for clarity.

Input Modality		IoU	pFM	pFM$_\Delta$
Init. Prediction [13]	-	32.86	49.28	–
Prediction	$\mathbf{Y}$	32.72	49.27	–
Interaction	$\mathbf{\Delta}$	35.83 ± .1	53.60 ± .2	+5.8 ± .23%
Both	$\mathbf{Y, \Delta}$	36.04 ± .2	53.44 ± .3	+5.5 ± .55%

leverage additional information on missing or superfluous parts, resulting in an increase of around +6% over $\mathbf{Y^\Delta}$. Finally, when utilizing both $\mathbf{Y}$ and $\mathbf{\Delta}$, we attain a comparable improvement over $\mathbf{Y^\Delta}$, with the difference deemed not statistically significant at a confidence level of 95%. However, the latter results in faster convergence, the reason for which we proceed with this option.

Stroke Widths: Next, we consider different options for the stroke width during the simulation of human interaction, namely: (a) keeping the stroke width constant at μ and (b) sampling it from $\mathcal{G}$. Our evaluation reveals that sampling does not significantly improve performance, with results showing +5.5 ± .55% improvement for fixed width versus +4.9 ± .35% for sampled one.

Table 2. Evaluating the impact of adding and erasing when refining mirror ANSA-1700: Employing both operations will result in the highest pFM$_\Delta$ of ca. +12%, where adding has a greater impact (ca. +8%) than erasing (ca. +2%); note that results are reported at the maximum pFM$_\Delta$.

Interaction		IoU	pFM	pFM$_\Delta$
Only Erasing	$\mathbf{\Delta^-}$	38.25 ± .06	58.92 ± .07	+1.9 ± .12%
Only Adding	$\mathbf{\Delta^+}$	55.19 ± .13	73.55 ± .11	+8.4 ± .16%
Both	$\mathbf{\Delta}$	58.41 ± .28	76.56 ± .16	+12.3 ± .17%

Operations: In order to illustrate the necessity of our two operations, namely adding $\mathbf{\Delta^+}$ and erasing $\mathbf{\Delta^-}$, we perform multiple interactions until convergence and report results at the maximum attained pFM$_\Delta$. For this, we inspect an entire mirror, ANSA-1700: Employing both operations jointly yields the highest pFM$_\Delta$ of approximately +12% (redline in Fig. 5). Notably, *add* has a more significant impact (ca. +8%) compared to *erase* (ca. +2%). However, this is very dependent on the initial prediction, thus only demonstrating that one operation supplements the other.

In summary, compared to the initial prediction $\mathbf{Y}$, stemming from f_{init}, providing human guidance via $\mathbf{\Delta}$ will yield improvements exceeding $\mathbf{Y^{\Delta}}$, utilizing both operations is beneficial, and augmenting stroke widths by random sampling performs worse than leaving it constant.

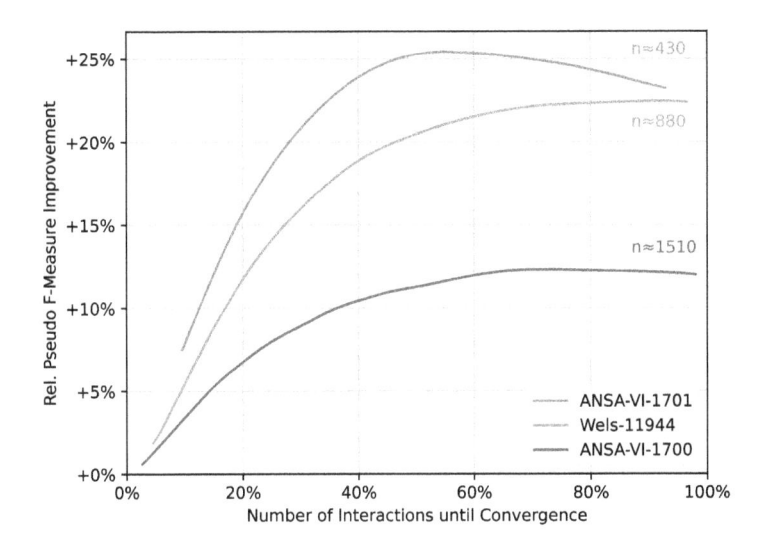

Fig. 5. An illustration of pFM_{Δ}, i.e., the relative pFM improvement of our method over pure manual refinement; n denotes the maximum number of human interactions within ten runs: With the relative improvement peaking at values between ca. $+12\%$ and $+26\%$, our human-in-the-loop approach immediately overtakes manual labeling, leading to drastically better annotations earlier.

5 Results

After verifying the effectiveness of our methodology, we pick the three mirrors from our validation/test set, namely ANSA-1700, ANSA-1701, and Wels-11944, and evaluate our human-in-the-loop approach on whole mirrors, performing multiple interactions (limited to 3,000; typically requiring much less). Again, due to the introduced randomness, we repeat this process ten times and report the average result, skipping the variation as it is negligible. As described in Section 4.2, for this, we start greedily by selecting the patch with the lowest pFM, simulate adding missing/erasing superfluous parts, selecting the operation which yields a larger improvement, refine the prediction based on the additional human input, and proceed from there until *convergence*, i.e., when neither adding nor erasing by itself increases the metric. We report the results of this in two figures: Fig. 5, which illustrates the relative pFM improvement over-performed interactions, as well as Fig. 6, which depicts potential reduction in annotation workload when employing our proposed interactive refinement paradigm.

Inspecting Fig. 5, we observe for all three mirrors significant relative improvements over the purely manual annotation baseline $\mathbf{Y}^\Delta$ when employing our method, with maximum improvements ranging from +12% to +26% depending on the mirror under consideration. Based on these results, we conclude that our human-in-the-loop approach quickly overtakes manual labeling, leading to drastically better annotations at an earlier stage. Interestingly, towards the end, relative improvement starts to decrease slightly before convergence (most notably for ANSA-1701), showcasing that at a point, the network may occasionally undo parts it had previously annotated correctly; human annotations are not impacted by this.

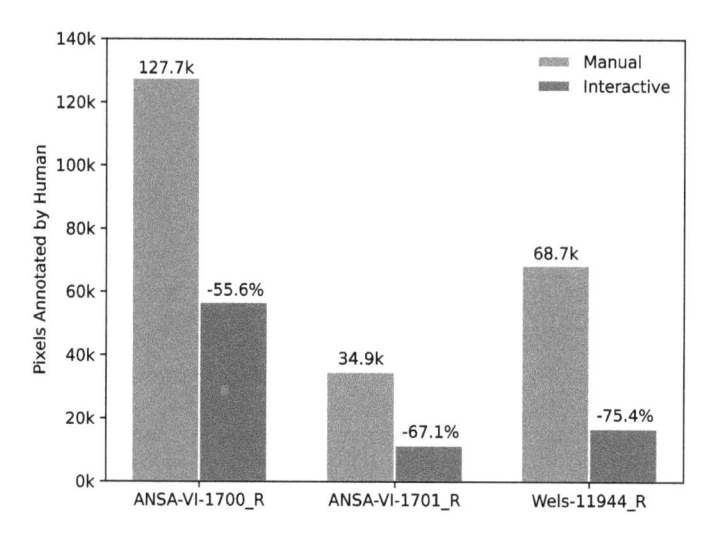

Fig. 6. An illustration of reduced workload: At convergence, our interactive approach requires drastically fewer annotated pixels to reach equal performance in pFM, resulting in a reduction of annotation effort ranging from 56% to 75%.

In Fig. 6, we directly contrast pure manual refinement against our interactive approach. For this, we determine the maximum attained pFM, calculated using $\mathbf{Y}^\Delta$, which corresponds to the final simulated human interaction. We then compare the amount of required human input to the human input necessary to reach equal or higher performance using our proposed method. By doing this, we are able to report a notion of workload reduction: Depending on the mirror under inspection, annotation requirements will experience a reduction ranging from around -56% to -75%, positioning our model well to be employed for simplifying the task of correcting erroneous segmentation masks.

6 Limitations and Future Work

While our proposed method shows promising results, it is important to acknowledge its limitations: At the moment human guidance aids refinement only locally, i.e., modifications happen just in the vicinity of the provided annotation. Moving forward, one could focus on further refining our methodology by exploring additional techniques to enhance efficiency. For instance, it would be meaningful to investigate the integration of quickly trainable learning algorithms, such as Gaussian processes which can immediately be adapted to newly provided annotation and thus allow for global adjustments, potentially further reducing the amount of human input required. Additionally, leveraging Gaussian processes is accompanied by the option of active learning strategies, which could allow the identification and annotation of patches where the model is most uncertain with the chance of expediting refinement further.

7 Conclusion

In summary, our research addresses the labor-intensive process of manually tracing intricate figurative illustrations found, for instance, on ancient Etruscan mirrors. In an attempt to automate this process, previous work has proposed the use of photometric-stereo scanning in conjunction with deep neural networks. By doing so, quantitative performance comparable to expert annotators has been achieved; however, in some instances, they still lack precision, necessitating correction through human labor. In response to the remaining resource intensity, we proposed a human-in-the-loop approach that streamlines the annotation process by training a deep neural network to interactively refine existing annotations based on human guidance. For this, we first developed a methodology to mimic human annotation behavior: We began by analyzing annotation statistics to capture stroke widths accurately and proceeded by introducing algorithms to select erroneous patches, identify false positives and negatives, as well as correct them by erasing superfluous or adding missing parts. Next, we verified our design choices by conducting an ablation study; its results showed that providing human guidance will yield improvements exceeding pure manual annotation, utilizing both operations is beneficial, and augmenting stroke widths by random sampling performs worse than leaving it constant. Finally, we evaluated our method by considering mirrors from our test and validation set. Here, we achieved equal quality annotations with up to 75% less manual input required. Moreover, the relative improvement over pure manual labeling reached peak values of up to 26%, highlighting the efficacy of our approach in reaching drastically better results earlier.

References

1. Amrehn, M., Gaube, S., Unberath, M., Schebesch, F., Horz, T., Strumia, M., Steidl, S., Kowarschik, M., Maier, A.K.: UI-Net: Interactive Artificial Neural Networks for Iterative Image Segmentation Based on a User Model. arXiv preprint arXiv:1709.03450 (2017)
2. Chen, L.C., Zhu, Y., Papandreou, G., Schroff, F., Adam, H.: Encoder-Decoder with Atrous Separable Convolution for Semantic Image Segmentation, p. 833-851. Springer International Publishing (2018)
3. Kingma, D., Ba, J.: Adam: A Method for Stochastic Optimization. In: International Conference on Learning Representations (ICLR). San Diega, CA, USA (2015)
4. Kirillov, A., Mintun, E., Ravi, N., Mao, H., Rolland, C., Gustafson, L., Xiao, T., Whitehead, S., Berg, A.C., Lo, W.Y., Dollar, P., Girshick, R.: Segment Anything. In: Proceedings of the IEEE/CVF International Conference on Computer Vision (ICCV). pp. 4015–4026 (Oct 2023)
5. Kluge, S.: Through the Looking-Glass: Der etruskische Spiegel VI 2627 aus dem Bestand des Kunsthistorischen Museums als Fallbeispiel für die rituelle Manipulation und Defunktionalisierung im Kontext der Grablege in der etruskischen Kultur. In: Weidinger, A., Leskovar, J. (eds.) Interpretierte Eisenzeiten. Fallstudien, Methoden, Theorie. Tagungsbeiträge der 10. Linzer Gespräche zur interpretativen Eisenzeitarchäologie, pp. 241–254. Studien zur Kulturgeschichte von Oberösterreich, Folge 55, Linz (2024)
6. Landstrom, A., Thurley, M.J., Jonsson, H.: Sub-Millimeter Crack Detection in Casted Steel Using Color Photometric Stereo. In: 2013 International Conference on Digital Image Computing: Techniques and Applications (DICTA). IEEE (Nov 2013)
7. Li, H., Xiong, P., An, J., Wang, L.: Pyramid Attention Network for Semantic Segmentation. In: British Machine Vision Conference 2018, BMVC 2018, Newcastle, UK, September 3-6, 2018. p. 285. BMVA Press (2018)
8. Li, L., Verma, M., Nakashima, Y., Nagahara, H., Kawasaki, R.: IterNet: Retinal Image Segmentation Utilizing Structural Redundancy in Vessel Networks. In: 2020 IEEE Winter Conference on Applications of Computer Vision (WACV). IEEE (Mar 2020)
9. Mahadevan, S., Voigtlaender, P., Leibe, B.: Iteratively trained interactive segmentation. In: British Machine Vision Conference 2018, BMVC 2018, Newcastle, UK, September 3-6, 2018. p. 212. BMVA Press (2018)
10. McGunnigle, G., Chantler, M.: Resolving handwriting from background printing using photometric stereo. Pattern Recogn. **36**(8), 1869–1879 (2003)
11. Pratikakis, I., Gatos, B., Ntirogiannis, K.: ICFHR 2012 Competition on Handwritten Document Image Binarization (H-DIBCO 2012). In: 2012 International Conference on Frontiers in Handwriting Recognition. IEEE (Sep 2012)
12. Ronneberger, O., Fischer, P., Brox, T.: U-Net: Convolutional Networks for Biomedical Image Segmentation, p. 234-241. Springer International Publishing (2015)
13. Sterzinger, R., Brenner, S., Sablatnig, R.: Drawing the Line: Deep Segmentation for Extracting Art from Ancient Etruscan Mirrors. In: 2024 ICDAR International Conference on Document Analysis and Recognition, submitted (2024)
14. Sudre, C.H., Li, W., Vercauteren, T., Ourselin, S., Jorge Cardoso, M.: Generalised Dice Overlap as a Deep Learning Loss Function for Highly Unbalanced Segmentations, p. 240-248. Springer International Publishing (2017)

15. Tan, M., Le, Q.V.: EfficientNet: Rethinking Model Scaling for Convolutional Neural Networks. In: Chaudhuri, K., Salakhutdinov, R. (eds.) Proceedings of the 36th International Conference on Machine Learning, ICML 2019, 9-15 June 2019, Long Beach, California, USA. Proceedings of Machine Learning Research, vol. 97, pp. 6105–6114. PMLR (2019)
16. Tao, J., Gong, H., Wang, F., Luo, X., Qiu, X., Huang, Y.: Automated image segmentation of air voids in hardened concrete surface using photometric stereo method. Int. J. Pavement Eng. **23**(14), 5168–5185 (2021)
17. Woodham, R.J.: Photometric method for determining surface orientation from multiple images. Optical Engineering **19**(1) (Feb 1980)
18. Xu, N., Price, B.L., Cohen, S., Yang, J., Huang, T.S.: Deep Interactive Object Selection. In: 2016 IEEE Conference on Computer Vision and Pattern Recognition, CVPR 2016, Las Vegas, NV, USA, June 27-30, 2016. pp. 373–381. IEEE Computer Society (2016)
19. Zhang, T.Y., Suen, C.Y.: A fast parallel algorithm for thinning digital patterns. Commun. ACM **27**(3), 236–239 (1984)
20. Zhang, W., Hansen, M.F., Smith, M., Smith, L., Grieve, B.: Photometric stereo for three-dimensional leaf venation extraction. Comput. Ind. **98**, 56–67 (2018)

Specular Region Detection and Covariant Feature Extraction

D. M. Bappy[1], Donghwa Kang[1], Jinkyu Lee[2], Youngmoon Lee[3], Minsuk Koo[1], and Hyeongboo Baek[4(✉)]

[1] Department of Computer Science and Engineering, Incheon National University, Incheon, Republic of Korea
[2] Department of Computer Science and Engineering, Sungkyunkwan University, Suwon, Republic of Korea
[3] Department of Robotics, Hanyang University, Ansan, Republic of Korea
[4] Department of Artificial Intelligence, University of Seoul, Seoul, Republic of Korea
hbbaek359@gmail.com

Abstract. Endoscopy images pose a distinct set of challenges, such as specularity, uniformity, and deformation, which can obstruct surgeons' observations and decision-making processes. These hurdles complicate feature extraction and may ultimately lead to the failure of a surgical navigation system. To tackle these obstacles, we introduce a Modified Maximal Stable Extremal Region (MMSER) detector that specifically targets fine specular regions. Subsequently, we ingeniously fuse the capabilities of MMSER and saturation region properties to precisely identify specular regions within endoscopy images. Furthermore, our approach harnesses the shared properties of covariant features and endoscopic imaging to detect features in intricate regions, such as low-textured and deformed areas. Surpassing contemporary methods, our proposed technique demonstrates remarkable performance when evaluated on the available CVC-ClinicSpec datasets. Our method has shown improvements in accuracy, recall, f1-score, and Jaccard index by 0.21%, 25.42%, 7.77% snd 11.77%, respectively, when compared to recent techniques. Owing to its exceptional ability to accurately pinpoint specular regions and extract features from complex areas, our approach holds the potential to significantly advance surgical navigation.

Keywords: Endoscopy Imaging · Specular Region · Saturation Region · Feature Extraction · Feature Matching

1 Introduction

Endoscopic imaging systems have revolutionized medical procedures, enabling quicker recovery times and less invasive surgeries compared to traditional methods. Doctors use their experience to estimate spatial relationships and distances within the surgical environment [30]. However, the narrow field of view in endoscopy images often forces surgeons to perform multiple observations to gather information about the same area, which increases the risk and duration of the operation [21]. Additionally, the 2D images lack depth information, making it difficult for doctors to accurately determine the movements of surgical instruments [31]. Therefore, recognizing the 3D structure during the operation is pivotal for doctors and correct feature matching is essential to achieve this. However, endoscopy images present difficulties for feature extraction due to the presence of specular and uniform regions.

Specularity is a constant challenge in endoscopic images, as the angles of the lighting source and camera are nearly identical, causing valuable information like vessels and lesions to be concealed. Specular reflections lead to significant discontinuities, resulting in lost image texture and color information, which hinders the surgeon's observation and judgment [29]. Several existing studies focus on specularity detection in endoscopy imaging. Endoscopic image specularity detection methods can be broadly categorized into those based on different color spaces [12] and those employing classifiers [1].

Oh et al. [24] defined specular reflection areas as absolute bright areas and relative bright areas, determined through outlier detection. However, the detected relative bright areas may include not only specular highlights but also white tissues. Shen et al. [28] transformed endoscopic images into grayscale and detected specular regions using an empirical grayscale threshold, followed by mask region expansion through morphological techniques. However, this method is only suitable for endoscopic images with uniform brightness. Asif et al. [5] employed the Intrinsic Image Layer Separation (IILS) technique to identify specular regions, but this approach misidentifies edges and highly saturated areas as highlights in highly saturated, high-resolution images. Nie et al. [23] suggested a technique for detecting specular regions through brightness classification, enhancement, and thresholding. Although the concept of brightness enhancement is promising, the technique's reliance on different fixed thresholds based on image brightness is not ideal. This approach may fail to detect complex specular regions, such as larger white tissue regions containing specularity.

Extracting features from endoscopy images can be a daunting task, especially when the scene contains specularity, deformation, and low texture [4]. Existing feature extraction methods, such as Scale-Invariant Feature Transform (SIFT) [20], Speeded Up Robust Features (SURF) [8], Oriented Fast and Rotated BRIEF (ORB) [26] and Harris [34] are typically used in 3D reconstruction but are unable to compute enough good feature points from endoscopy images. As such, finding enough good features and correct matching in continuous endoscopy image frames is a critical aspect of recognizing the 3D structure during surgery.

There are only a few works that attempt to extract features in endoscopy imaging. Yan et al. [19] proposed using SIFT for feature extraction and improving the matching process through feature-point pair purification. Although this technique enhances matching performance, it overlooks the fact that having enough available features is crucial for improved matching. In a recent study, Barbed et al. [7] introduced a self-supervised SuperPoint [13] adaptation for the endoscopic domain. However, this learning-based technique has computational complexity, and the adapted model avoids features within specular regions.

Recent advances in specularity removal have yielded promising results. Pan et al. [25] introduced an accelerated adaptive non-convex robust principal component analysis (AANC-RPCA) method that enhances the efficiency and accuracy of highlight removal through adaptive threshold segmentation and quasi-convex function approximation. Zhang et al. [35] developed a partial attention network (PatNet) that employs highlight segmentation and image inpainting, significantly improving the visual quality of endoscopic images. Another innovative approach by Joseph et al. [17] presents a parameter-free matrix decomposition technique that decomposes the original image into a highlight-free pseudo-low-rank component and a highlight component, effectively removing specular reflections and boundary artifacts. These methods demonstrate significant progress in addressing the challenges of specular highlight removal in endoscopic imaging.

Existing techniques primarily focus on detecting specular regions. However, these methods often suffer from high computational complexity and insufficient detection in complex situations, such as when large white tissues overlap with densely specular regions, as shown in Figure 1. It displays white tissue regions enclosed by red boundaries and specular regions enclosed by blue boundaries from publicly available datasets CVC-ClinicSpec [27], Kvasir-Seg [16], Hyper-Kvasir [11], and CVC-ClinicDB [9]. Both these regions exhibit similar properties, including high intensity and low saturation, which often causes existing techniques to misidentify them as a single specular region, known as a false specular region. These false regions are generally larger than the actual specular regions because they include a combination of specular and white tissue regions. Saturation detectors can effectively identify these false specular regions, as they detect regions as specular if they possess high intensity and low saturation. However, the inadequate detection of false regions by existing methods makes their removal challenging, subsequently impacting feature detection. Moreover, covariant detectors can consistently detect affine-invariant frames in deformed and textureless regions, from which distinctive SIFT descriptors can be extracted for reliable matching.

Our technical contributions are as follows:

- We introduce MMSER as a method for identifying fine specular regions.
- By integrating the specular regions identified using MMSER with those detected by the saturation detector, we effectively recognize complex specular regions and eliminate false regions.
- To tackle the complexity of feature extraction in deformed and low-texture regions of endoscopy images, we utilize the affine invariance properties of covariant detectors and the distinctiveness of SIFT descriptors.
- By employing adaptive distance thresholding and outlier rejection, we enhance the accuracy of matching.
- Our technique has shown improvements compared to recent techniques in accuracy 0.21%, recall 25.42%, f1-score 7.77%, and jaccard 11.77%.

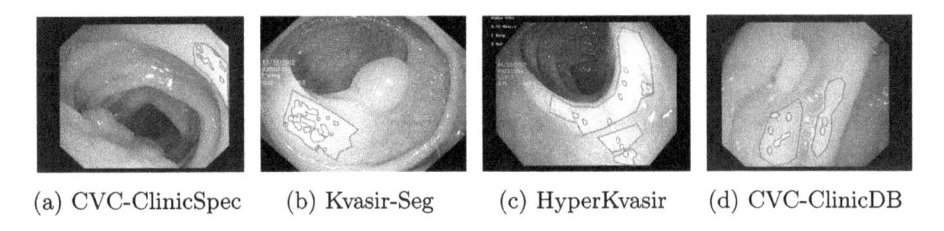

(a) CVC-ClinicSpec (b) Kvasir-Seg (c) HyperKvasir (d) CVC-ClinicDB

Fig. 1. Images with overlapping white tissue and specular regions from available public datasets.

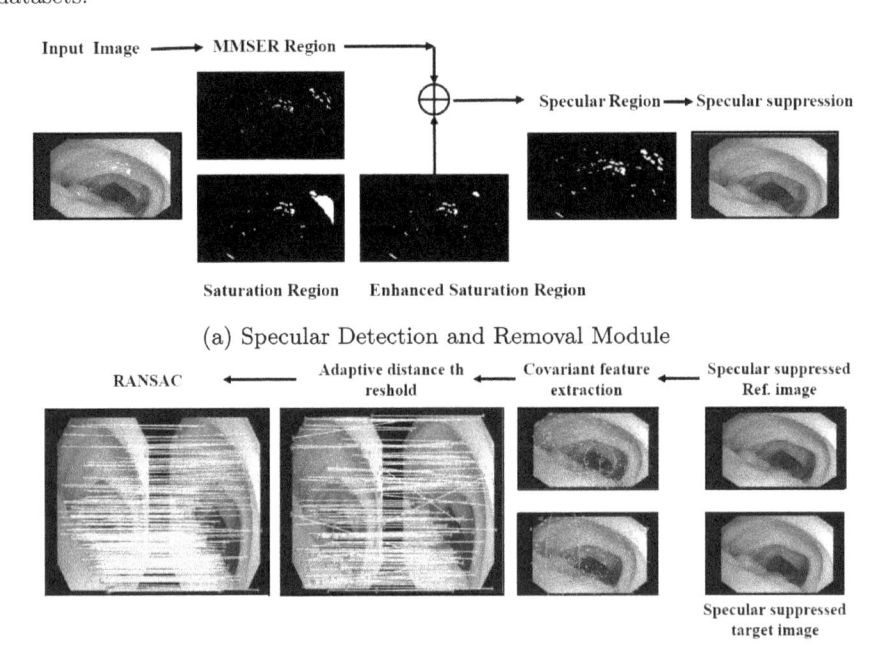

(a) Specular Detection and Removal Module

(b) Covariant Feature Extraction and Matching Module

Fig. 2. Schematic of the proposed technique

2 Proposed Method

Capitalizing on the synergistic combination of MMSER and saturation regions, our proposed technique offers significant advantages over existing methods, as it can adeptly handle complex situations such as false, low-texture, and deformed regions. Furthermore, our affine adaptation of covariant features enables the detection of features within low-textured and deformed areas. The affine invariance properties of these features also enhance matching accuracy. With the capacity to extract features from intricate regions and improve matching accuracy, our method paves the way for precise 3D dense reconstructions in endoscopy imaging, ultimately contributing to significant advancements in surgical navigation.

In essence, our proposed technique consists of the following steps: First, we separately detect MMSER and saturation regions. Next, we enhance the saturation region by suppressing false region areas, which typically correspond to white tissue regions containing specularity. We then merge the MMSER and the enhanced saturation regions to obtain our final detected specular region. Subsequently, we remove the detected specular region using existing techniques. Afterward, we extract covariant feature frames and SIFT descriptors with affine adaptation and match features between two images. Although there are some mismatched features, we employ adaptive distance thresholding to eliminate them. However, some mismatches may persist, so we use RANSAC to obtain accurate matches between features. The flowchart of our proposed technique is illustrated in Figure 2.

We can define the intensity function $\boldsymbol{I}$ using red (r), green (g) and blue (b) component of an image as follows:

$$\boldsymbol{I} = \frac{1}{3}(r + g + b). \tag{1}$$

2.1 Modified Maximally Stable Extremal Regions Detection

The MSER algorithm [22] identifies extremal regions (maximum or minimum intensity) as connected components within the level sets of an image. Among these extremal regions, locally maximally stable ones are chosen. The absence of smoothing enables the detection of both fine and large structures. While MSER's properties are valuable for extracting extremal regions and varying region sizes, our goal is to detect only maximum intensity and fine structures. Consequently, we modify the MSER detector's properties to suit our specific requirements for capturing higher intensity and fine regions in endoscopy images. The steps of our proposed MMSER detection method can be outlined as follows.

- We assume $\mathbf{I}(x,y)$ represent the intensity of an image at pixel location (x,y).
- We define an extremal region $\boldsymbol{M_R}$ as a connected component of an image $\boldsymbol{I}$, such that $\forall (x,y) \in \boldsymbol{M_R}$ and $\forall (x',y') \in \partial \boldsymbol{M_R}$ (the boundary of $\boldsymbol{M_R}$):

$$I(x,y) \geq I(x',y') + t, \tag{2}$$

where, t is intensity threshold.
- We compute the area of the extremal region $\boldsymbol{M_R}$ for a range of intensity thresholds $t \in [0, 255]$. Let $A(t)$ denote the area of $\mathbf{R}$ at threshold t.
- Calculate the stability score $S(\boldsymbol{M_R})$ for each extremal region $\boldsymbol{M_R}$ as the absolute difference in areas over a range of intensity thresholds $\triangle t$:

$$S(\boldsymbol{M_R}) = A(t + \triangle t) - A(t), \tag{3}$$

where $\triangle t$ is the sensitivity of stability.

- We consider an extremal region M_R is maximally stable M_{RS} if its stability score $S(M_R)$ is locally minimal compared to its neighboring regions in the intensity range, and its area $A(t)$ is smaller than a predefined maximum area A_{max}. Mathematically, this can be expressed as:

$$M_{RS} = \begin{cases} M_R & \text{if } S(\mathbf{R}) < S(\mathbf{R'}) \text{ and } A(t) < A_{max} \\ 0 & \text{otherwise} \end{cases} \quad (4)$$

where $\mathbf{R'}$ is a set of neighbouring regions.

2.2 Intensity-Saturation Regions Detection

Previously, in [32], the author established a correlation between intensity and saturation. In this study, specular regions in images were detected using the bidirectional histogram concept. The approach was based on the observation that specular areas exhibit higher brightness and lower saturation than surrounding regions. Although this technique was effective in detecting specular regions, it had limitations. For instance, it often incorrectly identified the white tissue region as specular because of its high intensity and low saturation, and it could not detect small regions. In this study, we leverage the intensity-saturation technique to detect the presence of white tissue in endoscopy images. Once we identify the white tissue region, we apply morphological operations to suppress it, resulting in an enhanced saturation region. The enhanced intensity-saturation regions IS_{en} are calculated as follows:

- We denote S as the saturation of an image, which can be expressed as:

$$S = \begin{cases} \frac{1}{2}(2r - g - b) = \frac{3}{2}(r - m) & , if \ (b + r) \geq 2g \\ \frac{1}{2}(r + g - 2b) = \frac{3}{2}(m - b) & , if \ (b + r) < 2g \end{cases} \quad (5)$$

- To identify the specular region IS in the image, we consider each pixel p and check if it satisfies the following conditions:

$$IS = \begin{cases} I_p \geq \frac{1}{2}I_{max} \\ S_p \leq \frac{1}{3}S_{max} \end{cases} \quad (6)$$

- We compute the connected component c in IS.
- We compute the area $A_S(i)$ for each connected component.
- Next, we compute the enhanced intensity-saturation regions IS_{en} as follows:

$$IS_{en} = \begin{cases} I_m & , if \ A_S(i) \leq t_h \ \forall c \\ 0 & otherwise \end{cases} \quad (7)$$

where t_h is an adaptive threshold that depends on the size of the image. For the CVC-ClinicSpec dataset, we evaluated with various threshold values and determined that the appropriate threshold value is 20.

2.3 Integration of Detected Regions

We combine the modified maximally stable extremal regions M_{RS} with the enhanced intensity-saturation regions IS_{en} to obtain the final specular regions S_R.

$$S_R = M_{RS} + IS_{en}. \tag{8}$$

Figure 2 top row appropriately illustrates the steps of our proposed technique for detecting specular regions, showcasing the effectiveness of our approach in complex situations. To highlight the need for our method, we carefully selected an image including false regions from the CVC-ClinicSpec dataset. Our process begins by converting the colored image to a grayscale image, followed by computing the MMSER (M_{RS}) containing relatively fine specular regions. Subsequently, we compute the intensity-saturation region IS, which encompasses relatively larger regions, including the crucial false regions (if present in the image). We refine the false regions based on their existence to obtain the enhanced intensity-saturation region IS_{en}. Finally, we integrate the M_{RS} with the IS_{en} to obtain the final specular region S_R with remarkable precision and accuracy.

2.4 Specular Region Suppress

We use the technique [10] to remove specularity and achieve a clean image without specularity after detecting the specular region S_R in the previous step.

$$I_{Refine} = I \otimes S_R, \tag{9}$$

where the operator $\otimes$ represents the specular removal operation implemented in [10].

2.5 Affine-Invariant Feature Detection

In the previous step, we obtained the specular-removed image I_{Refine}, which we use to extract affine-invariant features. Using a covariant detector [33], we detect feature frames F that are defined by an affine matrix comprising a translation vector t_r and a linear map L. These feature frames define elliptical regions in the image. Next, we extract description vectors d_v from these regions. An affine-invariant feature is associated with a matrix F and a vector d_v.

$$F = \begin{vmatrix} L & t_r \end{vmatrix} = \begin{vmatrix} l_{11} & l_{12} & t_{r1} \\ l_{21} & l_{22} & t_{r2} \end{vmatrix} \tag{10}$$

where the translation vector t_r represents the location of an image, while the linear map L represents the shape and orientation of the local features.

Following this, we extract a descriptor vector of dimension $128 - D$ from the detected region using the SIFT method. This descriptor will be used to calculate

Algorithm 1. Detection of specular highlights, extraction of affine-invariant features, and improvement of matching accuracy in endoscopy image pairs.

Require: Endoscopy image pair.
1: **for** each image **do**
2:     Compute the modified MSE regions M_{RS}.
3:     Compute the saturation regions IS.
4:     Compute the enhanced saturation regions IS_{en}.
5:     Integrate M_{RS} and IS_{en} to get S_R.
6:     Compute I_{refine} using Arnold's method [30] to suppress specularity.
7:     Extract the Affine-invariant features from I_{refine} by computing:
8:         (i) Feature frame F
9:         (ii) Feature descriptor d_v
10: **end for**
11: Compute the initial match between the image pair using the descriptors d_v and the Brute-Force technique [15].
12: Apply adaptive distance thresholding to improve the initial match.
13: Remove outliers using graph-cut RANSAC [6].

the matching confidence between two features by computing the distance ratio e_n.

$$e_n = \frac{d_{n,closest}}{d_{n,closest2}} = \frac{\| \boldsymbol{d}_{v,n} - \boldsymbol{d}_{v,m_{closest}} \|}{\| \boldsymbol{d}_{v,n} - \boldsymbol{d}_{v,m_{closest2}} \|} \quad (11)$$

In comparing features between two images, $\boldsymbol{d}_{v,n}$ and $\boldsymbol{d}_{v,m}$ denote the feature descriptors, while $d_{n,closest}$ is the descriptor distance between a particular feature and its nearest neighbor in the other image, and $d_{n,closest2}$ is the distance to the second-nearest neighbor. A smaller value of e_n indicates a greater similarity between the two features, increasing the likelihood that the correspondence is an inlier.

2.6 Feature Matching

Once the Affine-invariant features are computed, we can perform matching between corresponding images by comparing descriptors vector $\boldsymbol{d}_v$, using the Brute-force (BF) method [15].

Next, we will use the distance ratio e_n between the two best matching descriptors and only accept matches below an adaptive threshold th_{en}. After extensive evaluation of various endoscopy datasets, we determined the optimal threshold value to be 0.91.

Finally, we perform geometric verification on the previous matching results. The geometric verification technique Graph-Cut RANSAC [6] improves the matching from the previous step by removing outliers and mismatches that do not satisfy the geometric constraint.

3 Computational Procedures

Algorithm 1 outlines the complete steps of our proposed method for specular detection, affine-invariant feature extraction, and matching accuracy improvement.

4 Experimental Results and Analysis

4.1 Implementation and Data

Our study involves evaluating the effectiveness of the proposed technique in detecting specular regions and extracting affine-invariant features in endoscopy imaging. To accomplish this, we compare our approach to state-of-the-art specular detection and feature extraction techniques. The experiments are conducted on a Windows 10 x64 system, utilizing OpenCV 4.0 for image processing. The hardware consists of an Intel i5-9400K with a 2.90 GHzX2 processor and 16 GB of RAM. Implementation of the algorithm is carried out using both Matlab 2022 and Python 3.8. Our evaluation is conducted on the publicly available CVC-ClinicSpec dataset, which contains colonoscopy images with annotated specular ground truth labels. To measure the effectiveness of our proposed approach, we use gold standard metrics such as Accuracy, Precision, Recall, F1-score, and Jaccard. We also evaluated the performance of affine-invariant feature extraction and matching in challenging conditions using the Hyper-Kvasir dataset. Specifically, we used the labeled videos in "lower-gi-tract/quality-of-mucosal-view/BBPS-2-3" from the dataset. We extracted and matched features across pairs of frames taken 1 second apart from each other within the sequences of the Hyper-Kvasir dataset (where 1 second $\bar{2}5$ frames). To evaluate the result, we used 10% of the frames from the videos. Our assessment of the detection algorithm's ability to segment the specular region involves the use of five metrics: Accuracy, Precision, Recall, F1-score, and Jaccard.

Our main contribution in this research is the detection of specular regions and the extraction of covariant features. Specifically, we focus on extracting covariant features, as specular regions often mislead feature extraction processes, complicating surgical navigation. We utilize an existing technique to suppress the detected specular regions, with the source code publicly available. In the following sections, we will evaluate our detection and feature extraction performance both qualitatively and quantitatively.

4.2 Visual Evaluation

In this part we will visually asses the specular region detection, covariant feature extraction and matching ability in challenging endoscopy imaging condition.

Specular Region Detection and Suppression Figure 3 presents a comparison between recent state-of-the-art techniques and our proposed method for detecting specular regions. The input image and ground truth from the CVC-ClinicSpec dataset are shown. Specifically, Arnold's method [3] fails to detect

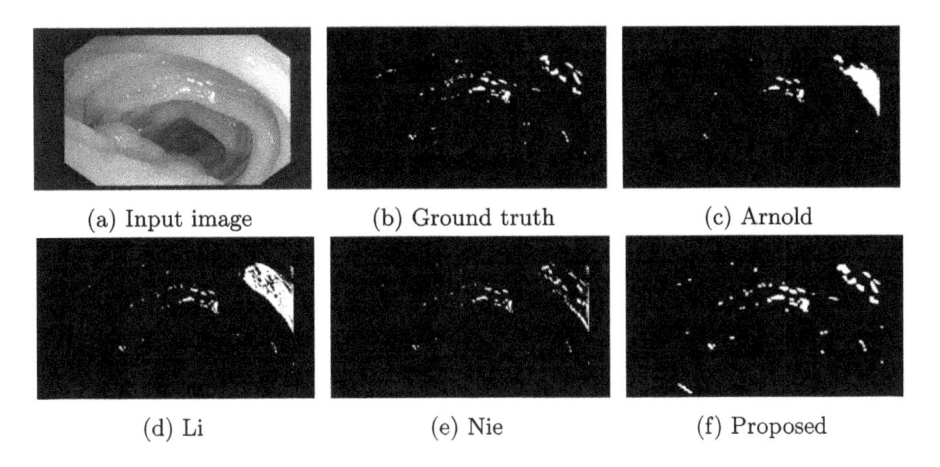

(a) Input image (b) Ground truth (c) Arnold

(d) Li (e) Nie (f) Proposed

Fig. 3. Comparison between different detection methods

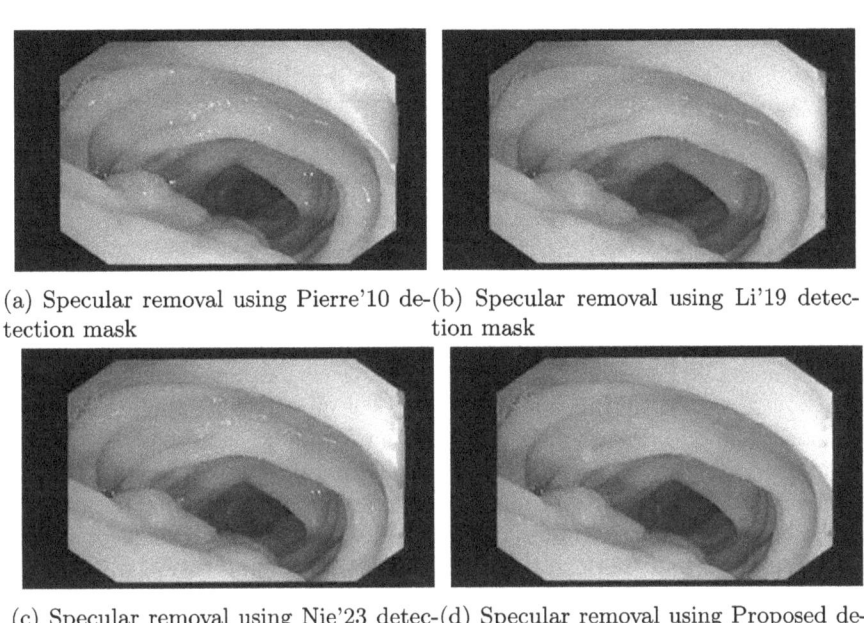

(a) Specular removal using Pierre'10 detection mask (b) Specular removal using Li'19 detection mask

(c) Specular removal using Nie'23 detection mask (d) Specular removal using Proposed detection mask

Fig. 4. Specularity removal using detected region from different techniques

fine specular regions and misidentifies the false region as a specular region. Li's method [18] is better at detecting fine specular regions, but also misclassifies the false region. Nie's method [23] can detect fine regions and specular regions on white tissue, but produces false edges that may affect the specular suppression process. In contrast, our proposed technique effectively detects fine specular regions and also accurately extracts specular regions from white tissue.

(a) Harris Features on Specular image
(b) ORB Features on Specular image
(c) SIFT Features on Specular image
(d) SURF Features on Specular image
(e) Affine-Invariant Features on Specular image

(f) Harris Features on filtered image
(g) ORB Features on filtered image
(h) SIFT Features on filtered image
(i) SURF Features on filtered image
(j) Affine-Invariant Features on filtered image

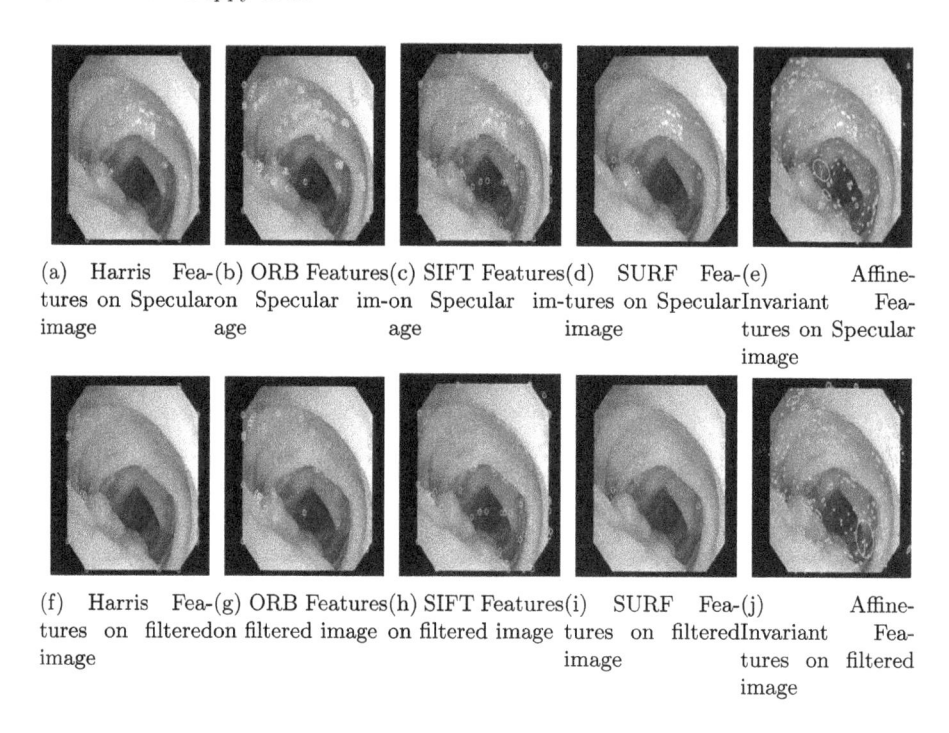

Fig. 5. Comparison between feature detectors

To suppress specular regions, we utilized Arnold's technique [3]. Figure 4 depicts the results of specular suppression using the detected specular region mask discussed in Figure 4. The results indicate that using Arnold's technique and regions detected by Li's method [18] do not effectively suppress the specular regions, and additionally introduce artifacts around the white tissue region. On the other hand, Nie's detected regions [23] perform better in removing the specular region compared to Arnold's and Li's methods, but still contain some specular regions and artifacts near the white tissue region. In contrast, our proposed detected region performs exceptionally well in suppressing the specular regions, with ignorable artifacts on the white tissue region.

Feature Detection and Matching The effectiveness of covariant detectors compared to commonly used detectors in endoscopy imaging is demonstrated in Figure 5. In Figure 5 top row, where the image contains specular regions, we observe that Harris, ORB, SIFT, and SURF detectors mostly detect features in specular highlight and corner-like areas, while avoiding low-texture areas. In contrast, covariant detector [33] detects affine-invariant frames throughout all the regions. Furthermore, in the filtered image in Figure 5 bottom row, we see a significant reduction in the number of features detected by Harris, ORB, SIFT, and SURF detectors, due to the absence of specular regions, while covariant detector detects a similar number of features throughout all the regions.

Specular Region Detection and Covariant Feature Extraction 181

(a) Image pair

(b) Initial match

(c) Distance threshold match

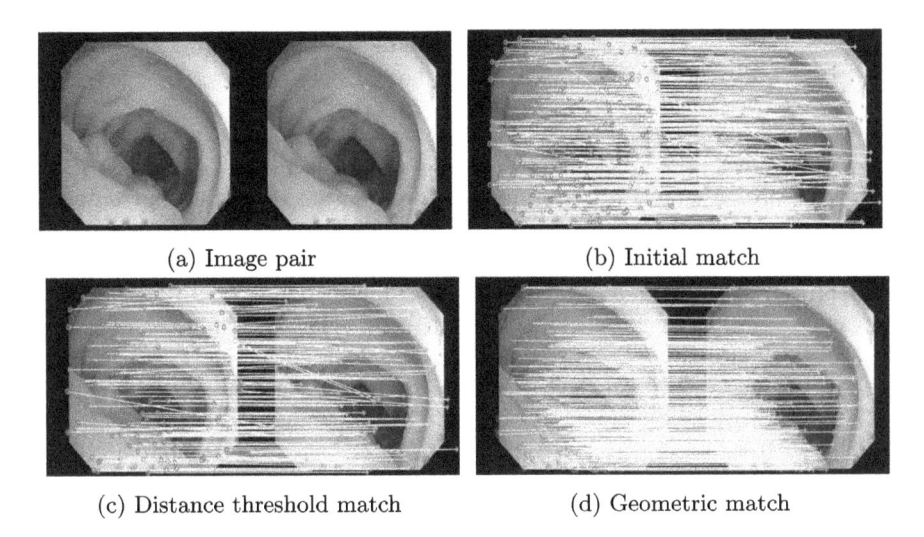

(d) Geometric match

Fig. 6. Steps for improving matching accuracy

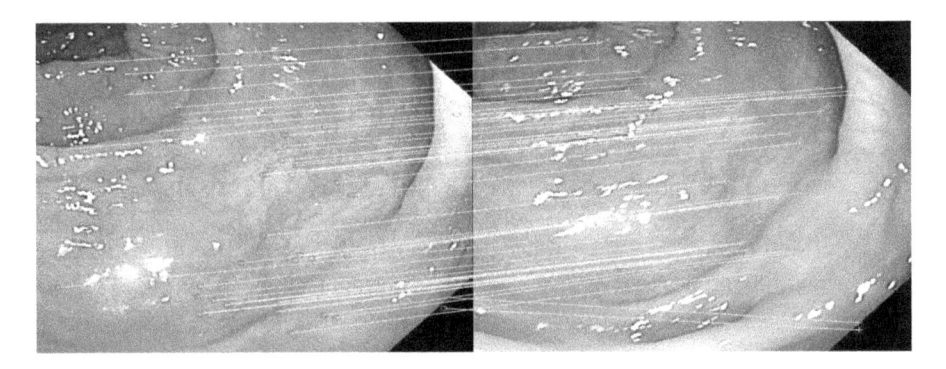

Fig. 7. Pair of 1-second-apart frames were used from the Hyper-Kvasir dataset to extract affine-invariant features and matches of inliers.

The matching process of affine-invariant features is illustrated in Figure 6. The filtered image sequence is represented in Figure 6(a). The initial matching results with numerous mismatches are depicted in Figure 6(b). By applying adaptive distance thresholding, the number of mismatches is significantly reduced, as shown in Figure 6(c). However, some mismatches still exist. Finally, Figure 6(d) demonstrates that there are no mismatches after applying the graph-cut RANSAC geometric verification technique.

Our evaluation uses the Hyper-Kvasir dataset with specularity, which allows us to compare our results with those in [7]. As shown in Figure 7, the detected affine-invariant features from the Hyper-Kvasir dataset are not confined to specific areas such as corners or specular regions but are also present in low-texture areas. Additionally, the figure demonstrates the feature matching between two pairs of frames taken 1 second apart from the dataset.

4.3 Quantitative Evaluation

Table 1 presents a quantitative comparison of different techniques for detecting specular regions on the CVC-ClinicSpec dataset. The bold values indicate the best detection performance. Our proposed method outperforms the other methods in terms of Accuracy, Recall, F1-score, and Jaccard. Specifically, compared to the best-performing method Nie [23], our method achieves higher Accuracy, F1-score, and Jaccard by 0.21, 7.77, and 11.77 percent respectively. Additionally, our Recall value is 1.97 percent higher than that of Meslouhi [14].

These improvements are primarily due to our innovative combination of MMSER with saturation region properties, which enhances the detection and management of specular regions in endoscopy images. This dual approach allows for more precise identification of true positives, especially in low-textured and deformed areas, leading to significantly higher recall and better alignment with the ground truth, as reflected in the Jaccard Index. Although our precision (0.8516) is slightly lower than that of Nie et al., the substantial gains in recall and the balanced F1-Score underscore the robustness and effectiveness of our method in accurately detecting and managing specular regions, which is crucial in medical imaging for ensuring comprehensive and reliable feature extraction.

Table 1. Quantitative comparison between state-of-art techniques

Methods	Accuracy	Precision	Recall	F1-Score	Jaccard
Arnold et al. [3]	0.9837	0.8938	0.6739	0.7684	0.6589
Meslouhi et al. [14]	0.9920	0.3744	0.8961	0.5281	0.6616
Alsaleh et al. [2]	0.9699	0.6016	0.8020	0.6875	0.6016
Shen et al. [28]	0.9767	0.9064	0.6683	0.7694	0.6518
Asif et al. [5]	0.9584	0.6972	0.7151	0.6489	0.6333
Nie et al. [23]	0.9932	**0.9083**	0.7286	0.8085	0.7153
Proposed	**0.9953**	0.8516	**0.9138**	**0.8713**	**0.7995**

Table 2 provides a quantitative comparison between commonly used detectors and the covariant detector, demonstrating the efficacy of the latter in extracting features from complex, low-texture, and deformed regions, particularly in endoscopy imaging. As shown in the table, the covariant detector is capable of extracting a very large number of features, 4907 and 4847 in the specular and filtered image, respectively, outperforming ORB, SIFT, Harris, and SURF. Furthermore, the table shows that our method achieves a greater number of inliers, 1274, compared to ORB, SIFT, Harris, and SURF after applying distance thresholding and Graph-cut RANSAC on initial Brute-Force matching to the extracted affine-invariant features.

The performance of various state-of-the-art techniques is presented in Table 3 using the Hyper-Kvasir dataset. The proposed technique outperforms other techniques in terms of feature extraction and inlier matching and is not limited to specific areas such as corners or specular regions. The table shows that the proposed technique achieves superior quantitative results compared to other techniques.

Table 2. Quantitative comparison between feature detectors

Methods	Specular image		Filtered image	
	Features	Inliers	Features	Inliers
Harris [34]	59	2	19	1
ORB [26]	447	29	80	6
SIFT [20]	140	22	101	17
SURF [8]	40	14	23	12
Affine-Invariant	4907	1344	4847	1274

Table 3. Matching quality metrics for state-of-the-art techniques using the Hyper-Kvasir dataset [4].

Methods	Feat/Img	F Inl.
SIFT [20]	825.7	151.3
ORB [26]	361.3	137.2
SP Base [13]	211.8	51.3
E-SP [7]	591.3	200.4
Affine-Invarinat	**1125.7**	**281.3**

5 Conclusions and Future Work

Our study proposes a novel specular detection technique that leverages the affine invariance properties of covariant detectors to overcome challenges in complex endoscopy imaging scenarios. Our approach enhances the characteristics of the Maximal Stable Extremal Region (MSER) to detect fine specular regions, followed by the computation of saturation regions. By merging the Modified MSER (MMSER) and saturation regions, our method accurately pinpoints areas of specular reflection. Using an existing method for suppression of specular regions, we refine the image and compute affine-invariant features, resulting in the extraction of a significant number of high-quality features. Our technique excels at detecting intricate specular regions and extracting features from uniform and deformed areas in endoscopy images. This state-of-the-art approach has the potential to enhance surgical navigation precision significantly.

Our detection technique is evaluated on the CVC-ClinicSpec. Visual evaluation reveals that our approach can successfully extract specular regions in complex conditions where other state-of-the-art techniques fail. Furthermore, in terms of Accuracy, Recall, F-1 Score, and Jaccard, our technique outperforms other existing methods in quantitative evaluation. Our evaluation also shows that recent techniques fail to detect the false regions, leading to artifacts during specular region suppression, unlike our technique.

Our approach to specular detection can significantly impact surgical navigation by enhancing the clarity and reliability of endoscopic images through any efficient removal process. Specular reflections in these images can obscure critical anatomical details, leading to potential misinterpretations during surgical procedures. By accurately detecting and suppressing these specular regions, the proposed technique ensures that important features are preserved and accurately extracted. This improved image quality facilitates better spatial understanding and decision-making for surgeons, thereby increasing the safety and efficacy of minimally invasive surgeries. Enhanced feature extraction and matching also contribute to more precise 3D reconstructions, further aiding in navigation and reducing the risk of complications.

The most robust aspect of the method is its combination of MMSER with saturation region properties to accurately detect and manage specular regions in endoscopy images. This technique significantly enhances the detection and extraction of features from low-textured and deformed areas, demonstrating notable improvements in metrics such as accuracy, recall, F1-score, and Jaccard index compared to existing methods. Conversely, the least effective aspect is the specular removal method used, which does not perform optimally. In the future, we plan to develop our own specular removal technique.

Acknowledgement. This work was supported by the National Research Foundation of Korea (NRF) grant funded by the Korea government (MSIT) (RS-2023-00250742, 2022R1A4A3018824, RS-2024-00438248, RS-2022-00155885). This research was also supported by the MSIT(Ministry of Science and ICT), Korea under the ITRC(Information Technology Research Center) support program(IITP-2023-RS-2023-00259061) supervised by the IITP(Institute for Information & Communications Technology Planning & Evaluation).

References

1. Ali, S., Zhou, F., Bailey, A., Braden, B., East, J.E., Lu, X., Rittscher, J.: A deep learning framework for quality assessment and restoration in video endoscopy. Med. Image Anal. **68**, 101900 (2021)
2. Alsaleh, S.M., Aviles-Rivero, A.I., Hahn, J.K.: Retouchimg: Fusioning from-local-to-global context detection and graph data structures for fully-automatic specular reflection removal for endoscopic images. Comput. Med. Imaging Graph. **73**, 39–48 (2019)
3. Arnold, M., Ghosh, A., Ameling, S., Lacey, G.: Automatic segmentation and inpainting of specular highlights for endoscopic imaging. EURASIP Journal on Image and Video Processing **2010**, 1–12 (2010)
4. Asif, M., Chen, L., Song, H., Yang, J., Frangi, A.F.: An automatic framework for endoscopic image restoration and enhancement. Appl. Intell. **51**, 1959–1971 (2021)
5. Asif, M., Song, H., Chen, L., Yang, J., Frangi, A.F.: Intrinsic layer based automatic specular reflection detection in endoscopic images. Comput. Biol. Med. **128**, 104106 (2021)
6. Barath, D., Matas, J.: Graph-cut ransac. In: Proceedings of the IEEE conference on computer vision and pattern recognition. pp. 6733–6741 (2018)
7. Barbed, O.L., Chadebecq, F., Morlana, J., Montiel, J.M., Murillo, A.C.: Super-point features in endoscopy. In: MICCAI Workshop on Imaging Systems for GI Endoscopy. pp. 45–55. Springer (2022)
8. Bay, H., Tuytelaars, T., Van Gool, L.: Surf: Speeded up robust features. In: Computer Vision–ECCV 2006: 9th European Conference on Computer Vision, Graz, Austria, May 7-13, 2006. Proceedings, Part I 9. pp. 404–417. Springer (2006)

9. Bernal, J., Sánchez, F.J., Fernández-Esparrach, G., Gil, D., Rodríguez, C., Vilariño, F.: Wm-dova maps for accurate polyp highlighting in colonoscopy: Validation vs. saliency maps from physicians. Computerized medical imaging and graphics **43**, 99–111 (2015)
10. Bertalmio, M., Bertozzi, A.L., Sapiro, G.: Navier-stokes, fluid dynamics, and image and video inpainting. In: Proceedings of the 2001 IEEE Computer Society Conference on Computer Vision and Pattern Recognition. CVPR 2001. vol. 1, pp. I–I. IEEE (2001)
11. Borgli, H., Thambawita, V., Smedsrud, P.H., Hicks, S., Jha, D., Eskeland, S.L., Randel, K.R., Pogorelov, K., Lux, M., Nguyen, D.T.D., et al.: Hyperkvasir, a comprehensive multi-class image and video dataset for gastrointestinal endoscopy. Scientific data **7**(1), 283 (2020)
12. Chu, Y., Li, H., Li, X., Ding, Y., Yang, X., Ai, D., Chen, X., Wang, Y., Yang, J.: Endoscopic image feature matching via motion consensus and global bilateral regression. Comput. Methods Programs Biomed. **190**, 105370 (2020)
13. DeTone, D., Malisiewicz, T., Rabinovich, A.: Superpoint: Self-supervised interest point detection and description. In: Proceedings of the IEEE conference on computer vision and pattern recognition workshops. pp. 224–236 (2018)
14. El Meslouhi, O., Kardouchi, M., Allali, H., Gadi, T., Benkaddour, Y.A.: Automatic detection and inpainting of specular reflections for colposcopic images. Central European Journal of Computer Science **1**, 341–354 (2011)
15. Jakubović, A., Velagić, J.: Image feature matching and object detection using brute-force matchers. In: 2018 International Symposium ELMAR. pp. 83–86. IEEE (2018)
16. Jha, D., Smedsrud, P.H., Riegler, M.A., Halvorsen, P., De Lange, T., Johansen, D., Johansen, H.D.: Kvasir-seg: A segmented polyp dataset. In: MultiMedia modeling: 26th international conference, MMM 2020, Daejeon, South Korea, January 5–8, 2020, proceedings, part II 26. pp. 451–462. Springer (2020)
17. Joseph, J., George, S.N., Raja, K.: Parameter-free matrix decomposition for specular reflections removal in endoscopic images. IEEE Journal of Translational Engineering in Health and Medicine **11**, 360–374 (2023)
18. Li, R., Pan, J., Si, Y., Yan, B., Hu, Y., Qin, H.: Specular reflections removal for endoscopic image sequences with adaptive-rpca decomposition. IEEE Trans. Med. Imaging **39**(2), 328–340 (2019)
19. Liu, Y., Tian, J., Hu, R., Yang, B., Liu, S., Yin, L., Zheng, W.: Improved feature point pair purification algorithm based on sift during endoscope image stitching. Front. Neurorobot. **16**, 840594 (2022)
20. Low, D.G.: Distinctive image features from scale-invariant keypoints. Journal of Computer Vision **60**(2), 91–110 (2004)
21. Marmol, A., Banach, A., Peynot, T.: Dense-arthroslam: Dense intra-articular 3-d reconstruction with robust localization prior for arthroscopy. IEEE Robotics and Automation Letters **4**(2), 918–925 (2019)
22. Matas, J., Chum, O., Urban, M., Pajdla, T.: Robust wide-baseline stereo from maximally stable extremal regions. Image Vis. Comput. **22**(10), 761–767 (2004)
23. Nie, C., Xu, C., Li, Z., Chu, L., Hu, Y.: Specular reflections detection and removal for endoscopic images based on brightness classification. Sensors **23**(2), 974 (2023)
24. Oh, J., Hwang, S., Lee, J., Tavanapong, W., Wong, J., de Groen, P.C.: Informative frame classification for endoscopy video. Med. Image Anal. **11**(2), 110–127 (2007)
25. Pan, J., Li, R., Liu, H., Hu, Y., Zheng, W., Yan, B., Yang, Y., Xiao, Y.: Highlight removal for endoscopic images based on accelerated adaptive non-convex rpca decomposition. Comput. Methods Programs Biomed. **228**, 107240 (2023)

26. Rublee, E., Rabaud, V., Konolige, K., Bradski, G.: Orb: An efficient alternative to sift or surf. In: 2011 International conference on computer vision. pp. 2564–2571. Ieee (2011)
27. Sánchez, F.J., Bernal, J., Sánchez-Montes, C., de Miguel, C.R., Fernández-Esparrach, G.: Bright spot regions segmentation and classification for specular highlights detection in colonoscopy videos. Mach. Vis. Appl. **28**(8), 917–936 (2017)
28. Shen, D.F., Guo, J.J., Lin, G.S., Lin, J.Y.: Content-aware specular reflection suppression based on adaptive image inpainting and neural network for endoscopic images. Comput. Methods Programs Biomed. **192**, 105414 (2020)
29. Son, M., Lee, Y., Chang, H.S.: Toward specular removal from natural images based on statistical reflection models. IEEE Trans. Image Process. **29**, 4204–4218 (2020)
30. Song, J., Wang, J., Zhao, L., Huang, S., Dissanayake, G.: Mis-slam: Real-time large-scale dense deformable slam system in minimal invasive surgery based on heterogeneous computing. IEEE Robotics and Automation Letters **3**(4), 4068–4075 (2018)
31. Sui, C., Wu, J., Wang, Z., Ma, G., Liu, Y.H.: A real-time 3d laparoscopic imaging system: design, method, and validation. IEEE Trans. Biomed. Eng. **67**(9), 2683–2695 (2020)
32. Tchoulack, S., Langlois, J.P., Cheriet, F.: A video stream processor for real-time detection and correction of specular reflections in endoscopic images. In: 2008 joint 6th international IEEE northeast workshop on circuits and systems and TAISA conference. pp. 49–52. IEEE (2008)
33. Vedaldi, A., Fulkerson, B.: Vlfeat: An open and portable library of computer vision algorithms (2008) (2012)
34. Yuan, X.C., Pun, C.M.: Invariant digital image watermarking using adaptive harris corner detector. In: 2011 Eighth International Conference Computer Graphics, Imaging and Visualization. pp. 109–113. IEEE (2011)
35. Zhang, C., Liu, Y., Wang, K., Tian, J.: Specular highlight removal for endoscopic images using partial attention network. Physics in Medicine & Biology **68**(22), 225009 (2023)

Low-Rank Adaptation of Segment Anything Model for Surgical Scene Segmentation

Jay N. Paranjape[1]( ), Shameema Sikder[2,3], S. Swaroop Vedula[3], and Vishal M. Patel[1]

[1] Electrical and Computer Engineering Department, Johns Hopkins University, Baltimore, MD 21218, USA
jparanj1@jhu.edu
[2] Wilmer Eye Institute, Johns Hopkins School of Medicine, Baltimore, MD 21287, USA
[3] Malone Center for Engineering in Healthcare, Johns Hopkins University, Baltimore, MD 21218, USA

Abstract. Surgical scene segmentation is an important task in the automated analysis of surgical videos and medical images. Traditional techniques for this task suffer from severe data scarcity characteristic of the medical domain. To address this challenge, contemporary research utilizes pre-trained models, which are finetuned using available data. Nonetheless, this approach involves repeated training of models with millions of parameters whenever new data becomes available. A recent foundational model called Segment Anything (SAM) has demonstrated impressive generalization capability in natural images, offering a potential resolution to this challenge. However, its applicability to the medical domain is hindered by its compute-intensive training along with need for task-specific prompts. These prompts can be bounding-boxes or foreground/background points, mandating expert annotation of every image. Expert annotation is not feasible as data volume increases. In this study, we propose LoRASAM - a highly efficient adaptation of SAM that enables text-guided segmentation of medical images. LoRASAM uses the labelname for precise image segmentation. To facilitate finetuning of LoRASAM, we utilize low-rank adaptation, which reduces the number of training parameters by more than 99% as compared to SAM, while significantly improving performance (about 70%). Extensive experimentation over three public surgical-scene datasets validates the superiority of LoRASAM over existing state-of-the-art methodologies. In addition, we show similar gains for non-surgical modalities such as x-ray and ultrasound. Our approach has the potential to enhance the segmentation performance and curtail expert involvement while adapting SAM for niche applications. Code: https://github.com/JayParanjape/LoRASAM

Keywords: Surgical Scene Segmentation · Foundational Models · Low Rank Approximation · Endoscopy

1 Introduction

The segmentation of structures and instruments in images is critical to advance situation-aware surgical systems. In the context of human-guided surgeries, segmentation of anatomical structures holds substantial utility in the development of computer-assisted systems aimed at augmenting the capabilities of surgeons [26]. Moreover, robotic surgery is also one domain where semantic segmentation plays a pivotal role, where it serves as a fundamental component for precise tracking of instruments throughout the procedure [29]. Deep learning-based solutions have emerged as the latest and most prevailing approach for addressing this challenge. Among these solutions, U-Net [24] and its variants [15,31,34] have shown commendable performance in segmenting surgical scenes when appropriately trained. However, deep learning methods require retraining a large number of parameters whenever they need to be used for new datasets. While this is a common problem occurring in non-medical datasets, foundational models like Segment Anything (SAM) [16] have been proposed to mitigate it to a certain extent. Foundational models are large-scale models with billions of parameters that have been trained on a large amount of data and can generalize well for their tasks. Other notable examples include the Contrastive Language-Image Pre-training (CLIP) model [22], which can associate text and images.

SAM addresses the task of prompted segmentation i.e. given an image and a suitable prompt, it outputs the corresponding masks of interest. However, using SAM for segmenting surgical images poses two major problems. First, since SAM is trained on natural images, it has a poor representation of surgical images, resulting in poor performance [8,13]. Second, a good prompt is required for every image during training as well as testing, which must be provided by experts. This becomes a tedious task, especially when the dataset size increases. To address the first challenge, various methods have been proposed to adapt SAM for specific applications by tuning fewer parameters than the conventional methods [25,29,30]. They are described in detail in the next section. In this work, we propose low-rank adaptation of SAM (LoRASAM), which performs text-prompted surgical scene segmentation. Thus, it only requires the name of the label of interest instead of an expert-provided point or bounding box prompt. Further, while existing adaptation methods require prompts per image, our approach can process an entire batch of images with the same text prompt. At the same time, our model freezes the original parameters of SAM, including the parameter heavy encoder and decoder. Thus, our method is more efficient than existing methods.

Our approach is inspired by low-rank adaptation used in the natural language processing literature, which has been shown as an effective method to finetune large language models [14]. Here, we approximate the tuning required in the encoder by a combination of two low-rank trainable matrices. Tuning just these matrices gives us an impressive performance in our experiments. This method of adaptation allows SAM to be trained on new datasets by reducing the number of trainable parameters by more than 99%. Furthermore, our approach operates in the regime of text prompted segmentation, where only the label names are required for producing masks of interest, thereby removing the onus on experts. Thus, the main contributions of this work are as follows:

1. We propose LoRASAM - a technique for text-prompted surgical scene segmentation, which is a more efficient and practical method to adapt SAM than existing methods. The proposed method utilizes LoRA to adapt encoder weights, a Text Affine Layer for enabling text prompts and tuning norm layers for better learning of domain shift, while keeping the weights of SAM's encoders and decoder frozen. This makes our method significantly more compact than existing methods.
2. We evaluate our method on three publicly available datasets for surgical scene segmentation. Extensive experimentation shows that our method is not only more efficient, but it also outperforms SOTA adaptation methods and conventional segmentation methods.
3. We evaluate our method on two medical (nonsurgical) imaging modalities to test the generalizability of our approach.

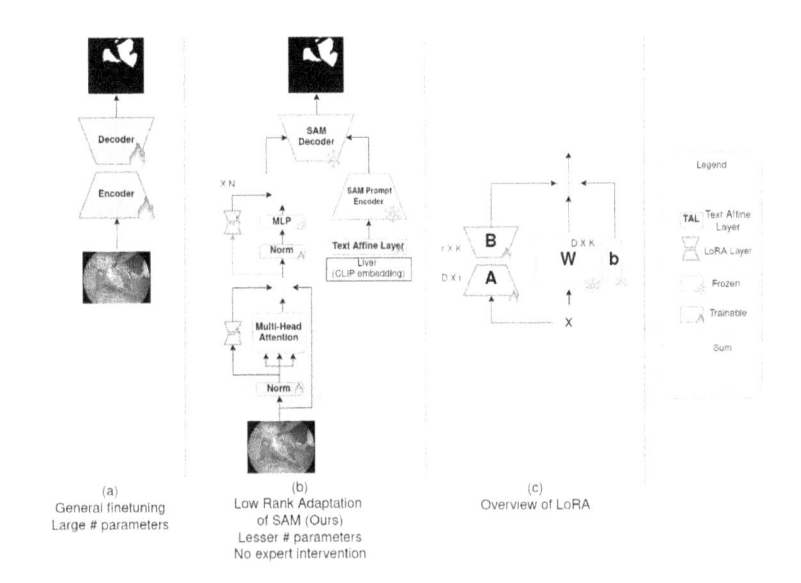

Fig. 1. Comparison of LoRASAM (b) with full finetuning (a). While full finetuning requires training of all the parameters of a network, our method adds very lightweight trainable low-rank matrices while keeping the original weights frozen. This is represented by LoRA layers in the diagram. At the same time, the Text Affine Layer is responsible for adapting the CLIP embeddings of surgical labels. (c) shows an overview of LoRA. A and B represent the low-rank matrices that transform the input to an r-dimensional subspace and back. The output from B is added to the original output, thereby modeling the required shift.

2 Related Work

SAM [16] is a recently released foundational model that proposes the task of promptable segmentation. Given an image and suitable prompts in the form of points, masks, or boxes, it can produce masks for the objects of interest corresponding to the prompt. SAM has been trained on 11 Million general images and 1 Billion masks over several GPUs. This makes it a very powerful model for natural image segmentation. However, the training corpus of SAM does not have surgical scenes. It is not surprising then, when recent literature showed that it fails for semantic segmentation in surgical images [8,19,21]. In such cases, models are often initialized with pretrained weights and then finetuned for the specific application. However, this is not possible in SAM due to the sheer number of parameters it has. Hence, many methods have been proposed that efficiently adapt SAM for medical segmentation. MedSAM Adapter [30] adds multiple adapter layers to all the blocks of SAM's encoder and decoder. Then, just these extra layers are trained while keeping the original SAM's weights frozen. While this significantly reduces the number of trainable parameters, its training still requires four GPUs. AutoSAM [25] replaces the prompt encoder of SAM with a convolutional layer and just finetunes that while keeping the encoders frozen. Some works also extend SAM for segmenting 3D medical images by adding extra trainable layers [5,10]. We refer readers to [17] for an extensive survey of SAM-based adaptation methods. However, all of the above mentioned methods are limited to using point or box-based prompts that need to be supplied by an expert surgeon for every image. Hence, deploying these models would be tedious and infeasible. In this paper, we do not use point or box-based expert level prompts. Instead, we explore prompting SAM using text, more specifically, the label names to segment out the object of interest. Thus, our methods do not require expert level prompts. We achieve this by adding a trainable module called TAL, and performing low-rank adaptation of the image encoder along with norm-layer tuning that can model the domain shift from natural images to surgical images better and more efficiently than the existing methods.

3 Methodology

In this section, we first describe the architecture of SAM, followed by the proposed low-rank modification that helps it train well on surgical data. Finally, we provide an intuition for why this method works.

3.1 Preliminaries: SAM Architecture

Given an image and set of prompts, SAM outputs masks that segment out the objects of interest as indicated by the input prompts. The prompts in question can be in the form of bounding boxes, foreground/background points, masks or text. This selective conditioning based on prompts is facilitated by having distinct encoders for the image and the prompt, followed by a decoder layer to fuse the encoded embeddings to produce the desired masks. Thus, SAM has three major components, namely the Image Encoder, the Prompt Encoder, and the Mask Decoder. The Image Encoder used by SAM uses a Vision Transformer (ViT) [9] and is pretrained with Mask AutoEncoder (MAE) strategy [11]. The prompt encoder also uses a ViT architecture and takes the encoded prompts as input. Foreground/Background point-based prompts are encoded using point embeddings. Bounding box-based prompts are encoded using the point embeddings of the top left and bottom right points of the boxes. Mask prompts are passed through convolutional layers and the output embeddings are used. On the other hand, text-based prompts are passed through a pre-trained CLIP [22] model to get embeddings. Note that the code support for text-based prompting is not available in SAM's codebase. Finally, the mask decoder is a lightweight module that fuses the outputs from the two encoders using a small ViT network. This is then upsampled to generate the final masks.

The SAM image encoder consists of N blocks, each one comprising of a self-attention module and a Multi-Layer Perceptron (MLP) layer. The first block operates on the input image which is divided into patches and fed to the block. Subsequent blocks act upon the outputs of the preceding block. For an input x to a block, it is passed through an affine transformation which is described as follows:

$$qkv = (W_{qkv}^n)^T x + b_{qkv}^n, \tag{1}$$

where, n is used to index the layers of the encoder. W, b denote the weights and biases of the affine transformation respectively. The output is further split into three different vectors, namely the key (k), query (q) and value (v). These are then used for the self-attention mechanism. The result is fed into an MLP, producing the final output of a block, denoted by o_n. This is also an affine transformation with weights W_{MLP} and biases b_{MLP} that can be denoted as follows:

$$o_n = (W_{MLP}^n)^T x + b_{MLP}^n \tag{2}$$

3.2 Low Rank Adaptation

The weight matrices as described previously are majorly responsible for the memory consumption of GPUs and hence, naively finetuning all of them is expensive. However, since SAM is already trained extensively on millions of images, we only need to adapt it for surgical scenes. We argue that tuning the full-rank weight matrix is not required for this task and that it can be achieved by tuning a separate, lower-ranked weight space and adding its output to the original weight matrix. More specifically, as shown in Figure 1 (c), consider any weight matrix $W \in \mathbb{R}^{DXK}$ in the image encoder of SAM, which works upon an input $x \in \mathbb{R}^{NXD}$. Here, N denotes the batch size, D denotes the input embedding size, and K denotes the output size of W. Then, to adapt it for representing surgical scenes, we can model the required change as follows:

$$W \leftarrow W + \Delta W. \tag{3}$$

However, ΔW is still a full rank matrix and hence, expensive to compute. In LoRA, we estimate ΔW as the multiplication of two trainable low rank matrices $A \in \mathbb{R}^{D \times r}$ and $B \in \mathbb{R}^{r \times K}$ as follows:

$$W + \Delta W \approx W + \lambda AB, \qquad (4)$$

where the ranks of the matrices A and B are $r << min(D, K)$, respectively, and λ is a hyperparameter. Now, the original weight matrix W is frozen and only A and B are learnt. This provides LoRA with a significant reduction in the number of training parameters (only $(r(D+K))$ as compared to general finetuning (DK). The low-rank adaptation is done for all the weight matrices of the image encoder. This includes the weights in the multi-head attention layers as well as MLP layers in all the transformer blocks. The prompt encoder and mask decoder remain unchanged.

While the above modifications can help adapt SAM to the surgical domain, it would still require precise bounding boxes or points as input since SAM does not understand medical terminology as is [21]. This is to be expected because the training data of SAM comprises of natural images and text descriptions. To mitigate this issue, we learn a Text Affine Layer (TAL) that transforms the text embeddings from CLIP as follows:

$$y = BatchNorm(ReLU(W_{TAL}^T X + b_{TAL})). \qquad (5)$$

The input to the TAL is the CLIP embedding X, which undergoes a learnable affine transformation to produce y. This is provided to the prompt encoder of SAM, which is frozen. We also finetune the layernorm layers and positional embeddings in the original SAM since they are highly biased towards a particular image size and distribution, similar to AdaptiveSAM [21]. However, AdaptiveSAM requires tuning the entire Mask Decoder in order to fuse the outputs from the two encoders. Instead, we argue that if the image encoder and the prompt encoder produce good embeddings, the pretrained mask decoder is an expert at fusing them to produce good masks. Hence, we freeze the entire Mask Decoder, unlike AdaptiveSAM.

In summary, the proposed modifications to SAM only include tuning the low-rank matrices in the image encoder as described in Equation 4, the Text Affine Layer, the positional embeddings, and the layernorm layers in the image encoder. This brings the total number of trainable parameters of our method to around 900,000, which is only 0.75% of SAM's total number of parameters and 37% of AdaptiveSAM's total number of parameters. The final model architecture for LoRASAM is illustrated in Figure 1.

3.3 Why should LoRA Work?

We provide intuitive reasoning behind the success of our method for adapting SAM to segment surgical scene images. SAM has an innate understanding of objects and boundaries in general, given the huge corpus of its training data. However, when there is a significant data distribution shift (eg. surgical scene), it might not perform as well. Hence, there is a need for finetuning the weights of the image encoder, the component of SAM responsible for encoding images. However, most of the components of the weight space are used for detecting general properties about the image that are not domain-specific and hence, need not be changed. The only change that is required to adapt to the domain shift is for that subspace of weights that are used for detecting features specific to the domain. By modeling the change as a multiplication of two low-rank matrices, the model learns the most suitable low-ranked subspace and the edits required to cater to the domain shift, and at the same time, reduces the number of parameters that need tuning. However, the choice of the rank of the learnable matrices is an important engineering decision and must be made using the validation dataset.

4 Experiments and Results

We evaluate our method on three widely used surgical scene segmentation datasets - Endovis 17 [2], Endovis 18 [1], and CholecSeg8k [12]. To showcase the generalizability of our method to segmentation tasks over other modalities, we also perform experiments on the ChestXDet (x-ray) [18] and abdominal ultrasound [28] datasets. For evaluation, we use the DICE score (DSC) and Intersection-over-Union (IoU) metrics as defined in existing literature [21,23] as follows:

$$DSC = \begin{cases} \frac{2*|Y \cap \hat{Y}|}{|Y|+|\hat{Y}|}, & \text{if } (|Y|+|\hat{Y}|) \neq 0 \\ 1, & \text{otherwise.} \end{cases} \quad (6)$$

4.1 Experimental Setup

In our experiments, we use the 'base' version of SAM for initializing weights. During training, we apply augmentations including random rotation ($\pm 10°$) with 0.5 probability, random saturation change with a scale of 2 with 0.2 probability and random brightness change with a scale of 2 with 0.5 probability. We set λ from Equation 4 to 1. The low rank r used in our experiments is 4. For all datasets, we use a batch size of 32, a focal loss function, and AdamW optimizer with a learning rate of 1e-4. All training is performed on a single Nvidia Quatro RTX 8000 GPU and requires less than 10 GB of memory.

4.2 Datasets

Endovis 17 [2] has eight training videos and ten testing videos from the Da-Vinchi robot system. We use two videos from the training dataset for validation. Endovis 17 has labels for six robotic instruments, namely Grasping Forceps, Bipolar Forceps, Large Needle Driver, Grasping Retractor and Monopolar Curved Scissors. Endovis 18 [1] comprises of sixteen training and four testing sequences with labels for different organs and surgical items. We use four sequences from the training set for validation. Finally, CholecSeg8k [12] provides labels for twelve objects including surgical tools, organs and tissues. We use the same train, val, test splits as [26] for our experiments.

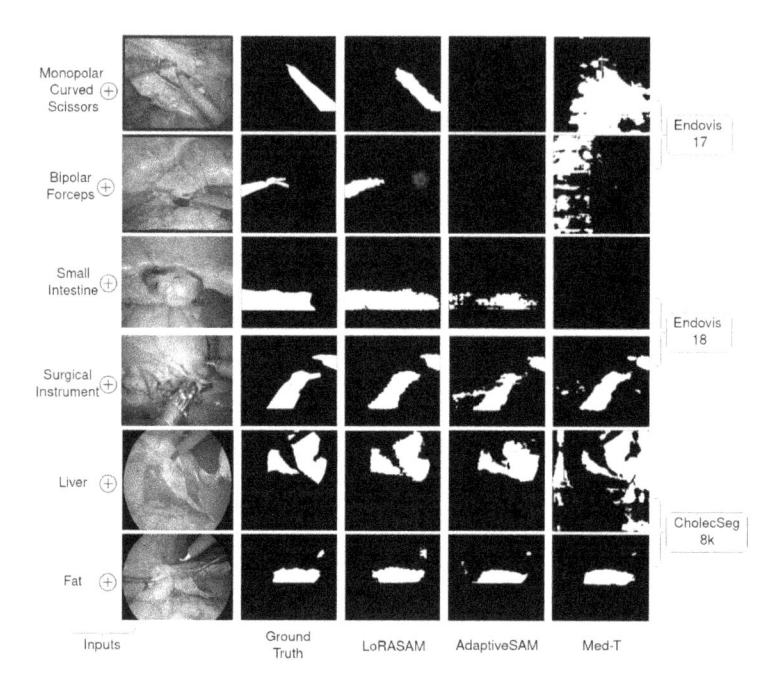

Fig. 2. Qualitative Results of LoRASAM on different surgical datasets. The inputs represent the image and the associated text which is the label of interest. With these two inputs, LoRASAM can produce good-quality masks.

4.3 Results on Surgical Datasets

Our approach performs on par if not better than the state-of-the-art methods on all three datasets while being significantly more efficient. As compared to AdaptiveSAM, we use 63% lesser parameters while matching its performance on Endovis 18 and outperforming it on the other two datasets. These results are tabulated in Table 1 for Endovis 17, Table 2 for Endovis 18, and Table 3 for CholecSeg8k datasets. A possible reason for the improved performance can be the fact that AdaptiveSAM only tunes the biases of the network, which might not be enough to model the shift required in Equation 3. In contrast, our approach does a better job by modeling this shift as a more complex operation (multiplication of two matrices). Here, SAM-ZS represents the zero-shot performance of the SAM model using the same label prompt as used for our method. We see a significant improvement of around 70% over this baseline.

Furthermore, we outperform state-of-the-art models in medical image segmentation. This includes U-Net [24], TransUNet [6] and Med-T [27] which were proposed as effective solutions against the data scarcity regime of the medical domain. These methods are highly prone to produce noisy masks which is mitigated to a good extent by our method. This can be attributed to the highly generalizable SAM backbone that is good at preserving the "objectness" of output masks and produces closed masks. Thus, our method combines the strengths of SAM while addressing its major weakness of requiring expert prompts for every image by enabling text prompts. Visual results on these datasets are shown in Figure 2.

Table 1. Results on Endovis17. PF - Prograsp Forceps, BF - Bipolar Forceps, LND - Large Needle Driver, GR - Grasping Retractor, VS - Vessel Sealer, MCS - Monopolar Curved Scissors.

Method	\multicolumn{7}{c}{Object wise DSC}
	PF	BF	LND	GR	VS	MCS	Avg.
Traditional DL methods							
UNet [24]	0.03	0.07	0.28	0.19	0.02	0.08	0.11
TransUNet [6]	0.08	0.10	0.20	0	0.03	0.10	0.08
MedT [27]	0.29	0.21	0.31	0.61	0.36	0.06	0.31
S3Net [3]	0.54	0.75	0.62	0.27	0.36	0.43	0.50
TraSeTR [33]	0.57	0.45	0.56	0.11	0.39	0.31	0.34
Mask2Former [7]	0.20	0.20	0.45	0	0.12	0.01	0.14
SAM based methods							
SAM w/ text prompt [16]	0.03	0.04	0.08	0	0.07	0.11	0.06
SAM w/ point prompt [16]	0	0.54	0	0.01	0.83	0.80	0.36
SAM w/ point and text prompt [16]	0.04	0.54	0.06	0.02	0.85	0.83	0.39
MedSAM [20]	0.01	0.54	0	0.01	0.84	0.80	0.36
SAMed [32]	0.33	0.27	0.40	0.73	0.66	0.72	0.52
AutoSAM [25]	0.56	0.54	0.25	0.98	0.83	0.80	0.66
AdaptiveSAM [21]	0.64	0.54	0.71	0.91	0.81	0.82	0.74
(Ours)	0.60	0.63	0.78	0.92	0.80	0.80	**0.76**

4.4 Ablation Study

We perform an ablation over the value of the rank r of the learnable matrices A and B. Here, we vary the value of r during training and evaluate the model on the Endovis 17 dataset. The results are tabulated in Table 4. A higher value of r would lead to a higher number of learnable parameters, which can improve the model, but can also lead to overfitting on the training data. This can be seen in the table, where increasing r from 2 to 4 increases the dice score by 7% but also increases the number of trainable parameters significantly. However, increasing the value to 8 does not produce significant improvement, and increasing it to 16 decreases the average dice score. However, if the value is too small, then the number of parameters is lower, leading to the model underfitting on the data, which can be seen for $r = 2$. Thus, there is a trade-off between the performance of the model and the memory consumption, making the choice of r an important engineering decision. We additionally plot the training curves for the cases of $r = 2, 4, 8$ in Figure 3. Here, for $r = 2$, the validation performance is significantly lower than the other two cases, indicating underfitting. For $r = 8$, even though the number of parameters is higher, the performance does not increase significantly over $r = 4$. Thus, we choose $r = 4$ for our experiments since it has a high performance with lesser trainable parameters.

We also conduct an ablation study over the role of each of the added modifications in LoRASAM in Table 5 on CholecSeg8k. In the first row, we start with zero-shot performance of SAM. Next, we only tune the positional embeddings, which does not improve performance. In the next row, we tune the layernorm layers and see a significant improvement in performance. Then, we add TAL which improves the model performance further and allows processing text queries. Finally, we add the LoRA layers, which give a boost to the model performance. Thus, each component in LoRASAM is necessary for improving model performance over SAM.

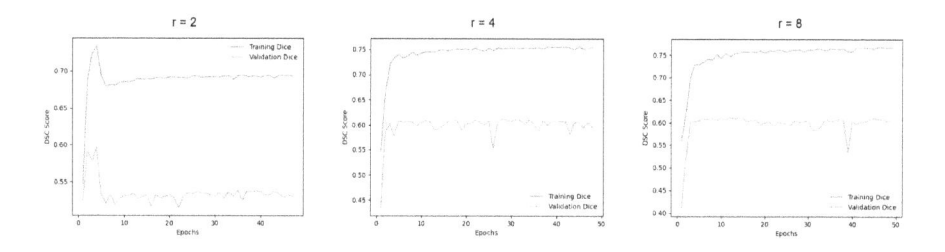

Fig. 3. Training and validation progress for different values of r. $r = 4$ provides a similar performance to r = 8 with fewer parameters.

4.5 Comparison for Number of Trainable Parameters

A visual representation of the number of trainable parameters can be found in Figure 4. Note that foundation models like SAM and MedSAM, as well as certain traditional methods like UNet++ need to train parameters on the order of 10^8. However, some recent methods like MedT and SAM adaptation methods require a significantly lesser number of trainable parameters. However, these are still around 6 million in number. Our method significantly reduces the number of trainable parameters to around 0.8 million, making it much more efficient than other methods.

4.6 Results on Non-Surgical Datasets

To evaluate the generalizability of our method on modalities other than surgical scene images, we evaluate our method on two nonsurgical datasets - ChestXDet [18] and Abdominal UltraSound [28]. ChestXDet has chest X-ray images and annotations for thirteen classes denoting abnormalities while Abdominal Ultrasound has labels for eight classes denoting different organs. The training data in this dataset consists of synthetic data only while the testing dataset has both real and synthetic ultrasounds. Table 6 shows the results on ChestXdet while Table 7 shows results on the Abdominal Ultrasound dataset. Our approach performs better than existing SOTA methods as well as the zero-shot implementation of SAM. Other medical segmentation methods like UNet, TransUNet and MedT perform poorly on these modalities owing to increased domain shift. However, our approach makes use of the generalization capability of a foundational model like SAM to perform well.

Table 2. Results on Endovis 18. BgT - Background Tissue, RI - Robotic Instrument, KP - Kidney Parenchyma, CK - Covered Kidney, SN - Suturing Needle, Su. I - Suction Instrument, SI - Small Intestine, UP - Ultrasound Probe, DSC - Dice Score

Method	\multicolumn{10}{c}{Object wise DSC}
	BgT	RI	KP	CK	SI	Thread	SN	Clamps	Su. I	UP	Avg.
Traditional DL methods											
LinkNet34[4]	0.77	0.87	0.23	0.23	0.23	0.74	0.74	0.33	0.33	0.33	0.59
LinkNet50[4]	0.76	0.87	0.21	0.21	0.21	0.73	0.73	0.37	0.37	0.37	0.59
UNet [24]	0.64	0.74	0.34	0.16	0.9	0.01	0.91	0.72	0.36	0.04	0.48
TransUNet [6]	0.73	0.70	0.49	0.33	0.63	0.01	0.04	1	0.66	0.62	0.52
MedT [27]	0.56	0.66	0.26	0.44	0.78	0.82	0.90	0.96	0.62	0.84	0.68
SAM based methods											
SAM w/ text prompt [16]	0.41	0.17	0.25	0.08	0	0	0	0	0.05	0.01	0.10
SAM w/ point prompt [16]	0.02	0.13	0.05	0.51	1	0.89	0.91	0.93	1	0.85	0.63
SAM w/ point and text prompt [16]	0.42	0.16	0.25	0.48	0.95	0.86	0.90	0.94	1	0.81	0.67
MedSAM [20]	0	0.12	0.05	0.51	0.77	0.90	0.91	0.93	1	0.85	0.61
SAMed [32]	0.52	0.47	0.02	0.48	0.74	0.86	0.89	0.90	0.99	0.81	0.67
AutoSAM [25]	0.57	0.57	0.05	0.50	0.77	0.87	0.89	0.91	0.98	0.83	**0.69**
AdaptiveSAM [21]	0.66	0.68	0.31	0.33	0.57	0.88	0.91	0.82	0.86	0.85	**0.69**
(Ours)	0.67	0.69	0.17	0.39	0.52	0.88	0.91	0.88	0.95	0.84	**0.69**

Table 3. Results on Choec8k. GB - Gall Bladder, AW - Abdominal Wall, GT - Gastrointestinal Tract, CD - Cystic Duct, LHEC - L Hook Electrocautery, HV - Hepatic Vein, CT - Connective Tissue, LL - Liver Ligament, DSC - Dice Score.

Method	\multicolumn{13}{c}{Object wise DSC}
	Fat	Liver	GB	AW	GT	Grasper	LHEC	Blood	HV	CT	LL	CD	Avg.
Traditional DL methods													
U-Net[24]	0.87	0.52	0.40	0.73	0.26	0.51	0.53	0.08	0.08	0.06	0.10	0.08	0.48
U-Net++[34]	0.91	0.75	0.63	0.83	0.11	0.61	0.60	0.14	0.16	0.15	0.14	0.17	0.61
TransUNet [6]	0.83	0.43	0.77	0.35	0.43	0.70	0.55	0.61	0.82	0.57	0.72	0.64	0.62
MedT [27]	0.81	0.39	0.56	0.34	0.25	0.48	0.71	1	0.70	0.69	0	0.89	0.57
SAM based methods													
SAM w/ text prompt [16]	0.05	0	0.02	0	0	0.01	0.04	0.01	0.14	0.01	0.14	0.01	0.04
SAM w/ point prompt [16]	0.17	0.23	0.07	0.30	0.10	0.22	0.63	1	0.69	0.43	1	1	0.49
SAM w/ point and text prompt [16]	0.19	0.20	0.07	0.28	0.10	0.26	0.65	0.99	0.70	0.45	1	1	0.49
MedSAM [20]	0	0	0.02	0	0.08	0.15	0.46	1	0.69	0.39	1	1	0.40
SAMed [32]	0	0	0.02	0	0.09	0.16	0.46	1	0.71	0.38	1	1	0.40
AutoSAM [25]	0.92	0.82	0.02	0.86	0.08	0.15	0.46	1	0.68	0.38	1	1	0.61
AdaptiveSAM	0.85	0.71	0.37	0.80	0.10	0.20	0.70	1	0.70	0.38	1	1	0.64
(Ours)	0.87	0.72	0.45	0.76	0.42	0.20	0.48	0.97	0.70	0.70	0.6	0.97	**0.65**

Table 4. Ablation Study over r on Endovis 17 Dataset. $r = 4$ has high performance and a lower number of trainable parameters. Thus, we use it for all our experiments.

Low Rank Parameter r	Number of Parameters	Average DSC
2	672000	0.69
4	974080	0.76
8	1578240	0.77
16	2786560	0.74

Table 5. Ablation Study over the role of various components in model performance on CholecSeg8k

Tuning Pos. Embeds	Tuning LayerNorm	TAL	LoRA	Trainable Parameters	Average DSC
				-	0.04
✓				196608	0.04
✓	✓			238080	0.50
✓	✓	✓		369920	0.52
✓	✓		✓	842240	0.59
✓	✓	✓	✓	974080	0.65

Table 6. Results on the ChestXDet dataset. SAM-ZS denotes zero-shot performance of the original SAM on the dataset. Ef - Effusion, No - Nodule, Cm - Cardiomegaly, Fb - Fibrosis, Co - Consolidation, Em - Emphysema, Ma - Mass, Ca - Calcification, Pt - Pleural Thickening, Pn - Pneumothorax, Fr - Fracture, At - Atelectasis, Dn - Diffuse Node

Method	\multicolumn{13}{c}{Object wise DSC}
	Ef	No	Cm	Fb	Co	Em	Ma	Ca	Pt	Pn	Fr	At	Dn	Avg.
SAM-ZS	0.05	0.13	0.53	0.36	0.15	0.28	0.23	0.10	0.37	0.07	0.40	0	0.26	0.22
UNet	0.15	0.08	0.06	0	0.13	0.02	0.95	0	0.08	0	0.50	0.02	0.02	0.15
TransUNet	0.06	0.87	0.06	0.59	0.13	0.01	0.89	0	0.74	0	0.08	0	0	0.26
MedT	0.06	0.75	0.08	0.01	0.10	0.03	0.12	0	0.91	0	0	0.37	0.07	0.19
AdaptiveSAM	0.52	0.88	0.86	0.86	0.43	0.93	0.95	0.91	0.84	0.93	0.86	0.94	0.93	0.83
(Ours)	0.50	0.89	0.90	0.83	0.42	0.93	0.96	0.91	0.84	0.93	0.86	0.94	0.95	**0.84**

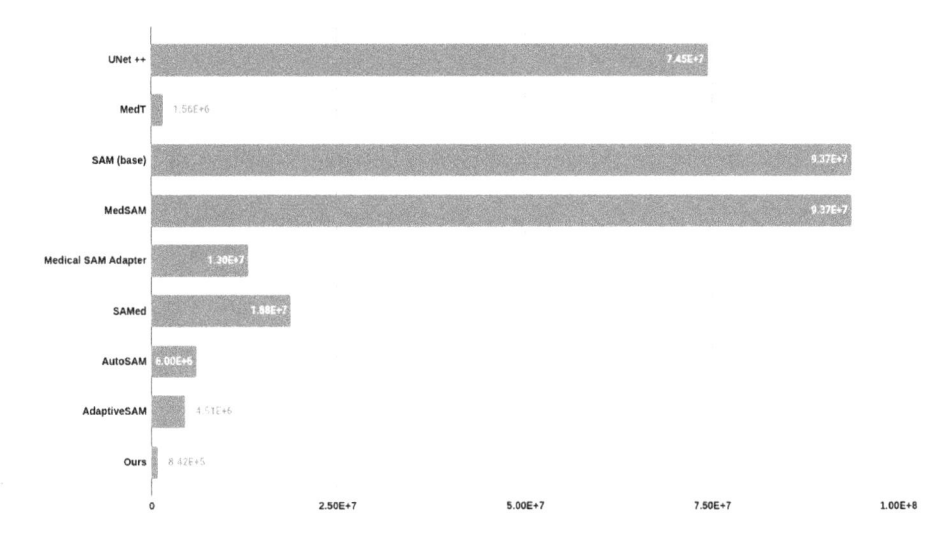

Fig. 4. Comparison of our method with others based on the number of trainable parameters.

Table 7. Results on the Ultrasound dataset. SAM-ZS denotes the zero-shot performance of the original SAM on the dataset

Method	\multicolumn{9}{c}{Objectwise DSC}
	Liver	Kidney	Pancreas	Vessels	Adrenals	Gall Bladder	Bones	Spleen	Avg.
SAM-ZS	0.17	0.20	0.72	0.21	0.44	0.65	0.67	0.63	0.46
UNet	0.28	0.37	0.11	0.16	0.85	0.08	0.17	0.14	0.27
TransUNet	0.18	0.09	0.03	0.03	0	0.11	0.05	0.02	0.08
MedT	0.18	0.03	0.27	0.10	0.85	0.15	0.02	0.08	0.21
AdaptiveSAM	0.36	0.30	0.50	0.40	0.86/0.86	0.63	0.67	0.54	0.53
(Ours)	0.43	0.35	0.45	0.61	0.90	0.59	0.67	0.67	**0.58**

5 Conclusion

In this paper, we propose LoRASAM - a highly efficient and well-performing adaptation of SAM for text-prompted surgical scene segmentation. LoRASAM employs a technique called low-rank adaptation that allows it to adapt the image encoder of SAM while freezing a high majority of its original parameters. We conduct experiments over three widely used surgical segmentation datasets and show the effectiveness of our approach over existing state-of-the-art adaptation and segmentation methods. Finally, we evaluate the generalized nature of LoRASAM for non-surgical modalities.

Acknowledgements. This research was supported by a grant from the National Institutes of Health, USA; R01EY033065. The content is solely the responsibility of the authors and does not necessarily represent the official views of the National Institutes of Health.

References

1. Allan, M., Kondo, S., Bodenstedt, S., Leger, S., Kadkhodamohammadi, R., Luengo, I., Fuentes, F., Flouty, E., Mohammed, A., Pedersen, M., Kori, A., Alex, V., Krishnamurthi, G., Rauber, D., Mendel, R., Palm, C., Bano, S., Saibro, G., Shih, C.S., Chiang, H.A., Zhuang, J., Yang, J., Iglovikov, V., Dobrenkii, A., Reddiboina, M., Reddy, A., Liu, X., Gao, C., Unberath, M., Kim, M., Kim, C., Kim, C., Kim, H., Lee, G., Ullah, I., Luna, M., Park, S.H., Azizian, M., Stoyanov, D., Maier-Hein, L., Speidel, S.: 2018 robotic scene segmentation challenge (2020)
2. Allan, M., Shvets, A., Kurmann, T., Zhang, Z., Duggal, R., Su, Y.H., Rieke, N., Laina, I., Kalavakonda, N., Bodenstedt, S., Herrera, L., Li, W., Iglovikov, V., Luo, H., Yang, J., Stoyanov, D., Maier-Hein, L., Speidel, S., Azizian, M.: 2017 robotic instrument segmentation challenge (2019)
3. Baby, B., Thapar, D., Chasmai, M., Banerjee, T., Dargan, K., Suri, A., Banerjee, S., Arora, C.: From forks to forceps: A new framework for instance segmentation of surgical instruments (2023)
4. Chaurasia, A., Culurciello, E.: LinkNet: Exploiting encoder representations for efficient semantic segmentation. In: 2017 IEEE Visual Communications and Image Processing (VCIP). IEEE (dec 2017)
5. Chen, C., Miao, J., Wu, D., Yan, Z., Kim, S., Hu, J., Zhong, A., Liu, Z., Sun, L., Li, X., Liu, T., Heng, P.A., Li, Q.: Ma-sam: Modality-agnostic sam adaptation for 3d medical image segmentation (2023)
6. Chen, J., Lu, Y., Yu, Q., Luo, X., Adeli, E., Wang, Y., Lu, L., Yuille, A.L., Zhou, Y.: Transunet: Transformers make strong encoders for medical image segmentation. arXiv preprint arXiv:2102.04306 (2021)
7. Cheng, B., Misra, I., Schwing, A.G., Kirillov, A., Girdhar, R.: Masked-attention mask transformer for universal image segmentation (2022)
8. Deng, R., Cui, C., Liu, Q., Yao, T., Remedios, L.W., Bao, S., Landman, B.A., Wheless, L.E., Coburn, L.A., Wilson, K.T., Wang, Y., Zhao, S., Fogo, A.B., Yang, H., Tang, Y., Huo, Y.: Segment anything model (sam) for digital pathology: Assess zero-shot segmentation on whole slide imaging (2023)
9. Dosovitskiy, A., Beyer, L., Kolesnikov, A., Weissenborn, D., Zhai, X., Unterthiner, T., Dehghani, M., Minderer, M., Heigold, G., Gelly, S., Uszkoreit, J., Houlsby, N.: An image is worth 16x16 words: Transformers for image recognition at scale (2021)
10. Gong, S., Zhong, Y., Ma, W., Li, J., Wang, Z., Zhang, J., Heng, P.A., Dou, Q.: 3dsam-adapter: Holistic adaptation of sam from 2d to 3d for promptable medical image segmentation (2023)
11. He, K., Chen, X., Xie, S., Li, Y., Dollár, P., Girshick, R.: Masked autoencoders are scalable vision learners (2021)
12. Hong, W.Y., Kao, C.L., Kuo, Y.H., Wang, J.R., Chang, W.L., Shih, C.S.: Cholecseg8k: A semantic segmentation dataset for laparoscopic cholecystectomy based on cholec80 (2020)
13. Hu, C., Xia, T., Ju, S., Li, X.: When sam meets medical images: An investigation of segment anything model (sam) on multi-phase liver tumor segmentation (2023)
14. Hu, E.J., Shen, Y., Wallis, P., Allen-Zhu, Z., Li, Y., Wang, S., Wang, L., Chen, W.: Lora: Low-rank adaptation of large language models (2021)

15. Isensee, F., Jaeger, P.F., Kohl, S.A.A., Petersen, J., Maier-Hein, K.H.: nnu-net: a self-configuring method for deep learning-based biomedical image segmentation. Nat. Methods **18**, 203–211 (2021)
16. Kirillov, A., Mintun, E., Ravi, N., Mao, H., Rolland, C., Gustafson, L., Xiao, T., Whitehead, S., Berg, A.C., Lo, W.Y., Dollár, P., Girshick, R.: Segment anything. arXiv:2304.02643 (2023)
17. Lee, H.H., Gu, Y., Zhao, T., Xu, Y., Yang, J., Usuyama, N., Wong, C., Wei, M., Landman, B.A., Huo, Y., Santamaria-Pang, A., Poon, H.: Foundation models for biomedical image segmentation: A survey (2024)
18. Lian, J., Liu, J., Zhang, S., Gao, K., Liu, X., Zhang, D., Yu, Y.: A structure-aware relation network for thoracic diseases detection and segmentation (2021)
19. Ma, J., He, Y., Li, F., Han, L., You, C., Wang, B.: Segment anything in medical images (2023)
20. Ma, J., He, Y., Li, F., Han, L., You, C., Wang, B.: Segment anything in medical images. Nat. Commun. **15**, 654 (2024)
21. Paranjape, J.N., Nair, N.G., Sikder, S., Vedula, S.S., Patel, V.M.: Adaptivesam: Towards efficient tuning of sam for surgical scene segmentation (2023)
22. Radford, A., Kim, J.W., Hallacy, C., Ramesh, A., Goh, G., Agarwal, S., Sastry, G., Askell, A., Mishkin, P., Clark, J., Krueger, G., Sutskever, I.: Learning transferable visual models from natural language supervision (2021)
23. Rahman, A., Valanarasu, J.M.J., Hacihaliloglu, I., Patel, V.: Ambiguous medical image segmentation using diffusion models. ArXiv **abs/2304.04745** (2023)
24. Ronneberger, O., Fischer, P., Brox, T.: U-net: Convolutional networks for biomedical image segmentation. vol. 9351 (2015)
25. Shaharabany, T., Dahan, A., Giryes, R., Wolf, L.: Autosam: Adapting sam to medical images by overloading the prompt encoder (2023)
26. Silva, B., Oliveira, B., Morais, P., Buschle, L.R., Correia-Pinto, J., Lima, E., Vilaça, J.L.: Analysis of current deep learning networks for semantic segmentation of anatomical structures in laparoscopic surgery. vol. 2022-July (2022)
27. Valanarasu, J.M.J., Oza, P., Hacihaliloglu, I., Patel, V.M.: Medical transformer: Gated axial-attention for medical image segmentation. In: Medical Image Computing and Computer Assisted Intervention – MICCAI 2021. pp. 36–46. Springer International Publishing, Cham (2021)
28. Vitale, S., Orlando, J., Iarussi, E., Larrabide, I.: Improving realism in patient-specific abdominal ultrasound simulation using cyclegans. International Journal of Computer Assisted Radiology and Surgery (07 2019)
29. Wang, A., Islam, M., Xu, M., Zhang, Y., Ren, H.: Sam meets robotic surgery: An empirical study in robustness perspective (2023)
30. Wu, J., Zhang, Y., Fu, R., Fang, H., Liu, Y., Wang, Z., Xu, Y., Jin, Y.: Medical sam adapter: Adapting segment anything model for medical image segmentation (2023)
31. Zeng, Y., Chen, X., Zhang, Y., Bai, L., Han, J.: Dense-u-net: densely connected convolutional network for semantic segmentation with a small number of samples. In: International Conference on Graphic and Image Processing (2019)
32. Zhang, K., Liu, D.: Customized segment anything model for medical image segmentation (2023)
33. Zhao, Z., Jin, Y., Heng, P.A.: Trasetr: Track-to-segment transformer with contrastive query for instance-level instrument segmentation in robotic surgery (2022)

34. Zhou, Z., Rahman Siddiquee, M.M., Tajbakhsh, N., Liang, J.: Unet++: A nested u-net architecture for medical image segmentation. In: Stoyanov, D., Taylor, Z., Carneiro, G., Syeda-Mahmood, T., Martel, A., Maier-Hein, L., Tavares, J.M.R., Bradley, A., Papa, J.P., Belagiannis, V., Nascimento, J.C., Lu, Z., Conjeti, S., Moradi, M., Greenspan, H., Madabhushi, A. (eds.) Deep Learning in Medical Image Analysis and Multimodal Learning for Clinical Decision Support, pp. 3–11. Springer International Publishing, Cham (2018)

DRIVPocket: A Dual-stream Rotation Invariance in Feature Sampling and Voxel Fusion Approach for Protein Binding Site Prediction

Bowen Deng[1], Yang Hua[1], Wenjie Zhang[1], Xiaoning Song[1,2(✉)], and Xiao-jun Wu[1]

[1] School of Artificial Intelligence and Computer Science, Jiangnan University, Wuxi 214122, China
{6223152013,7211905018,wenjie.zhang}@stu.jiangnan.edu.cn,
{x.song,wu_xiaojun}@jiangnan.edu.cn
[2] DiTu (Suzhou) Biotechnology Co., Ltd., Suzhou 215000, China

Abstract. Protein binding site prediction is crucial for drug design, but it is challenging due to the small size of the pockets and the complex interactions of the amino acids involved. Many existing methods use a 3D voxel U-Net to extract single-scale samples. However, this approach may overlook the structured or chemical information of the protein and fail to consider the impact of nearby atoms in the pocket. To tackle these issues, we propose a new protein binding site prediction model (DRIVPocket) based on dual-stream rotational invariance and voxel feature fusion. Specifically, DRIVPocket uses a dual-stream framework consisting of a 3D voxel network and an atomic point cloud network to predict the basic pockets and binding atoms, respectively. In addition, we present a novel feature extraction backbone based on dual rotational invariance attention (DRIA), which combines the advantages of shared dual attention information and the point cloud rotational invariance features. This module can extract the detailed spatial and hidden chemical information of the sample. Finally, we achieve a more robust predicted site by integrating the predictions of the above two networks through a semantic fusion module. Extensive experimental results obtained on four benchmarks demonstrate the merits and superiority of DRIVPocket over the existing state-of-the-art approaches. Our code has been released at https://github.com/lv5misaki/DRIVPocket.

Keywords: Binding site Detection · Drug Design · Deep Learning

1 Introduction

The prediction of protein binding sites based on the 3D structure is one of the critical steps in drug design [2,21]. However, the prediction task is challenging because of the small size of the binding sites and the large variation in size between different proteins. As shown in Fig. 1, the binding site (also known as the pocket or cavity) is a hole or tunnel on the surface of the protein. The properties and function of the pocket can be influenced by the surrounding amino acids, leading to variations in the shape of the pocket between different proteins and increasing the difficulty of the prediction task.

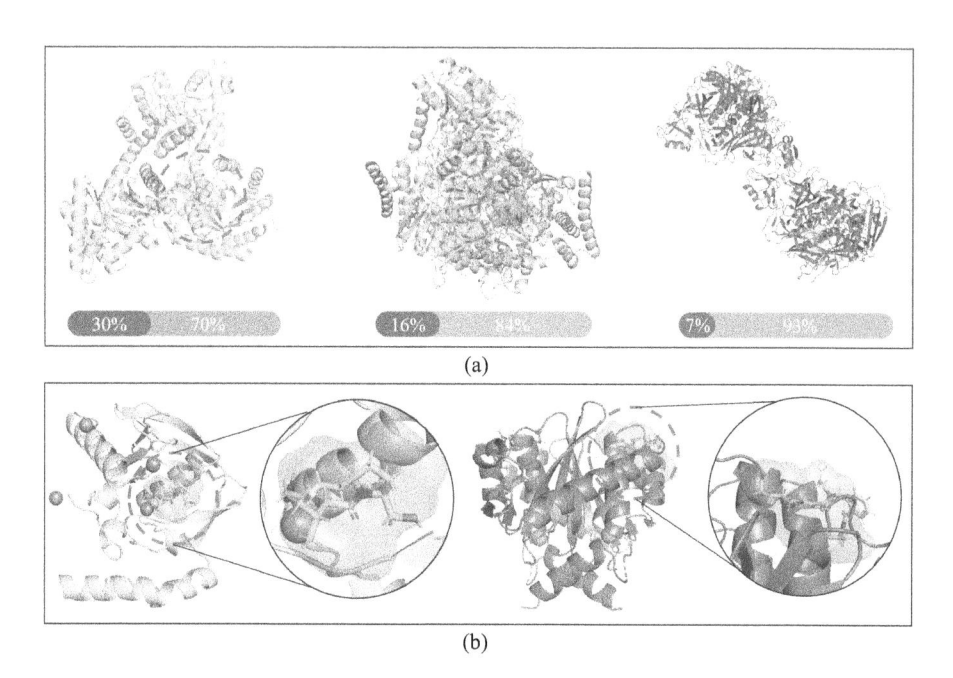

Fig. 1. Sub-figure (a) shows the difference in the size of binding sites in different proteins. Compared to the proteins, the size of the binding sites is significantly smaller than that of the proteins. Sub-figure (b) shows the different shapes of the different pockets, which challenges the accurate prediction of binding sites. The rod objects represent ligands, and the transparent bodies illustrate the cavity they occupy.

Many methods for this challenging task have been explored in the recent past. At first, the traditional methods require multiple matches or a massive collection of templates to generate predictions. The former appears mainly in some geometric-based [11,15] and energy-based methods [26,35], while the latter appears in template-based methods [3,34]. As deep learning progresses, more and more deep-neural-network-based approaches have been studied for binding site detection. Pointsite [36] provides a fine-grained representation of the pocket at the atomic level, ensuring that appropriate binding site predictions are obtained through point sampling of protein atoms. DeepPocket [1] reconstructs the target pocket from voxel information derived from the candidate protein based on a 3D U-network. On this basis, GLPocket [17] uses the attention structure of the transformer [31] to select the local regions of high response from the global representations.

Although existing methods have made significant progress in predicting binding sites, they still have shortcomings that need to be improved. The voxel-based methods, which are similar to segmentation models, tend to extract global features of proteins and use them to predict the entire pocket region. However, the predicted results lack generalizability because they cannot extract the complex details of proteins. In contrast, point-based methods focus on predicting the position of binding atoms, which improves their ability to represent detailed information. However, understanding the 3D global spatial information of the pocket directly is a challenge for them. Both of the above two groups of methods have their own specific problems, which make them unable to explore the entire pocket information and the relationship between the binding atoms simultaneously. Therefore, it is more difficult for them to capture the deeper spatial and chemical properties of the samples.

To address the above issues, we proposed a new protein binding site prediction model, called DRIVPocket, based on dual-stream rotational invariance and voxel feature fusion. Remarkably, DRIVPocket could predict the entire pocket region and atoms near the cavity simultaneously. In particular, we first represent the protein in two modalities, voxel and point cloud, and extract the relevant features via Dual Rotational Invariance Attention (DRIA) feature extraction and Rotational Invariance Down-Up (RID/RIU) sampling modules, respectively. We also fuse the point cloud feature into voxel features via DRIA, which is based on shared channel attention and spatial attention. By design, DRIVPocket could better understand the chemical properties and structural features of the protein according to the fusion features. In addition, DRIVPocket predicts the binding regions and binding atoms from the voxel features and the point cloud features, respectively. Finally, a more accurate segmentation prediction is obtained by integrating the two predictions.

Through extensive experiments conducted on various benchmarks, it shows that our proposed method significantly improves the performance of binding site prediction models. The results also verify that the voxel and point cloud features could complement each other, and the fusion result could provide a better option to represent the binding site.

Our contribution can be summarized as follows:

- We propose a novel dual-stream network method for predicting protein binding sites. This is the first method that predicts protein binding sites using point cloud and voxel information.
- A dual rotational invariance attention (DRIA) module is proposed for the fusion of protein voxel and point cloud information.
- Rotational invariance down-up (RID/RIU) sampling modules are introduced to extract the point cloud feature of the binding atoms.
- We evaluate the proposed method on four datasets and present a significant improvement in the DCC, DCA, and DVO metrics compared to the existing advanced methods.

2 Related Work

2.1 Traditional Binding Site Prediction methods

In this field, researchers have explored a series of traditional methods based on three paradigms, namely geometry-based, template-based, and energy-based approaches. The typical **geometric-based methods** include Fpocket [15], Ligsite-series [6,8], and CB-Dock [18] and CriticalFinder [4]. These methods predict pockets based on geometric structure analysis of proteins. The **template-based methods**, such as FINDSITE [3] and 3DLigandSite [34], work by finding proteins in a database that are closely resemble the target protein, and then projecting the outcomes from these similar proteins onto the target protein. These methods are dependent on a significant quantity of samples in the database and similar existing knowledge about the protein binding sites. Furthermore, the **energy-based methods**, such as AutoSite [26], FTSite [24], and Q-SiteFinder [14], focus on finding the ligand that requires the lowest interaction energy to bind to a protein. To find the best fit, these methods usually require multiple matching attempts.

In the past, traditional methods also used machine learning to predict binding sites. With the increase in labeled data, machine learning could learn and understand more complex relationships between data observations and results through an iterative process. Typical machine learning methods, such as Prank [12], use the random forest algorithm to calculate the physical and chemical properties of points and their surrounding neighborhoods. It can also rank the ligandability probability of potential pockets predicted by the Fpocket and Concavity tools [4], which is used until now.

2.2 Deep-learning-based Binding Site Prediction methods

Recently, several studies have employed deep learning technique to detect binding sites with remarkable results. The current research methods typically use pocket prediction as a computerized binary classification task. DeepSite [9] utilizes a 3D voxel-based Convolutional Neural Network (CNN) to predict pockets. A sliding window $16 \times 16 \times 16$ is used to evaluate the grid scores. Kalasanty [28] for the first time uses the U-Net [27] model to predict the pocket probability of each voxel.

Some of these networks replace global information with partial information for pocket detection. By using local grids as input, it is possible to ensure that smaller pockets receive more focused attention. Deepsurf [22] predicts the score of each local grid using LDS-Resnet, exploiting protein surface information, and finally groups high scores to form binding sites. However, this method requires traversing all surface points and has a high computational complexity.

The pointsite [36] converts the initial 3D protein structure into point clouds. The segmentation is performed using the U-Net architecture with Submanifold Sparse Convolution. The target cropping block was developed by GLPocket [17] specifically to capture localized features of interest, ensuring precise information on binding pockets. GLPocket also implemented the transformer block to create connections between patches in the local area, thereby establishing dependency relationships.

The rest of the algorithms use Fpocket [15] for auxiliary operations. Fpocket is a tool designed for predicting pocket based on geometry information. It achieves high recall but suffers from low precision. DeepPocket [1] is the first two-step pocket prediction model using Fpocket that predicts many candidate pockets. RecurPocket [16] uses feedback links and a mask to filter the background and noise region. RefinePocket [19] uses the dual attention block to obtain global information, while a refine block is used to improve segmentation within the decoder.

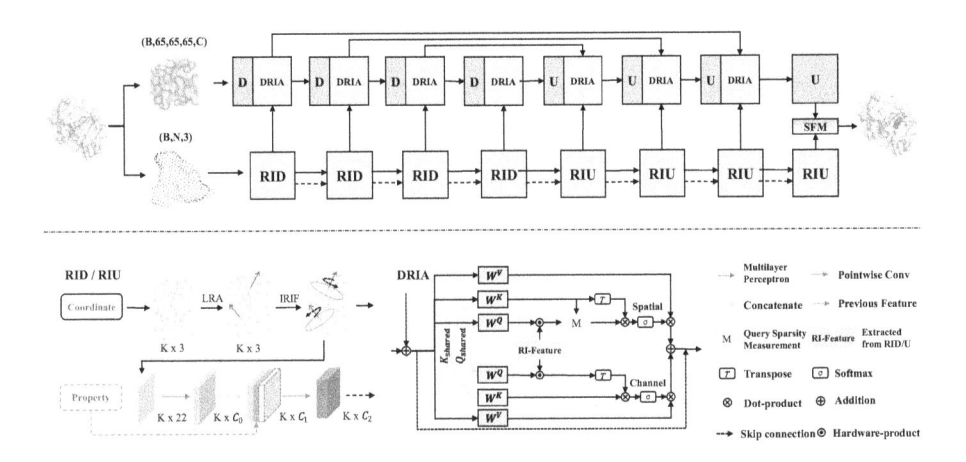

Fig. 2. An overview of the proposed DRIVPocket, a dual-stream based on rotational invariance pocket prediction network. Rotational Invariance Downsampling and Upsampling (RID/RIU) is responsible for extracting point cloud information by rotational invariance. In the RIU dashed section, we concatenate the residual features from the corresponding layers of the RID. Modules D and U are 3D downsampling and upsampling convolution, respectively. Pocket information is extracted via a shared Dual Rotaion Invariance based Attention (DRIA) module. With two outputs, the Semantic Fusion Module (SFM) is used to fuse the predictions.

3 Method

Firstly, we will present the dual-stream architecture of DRIVPocket in this section. Subsequently, we introduce the Rotation Invariant which is based Down-Up sampling modules(RID/RIU), the Dual Rotational Invariance based Attention(DRIA) module and the Semantic Fusion Module (SFM) in details.

3.1 Overview of DRIVPocket Architeture

As shown in Fig. 2, we present the pocket prediction pipeline in DRIVPocket, which consists of two parts: the point cloud network and the voxel network. In terms of the point cloud network, we follow the extraction approach of Libmolgrid [29], which produces a voxellized protein denoted as $\mathbf{X}_v \in \mathbb{R}^{HWD \times C}$. Meanwhile, we add its corresponding point cloud expression, $\mathbf{X}_{pc} \in \mathbb{R}^{N \times C}$, and extract the associated features using two backbones, the Dual Rotational Invariance Attention (DRIA) module and the Rotational Invariance Down-Up (RID/RIU) sampling module. The RID/RIU modules are used in the point cloud network to gather similar features of comparable proteins. This backbone consists of four downsampling and four upsampling modules and is described in detail in section 3.2.

The voxel network uses four encoder layers to transform inputs into refined protein features. Each encoder layer contains a convolutional block and a DRIA module. In the convolutional block, we use $3 \times 3 \times 3$ sized kernels, followed by three-dimensional Group Normalization (GN) and Rectified Linear Unit (ReLU) activation. A nonoverlapping convolution is also used to halve the resolution of the voxel feature after each encoder. We then use four decoder layers to reconstruct the protein feature, with the exception of DRIA in the last layer due to memory limitations. To facilitate the reconstruction of protein features with high-resolution detail, we link encoder blocks to their matching decoder blocks with skip connections.

Finally, after completion of the predictions for the binding atom, illustrated as $\mathbf{P}_{pc} \in \mathbb{R}^{N \times 1}$ and the voxel result $\mathbf{P}_v \in \mathbb{R}^{HWD \times 1}$, the result obtained from the prediction of point cloud can be utilized to refine the prediction of voxel. Therefore, the implementation of a Semantic Fusion Module (SFM) generates an accurate and robust final output $\mathbf{P}_{fuse} \in \mathbb{R}^{HWD \times 1}$ based on the two prediction results above.

3.2 Rotational Invariance Down-Up Sampling Module

The complicated and variable spatial structures of proteins make it difficult to extract spatially consistent information from proteins using translation-invariant convolution. To address this, we extracted the unique information for proteins using a rotation-invariant approach. Given a set of coordinate points for protein atoms, a uniformly distributed subset N_{sub} within the 3D coordinates is initially obtained by using the farthest point sampling method. The 3D coordinate subset will be used as input for the next layer. The Local Reference Axis (LRA) information and the essential reference atoms are extracted from these 3D coordinates. The LRA, based on the concept of the Local Reference Frame (LRF), serves as a more stable reference vector for rotation invariant shape descriptors [37]. At the reference atom x, we determine the LRA by finding the eigenvector that is associated with the smallest eigenvalue of the covariance matrix. By using the LRA, it becomes possible to identify the corresponding neighbor p_i of the atom

x, thus facilitating the extraction of rotation invariant features from the atoms. The LRA equation is given below:

$$LRA = \sum_{i=1}^{N_{sub}} l_i \left(p_i - x\right) \left(p_i - x\right)^\top, \quad (1)$$

where N_{sub} is the set of points in the neighborhood of the atom x, and

$$l_i = \frac{m - \|p_i - x\|}{\sum_{i=1}^{N} m - \|p_i - x\|}, \quad (2)$$

$$m = \max_{i=1..N_{sub}} \left(\|p_i - x\|\right). \quad (3)$$

After constructing the LRA, the unique Informative Rotation Invariant Feature (IRIF) is extracted for each atom x by computing Euclidean distances and angles. It consists of eight variables including radial distance, azimuthal angle, and six dimensions around the polar angle between the reference atom x and the neighboring points p_i and p_{i+1} in the clockwise direction, respectively. They uniquely define a 3D point in the local system [37]. In addition, we concatenate 14 types of input atoms, bringing the total to 22 unique dimensions.

To keep the size of the rotation invariant feature maps consistent, IRIF downsamples and upsamples using a MultiLayer Perceptron (MLP) and an activation function σ. The rotation invariant feature sampling equation is given below:

$$\mathbf{F}_i = \sigma \left(MLP\left(\mathbf{F}_{x_i}\right)\right), \quad (4)$$

$$\widetilde{\mathbf{F}}_i = \mathbf{W}_i \cdot concat \left(\mathbf{F}_i, \mathbf{F}_{prev}\right), \quad (5)$$

where $\mathbf{F}_{x_i}$ is the input feature of informative rotation invariant features $\mathbf{F}_{x_i} = IRIF(x_i)$. In the RIU module, $\mathbf{F}_i$ and $\mathbf{F}_{prev}$ are additionally concatenated with the residual features from the corresponding layers of the RID. Similar to PointNet [25], $\mathbf{W}_i$ is the weight parameter that the network learns, and · is the product per element.

A protein binding site is defined by its remarkable chemical specificity and affinity, which allows it to efficiently bind to a ligand. Only a small number of residues within the functional pocket are involved in ligand binding, and their spatial arrangement is often conserved by evolution [30]. Therefore, we believe that the rotation invariant features of the point cloud can compensate for some of the voxel information lost due to coarse detail during sampling. Meanwhile, pocket prediction in proteins differs from other segmentation prediction tasks by focusing on the concave hole portion instead of the protein entity itself. To tackle this task, we employ the DRIA module, capable of integrating features across both spatial and channel dimensions through the application of a self-attention mechanism. Next, we will present our DRIA block in detail.

3.3 DRIA module

The DRIA module includes the ProbSparse spatial and channel attention modules to efficiently capture spatial attention and efficiently perform chemical channel attention on the input features. Remarkably, ProbSparse self-attention can reduce the complexity of spatial attention from $O(L^2)$ to $O(LlnL)$ [38]. Previous work[7] has shown its effectiveness in bioinformatics.

In specific, as shown in the DRIA block of Fig. 2, the input features are transformed into the Queries (Q) and Keys (K) via the linear layer, and they are shared between the two attention modules. Besides, the input features are also transformed into two different Values (V), which are used for the spatial and channel attention modules, respectively. Before computation, we fuse feature with rotationally invariant information from the rotational invariance down-up sampling module. A dot product fusion of $\mathbf{Q}_{\text{shared}}$ with the voxelized features $\mathbf{F}_{RI}$, extracted from the RID/U module, produces a new expression for $\widetilde{\mathbf{Q}}_{\text{shared}}$:

$$\widetilde{\mathbf{Q}}_{\text{shared}} = \mathbf{Q}_{\text{shared}} \cdot \mathbf{F}_{RI}. \tag{6}$$

The two attention modules can be formulated as follows:

$$\hat{\mathbf{X}}_s = \text{SA}\left(\widetilde{\mathbf{Q}}_{\text{shared}}, \mathbf{K}_{\text{shared}}, \mathbf{V}_{\text{spatial}}\right), \tag{7}$$

$$\hat{\mathbf{X}}_c = \text{CA}\left(\widetilde{\mathbf{Q}}_{\text{shared}}, \mathbf{K}_{\text{shared}}, \mathbf{V}_{\text{channel}}\right), \tag{8}$$

where $\hat{\mathbf{X}}_s$ and $\hat{\mathbf{X}}_c$ are ProbSparse spatial and channel attention matrices, respectively. SA and CA are the global ProbSparse spatial and channel attention modules, respectively. The $\mathbf{K}_{\text{shared}}$ correspond to shared keys, while $\mathbf{V}_{\text{spatial}}$ and $\mathbf{V}_{\text{channel}}$ represent the spatial and channel values, respectively.

More specifically, the **ProbSparse Spatial attention** aims to effectively reduce the complexity of learning global information computation. Therefore, we take the dot product pairs from $\widetilde{\mathbf{Q}}_{\text{shared}}\mathbf{K}_{\text{shared}}^\top$ and measure their similarity using the Kullback-Leibler divergence, based on long-tailed distribution principles [38] that the more similar the parts are, the more important they are to the final computation result. For the i-th query, we can define the equation as follows:

$$M\left(\mathbf{q}_i, \mathbf{K}_{\text{shared}}\right) = \ln \sum_{j=1}^{L} e^{\frac{\mathbf{q}_i \mathbf{k}_j^\top}{\sqrt{d}}} - \frac{1}{L} \sum_{j=1}^{L} \frac{\mathbf{q}_i \mathbf{k}_j^\top}{\sqrt{d}}, \tag{9}$$

where L denotes the length of queries. Previous work[38] has shown that the boundaries of M in

$$\ln L < M\left(\mathbf{q}_i, \mathbf{K}_{\text{shared}}\right) < \max_j \left\{\frac{\mathbf{q}_i \mathbf{k}_j^\top}{\sqrt{d}}\right\} - \frac{1}{L} \sum_{j=1}^{L} \frac{\mathbf{q}_i \mathbf{k}_j^\top}{\sqrt{d}} + \ln L. \tag{10}$$

Depending on the boundaries, we can get the max-mean measurement as

$$\bar{M}\left(\mathbf{q}_i, \mathbf{K}_{\text{shared}}\right) = \max_j \left\{ \frac{\mathbf{q}_i \mathbf{k}_j^\top}{\sqrt{d}} \right\} - \frac{1}{L} \sum_{j=1}^{L} \frac{\mathbf{q}_i \mathbf{k}_j^\top}{\sqrt{d}}. \tag{11}$$

We use the randomly sample $U = L \cdot lnL$ dot-product pairs to calculate the $\bar{M}$ under the long tail distribution. Then, from the set $\bar{M}$, the sparse Top-u is extracted as $\overline{\mathbf{Q}}$. We set $u = c \cdot lnL$, which is controlled by a constant constant c. The spatial attention map is calculated by multiplying the $\overline{\mathbf{Q}}$ layer with the transpose of $\mathbf{K}_{\text{shared}}$. Next, we employ the Softmax operation to assess the similarity between each feature and the others. This approach reduces the time and space complexity of ProbSparse attention to $O\left(LlnL\right)$. These similarities are multiplied by the $\mathbf{V}_{\text{spatial}}$ layer to generate the final spatial global attention map $\hat{\mathbf{X}}_s \in \mathbb{R}^{HWD \times C}$. The ProbSparse attention equation is formulated as follows:

$$\hat{\mathbf{X}}_s = \text{Softmax}\left(\frac{\overline{\mathbf{Q}} \mathbf{K}_{\text{shared}}^\top}{\sqrt{d}}\right) \cdot \mathbf{V}_{\text{spatial}}, \tag{12}$$

where, $\overline{\mathbf{Q}}$, $\mathbf{K}_{\text{shared}}$, $\mathbf{V}_{\text{spatial}}$ denote shared sparse queries, shared keys, and global spatial values, respectively, and d is the size of each vector.

The **Channel Attention** can effectively capture the interdependencies among feature chemical channels by executing the dot product operation in the channel dimension between the channel values and the attention maps. The value of the channel attention module is the same as the spatial attention module for $\widetilde{\mathbf{Q}}_{\text{shared}}$ and $\mathbf{K}_{\text{shared}}$. We compute complementary features through a linear layer to obtain the value matrix $\mathbf{V}_{\text{channel}}$. The channel attention equation is formulated as follows:

$$\hat{\mathbf{X}}_c = \mathbf{V}_{\text{channel}} \cdot \text{Softmax}\left(\frac{\widetilde{\mathbf{Q}}_{\text{shared}}^\top \mathbf{K}_{\text{shared}}}{\sqrt{d}}\right), \tag{13}$$

where, $\mathbf{V}_{\text{channel}}$, $\widetilde{\mathbf{Q}}_{\text{shared}}$, $\mathbf{K}_{\text{shared}}$ denote channel value layer, shared RIfeature fused queries, and shared keys, respectively, and d is the size of each vector. Finally, the sum fusion result $\hat{\mathbf{X}}$ is performed by adding the outputs from the two attention modules.

$$\hat{\mathbf{X}} = \hat{\mathbf{X}}_s + \hat{\mathbf{X}}_c. \tag{14}$$

3.4 Semantic Fusion Module and Loss Function

We will combine the obtained outputs of the two networks as the final predicted result based on the semantic fusion module. Specifically, the output of the point cloud network is to determine if input points are binding atoms $\mathbf{P}_{pc} \in \mathbb{R}^{N \times 1}$. The voxel network, like the DeepPocket [1], is a prediction of the pocket $\mathbf{P}_v \in \mathbb{R}^{HWD \times 1}$. First, a sphere is drawn centered on the prediction point of the point cloud, with a radius equal to the distance from the point cloud prediction point to the voxel prediction center. Then, the part of the voxel facing this sphere is cropped. This part of the information represents the potential boundaries of the pocket. All predicted sums are aggregated to produce results of the same dimension, $\hat{\mathbf{P}}_{pc} \in \mathbb{R}^{HWD\times 1}$, as the voxel output. Finally, the transformed point cloud predictions are combined with the voxel output. The fuse prediction equation is given below:

$$\mathbf{P}_{fuse} = \mathbf{P}_v + \hat{\mathbf{P}}_{pc} \cdot \mathbf{P}_v + \hat{\mathbf{P}}_{pc}, \tag{15}$$

where $\mathbf{P}_{fuse}$ is the final fused prediction output, $\hat{\mathbf{P}}_{pc}$ is the transformed point cloud prediction, and $\mathbf{P}_v$ is the output of the voxel network.

Predicting protein binding sites is a binary classification challenge at the voxel level. Furthermore, we labeled atoms with euclidean distances less than three as binding atoms and converted them to a point-level classification challenge. Here, the network has been optimized using binary cross entropy loss:

$$\begin{aligned}\mathcal{L} = &\sum_t \sum_l - \left[y_{p_l}^{(t)} \log\left(\hat{p}_l^{(t)}\right) + \left(1 - y_{p_l}^{(t)}\right) \log\left(1 - \hat{p}_l^{(t)}\right) \right] \\ &+ \sum_t \sum_{i,j,k} - \left[y_{v_{ijk}}^{(t)} \log\left(\hat{v}_{ijk}^{(t)}\right) + \left(1 - y_{v_{ijk}}^{(t)}\right) \log\left(1 - \hat{v}_{ijk}^{(t)}\right) \right], \end{aligned} \tag{16}$$

where $v_{ijk}^{(t)} \in [0,1]$ represents the binary classification of voxel in the t-th binding pocket. Here, labels 1 and 0 signify whether the voxel is part of the cavity or not. Similarly, y_{p_l} represents the binary classification of point l, where 1 and 0 indicate whether the point is the binding atom, respectively. The $\hat{p}_l^{(t)} \in [0,1]$ and $\hat{v}_{ijk}^{(t)} \in [0,1]$ represents the predictions, indicating the probability that points belong to binding atoms and that the voxel belongs to the cavity.

4 EXPERIMENT

4.1 Dataset and Implementation

In this paper, we use five publicly available datasets in our training and evaluation approach. The ScPDB v2017 dataset [5] is used for both training and validation purposes, while our testing involves four datasets: HOLO4K [13], COACH420 [13], SC6K [1], and PDBbind [32]. ScPDB is one of the most extensive datasets and is commonly employed for predicting binding sites [10,20,23,33]. The provided resource presents a comprehensive depiction of protein-ligand pairs and their binding sites, covering 16,612 proteins and 17,594 binding sites at an all-atom level. Furthermore, we use the Libmolgrid [29] tools to provide datasets. Libmolgrid is a library capable of voxelizing three-dimensional molecules into multidimensional arrays as $\mathbf{X}_v \in \mathbb{R}^{HWD\times C}$, where each channel represents a receptor atom type, totaling 14 types. Meanwhile, Libmolgrid will provide the coordinates of these atoms. All settings and 3D inputs are consistent with DeepPocket.

The test datasets used include HOLO4K, COACH420, and SC6K, which were also used in DeepPocket, the preprocessing procedures remaining consistent between them. For PDBbind, the refined set from the 2020 version was employed in the experiments. The final counts for proteins and binding sites were 207 and 248 for COACH420, 2752 and 3449 for HOLO4K, 2378 and 6388 for SC6K, and 1113 and 1113 for PDBbind, respectively.

DRIVPocket was implemented in PyTorch and trained for 100 epochs. The training batch size is 8 across four 1080Ti GPUs. The SGD optimizer, along with the StepLR scheduler, was used for training, with the learning rate initially set to 0.001.

4.2 Evaluation Metrics

Consistent with previous studies, three types of metrics are used to evaluate the performance of the models, and all thresholds are assigned a value of 4τ, consistent with other work.

DCA (Distance to any atom of the ligand): The DCA quantifies the distance between the ligand's nearest atom and the predicted pocket center. If it's below threshold, the prediction is accurate.

DCC (Distance between Centers of the biding site): The DCC calculates the distance between the centers of the predicted and ground-truth pockets. A prediction is considered correct if the distance fall below the threshold.

DVO (Distance Volume Overlap): The DVO evaluates the IOU for accurately matched pairs. Pockets that have been inaccurately predicted within the DCC are set to zero.

4.3 Ablation Study

In this part, we evaluate the effectiveness of our method with each novel component on the Coach420 and SC6K datasets, and report the result in Table. 1. We use 3D U-Net as a baseline (Model-1) and compare it with models that add different attention architectures and different point-cloud and voxel-fusion strategies. The effectiveness of our proposed point cloud voxel fusion module SFM is evaluated using Models 2-6, respectively. Model-2 proves that the combination of voxelization of point cloud prediction and voxel prediction by addition could enable the predicted results to be closer to the real pocket center, demonstrating that fusion prediction is effective. However, Model-2 and Model-5 have worse DVO metrics than the baseline approach in both datasets, indicating that simple addition is limited in improving the overlap between predicted and true pockets. In fact, simple addition introduces redundant information, which has a negative impact on prediction. However, our SFM can effectively address this issue by merging more precise information, and Model-6 shows an improvement of 3.15% and 3.78% in both metrics compared to Model-5.

We also compare Model-3 and Models 7-9 to verify the effectiveness of the ProbSparse Spatial Attention and Channel Attention modules. The experiments

Table 1. We evaluate the performance of DRIVPocket with different modules, including ProbSparse Spatial (PSS) Attention, Channel Attention and Shared Dual Attention, on COACH420 and SC6K datasets. Results in bold indicate best performance.

Model	Attention	Fusion	COACH420 DCC DVO	SC6K DCC DVO	Params	FLOPs
Model-1	None	None	85.08 53.89	84.20 50.31	26.37M	49.57G
Model-2		Addition	86.24 53.15	86.14 50.02	27.30M	50.15G
Model-3		SFM	88.31 55.04	87.46 51.89	27.30M	50.15G
Model-4	Shared Dual Attention	None	89.92 56.15	90.67 54.24	28.96M	51.03G
Model-5		Addition	90.52 54.65	91.05 52.79	30.21M	51.82G
Model-6		SFM	**93.68 59.31**	**94.76 57.94**	30.21M	51.82G
Model-7	PSS Attention	SFM	91.09 56.98	92.47 55.47	28.52M	50.56G
Model-8	Channel Attention		91.39 57.29	92.65 55.66	28.52M	50.98G
Model-9	Shared Dual Attention		**93.68 59.31**	**94.76 57.94**	30.21M	51.82G

show that attention is effective in improving atomic feature extraction. In particular, Model-7 and Model-8 improve the predictions when a single attention mechanism is used. The ProbSparse Spatial Attention and Channel Attention approaches can optimize Model-3 due to their enriched protein spital and atomic chemical information, respectively. Moreover, when we fuse the two attention approaches as Shared Dual Attention, the performance of the model is further improved. Specifically, Model-9 improved by 2.63% on DCC and 2.22% on DVO compared to Model-7 and Model-8. Notably, we used the shared parameter strategy, which adds only a few parameters and has a faster computation compared to self-attention.

Table 2. Different models are evaluated based on the DVO and DCC metrics. For RecurPocket, the results are obtained under the condition where τ equals 2, employing a voxel-level mask. The results highlighted in bold signify the best performance.

Method	COACH420 DCC DVO	HOLO420 DCC DVO	SC6K DCC DVO	PDBbind DCC DVO
Kalasanty [28]	56.85 24.49	51.08 21.53	91.94 48.24	42.40 22.69
DeepPocket [1]	85.08 54.12	83.62 51.82	84.03 50.22	63.96 36.11
RecurPocket [16]	89.91 53.19	89.94 53.43	92.77 54.22	70.85 36.49
GLPocket [17]	92.74 55.18	90.20 54.21	92.50 52.67	77.14 38.51
DRIVPocket(Ours)	**93.68 59.31**	**91.32 55.71**	**94.76 57.94**	**77.77 40.32**

4.4 Comparison to State-of-the-art Methods

We compare our method with the state-of-the-art method on four datasets, and report the result in Table 2. Multiple pockets within proteins are predicted, and each pocket is ranked based on its probability. These probabilities are then sorted in descending order. DCC and DVO measure the degree of overlap between predicted pockets and real labels. Compared to DCC, DCA evaluates the model prediction for the degree of atomic and predicted pocket centering misalignment.

Table 3. We evaluate different models using the DCA Top-(n) and Top-$(n+2)$ metrics for comparison. The "-" symbol indicates that the datasets are not mentioned in the related paper. In this table, we use the DeepPocket classification model for evaluation. The results highlighted in bold indicate the best performance. Underlined values are the second best performance.

Method	COACH420 Top-n	Top-$(n+2)$	HOLO4K Top-n	Top-$(n+2)$	SC6K Top-n	Top-$(n+2)$	PDBbind Top-n	Top-$(n+2)$
Fpocket [15]	22.06	44.48	21.97	29.87	7.73	17.30	19.14	40.70
Kalasanty [28]	63.52	65.18	61.21	62.63	61.75	61.75	47.07	51.21
P2Rank [13]	68.24	75.48	70.60	80.05	62.80	75.74	-	-
DeepSite [9]	53.07	53.07	51.65	51.67	52.94	65.41	-	-
DeepPocket [1]	71.53	76.87	73.36	82.97	64.58	83.01	40.61	52.74
Pointsite [36]	72.12	76.34	80.42	86.21	-	-	-	-
RecurPocket [16]	72.95	<u>80.42</u>	81.12	<u>89.59</u>	67.28	85.84	<u>57.14</u>	<u>78.71</u>
GLPocket [17]	**75.30**	80.23	<u>81.48</u>	88.46	<u>67.70</u>	<u>86.42</u>	56.50	76.87
DRIVPocket(Ours)	<u>74.71</u>	**81.12**	**82.24**	**91.27**	**68.45**	**87.44**	**57.71**	**79.34**

The result proves that the proposed DRIVPocket outperforms other methods on four datasets, which can be attributed to several merits. First, unlike DeepPocket [1] and RecurPocket [16], our proposed Shared Dual Attention can capture the spatial and chemical information of proteins to predict binding sites more accurately. Therefore, DRIVPocket improves DCC and DVO by an average of 9.46% and 5.27%, respectively, compared to DeepPocket (baseline). Notably, the performance of our method in terms of DVO on the coach420 dataset is exactly, which gains 5% improvement compared to the second best method, GLPocket [17]. This is because GLPocket [17] only adopt voxel-based method, which loses structural protein information in their approach. However, our proposed point cloud fusion prediction method effectively overcomes this problem.

Following previous works [1,16,17], we use the DeepPocket classification network to select n and $n+2$ candidate pocket centers as input to calculate DCA metrics. As shown in Table 3, the proposed DRIVPocket shows better performance than other methods for the DCA metric. In particular, with Dual-stream

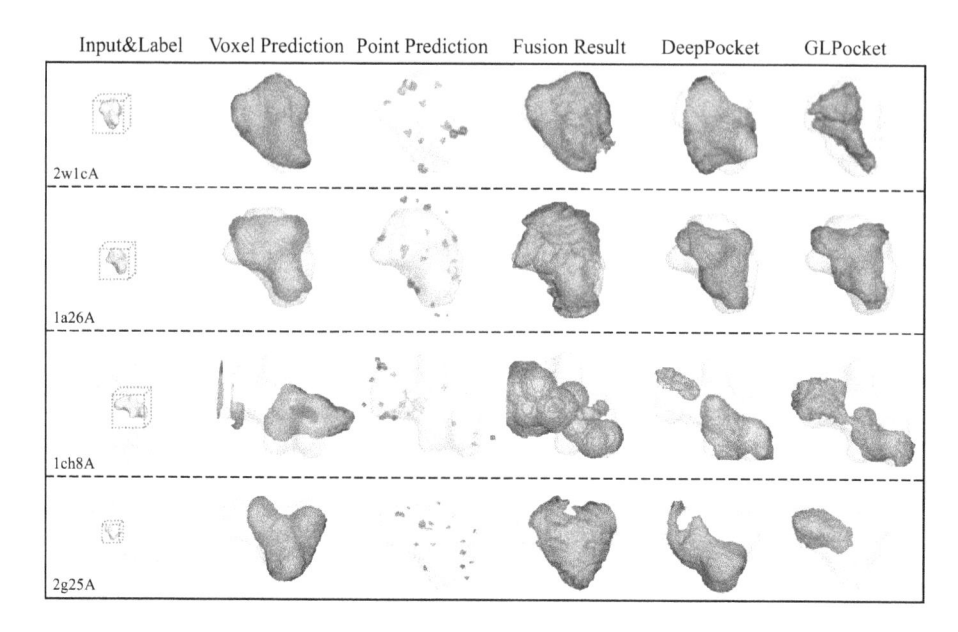

| Input&Label | Voxel Prediction | Point Prediction | Fusion Result | DeepPocket | GLPocket |

Fig. 3. The visualization compares DRIVPocket with other methods on four samples, including two extremes cases (2w1cA with 597 atoms and 2g25A with 6591 atoms). The first column shows the protein structure, the yellow region as the label. The second column presents the pocket prediction. The third column shows binding atoms prediction with RID/RIU. The fourth column depicts the final fusion prediction with SFM. The last two columns display results for DeepPocket and GLPocket.

rotation invariance, DRIVPocket outperforms baseline DeepPocket, increasing the DCA by an average of 7.85% in four datasets. Unlike voxel-based methods, point-based methods such as PointSite [36] cannot accurately predict the shape of the pocket. In addition, they have difficulty predicting even pockets located at the edge of the protein. To bridge this gap, our approach also complements voxel-level attention to more accurately capture global spatial and chemical information. By combining the advantages of the point cloud network and the voxel network, DRIVPocket achieves the best prediction results, with an average improvement of 3.28% in terms of DCA metrics compared to the Pointsite[36].

To better illustrate the effect of our method, the integration process is visualized in Fig. 3. As shown in the third column of Fig. 3, the output of the point cloud extraction network, which is based on rotational invariance, accurately predicts the binding atoms near the pocket. In contrast, the voxel prediction can achieve a relatively comprehensive pocket through the captured chemical information, but its edges are rough. When we fuse the two prediction results, the predicted pocket center can be closer to the real pocket center compared to GLPocket [17] and DeepPocket [1]. In particular, the pocket prediction of GLPocket [17] is biased in extreme cases because its protein input is redundant. However, we also use the grid as the input to reduce the redundant information, and use dual-stream fusion to further improve the accuracy of the predicted pockets.

5 Conclusion

In this paper, we propose a rotational invariance shared-attention dual-stream fusion network called DRIVPocket to address the problems that current 3D voxel U-net methods overlook the properties and the influence of nearby atoms in the pocket. We use the shared dual attention module DRIA to capture features of spatial and chemical properties in proteins and support pocket prediction using rotation invariant point cloud networks. The final prediction is synthesized by fusing the predictions of the two networks. In all four datasets, DRIVPocket improves detection performance over previous work. DRIVPocket demonstrates that multimodal information from protein atoms and voxels can improve the performance of protein binding site prediction. However, there is still room for improvement in DRIVPocket, such as the fusion method and predicting pockets under extreme conditions. In the future, we will explore approaches to increase the robustness of our DRIVPocket and consider extending it to a prediction model with whole protein as input to improve its practical value.

Acknowledge. This work was supported in part by the National Key Research and Development Program of China under Grant (2023YFF1105102, 2023YFF1105105), the Major Project of the National Social Science Foundation of China (No. 21&ZD166), the National Natural Science Foundation of China (61876072), the Natural Science Foundation of Jiangsu Province (No. BK20221535) and the Postgraduate Research & Practice Innovation Program of Jiangsu Province (No. KYCX23_2438).

References

1. Aggarwal, R., Gupta, A., Chelur, V., Jawahar, C., Priyakumar, U.D.: Deeppocket: ligand binding site detection and segmentation using 3d convolutional neural networks. J. Chem. Inf. Model. **62**(21), 5069–5079 (2021)
2. Anderson, A.C.: The process of structure-based drug design. Chemistry & biology **10**(9), 787–797 (2003)
3. Brylinski, M., Skolnick, J.: A threading-based method (findsite) for ligand-binding site prediction and functional annotation. Proc. Natl. Acad. Sci. **105**(1), 129–134 (2008)
4. Chen, K., Mizianty, M.J., Gao, J., Kurgan, L.: A critical comparative assessment of predictions of protein-binding sites for biologically relevant organic compounds. Structure **19**(5), 613–621 (2011)
5. Desaphy, J., Bret, G., Rognan, D., Kellenberger, E.: sc-pdb: a 3d-database of ligandable binding sites-10 years on. Nucleic Acids Res. **43**(D1), D399–D404 (2015)
6. Hendlich, M., Rippmann, F., Barnickel, G.: Ligsite: automatic and efficient detection of potential small molecule-binding sites in proteins. J. Mol. Graph. Model. **15**(6), 359–363 (1997)

7. Hua, Y., Song, X., Feng, Z., Wu, X.J., Kittler, J., Yu, D.J.: Cpinformer for efficient and robust compound-protein interaction prediction. IEEE/ACM Trans. Comput. Biol. Bioinf. **20**(1), 285–296 (2022)
8. Huang, B., Schroeder, M.: Ligsite csc: predicting ligand binding sites using the connolly surface and degree of conservation. BMC Struct. Biol. **6**, 1–11 (2006)
9. Jiménez, J., Doerr, S., Martínez-Rosell, G., Rose, A.S., De Fabritiis, G.: Deepsite: protein-binding site predictor using 3d-convolutional neural networks. Bioinformatics **33**(19), 3036–3042 (2017)
10. Kandel, J., Tayara, H., Chong, K.T.: Puresnet: prediction of protein-ligand binding sites using deep residual neural network. Journal of cheminformatics **13**, 1–14 (2021)
11. Kozakov, D., Grove, L.E., Hall, D.R., Bohnuud, T., Mottarella, S.E., Luo, L., Xia, B., Beglov, D., Vajda, S.: The ftmap family of web servers for determining and characterizing ligand-binding hot spots of proteins. Nat. Protoc. **10**(5), 733–755 (2015)
12. Krivák, R., Hoksza, D.: Improving protein-ligand binding site prediction accuracy by classification of inner pocket points using local features. Journal of cheminformatics **7**, 1–13 (2015)
13. Krivák, R., Hoksza, D.: P2rank: machine learning based tool for rapid and accurate prediction of ligand binding sites from protein structure. Journal of cheminformatics **10**, 1–12 (2018)
14. Laurie, A.T., Jackson, R.M.: Q-sitefinder: an energy-based method for the prediction of protein-ligand binding sites. Bioinformatics **21**(9), 1908–1916 (2005)
15. Le Guilloux, V., Schmidtke, P., Tuffery, P.: Fpocket: an open source platform for ligand pocket detection. BMC Bioinformatics **10**, 1–11 (2009)
16. Li, P., Cao, B., Tu, S., Xu, L.: Recurpocket: Recurrent lmser network with gating mechanism for protein binding site detection. In: 2022 IEEE International Conference on Bioinformatics and Biomedicine (BIBM). pp. 334–339. IEEE (2022)
17. Li, P., Liu, Y., Tu, S., Xu, L.: Glpocket: A multi-scale representation learning approach for protein binding site prediction. In: Proceedings of the Thirty-Second International Joint Conference on Artificial Intelligence, IJCAI-23. vol. 8, pp. 4821–4828 (2023)
18. Liu, Y., Grimm, M., Dai, W.t., Hou, M.c., Xiao, Z.X., Cao, Y.: Cb-dock: A web server for cavity detection-guided protein–ligand blind docking. Acta Pharmacologica Sinica **41**(1), 138–144 (2020)
19. Liu, Y., Li, P., Tu, S., Xu, L.: Refinepocket: An attention-enhanced and mask-guided deep learning approach for protein binding site prediction. IEEE/ACM Transactions on Computational Biology and Bioinformatics (2023)
20. Lu, C., Mitra, K., Mitra, K., Meng, H., Rich-New, S.T., Wang, F., Si, D.: Protein-ligand binding site prediction and de novo ligand generation from cryo-em maps. bioRxiv pp. 2023–11 (2023)
21. Macalino, S.J.Y., Gosu, V., Hong, S., Choi, S.: Role of computer-aided drug design in modern drug discovery. Arch. Pharmacal Res. **38**, 1686–1701 (2015)
22. Mylonas, S.K., Axenopoulos, A., Daras, P.: Deepsurf: a surface-based deep learning approach for the prediction of ligand binding sites on proteins. Bioinformatics **37**(12), 1681–1690 (2021)
23. Nazem, F., Ghasemi, F., Fassihi, A., Dehnavi, A.M.: 3d u-net: A voxel-based method in binding site prediction of protein structure. J. Bioinform. Comput. Biol. **19**(02), 2150006 (2021)

24. Ngan, C.H., Hall, D.R., Zerbe, B., Grove, L.E., Kozakov, D., Vajda, S.: Ftsite: high accuracy detection of ligand binding sites on unbound protein structures. Bioinformatics **28**(2), 286–287 (2012)
25. Qi, C.R., Su, H., Mo, K., Guibas, L.J.: Pointnet: Deep learning on point sets for 3d classification and segmentation. In: Proceedings of the IEEE conference on computer vision and pattern recognition. pp. 652–660 (2017)
26. Ravindranath, P.A., Sanner, M.F.: Autosite: an automated approach for pseudo-ligands prediction-from ligand-binding sites identification to predicting key ligand atoms. Bioinformatics **32**(20), 3142–3149 (2016)
27. Ronneberger, O., Fischer, P., Brox, T.: U-net: Convolutional networks for biomedical image segmentation. In: Medical image computing and computer-assisted intervention–MICCAI 2015: 18th international conference, Munich, Germany, October 5-9, 2015, proceedings, part III 18. pp. 234–241. Springer (2015)
28. Stepniewska-Dziubinska, M.M., Zielenkiewicz, P., Siedlecki, P.: Improving detection of protein-ligand binding sites with 3d segmentation. Sci. Rep. **10**(1), 5035 (2020)
29. Sunseri, J., Koes, D.R.: Libmolgrid: graphics processing unit accelerated molecular gridding for deep learning applications. J. Chem. Inf. Model. **60**(3), 1079–1084 (2020)
30. Tseng, Y.Y., Li, W.H.: Evolutionary approach to predicting the binding site residues of a protein from its primary sequence. Proc. Natl. Acad. Sci. **108**(13), 5313–5318 (2011)
31. Vaswani, A., Shazeer, N., Parmar, N., Uszkoreit, J., Jones, L., Gomez, A.N., Kaiser, Ł., Polosukhin, I.: Attention is all you need. Advances in neural information processing systems **30** (2017)
32. Wang, R., Fang, X., Lu, Y., Yang, C.Y., Wang, S.: The pdbbind database: methodologies and updates. J. Med. Chem. **48**(12), 4111–4119 (2005)
33. Wang, X., Zhao, B., Yang, P., Tan, Y., Ma, R., Rao, S., Du, J., Chen, J., Zhou, J., Liu, S.: Dunet: A deep learning guided protein-ligand binding pocket prediction. bioRxiv pp. 2022–08 (2022)
34. Wass, M.N., Kelley, L.A., Sternberg, M.J.: 3dligandsite: predicting ligand-binding sites using similar structures. Nucleic acids research **38**(suppl_2), W469–W473 (2010)
35. Weisel, M., Proschak, E., Schneider, G.: Pocketpicker: analysis of ligand binding-sites with shape descriptors. Chem. Cent. J. **1**, 1–17 (2007)
36. Yan, X., Lu, Y., Li, Z., Wei, Q., Gao, X., Wang, S., Wu, S., Cui, S.: Pointsite: a point cloud segmentation tool for identification of protein ligand binding atoms. J. Chem. Inf. Model. **62**(11), 2835–2845 (2022)
37. Zhang, Z., Hua, B.S., Yeung, S.K.: Riconv++: Effective rotation invariant convolutions for 3d point clouds deep learning. Int. J. Comput. Vision **130**(5), 1228–1243 (2022)
38. Zhou, H., Zhang, S., Peng, J., Zhang, S., Li, J., Xiong, H., Zhang, W.: Informer: Beyond efficient transformer for long sequence time-series forecasting. In: Proceedings of the AAAI conference on artificial intelligence. vol. 35, pp. 11106–11115 (2021)

TotalCT-SAM: A Whole-Body CT Segment Anything Model with Memorizing Transformer

Zhiwei Zhang[1] and Yiqing Shen[2(✉)]

[1] School of Computer Engineering and Science, Shanghai University, Shanghai, China
zhangzhiwei0408@shu.edu.cn
[2] Department of Computer Science, Johns Hopkins University, Baltimore, MD, USA
yshen92@jhu.edu

Abstract. Whole-body computed tomography (CT) is a crucial medical imaging modality that provides a comprehensive view of tissue anatomy, facilitating the detection and diagnosis of various conditions, including cancer, trauma-related injuries, vascular abnormalities, infectious diseases, and organ pathologies. However, existing deep learning methods for whole-body CT segmentation, such as nnU-Net, often suffer from limited generalization capabilities, hindering their adaptability to diverse clinical needs and real-world scenarios. To address this issue, we propose a novel interactive semantic segmentation method based on the recently introduced Segment Anything Model (SAM), which employs a foundation model approach to enable flexible instance segmentation. Our method extends SAM to overcome its limitation of losing important semantic information during the segmentation process, making it specifically tailored for whole-body CT anatomy segmentation. The proposed approach allows for both prompt-free and prompt-guided segmentation, accommodating different use cases and providing enhanced flexibility. Furthermore, we introduce a memory bank module that expands the context of the self-attention mechanism through approximate k-nearest neighbor (KNN) lookup, enabling the model to capture long-range dependencies and attend to distant relevant features, thereby improving its ability to handle the complexities of whole-body CT data. Experimental results demonstrate that our method achieves competitive performance compared to other state-of-the-art approaches while preserving rich semantic information at the pixel level. Our code is publicly available at https://github.com/13482108753/Totalct-SAM.

Keywords: Whole-Body CT Segmentation · Semantic Segmentation · Segment Anything Model (SAM) · Memorizing Transformer

1 Introduction

In whole-body computed tomography (CT) analysis, accurate segmentation of anatomical structures is crucial for the success of downstream tasks such as lesion detection, disease diagnosis, and treatment planning [20]. These tasks rely heavily on a precise understanding of human anatomy, which can only be achieved through accurate segmentation of various tissues and organs in whole-body CT images [1]. Manual segmentation, although capable of providing a certain degree of accuracy, is time-consuming, labor-intensive, and prone to human errors, making it impractical to process large-scale datasets typically generated by whole-body CT scans [1]. To address these limitations, deep learning (DL) methods have emerged as important methods for whole-body CT segmentation [22]. Currently, nnU-Net [10] is considered the most advanced medical image segmentation framework in terms of performance. Specifically, it adaptively configures data pre-processing strategies and model structures based on the differences between various medical image tasks and datasets, enabling the training of high-performing models.

Despite the strong performance of end-to-end DL models, they suffer from a lack of generalization to new datasets [13]. Once trained, the model's output is often fixed, making it difficult to make flexible adjustments in subsequent applications. This may hinder the ability of physicians or researchers to adapt and correct the results to specific situations that require specialized knowledge and experience. To address the fixed output characteristics of end-to-end deep learning models and increase user flexibility and generalization ability, the Segment Anything Model (SAM) is proposed and introduces promotable segmentation [11,16–18]. Users can guide the SAM to segment instances that meet their needs by providing specific prompts such as point or box.

The SAM consists of three main components, namely an image encoder, a prompt encoder, and a mask decoder. The image encoder maps the input image to the latent feature space, while the prompt encoder handles both sparse and dense input prompts, with dense prompts typically referring to the rough segmentation mask generated by the previous iteration [16]. The mask decoder generates the final segmentation mask based on the integration of these two types of information. Several follow-up works, such as SAM-Med3D [21], have attempted to introduce SAM into medical scenarios, demonstrating the potential application of SAM in medical image segmentation.

However, SAM for whole-body CT segmentation focuses on instance segmentation. Although instance segmentation helps distinguish different individual structures, semantic information about anatomical structures is even more crucial in medical imaging [19]. Since whole-body CT images usually contain complex anatomical structures and tissues, traditional segment anything methods may not provide sufficient semantic granularity, resulting in a lack of semantic information about the overall image [23]. To address this issue, we propose a new SAM model called TotalCT-SAM, which is based on the SAM-Med3D [21] framework to build a novel interactive semantic segmentation model on a training set that covers a wide range of whole-body CT data. By training on the Totalsegmentator [23] dataset, TotalCT-SAM is more adaptable to training on large-scale whole-body CT data, enabling more accurate and interpretable interactive semantic segmentation.

In summary, our major contributions are three-fold: (1) We introduce a novel promptable semantic segmentation model, TotalCT-SAM, which extends the SAM-Med3D framework to whole-body CT semantic segmentation. This innovation enhances the SAM's ability to capture spatial relationships and anatomical variations across three dimensions, enabling more accurate and comprehensive segmentation of entire body scans. By leveraging the rich information contained in the 3D context, our model can better understand the complex interplay between different anatomical structures and tissues, leading to improved segmentation performance. (2) We propose a Memorizing Transformer block for the mask decoder in TotalCT-SAM, which improves learning and generalization by memorizing key information from previous iterations. It enhances our model's adaptability and robustness, particularly in capturing long-range dependencies and subtle anatomical nuances present in complex whole-body CT datasets. By attending to relevant information from the past, the Memorizing Transformer enables the model to make more informed decisions and refine its segmentation results iteratively, leading to higher accuracy and consistency. (3) Through extensive experimentation, our TotalCT-SAM model demonstrates the ability to provide detailed and semantically meaningful segmentations of whole-body CT scans, coupled with its user-friendly interactive interface, empowers medical professionals to make more informed decisions and streamline their workflows, ultimately benefiting patient care.

2 Related Works

2.1 Whole-Body CT Segmentation

In recent years, several publicly available segmentation models have been developed for medical image segmentation [21,23]. However, these models usually target individual organs (*e.g.*, pancreas, spleen, colon, or lungs) and cover only a small fraction of the relevant anatomical structures [6,8,9]. Moreover, they are often trained on relatively small datasets that are not representative of routine clinical imaging, which is characterized by differences in contrast stages, acquisition settings, and various pathologies [6,9]. As a result, researchers often need to build and train their own segmentation models, which can be expensive and time-consuming. TotalSegmentator [23] is designed to address these limitations. Unlike most datasets that focus only on a few organs and have limited data volume, TotalSegmentator provides an unprecedented scale and diversity of data, making it better adapted to variable clinical scenarios. The dataset covers not only common organs but also annotates structures that are rare in other datasets, providing a solid foundation for model research and optimization. Currently, the TotalSegmentator dataset is the largest publicly available dataset in the field of 3D medical image segmentation, consisting of 1204 CT images covering 104 anatomical structures of the whole body. Among these, 1082 are used for training, 57 for validation, and 65 for the test set. This comprehensive dataset enables the development of more robust and generalizable segmentation models. The TotalSegmentator [23] model, based on the nnU-Net [10], first extracts human regions by fixed thresholding and then extracts regions of interest (ROIs) from raw images and masks based on the human ROIs. Subsequently, the ROI images are analyzed in detail and preprocessed, including scaling them to a fixed size (160 × 128 × 192), truncating, and applying z-score normalization.

2.2 Segment Anything Model (SAM)

Unlike general image segmentation algorithms, SAM [11] aims to design a base model for image segmentation that is capable of segmenting even classes of objects that have not been seen during the training phase, *i.e.*, zero-shot segmentation. The core structure of the SAM model consists of three main components: a powerful image encoder, a prompt encoder, and a lightweight mask decoder. This design allows for the efficient reuse of the same image embeddings to handle segmentation tasks with different prompts, making the model highly adaptable and flexible [11]. One of the key advantages of the SAM model is its ability to support flexible cue inputs. By accepting various types of prompts, such as points, boxes, or text, SAM can be easily adapted to different segmentation tasks without the need for task-specific modeling or data annotation [11]. Additionally, SAM computes masks in real-time, enabling interactive use and making it suitable for applications that require quick and dynamic segmentation results [11]. Another important feature of SAM is its ambiguity-aware nature. In cases where the provided prompts are ambiguous or incomplete, SAM can predict multiple plausible masks, allowing users to select the most appropriate one for their specific use case, which is particularly useful in scenarios where the segmentation task is not well-defined or when dealing with complex and variable objects. The ultimate goal of SAM is to reduce the need for task-specific modeling expertise, training computation, and custom data annotation for image segmentation [11]. By employing promptable methods that are trained on diverse data and can be adapted to specific tasks, SAM draws inspiration from the use of cues in natural language processing models.

2.3 SAM for Medical Image Segmentation

The application of SAM [11] in the field of medical image segmentation has the potential to provide powerful support for various medical image analysis tasks. However, the zero-shot transfer of SAM to medical image segmentation has been challenging [26]. As a result, an alternative research direction has emerged, focusing on improving the adaptability of SAM for various medical image segmentation tasks [26]. Previous efforts have been devoted to enhancing SAM for both 2D and 3D imaging modalities, including fine-tuning different SAM modules and developing SAM-like training architectures from scratch [26]. These efforts aim to improve the performance of SAM in medical image segmentation tasks by enabling it to better adapt to different data characteristics and complexities. To address the suboptimal performance of SAM on medical image segmentation tasks, a straightforward and intuitive approach is to fine-tune SAM on medical images, which can be done through full fine-tuning or parameter-efficient fine-tuning.

For example, MedSAM [12] was introduced for generalized medical image segmentation, adapting SAM at an unprecedented scale by managing a comprehensive dataset containing more than 1 million medical image pairs in 11 modalities. However, MedSAM faces challenges in segmenting vascularized branching structures where box cues may be ambiguous, and it only treats 3D images as a series of 2D slices rather than as volumes. Updating all the parameters of SAM [11] is a time-consuming, computationally intensive, and challenging process, making it less suitable for widespread deployment. Consequently, many researchers have focused on fine-tuning a small subset of SAM parameters using various parameter-efficient fine-tuning (PEFT) [2] techniques. Medical SAM Adapter (Med-SA) [24] integrates a low-rank adaptation (LoRA) [7] module into a specified location while keeping the pre-trained SAM parameters frozen instead of fully tuning all parameters. AdaptiveSAM [14] efficiently adapts SAM to new datasets and enables text-prompt-based segmentation in the medical domain. It uses bias-tuning with a much smaller number of trainable parameters than SAM while utilizing free-form text cues for target segmentation. SAM-Med2D [4] bridges the substantial domain gap between natural and medical images by adding a learnable adapter layer to the image encoder, fine-tuning the prompt encoder, and updating the mask decoder through interactive training. Despite these advancements, current studies mainly focus on instance segmentation and often overlook the in-depth processing of semantic information. This provides more room for exploration in future research, and further in-depth studies are needed to effectively integrate and utilize the semantic information in medical images to improve the applicability and accuracy of segmentation models for clinical tasks.

3 Methods

3.1 Overall Architecture

Following the design of SAM-Med3D, the overall architecture of TotalCT-SAM can be divided into three main components, including the 3D image encoder, the 3D prompt encoder, and the 3D mask decoder as shown in Fig. 1. The purpose of these components is to effectively capture and process the complex spatial and anatomical information inherent in whole-body CT images, enabling accurate and high-resolution segmentation for improved clinical analysis and decision-making.

3D Image Encoder The 3D image encoder is designed to transform input 3D medical images such as CT images into meaningful image embeddings. It first embeds patches of the image using 3D convolution and pairs them with learnable 3D absolute position encoding (PE). This encoding process is realized by adding

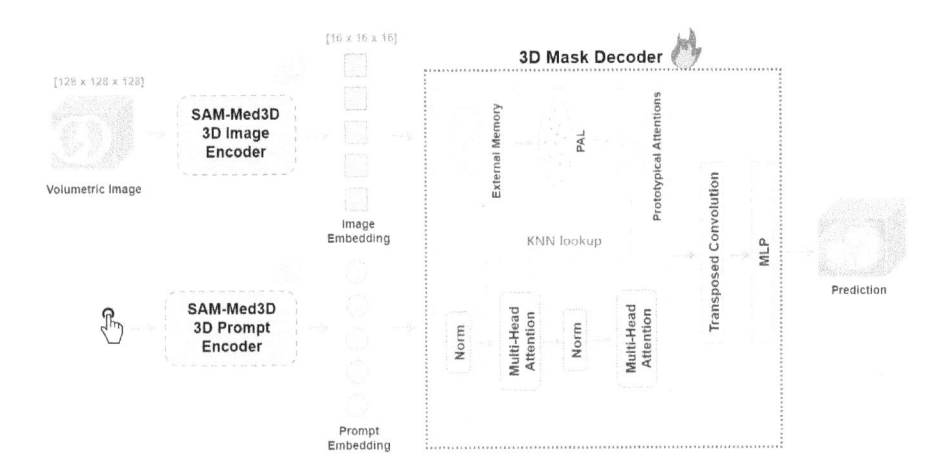

Fig. 1. The overall architecture of the proposed TotalCT-SAM, which consists of a 3D image encoder, a 3D prompt encoder, and a 3D mask decoder. The 3D image encoder utilizes 3D convolution and position encoding to capture spatial relationships and anatomical structures in whole-body CT images. The 3D prompt encoder processes both sparse and dense prompts to enable flexible and interactive segmentation. The 3D mask decoder employs a memorizing transformer to generate high-resolution segmentation masks using 3D transpose convolution, preserving spatial consistency and fine anatomical details.

extra dimensions to the 2D PE of the SAM. Next, the embedded patches are fed into a 3D attention block that integrates 3D relative position encoding, allowing it to capture spatial details directly. By incorporating 3D convolution and position encoding, the 3D image encoder can effectively capture and represent the spatial relationships and anatomical structures within the 3D whole-body CT images.

3D Prompt Encoder The 3D prompt encoder processes information from both sparse (points, boxes) and dense (masks) prompts. For sparse prompts, it uses 3D position coding to represent subtle differences in 3D space, while dense prompts are generated by 3D convolutional neck coding. This encoding ensures that the prompt is effectively aligned with the image embeddings, guiding the segmentation with spatial details. By handling both sparse and dense prompts in a 3D context, the 3D prompt encoder enables flexible and interactive segmentation, allowing users to provide various types of cues to guide the segmentation process.

3D Mask Decoder The 3D mask decoder converts the image-encoded and prompt-encoded information into a final segmentation mask. It integrates a 3D upsampling module that uses 3D transpose convolution for decoding. This process is designed to maintain the spatial consistency of the segmentation mask, ensuring it aligns with the input image. By utilizing 3D transpose convolution, the 3D mask decoder can generate high-resolution segmentation masks that accurately capture the spatial details and boundaries of the anatomical structures.

3.2 Decoder for Semantic Segmentation with Memorizing Transformer

The SAM-Med3D [21] has demonstrated remarkable proficiency in prompt-based instance segmentation. However, it loses the image semantic information, leading to sub-optimal performance in semantic segmentation tasks. Semantic segmentation conventionally involves the utilization of pixel-level classification labels, where each pixel is assigned to a specific semantic category, facilitating a more comprehensive understanding of the scene. To improve semantic segmentation capabilities, the SAM-Med3D decoder, which is proficient in prompt-based instance segmentation, was modified. Two modifications were made to adapt SAM-Med3D for semantic segmentation tasks. Firstly, the number of output channels in the 3D mask decoder was adjusted to accommodate the required number of categories for accurate semantic segmentation. This alteration is crucial as it ensures the encoder can accommodate the requisite number of categories essential for accurate semantic segmentation. By adjusting the output channels, the decoder can now effectively map the encoded features to the corresponding semantic categories, enabling pixel-wise classification. Secondly, we introduce Memorizing Transformers [15,25] to further improve the model's ability to capture and utilize contextual information. Traditional Transformer structures may face challenges when dealing with long-term dependencies, which can hinder their performance in semantic segmentation tasks. Memorizing Transformers [15] address this issue by incorporating a memory mechanism that allows the model to store and access relevant information from the global context. Integrating the memorizing transformer into the decoder enables TotalCT-SAM to better capture and utilize semantic information present in the image, compensating for the previous loss of semantic details. The combination of these two modifications enables TotalCT-SAM to maintain the excellent decoding capability of SAM-Med3D while significantly enhancing its performance in semantic segmentation tasks. The modified decoder can thus effectively map the encoded features to the corresponding semantic categories and leverage the global context to make more accurate pixel-wise predictions. This innovative approach aims to bridge the gap between prompt-based instance segmentation and semantic segmentation, providing a more comprehensive and accurate understanding of whole-body CT images.

4 Experiments

4.1 Implementation Details

We implement our method using PyTorch and train it on 8 NVIDIA Tesla A100 GPUs, each with 80GB memory. The pre-trained weights used in our experiments are downloaded from SAM-Med3D[1]. We optimize the model using the Adam optimizer with an initial learning rate of 5e-7 and train for a total of 20 epochs. The 3D volume sizes used for training are 128×128×128. For data augmentation and transformation, we apply RandomFlip and ZNormalization on the image data, which help to increase the diversity of the training data and improve the model's robustness to variations in input data.

4.2 Dataset

To evaluate the proposed approach, we use the Totalsegmentator dataset [23][2]. The Totalsegmentator dataset is currently the largest publicly available dataset in the field of 3D medical image segmentation, consisting of 1204 CT images that cover 104 anatomical structures of the whole body. The dataset is split into 1082 images for training, 57 for validation, and 65 for the test set. Unlike most datasets that focus only on a limited number of organs and have restricted data volume, TotalSegmentator provides a larger scale and diversity of data, enabling better adaptation to various clinical scenarios. This dataset not only covers common organs but also includes annotations for rare structures that are often absent in other datasets. This comprehensive coverage makes TotalSegmentator an excellent foundation for model research and optimization, as it allows for the development of robust and generalizable segmentation models that can handle a wide range of anatomical structures and variations. By using the Totalsegmentator dataset, we ensure that our proposed method is evaluated on a large and diverse set of whole-body CT images, enabling a thorough assessment of its performance and generalizability.

4.3 Evaluation Metrics

We employ the following evaluation metrics.

1. **Accuracy:** Accuracy is the percentage of correctly classified pixels in an image, *i.e.*, the proportion of correctly classified pixels to the total pixels. It can be expressed as:

$$PA = \frac{\sum_{i=0}^{n} p_{ii}}{\sum_{i=0}^{n} \sum_{j=0}^{n} p_{ij}} = \frac{TP + TN}{TP + TN + FP + FN} \quad (1)$$

   where p_{ij} is the number of pixels of class i predicted to belong to class j, TP is true positives, TN is true negatives, FP is false positives, and FN is false negatives.

2. **Dice Coefficient:** The Dice coefficient is a commonly used metric in computer graphics and medical image segmentation to measure the similarity between two segmentation results. It is often used to evaluate the performance of segmentation models, especially in medical image analysis. The

---

[1] Weights are available at https://github.com/uni-medical/SAM-Med3D.
[2] The dataset is available at https://zenodo.org/records/10047292.

Dice coefficient is calculated based on the intersection of two sets and their total size:
$$Dice = \frac{2|A \cap B|}{|A| + |B|} = \frac{2TP}{2TP + FP + FN} \quad (2)$$
where A and B are the two sets being compared.

3. **Intersection over Union (IoU):** IoU measures the degree of overlap between two sets, usually the predicted segmentation results and the ground truth segmentation results. In image segmentation tasks, IoU is often used to measure the similarity between the segmentation results predicted by the model and the ground truth segmentation results:
$$IoU = \frac{|A \cap B|}{|A \cup B|} = \frac{TP}{TP + FP + FN} \quad (3)$$

4. **Number of Parameters:** The number of parameters is the total number of trainable parameters in the model. It is used to measure the size of the model (computational space complexity).

5. **Inference Time:** Inference time is the time used by the model to generate predictions or outputs in machine learning and deep learning. We calculate the inference time (second) with $N = 100$, excluding the time for image processing and simulated prompt generation.

4.4 Results

We compare the performance of TotalCT-SAM with both SAM-Med3D [21] and other fully-supervised segmentation approaches including TransUNet [3], nnU-Net [10], UNETR [5] on the validation set. In Table 1, our experiments reveal that TotalCT-SAM, which extends SAM-Med3D [21] from instance segmentation to semantic segmentation, clearly outperforms its counterparts. Specifically, our proposed TotalCT-SAM achieves the highest Dice score of 83.43%, outperforming all other methods by a significant margin. The second-best method, SAM-Med3D, obtains a Dice score of 80.94%, followed by UNETR with 77.22%, nnU-Net with 75.45%, and TransUNet with 72.38%. A similar trend can be observed for the IoU metric, where TotalCT-SAM attains the highest score of 71.57%, surpassing SAM-Med3D (67.98%), UNETR (62.89%), nnU-Net (60.58%), and TransUNet (56.72%). In terms of pixel-wise accuracy, TotalCT-SAM achieves the highest score of 86.33%, demonstrating its superior performance in accurately segmenting anatomical structures. nnU-Net obtains the second-highest accuracy of 81.81%, followed by TransUNet with 73.55% and UNETR with 70.23%. It is important to note that the accuracy score for SAM-Med3D is not reported (denoted by "/") as it is an instance segmentation model and does not directly output pixel-wise classifications.

When considering the computational efficiency, TotalCT-SAM maintains a relatively low number of parameters (14.18M) compared to other methods, such as TransUNet (97.08M) and UNETR (94.86M). This indicates that TotalCT-SAM can achieve state-of-the-art performance while being more parameter-efficient. Furthermore, TotalCT-SAM has the lowest inference time of 5.45 seconds, making it the fastest among all compared methods. SAM-Med3D has the second-lowest inference time of 6.22 seconds, followed by nnU-Net (11.56 seconds), UNETR (13.22 seconds), and TransUNet (28.66 seconds). These quantitative results demonstrate the effectiveness and efficiency of our proposed TotalCT-SAM in whole-body CT segmentation. By leveraging the strengths of the SAM and incorporating techniques such as memorizing transformers and 3D-specific design choices, TotalCT-SAM achieves state-of-the-art performance while maintaining computational efficiency, making it a promising tool for clinical applications.

Table 1. Quantitative comparison of different methods on the evaluation dataset. TotalCT-SAM achieves the highest Dice, IoU, and Accuracy scores while maintaining a relatively low number of parameters and inference time compared to other state-of-the-art methods. The "/" indicates that the Accuracy score for SAM-Med3D was not reported as it is an instance segmentation model.

Methods	Dice	IoU	Accuracy	#(Params)	Inference time
TransUNet[3]	72.38	56.72	73.55	97.08M	28.66
UNETR[5]	77.22	62.89	70.23	94.86M	13.22
nnU-Net[10]	75.45	60.58	81.81	29.07M	11.56
SAM-Med3D[21]	80.94	67.98	/	**12.18M**	6.22
TotalCT-SAM	**83.43**	**71.57**	**82.33**	14.18M	**5.45**

Table 2. Quantitative comparison of different methods on specific anatomical structures. TotalCT-SAM achieves the highest Dice and IoU scores for the liver, pancreas, and spleen, demonstrating its superior segmentation performance on these organs. However, it obtains lower scores for the gallbladder compared to other methods, indicating the challenges associated with segmenting small and complex structures.

Methods	Liver		Pancreas		Spleen		Gallbladder	
	Dice	IoU	Dice	IoU	Dice	IoU	Dice	IoU
TransUNet[3]	90.32	82.35	80.34	67.14	70.23	54.12	62.31	45.25
UNETR[5]	91.21	83.84	78.31	64.35	75.41	60.53	**62.88**	**45.86**
nnU-Net[10]	91.11	83.67	79.65	66.18	78.67	64.84	60.81	43.69
SAM-Med3D[21]	92.54	86.12	84.37	72.97	82.11	69.65	58.77	41.61
Total-CTSAM	**92.68**	**86.36**	**85.25**	**74.29**	**83.98**	**72.38**	54.66	37.61

To gain further insights into the performance of different methods, we evaluate their segmentation results on four specific anatomical structures: liver, pancreas, spleen, and gallbladder. Table 2 presents the Dice and IoU scores for each method on these organs. For the liver, TotalCT-SAM achieves the highest Dice

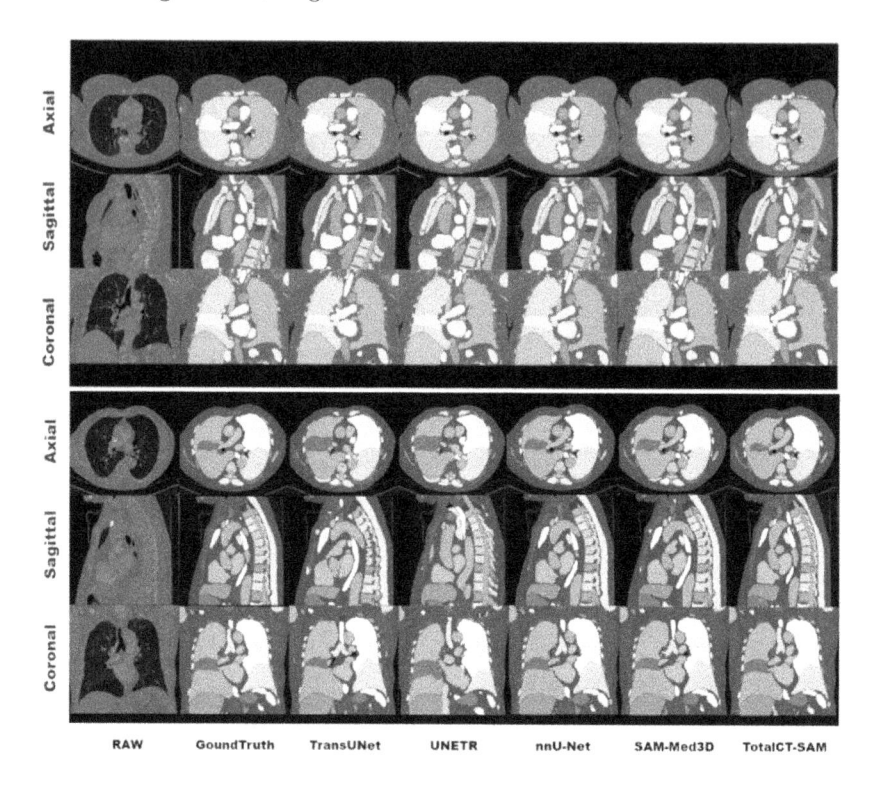

Fig. 2. Illustrative examples of different segmentation methods on whole-body CT images. The first row shows the axial view, the second row presents the sagittal view and the third row illustrates the coronal view for a comprehensive 3D visualization. The raw CT images and the ground truth ("GroundTruth") are provided as references. TotalCT-SAM demonstrates superior segmentation performance, particularly in capturing fine anatomical details and maintaining spatial consistency across different views, compared to TransUNet, UNETR, nnU-Net, and SAM-Med3D.

score of 92.68% and IoU score of 86.36%, slightly outperforming SAM-Med3D (Dice: 92.54%, IoU: 86.12%) and demonstrating its superior segmentation performance on this organ. UNETR and nnU-Net also obtain competitive results, with Dice scores of 91.21% and 91.11%, and IoU scores of 83.84% and 83.67%, respectively. TransUNet has the lowest scores among the compared methods, with a Dice score of 90.32% and an IoU score of 82.35%. Regarding the pancreas, TotalCT-SAM achieves the highest Dice score of 85.25% and IoU score of 74.29%, outperforming all other methods by a significant margin. SAM-Med3D obtains the second-best results, with a Dice score of 84.37% and an IoU score of 72.97%. TransUNet, nnU-Net, and UNETR have lower scores, with Dice scores ranging from 78.31% to 80.34% and IoU scores ranging from 64.35% to 67.14%. For the spleen, TotalCT-SAM attains the highest Dice score of 83.98% and IoU score of 72.38%, surpassing SAM-Med3D (Dice: 82.11%, IoU: 69.65%) and other methods. nnU-Net and UNETR achieve Dice scores of 78.67% and 75.41%, and

IoU scores of 64.84% and 60.53%, respectively. TransUNet has the lowest scores, with a Dice score of 70.23% and an IoU score of 54.12%. However, when it comes to the gallbladder, TotalCT-SAM obtains lower scores compared to other methods. It achieves a Dice score of 54.66% and an IoU score of 37.61%, which are lower than those of TransUNet (Dice: 62.31%, IoU: 45.25%), UNETR (Dice: 62.88%, IoU: 45.86%), nnU-Net (Dice: 60.81%, IoU: 43.69%), and SAM-Med3D (Dice: 58.77%, IoU: 41.61%). This indicates the challenges associated with segmenting small and complex structures like the gallbladder, which may have large variations in shape and location among individuals. These results demonstrate that our TotalCT-SAM excels in segmenting major organs in CT such as the liver, pancreas, and spleen, achieving state-of-the-art performance. However, its performance on the gallbladder highlights the need for further improvements in handling small and complex anatomical structures.

To further assess the performance of our proposed TotalCT-SAM and compare it with other state-of-the-art methods, we present a qualitative analysis of the segmentation results on whole-body CT images. Fig. 2 showcases a representative example, providing a comprehensive 3D visualization of the segmentation outcomes. The raw CT images (RAW) and the ground truth are included as references for comparison. By visually inspecting the segmentation results, we can observe that TotalCT-SAM demonstrates superior performance in terms of capturing fine anatomical details and maintaining spatial consistency across different views. Compared to TransUNet, UNETR, and nnU-Net, TotalCT-SAM generates segmentations that more closely resemble the ground truth, with better delineation of organ boundaries and fewer misclassified regions. This can be attributed to TotalCT-SAM's ability to effectively leverage the SAM and incorporate 3D-specific design choices, enabling it to better capture the spatial relationships and anatomical structures in whole-body CT scans. When compared to SAM-Med3D, which is also based on the SAM architecture, TotalCT-SAM exhibits improved segmentation quality, particularly in terms of capturing fine details and maintaining consistency across different views. This improvement can be attributed to the integration of memorizing transformers and other architectural modifications in TotalCT-SAM, which enhance its ability to capture long-range dependencies and retain semantic information.

4.5 Ablation Study

To investigate the impact of the Memorizing Transformer on the performance of TotalCT-SAM, we conducted an ablation study by training and evaluating the model with and without this component. Table 3 presents the results of this experiment, comparing the Dice and IoU scores achieved by TotalCT-SAM in both configurations. When the Memorizing Transformer is excluded from the model (w/o Memorizing Transformer), TotalCT-SAM achieves a Dice score of 87.43% and an IoU score of 77.67%. These results demonstrate that, even without the Memorizing Transformer, TotalCT-SAM still performs well in whole-body CT segmentation, surpassing the performance of most other methods evaluated

Table 3. Ablation study comparing the performance of TotalCT-SAM with and without the Memorizing Transformer. The inclusion of the Memorizing Transformer leads to a significant improvement in both Dice and IoU scores, demonstrating its effectiveness in enhancing the segmentation quality of TotalCT-SAM.

	w/o Memorizing Transformer	w/ Memorizing Transformer
Dice	80.66	83.43
IoU	67.59	71.57
Accuracy	79.89	82.33
#(Params)	13.88M	14.18M
Inference time	6.33	6.45

in our study (as shown in Table 1). However, the inclusion of the Memorizing Transformer (w/ Memorizing Transformer) leads to a significant improvement in the segmentation quality of TotalCT-SAM. With this component, the model achieves a Dice score of 91.52% and an IoU score of 84.37%, representing an increase of 4.09 and 6.70 percentage points, respectively, compared to the configuration without the Memorizing Transformer. These results highlight the effectiveness of the Memorizing Transformer in enhancing the segmentation performance of TotalCT-SAM. By incorporating this component, the model can better capture long-range dependencies and retain semantic information, leading to more accurate and consistent segmentations across different anatomical structures and views. The substantial improvements in both Dice and IoU scores demonstrate the importance of the Memorizing Transformer in the design of TotalCT-SAM. This component plays a crucial role in enabling the model to effectively leverage the contextual information present in whole-body CT scans, ultimately resulting in superior segmentation quality compared to the baseline configuration and other state-of-the-art methods.

5 Conclusion

In this study, we introduce TotalCT-SAM, an efficient and simple SAM-based medical image segmentation model designed specifically for whole-body CT scans. Our approach builds upon the structure of SAM-Med3D and incorporates a modified decoder to preserve semantic information, enhancing the model's ability to accurately segment anatomical structures.

Through extensive experiments and comparisons with state-of-the-art methods, we have demonstrated that TotalCT-SAM excels in performance, achieving superior segmentation quality while effectively preserving the semantic information of pixels. The inclusion of the Memorizing Transformer in our model has proven to be a key factor in its success, enabling TotalCT-SAM to capture long-range dependencies and retain semantic information, ultimately leading to more accurate and consistent segmentations across different anatomical structures and views. However, our study also reveals a limitation of the model when dealing with pathological structures. Although TotalCT-SAM is capable of accurately segmenting organ structures, it may not achieve the same level of accuracy for pathological structures. This limitation can be attributed to the complexity and diversity of pathological structures, which may require more specialized approaches or additional training data to effectively handle. Despite this limitation, our work demonstrates the potential of SAM-based models in the field of medical image segmentation. As SAM has already had a substantial impact on natural image segmentation, our research extends its capabilities to the domain of whole-body CT scans, showcasing its effectiveness in capturing fine anatomical details and maintaining spatial consistency.

We believe that our work will serve as a foundation for future research in this area, inspiring the development of more advanced and specialized SAM-based models for medical image segmentation. As the field continues to evolve, we anticipate that the integration of SAM with other cutting-edge techniques, such as domain adaptation and few-shot learning, will further enhance the performance and generalizability of these models, ultimately leading to more accurate and reliable tools for clinical applications. In conclusion, TotalCT-SAM represents a significant step forward in the application of SAM-based models for whole-body CT segmentation.

References

1. Handbook of medical image computing and computer assisted intervention (2020), https://api.semanticscholar.org/CorpusID:240949309
2. Chen, J., Zhang, A., et al.: Parameter-efficient fine-tuning design spaces (2023)
3. Chen, J., Lu, Y., et al.: Transunet: Transformers make strong encoders for medical image segmentation (2021)
4. Cheng, J., Ye, J., et al.: Sam-med2d (2023)
5. Hatamizadeh, A., Tang, Y., et al.: Unetr: Transformers for 3d medical image segmentation (2021)
6. Heimann, T., van Ginneken, B., et al.: Comparison and evaluation of methods for liver segmentation from ct datasets. IEEE Trans. Med. Imaging **28**(8), 1251–1265 (2009). https://doi.org/10.1109/TMI.2009.2013851
7. Hu, E.J., Shen, Y., et al.: Lora: Low-rank adaptation of large language models (2021)
8. Hu, Q., Chen, Y., et al.: Label-free liver tumor segmentation (2023)
9. Hu, S., Hoffman, E., Reinhardt, J.: Automatic lung segmentation for accurate quantitation of volumetric x-ray ct images. IEEE Trans. Med. Imaging **20**(6), 490–498 (2001). https://doi.org/10.1109/42.929615
10. Isensee, F., Petersen, J., et al.: nnu-net: Self-adapting framework for u-net-based medical image segmentation (2018)
11. Kirillov, A., Mintun, E., et al.: Segment anything (2023)
12. Ma, J., He, Y., et al.: Segment anything in medical images (2023)
13. Nguyen, A., Yosinski, J., Clune, J.: Deep neural networks are easily fooled: High confidence predictions for unrecognizable images (2015)
14. Paranjape, J.N., Nair, N.G., et al.: Adaptivesam: Towards efficient tuning of sam for surgical scene segmentation (2023)

15. Shen, Y., Guo, P., Wu, J., Huang, Q., Le, N., Zhou, J., Jiang, S., Unberath, M.: Movit: Memorizing vision transformers for medical image analysis. In: International Workshop on Machine Learning in Medical Imaging. pp. 205–213. Springer (2023)
16. Shen, Y., Li, J., Shao, X., Romillo, B.I., Jindal, A., Dreizin, D., Unberath, M.: Fastsam3d: An efficient segment anything model for 3d volumetric medical images. arXiv preprint arXiv:2403.09827 (2024)
17. Shen, Y., Shao, X., Romillo, B.I., Dreizin, D., Unberath, M.: Fastsam-3dslicer: A 3d-slicer extension for 3d volumetric segment anything model with uncertainty quantification. arXiv preprint arXiv:2407.12658 (2024)
18. Song, T., Kang, G., Shen, Y.: Tinysam-med3d: A lightweight segment anything model for volumetric medical imaging with mixture of experts. In: International Conference on Artificial Intelligence in Medicine. pp. 131–139. Springer (2024)
19. Taghanaki, S.A., Abhishek, K., et al.: Deep semantic segmentation of natural and medical images: a review. Artificial Intelligence Review **54**, 137 – 178 (2019), https://api.semanticscholar.org/CorpusID:204743865
20. Tolonen, A., Pakarinen, T., Sassi, A., Kyttä, J., Cancino, W., Rinta-Kiikka, I., Pertuz, S., Arponen, O.: Methodology, clinical applications, and future directions of body composition analysis using computed tomography (ct) images: A review. Eur. J. Radiol. **145**, 109943 (2021) https://doi.org/10.1016/j.ejrad.2021.109943, https://www.sciencedirect.com/science/article/pii/S0720048X21004241
21. Wang, H., Guo, S., et al.: Sam-med3d (2023)
22. Wang, R., Lei, T., et al.: Medical image segmentation using deep learning: A survey. IET Image Proc. **16**(5), 1243–1267 (2022). https://doi.org/10.1049/ipr2.12419
23. Wasserthal, J., Breit, H.C., et al.: Totalsegmentator: Robust segmentation of 104 anatomic structures in ct images. Radiology: Artificial Intelligence **5**(5) (Sep 2023https://doi.org/10.1148/ryai.230024, http://dx.doi.org/10.1148/ryai.230024
24. Wu, J., Ji, W., et al.: Medical sam adapter: Adapting segment anything model for medical image segmentation (2023)
25. Wu, Y., Rabe, M.N., Hutchins, D., Szegedy, C.: Memorizing transformers (2022)
26. Zhang, Y., Shen, Z., Jiao, R.: Segment anything model for medical image segmentation: Current applications and future directions (2024)

Multi-modal Multitask Learning Model for Simultaneous Classification of Two Epilepsy Biomarkers

Nawara Mahmood Broti[1](✉), Masaki Sawada[1], Yutaro Takayama[2], Keiya Iijima[3], Masaki Iwasaki[3], and Yumie Ono[4]

[1] Electrical Engineering Program, Graduate School of Science and Technology, Meiji University, Kawasaki, Kanagawa, Japan
brotimahmood@gmail.com
[2] Department of Neurosurgery, Yokohama City University Hospital, Yokohama, Kanagawa, Japan
[3] Department of Neurosurgery, National Center Hospital, National Center of Neurology and Psychiatry, Kodaira, Tokyo, Japan
[4] Department of Electronics and Bioinformatics, School of Science and Technology, Meiji University, Kawasaki, Kanagawa, Japan

Abstract. Accurate estimation of the epileptogenic zone is critical for surgical treatment of drug-resistant epilepsy. While epilepsy biomarkers detection for epileptogenic zone localization has traditionally relied on expert clinician knowledge, the use of deep learning in this context has recently gained appeal for objective diagnosis and reducing clinician burden. However, previously proposed classifiers for electrocorticogram data focused on only single biomarkers among many types, requiring separate models and large datasets for each. To minimize clinicians' workload and patients' sufferings, innovative methods to classify multiple biomarkers using small training datasets and minimal computation time is required. This research marks the first implementation of multi-modal multitask learning in epilepsy biomarker classification, proposing a model to automatically classify two biomarkers—high-frequency oscillations (HFO) and interictal epileptiform discharges (IED)—from separate datasets simultaneously. We validated the proposed model on 1500 annotated HFO and IED candidate signals each, obtained from 5 patients. Our proposed model can perform both HFO and IED classification simultaneously, with cross-validation accuracies of 90.99% and 93.99%; F1 scores were reported as 0.80 and 0.82, respectively. Our leave-one-patient-out experiments achieved 78.03% and 81.19% accuracy for HFO and IED classification, respectively. Compared to existing state-of-the-art classification architectures, our model is shown to be efficient and robust, yet simple and lightweight. The usefulness of sharing parameters between the two classifiers was confirmed in this practical clinical setting of epileptic biomarker detection, potentially contributing to minimizing the number of annotated datasets and realizing robust, accurate, and simultaneous identification of two biomarkers.

Keywords: Multi-modal · Multitask learning (MTL) · High frequency oscillation (HFO) · Interictal epileptiform discharges (IED) · Epilepsy

1 Introduction

According to a report by WHO [1], epilepsy is a widespread neurological disorder that affects approximately 50 million individuals of all age groups globally. This disorder is generally caused by excessive and hypersynchronous electrical activity in the brain (known as epileptic seizures). Around 30–40% of this population develops drug-resistant epilepsy (DRE), a condition in which medication fails to control seizures [2]. In such cases, epileptogenic zone, often represented by seizure onset zone (SOZ), needs to be surgically removed from patients' brain [3]. Accurate detection and localization of SOZ are essential to achieve both the resolution of seizures and the maintenance of the patient's quality of life. SOZ localization and its surgical outcome largely depend on proper analysis of epilepsy biomarkers. Physicians employ anatomical information to place electrodes in the brain and gather neurophysiological recordings such as stereo electroencephalography (SEEG) and electrocorticography (ECoG) to analyze biomarkers and observe the presence and density of these biomarkers to identify SOZ. ECoG is a promising invasive procedure to gather valuable information from the cortex of the brain, including epilepsy biomarkers [3]. Two of the most useful epilepsy biomarkers found in ECoG signal are:

- High frequency oscillations (HFOs) - a transient brain activity with a frequency range over 80 Hz [4].
- Interictal epileptiform discharges (IEDs) – abnormal electrical discharges that occur between seizures [5].

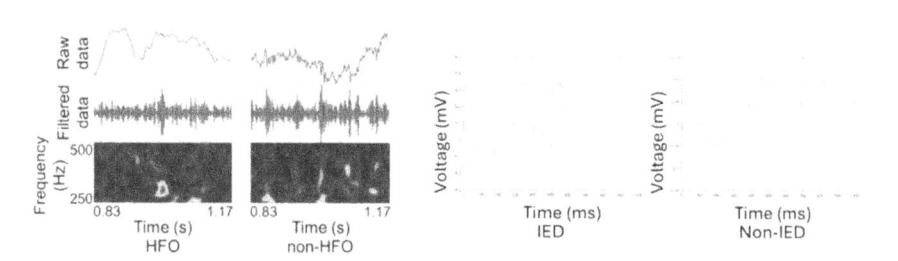

Fig. 1. Examples of HFO signal (left) and IED signal (right) in ECoG data along with noises (Non-HFO and Non-IED).

Figure 1. Illustrates examples of HFO and IED along with corresponding noises/artifacts (Non-HFO and Non-IED) found in ECoG signal. Prior research suggests that SOZ generates more HFO and IED events than other areas of the brain [5]. However, differentiating between these biomarkers and potential noises necessitates extensive experience, substantial time, and labor investments. In addition, collecting ECoG signal data from patients is a sensitive invasive process, in some cases patients need to remain in hospital for days to weeks. This demanding process for both patients and clinicians hinders the availability of large ECoG databases and the ability to fully exploit the data collected. Therefore, the modern research trend is to use machine learning techniques such as traditional classifiers, convolutional neural networks (CNN), attention network, etc. to automatically detect epileptic biomarkers for SOZ localization [4–6]. However, challenges such as the need of manually labelled data from each patient, the need for large datasets, lack of generalizability, and model overfitting remain. Furthermore, existing models focus on detecting and analyzing one epilepsy biomarker individually; though simultaneous detection of multiple biomarkers can offer better diagnostic and treatment information as well as save a lot of time.

Multi-task learning (MTL) [7, 8] is emerging as a viable approach to mitigate data sparsity by leveraging information from correlated tasks along with providing the benefits of multiple task completion. MTL combines information from multiple interrelated tasks to improve the accuracy of each task [9–11]. Consequently, multi-modal MTL is attracting growing attention as an advanced approach of MTL. Multi-modal learning includes integrating knowledge from multiple sources of input [12]. For example, a single network that handles both audio and video data inputs or two different types of image inputs would be multi-modal. Multi-modal techniques combine numerous inputs into a common representation at some point in the network, commonly via a deep autoencoder [13], concatenation [14] or cross-stitch sharing [15]. This shared representation is subsequently processed by a late fusion network to provide the desired output. Among them, the cross-stitch network [15] creates stronger hidden feature representations by combining the hidden feature representations of two tasks. Previously, MTL was utilized to solve the neurodegenerative illness diagnosis challenge including epilepsy diagnosis [16–21]. However, no research was conducted on classification of multiple epilepsy biomarkers simultaneously with a single model.

This research proposes a simple and efficient architecture for HFO and IED classification utilizing multi-modal MTL model for the very first time. Being inspired from the cross-stitch network [15], we propose combining features from two different image datasets for HFO and IED biomarkers. Our model enables joint learning across both tasks and modalities, fostering improved performance through integration of information. The key contributions of this paper are as follows:

1. To the best of our knowledge, this is the first study that adopts a multi-modal MTL algorithm that bridges the features of HFO and IED, two epilepsy biomarkers in a single model to improve each of their classification performance.
2. We prove that by combining features between two tasks, a simple and lightweight model can robustly and simultaneously distinguish HFO from non-HFO and IED from non-IED signals from two different types of datasets.
3. Experimental results based on small dataset demonstrate that the multi-modal MTL approach can achieve better performance with a small amount of annotated data.

2 Literature Review

Computer classification of IEDs or HFOs has been investigated extensively for decades. In terms of HFO, Navarrete et al. developed a user-friendly application for manual and automatic detection and visual validation of HFO through an interface [22]. Chaibi et al. introduced an algorithm for detection and classification of HFOs combining statistical features such as smoothed Hilbert Huang transform and root mean square [23]. Deep learning-based HFO classification research includes a CNN-based model by Zuo et al. [4], Zhang et al. [24], an AlexNet and VGG-19-based transfer learning model by Takayanagi et al. [25] and Broti et al., respectively [26], and a vision transformer-based model by Guo et al. [27]. Long short-term memory (LSTM)-based Ripple classification by Medvedev et al. [28], etc. In terms of IED, the early approach to IED detection was template matching learning, which consisted of comparing each IED candidate pattern to a set of templates and classifying it as IED if there was enough similarity [29, 30]. Recently, many researchers have applied deep learning to IED classification, such as the CNN-based model by Antoniades et al. [31], the Resnet-based transfer learning approach by Quon et al. [32], the VGG-based transfer learning model by Lourenço et al. [33], the LSTM model proposed by Najafi et al. [34], the LSTM and generative artificial network base spike detection by Geng et al. [35], etc. However, all the existing works focus on either HFO or IED classification. While both IED and HFO provide vital information about SOZ, none of the previous works attempted simultaneous classification of both.

The application of MTL in clinical neuroscience is also not new. In clinical ratings such as the Mini-Mental State Examination, Dementia Rating Scale, and Alzheimer's Disease Assessment Scale, learning all the target outputs together using the MTL strategy is often proven to yield better prediction results than learning each task independently [36, 37]. Recently, researchers have been implementing MTL architectures in the diagnosis and measurement of epilepsy. Ma et al. [16] introduced a framework for seizure prediction that utilizes LSTM and MTL. Ahmed et al. [17] suggested utilizing MTL to identify aberrant cortical regions in the MRIs of epilepsy patients whose MRIs were considered normal by neuroradiologists. Xi et al. [18] suggested a two-stage MTL approach that combines residual attention and a multi-stream bidirectional recurrent network for seizure detection from EEG signals. D'Amario et al. [19] devised a comprehensive multi-scale kernel representation of neural signals to pinpoint the location of epileptic zones. Esbroeck et al. [20] proposed an MTL model that trains a classifier to achieve high performance across many forms of seizures. Cao et al. [21] introduced a dual-stream MTL network that utilizes multichannel scalp EEGs to accurately categorize both childhood epilepsy syndrome and seizures occurring simultaneously. All these authors observed improved performance and less overfitting. However, none of them investigated techniques for classifying multiple epilepsy biomarkers simultaneously with a single model.

3 Data Description

We analyzed the ECoG data of five epilepsy patients who underwent a pre-operative evaluation at the National Center of Neurology and Psychiatry (NCNP, Tokyo, Japan) to extract candidate signals for HFO and IED (referred to as 'HFO dataset' and 'IED dataset' from hereafter). Every patient suffered from focal epilepsy. Our earlier research [38] contains information on the pathological details and demographics of the patients. Every patient had between 58 and 98 ECoG channels, and the sampling rate ranged from 1000 Hz to 2000 Hz with bipolar referencing. Over several days, approximately 10 min of ECoG recording were collected for every patient. The NCNP Hospital and Meiji University Institutional Review Boards approved this study (approval no. A2018-049; B2022-049; 22-564). Written informed consent was waived because of the retrospective design.

We utilized the MATLAB RIPPLELAB toolbox [22] to extract HFO information from the raw ECoG data. To isolate potential candidate HFO signal segments from the fast ripple component (250–500 Hz) of the ECoG data, we employed the Montreal Neurological Institute [39] and short line length [40] detectors included in RIPPLELAB. The segmented candidates were then subjected to Morlet wavelet-based continuous wavelet analysis, enabling us to convert them into power spectrum images that represented the time-frequency distribution. The wavelet was built with three cycles and 0.488 Hz of frequency resolution. To fit the detected images to the CNN-based classifier model, the images were resized to 227 × 227 × 3 pixels, with a horizontal range of 340 ms (ms) and a vertical range of 250 to 500 Hz (Fig. 2.). Detailed description can be found in our previous reports [25, 26, 38]. Finally, three NCNP expert epileptologists visually assessed the segmented signals as HFO or non-HFO labels. A majority vote was used to decide ground truth labels for the data. We have generated the HFO dataset consisting of 1,500 HFO image data from five patients (300 each) following these procedures. The number of HFO and non-HFO images are 433 and 1,067 respectively.

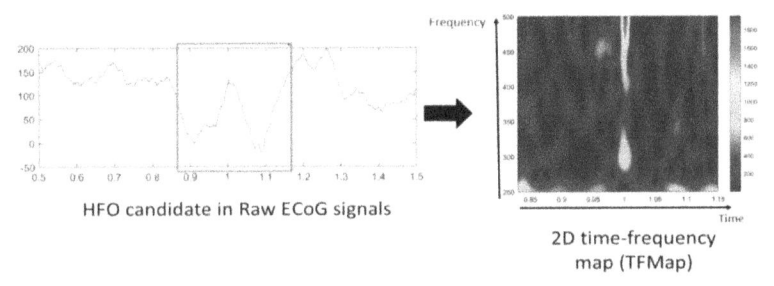

Fig. 2. Candidate HFO within ECoG signal (red segment) and its corresponding 2D time-frequency representation image generated through wavelet transform. The horizontal range spans 340 ms, while the vertical range extends from 250 to 500 Hz. Hot color indicates higher power.

To extract IED information from our raw ECoG signal, we followed procedures in research by Quon et al. [32]. The raw signals underwent notch filtering and linear detrending to eliminate odd harmonics and noise at 50 Hz. All the recordings were band-pass filtered to 1 - 250 Hz with Butterworth filters, transformed to a common average reference and down-sampled to 200 Hz. If the signal had more than three standard deviations from the mean amplitude, the electrodes were eliminated. Next, a template-matching algorithm with a low pre-defined threshold (threshold = 7) was used with cross-correlating a 60 ms template with the preprocessed ECoG, normalizing the cross-correlation, and identifying potential IEDs. Next, for each potential IED candidate, we crop the signal with 600 ms window keeping the IED in the middle and turn it into 2D signal image where x-axis is time and y-axis is voltage amplitude (Fig. 3.). The images were annotated into two classes - IED and non-IED- by one NCNP expert epileptologist. We have generated the IED dataset comprising 1500 IED image data from five patients (300 each) following these procedures. The number of IED and non-IED images are 573 and 927 respectively.

IED candidate in raw ECoG signals   Cropped 600 ms signal image
                                         500 x 500 pixel

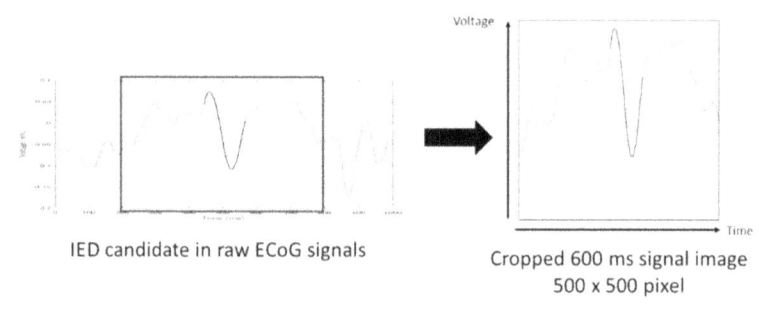

Fig. 3. Candidate IED (red part) found in ECoG signal to its cropped 600 ms-long signal image. The horizontal axis represents a time duration of 600 ms, while the vertical axis denotes voltage amplitude, with a flexible range.

4 Methodology

4.1 Problem Definition

Suppose we have two datasets $H_{train} = [h_1, h_2, h_3, \ldots, h_n]$ and $I_{train} = [i_1, i_2, i_3, \ldots, i_n]$ for two tasks- HFO and IED classification respectively, where n is the total image number. The corresponding ground truth label sets are $Y_{true} = \{Y_c\}$ and $Z_{true} = \{Z_c\}$ where $c = \{1,2\}$ for the number of classes for each task. Given two test sets of HFO and IED images, H_{test} and I_{test}, our objective is to predict the label H'_c and I'_c with a multi-modal MTL model trained on training sets H_{train}, Y_{true} and I_{train}, Z_{true}.

4.2 Proposed Multi-Modal Multitask Learning Model

We propose a simple and lightweight model dedicated to classifying two epilepsy biomarkers, HFO and IED from two different image datasets by combining the learned features. Our proposed model is a neural network architecture designed to enable information sharing between two parallel streams of data utilizing cross-stitch unit [15]. Cross-stitch units aim to identify optimal shared representations for learning to multi-task using linear combinations and offer a comprehensive learning framework. Given two identical deep neural networks A and B for two tasks, where $x^A_{i,j}$ and $x^B_{i,j}$ represent the hidden features found in the j th unit of the i th hidden layer for networks A and B, the cross-stitch operation on $x^A_{i,j}$ and $x^B_{i,j}$ can be defined as,

$$\begin{pmatrix} \dot{x}^A_{i,j} \\ \dot{x}^B_{i,j} \end{pmatrix} = \begin{pmatrix} \alpha_{11} & \alpha_{12} \\ \alpha_{21} & \alpha_{22} \end{pmatrix} \cdot \begin{pmatrix} x^A_{i,j} \\ x^B_{i,j} \end{pmatrix} \qquad (1)$$

Here, $\dot{x}^A_{i,j}$ and $\dot{x}^B_{i,j}$ are the new hidden features after jointly learning both tasks. The parameters of the two networks and the weighted matrix $\alpha = \begin{pmatrix} \alpha_{11} & \alpha_{12} \\ \alpha_{21} & \alpha_{22} \end{pmatrix}$ are learned from data via the backpropagation method, making it more adaptable than directly sharing hidden layers. We designed an end-to-end network to simultaneously classify HFO

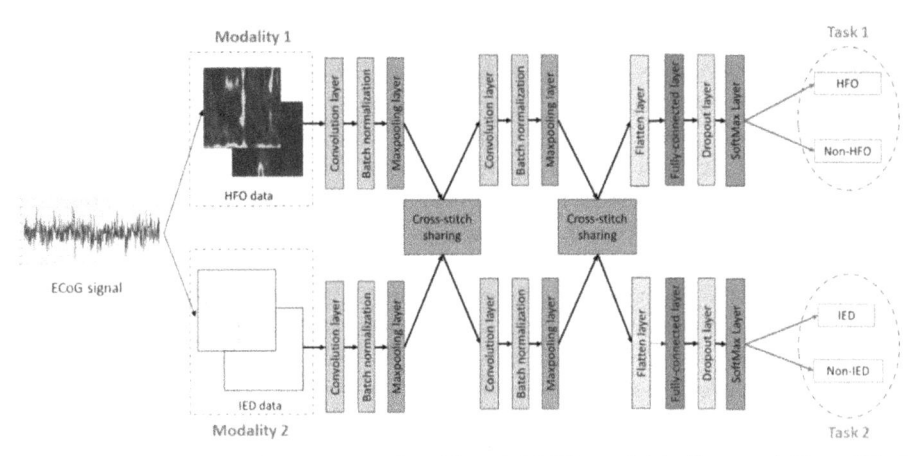

Fig. 4. The architecture of our proposed multi-modal MTL model is illustrated. From ECoG signals, we derive two different modalities: HFO data and IED data; the model receives these two distinct modal images. By integrating two identical CNN models through multiple cross-stitch units, the proposed end-to-end network concurrently classifies HFO and IED tasks.

and IED by merging two identical CNN models through multiple cross-stitch units. The schematic representation of our model design is depicted in Fig. 4.

Each CNN model comprises an input layer ($28 \times 28 \times 3$), two convolutional layers (with filter sizes of 8 and 16 respectively, kernel size of 3×3 and ReLU activation), each followed by batch normalization layer for feature regularization and Maxpooling layer (with a size of 3×3). Following the last Maxpooling layer, both CNNs feature a fully connected layer consisting of 2 units, followed by Dropout layers with a dropout rate of 0.8, and a final fully connected layer with 2 units corresponding to the respective classes, employing SoftMax activation. To merge the two CNNs, we incorporated cross-stitch units, enabling the combination of learned features after the first and second Maxpooling layers. Initially, the weighted matrix (α) for each cross-stitch unit is set to an identity matrix. We employed categorical cross-entropy loss to quantify the disparity between predicted and true class distributions. Subsequently, individual losses for both outputs were aggregated, resulting in a comprehensive measure that captures the overall deviation between predictions and true labels across both outputs. The model aims to enhance feature learning and representation by enabling cross-stream communication while maintaining individual stream characteristics. The process of model parameter selection, including hyperparameter tuning and ablation study experiments, can be found in the supplementary materials.

4.3 Experiment Setup

Our experiments were designed in two setups. In the first setup, we performed 5-fold cross validation with both HFO and IED image data; we trained our model with the training set of randomly selected 1200 images and assessed its performance on the remaining 300 test images. We report the evaluating metrics such as average accuracy, sensitivity, specificity, F1 score and area under the curve (AUC) over 5 repetitions of the experiment. This setup aims to gauge the MTL models' efficacy under challenging conditions characterized by limited annotated data availability.

Moving to the second experiment, we performed leave-one-patient-out, where we reserved one patient data for test and trained with the rest. Following a similar procedure, we trained the model with the 1200 training images from four patients and evaluated its performance using the test set from one patient, repeating the experiment 5 times to calculate average accuracy, sensitivity, specificity, F1 score and AUC. This experiment assesses the model's performance in scenarios where training data is scarce, and annotation for the test data is unavailable for training, essentially measuring the interpatient classification capabilities of our model under data scarcity.

In addition, we conducted a baseline experiment to demonstrate how feature sharing enhances the overall performance of HFO and IED classification. We recorded the performance under three conditions: 1) when HFO and IED are classified using two separate identical models without feature sharing (the baseline), 2) when HFO and IED are classified using a single model with one cross-stitch unit, and 3) when HFO and IED are classified using a single model with two cross-stitch units (the proposed model). For both IED and HFO, we compared the accuracy and F1 score obtained from each condition.

In all experiments with our proposed model, we maintained consistency by fixing the seed at 123, learning rate at 0.001 decayed by a factor of 0.95 every 500 steps., the number of epochs at 200, and utilizing the stochastic gradient descent optimizer. We used lasso regularization in each convolutional layer. Both HFO and IED images were resized to 28 × 28 pixels to align with the input layer of the model. The proposed model was built using the Python programming language, utilizing the TensorFlow and Keras libraries.

The objective of this study was to develop a robust model capable of effectively classifying HFO and IED using a small dataset. To evaluate the effectiveness of our proposed MTL model compared to established methods, we selected existing research that dealt with small quantities of HFO or IED image data for classification (up to 3000 images) including 2D-CNN by Zuo et al. [4], Resnet18 by Quon et al. [32], and VGG19 [26] as described in our previous study and compared their performance with ours. In addition, we compared our result with some state-of-the-art (SOTA) architectures including MobileNet-v2, Efficientnet-b0, NaseNet-mobile, DenseNet-201, Inception-v3, Xception and a simple Vision transformer (ViT) model [41]. We utilized MATLAB Deep learning toolbox to build and import the SOTA architectures and evaluate them.

All the experiments were conducted utilizing an Intel(R) Core (TM) i7-10750H processor, 16GB of RAM, and an NVIDIA GeForce RTX 2070 GPU.

5 Results

Table 1 presents the average accuracy, sensitivity, specificity, F1 score, and AUC obtained from both cross-validation and leave-one-patient-out experiments. The results reveal that HFO and IED classification achieved accuracies of 90.99% and 93.99% respectively in 5-fold cross-validation. The corresponding F1 scores are reported as 0.80 and 0.82 for HFO and IED respectively, indicating the successful discrimination between HFO and

Non-HFO, as well as IED and Non-IED, even with a limited number of annotated data and imbalanced class distribution.

In the leave-one-patient-out experiment, the mean accuracies for HFO and IED classification are 78.03% and 81.19% respectively, suggesting the model's capability in classifying HFO and IED from new patient data. However, the F1 scores of 0.42 and 0.65 for HFO and IED classification respectively imply that the uneven distribution of images per class may impact the model's generalization. Overall, the results suggest that the fundamental signal differences between patients and class imbalance can affect model performance.

Table 1. The proposed model's performance in cross-validation and leave-one-patient-out experiments.

Evaluation matrix	5-fold cross validation		Leave-one-patient-out	
	HFO	IED	HFO	IED
Mean Accuracy	90.99%	93.99%	78.03%	81.19%
Mean Sensitivity	78.55%	82.02%	43.17%	68.93%
Mean Specificity	91.93%	88.12%	90.63%	85.07%
Mean F1 score	0.80	0.82	0.42	0.65
Mean AUC	85.24%	85.39%	67.34%	74.27%

Figure 5. Presents the results of our baseline experiments using the proposed model to demonstrate how combining features impacts the classification performance of each task. By comparing the accuracy and F1 scores for both IED and HFO, we observed a gradual improvement in performance with an increasing number of sharing units. The classification accuracy for HFO and IED is 11.86% and 10.8% higher with our proposed model compared to the baseline model. Additionally, the F1 scores for HFO and IED increased by 0.2633 and 0.0457, respectively, confirming that sharing learned features between HFO and IED data enhances classification performance for both tasks.

To validate our proposed model performance, we compared our model's performance with three previous approaches for HFO/IED classification and seven modern SOTA architectures using our dataset, which includes 1500 HFO and 1500 IED images. The 5-fold cross-validation results, displayed in Table 2, indicate that our proposed model outperforms the others in terms of accuracy for both HFO and IED classification. Additionally, our model achieved the highest F1 scores for both HFO and IED classification (with scores tied with MobileNet-v2, EfficientNet-b0, and Xception for IED). This demonstrates our model's superior performance compared to the mentioned SOTA methods and its enhanced capability in classifying both positive and negative classes for HFO and IED under challenging conditions characterized by limited availability of annotated data.

Finally, to verify the efficiency of the proposed model, we compared it with the aforementioned models in terms of total parameter count (in millions), model size (in megabytes), and inference time (in milliseconds) required to classify 300 test images for

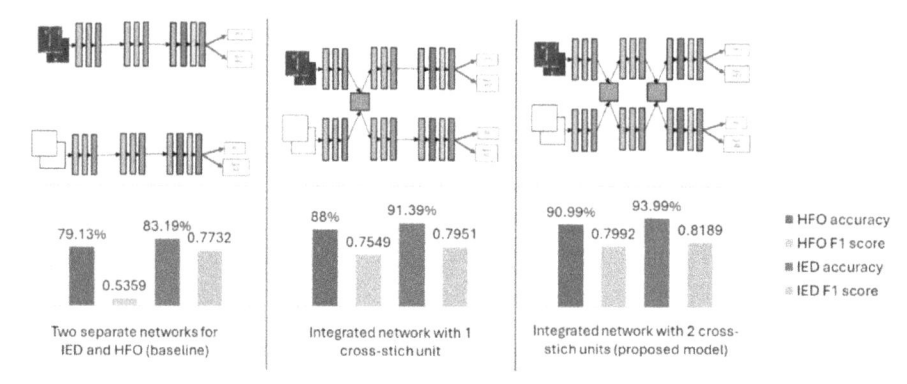

Fig. 5. Baseline experiment results. The leftmost architecture represents the baseline condition where HFO and IED are classified using separate models. The middle architecture is a single model for HFO and IED classification with one sharing unit. The rightmost architecture is our proposed model, which includes two feature-sharing units. Gradual improvement in performance is observed as the number of sharing units increases.

both HFO and IED (combined classification time) as shown in Table 3. Remarkably, our proposed model has the shortest inference time of 455 ms. It also has the second lowest number of parameters at 1.06 million and the second smallest size at 4.0 Megabytes. Compared to the SOTA architectures, these numbers are quite low. This demonstrates that by utilizing cross-sharing between related modal features, efficiency and robustness can be achieved in HFO and IED classification with a lightweight model such as our proposed one.

6 Discussion

The multi-modal MTL model approach outlined in this paper offers numerous advantageous qualities, including robustness, simplicity, lightweight design, and highly efficient performance. From a practical standpoint, the task of preparing labeled datasets for each patient before surgery poses significant challenges and consumes considerable time. In this context, the MTL-based HFO and IED classification approach proposed in this study presents a practical solution for detecting epilepsy biomarkers and localizing them, requiring only a few annotated datasets. By employing a straightforward MTL model, this approach circumvents the necessity for complex computations and facilitates prompt analysis. HFO and IED images, both derived from ECoG signals, contain complementary information. By sharing features, our proposed model could leverage common patterns and structures in both image types, leading to a more comprehensive data representation. Thus, shared feature extraction enhances the model's ability to identify subtle differences and similarities between HFO and IED events even with a lower number of learning parameters. This showcases efficiency in classification performance and supports the robustness of our proposed approach.

On another note, accurate biomarker detection plays a pivotal role in the precise localization of the surgical region within the brain. Previous studies show a strong correlation between biomarker populations and SOZ localization, suggesting SOZ electrodes

Table 2. Comparison between existing models and state-of-the-art models with the proposed model, experimented with our HFO and IED dataset.

Model name	Previous use	Accuracy HFO	Accuracy IED	Sensitivity HFO	Sensitivity IED	Specificity HFO	Specificity IED	F1 score HFO	F1 score IED
2D-CNN	HFO classification [4]	85.47%	86.27%	71.12%	**86.27%**	91.28%	88.02%	0.73	0.81
VGG19	HFO classification [26]	89.20%	87.24%	77.35%	81.51%	90.40%	90.03%	0.79	0.80
Resnet18	IED classification [32]	85.4%	83.40%	81.27%	84.63%	87.06%	82.61%	0.76	0.79
ViT	--	77.53%	74.47%	36.45%	52.30%	91.18%	88.12%	0.67	0.71
Mobile Net-V2	--	85.73%	86.27%	74.11%	83.41%	90.44%	88.02%	0.75	0.82
Efficient Net-b0	--	85.86%	86.73%	75.53%	80.29%	90.06%	90.72%	0.75	0.82
Nasenet-mobile	--	86.40%	88.80%	73.88%	78.46%	91.47%	91.98%	0.76	0.77
Densenet 201	--	87.73%	86.27%	**80.82%**	76.77%	90.53%	**92.12%**	0.78	0.81
Inception-V3	--	87.33%	87.73%	72.74%	85.87%	**93.25%**	88.89%	0.77	0.81
Xception	--	86.73%	87.33%	73.64%	82.72%	92.03%	90.18%	0.76	0.82
Proposed MTL model		**90.99%**	**93.99%**	78.55%	82.02%	91.93%	88.12%	**0.80**	**0.82**

Table 3. Comparison between different models with the proposed model in terms of total parameter number, model size, and inference time with 300 images for HFO and IED each.

Model name	Total parameter (Millions)	Size (MB)	Inference time (ms)
2D-CNN [4]	0.71	2.9	1532
VGG19 [26]	143.67	548.1	2138
Resnet18 [32]	11.69	44.7	1248
ViT	5.70	21.8	4934
MobileNet-V2	3.50	13.6	2430
EfficientNet-b0	5.29	20.5	4171
Nasenet-mobile	5.30	20.0	6542
Densenet 201	20.01	77.4	8988
Inception-V3	27.16	103.9	9372
Xception	22.90	85.0	7021
Proposed MTL model	**1.06**	**4.0**	**455**

typically contain more biomarker events than other brain regions [5, 38, 42]. By incorporating multiple biomarkers, the process of localization becomes more refined, offering enhanced precision and reliability. This refinement is beneficial for neurologists, who rely on accurate localization for clinical decision-making, also for deep learning models tasked with assisting in medical image analysis and diagnosis. Our proposed model can effectively detect both HFOs and IEDs at a time, allowing comprehensive assessment of the brain's activity, facilitating both overall evaluation and individual patient analysis. The practical implications of such a model are profound, in not only detecting critical biomarkers but also showcasing its potential to significantly improve the surgical localization process, thereby contributing to better patient outcomes and advancing the field of neurosurgery.

This study has a few shortcomings that should be addressed. To begin, our proposed model only works with an equal amount of HFO and IED images due to the linear feature combining used in cross-stitch units. In the future, we plan to address these two issues by modifying the cross-stitch unit. Next, performance comparisons with existing methods under the condition of a larger size of the training dataset are currently unavailable due to the limited number of annotated datasets available in some of the patients. In addition, due to restrictions and consent issues, HFO or IED annotated ECoG dataset are not to be found publicly, therefore we could not evaluate our model with other datasets. We hope to include more patients to validate our results and upon finding other online dataset, include that as well in future. In addition, we observed intrinsic variation between subjects and observed low sensitivity in patient-wise classification, indicating that the model struggled to correctly classify positive classes such as HFO and IED, especially when the number of positive and negative classes is largely imbalanced. Moving forward, we aim to focus more on this issue. Finally, this study only includes patients who had a successful surgical outcome; nevertheless, in some cases, individuals may still experience seizures following surgery. In future studies, we intend to study the effectiveness of our technique on such critical patients as well. Prior research on surgical region localization has indicated that the utilization of multi-modal biomarkers [42, 43] offers enhanced localization performance, accurate diagnosis and disease monitoring compared to single biomarker approaches. In this context, our proposed multi-modal MTL biomarker detection approach promises to contribute to improving diagnosis, treatment as well as patients' quality of life.

7 Conclusion

This paper introduces a novel multi-modal MTL model designed to automatically classify HFO and IED, two critical epilepsy biomarkers, simultaneously from distinct datasets. To the best of our knowledge, this is the very first attempt to imply MTL in classification of multiple epilepsy biomarkers. Leveraging the benefits of combining the hidden feature representations of two tasks, our proposed model demonstrates the capability to perform both HFO and IED classification efficiently from small datasets. Achieving cross-validation accuracies of 90.99% and 93.99% for HFO and IED respectively, alongside respective F1 scores of 0.80 and 0.82, highlights the effectiveness of our approach. Moreover, our leave-one-patient-out experiments yield accuracies of 78.03%

and 81.19% for HFO and IED classification, indicating the model's generalization ability. Comparative analysis proved the efficiency and robustness of our model, despite its simplicity and lightweight nature. By employing multi-modal MTL, our proposed framework offers promises to clinical research, potentially reducing the dependency on large number of annotated datasets while facilitating precise identification of both HFO and IED biomarkers. In the future, we envision including more patients in our study and creating a system that can pinpoint surgical regions in the brain automatically based on the population information of these biomarkers.

Acknowledgement. We acknowledge the contribution of National Center of Neurology and Psychiatry, Tokyo, Japan for assistance in data collection and annotation. This study was supported by JSPS KAKENHI (22K09296).

Conflicts of Interests. The authors declare that the research was conducted in the absence of any commercial or financial relationships that could be construed as a potential conflict of interest.

Supplementary Materials. Supplementary materials related to this research can be found at https://figshare.com/projects/Multi-modal_Multitask_Learning_Model_for_Simultaneous_Classification_of_Two_Epilepsy_Biomarkers-_Supplementary_Materials/212504.

References

1. World Health Organization. "Epilepsy: a public health imperative." (2019), https://www.who.int/publications/i/item/epilepsy-a-public-health-imperative, last accessed 2024/02/20
2. Guery, D., & Rheims, S. Clinical management of drug resistant epilepsy: a review on current strategies. Neuropsychiatric Disease and Treatment, 2229–2242 (2021)
3. Burns, S. P., Santaniello, S., Yaffe, R. B., Jouny, C. C., Crone, N. E., Bergey, G. K., Anderson, W.S. & Sarma, S. V. Network dynamics of the brain and influence of the epileptic seizure onset zone. Proceedings of the National Academy of Sciences, 111(49), E5321-E5330 (2014)
4. Zuo, R., Wei, J., Li, X., Li, C., Zhao, C., Ren, Z., Liang, Y., Geng, X., Jiang, C., Yang, X. and Zhang, X. Automated detection of high-frequency oscillations in epilepsy based on a convolutional neural network. Frontiers in computational neuroscience, 13, 6 (2019)
5. Selvitelli, M. F., Walker, L. M., Schomer, D. L., & Chang, B. S. The relationship of interictal epileptiform discharges to clinical epilepsy severity: a study of routine EEGs and review of the literature. Journal of clinical neurophysiology: official publication of the American Electroencephalographic Society, 27(2), 87 (2010)
6. Abbasi, B., & Goldenholz, D. M. Machine learning applications in epilepsy. Epilepsia, 60(10), 2037-2047 (2019)
7. Caruana, R. Multitask learning. Machine learning, 28, 41-75 (1997)
8. Zhang, Y., & Yang, Q. A survey on multi-task learning. IEEE Transactions on Knowledge and Data Engineering, 34(12), 5586-5609 (2021).
9. Baxter, J. A Bayesian/information theoretic model of learning to learn via multiple task sampling. Machine Learning, 28, 7–39 (1997).
10. Long, M., & Wang, J. Learning Multiple Tasks with Deep Relationship Networks. arXiv Preprint arXiv:1506.02117 (2015)
11. Lu, Y., Kumar, A., Zhai, S., Cheng, Y., Javidi, T., & Feris, R. Fully-adaptive Feature Sharing in Multi-Task Networks with Applications in Person Attribute Classification (2016)

12. Ngiam, J., Khosla, A., Kim, M., Nam, J., Lee, H., & Ng, A. Y. Multimodal deep learning. In Proceedings of the 28th international conference on machine learning (ICML-11) (pp. 689–696) (2011)
13. Hu, R., & Singh, A. Unit: Multimodal multitask learning with a unified transformer. In Proceedings of the IEEE/CVF International Conference on Computer Vision (pp. 1439–1449) (2021)
14. Chen, S., Jin, Q., Zhao, J. and Wang, S. Multimodal multi-task learning for dimensional and continuous emotion recognition. In Proceedings of the 7th Annual Workshop on Audio/Visual Emotion Challenge (pp. 19–26) (2017)
15. Misra, I., Shrivastava, A., Gupta, A. and Hebert, M. Cross-stitch networks for multi-task learning. In Proceedings of the IEEE conference on computer vision and pattern recognition. pp. 3994–4003 (2016)
16. Ma, X., Qiu, S., Zhang, Y., Lian, X., & He, H. Predicting epileptic seizures from intracranial EEG using LSTM-based multi-task learning. In Chinese Conference on Pattern Recognition and Computer Vision (PRCV) (pp. 157–167). Cham: Springer International Publishing (2018)
17. Ahmed, B., Thesen, T., Blackmon, K., Kuzniecky, R., Devinsky, O., Dy, J., & Brodley, C. Multi-task learning with weak class labels: Leveraging iEEG to detect cortical lesions in cryptogenic epilepsy. In Machine learning for healthcare conference (pp. 115–133). PMLR (2016)
18. Xi, H., Wang, Y., Niu, R., Hao, Y., & Chen, Y. Two-Stage Multi-task Learning for Automatic Epilepsy Detection. In Advances in Natural Computation, Fuzzy Systems and Knowledge Discovery: Proceedings of the ICNC-FSKD 2021 17 (pp. 866–873). Springer International Publishing (2022)
19. D'Amario, V., Tomasi, F., Tozzo, V., Arnulfo, G., Barla, A., & Nobili, L. Multi-task multiple kernel learning reveals relevant frequency bands for critical areas localization in focal epilepsy. In Machine Learning for Healthcare Conference (pp. 348–382). PMLR (2018)
20. Van Esbroeck, A., Smith, L., Syed, Z., Singh, S., & Karam, Z. Multi-task seizure detection: addressing intra-patient variation in seizure morphologies. Machine Learning, 102, 309-321 (2016)
21. Cao, J., Chen, Y., Zheng, R., Cui, X., Jiang, T., & Gao, F. DSMN-ESS: Dual-stream Multi-task Network for Epilepsy Syndrome Classification and Seizure Detection. IEEE Transactions on Instrumentation and Measurement (2023)
22. Navarrete, M., Alvarado-Rojas, C., Le Van Quyen, M., & Valderrama, M. RIPPLELAB: A comprehensive application for the detection, analysis and classification of high frequency oscillations in electroencephalographic signals. PloS one, 11(6), e0158276 (2016)
23. Chaibi, S., Sakka, Z., Lajnef, T., Samet, M. and Kachouri, A. Automated detection and classification of high frequency oscillations (HFOs) in human intracereberal EEG. Biomedical Signal Processing and Control, 8(6), pp.927-934 (2013)
24. Zhang, Y., Lu, Q., Monsoor, T., Hussain, S.A., Qiao, J.X., Salamon, N., Fallah, A., Sim, M.S., Asano, E., Sankar, R. and Staba, R.J. Refining epileptogenic high-frequency oscillations using deep learning: a reverse engineering approach. Brain communications, 4(1), p.fcab267 (2022)
25. Takayanagi, Y., Takayama, Y., Iijima, K., Iwasaki, M., & Ono, Y. Efficient Detection of High-frequency Biomarker Signals of Epilepsy by a Transfer-learning-based Convolutional Neural Network. Advanced Biomedical Engineering, 10, 158-165 (2021)
26. Broti, N. M., Sawada, M., Takayama, Y., Iwasaki, M., & Ono, Y. Detection of high-frequency biomarker signals of epilepsy by combined deep-learning feature selection and linear discrimination analysis. 37th annual conference of the Japanese society of artificial intelligence. 1L5OS18b03, 1–4 (2023)

27. Guo, J., Xiao, N., Li, H., He, L., Li, Q., Wu, T., He, X., Chen, P., Chen, D., Xiang, J. and Peng, X.. Transformer-based high-frequency oscillation signal detection on magnetoencephalography from epileptic patients. Frontiers in Molecular Biosciences, 9, 822810 (2022)
28. Medvedev, A.V., Agoureeva, G.I. and Murro, A.M.. A long short-term memory neural network for the detection of epileptiform spikes and high frequency oscillations. Scientific reports, 9(1), p.19374 (2019)
29. El-Gohary, M., McNames, J. and Elsas, S. User-guided interictal spike detection. In 2008 30th Annual International Conference of the IEEE Engineering in Medicine and Biology Society (pp. 821–824). IEEE (2008)
30. Vijayalakshmi, K. and Abhishek, A.M. Spike detection in epileptic patients EEG data using template matching technique. International Journal of Computer Applications, 2(6), pp.5-8 (2010)
31. Antoniades, A., Spyrou, L., Martin-Lopez, D., Valentin, A., Alarcon, G., Sanei, S. and Took, C.C. Detection of interictal discharges with convolutional neural networks using discrete ordered multichannel intracranial EEG. IEEE Transactions on Neural Systems and Rehabilitation Engineering, 25(12), pp.2285-2294 (2017)
32. Quon, R. J., Meisenhelter, S., Camp, E. J., Testorf, M. E., Song, Y., Song, Q., Culler, G.W., Moein, P. & Jobst, B. C. AiED: Artificial intelligence for the detection of intracranial interictal epileptiform discharges. Clinical neurophysiology, 133, 1-8 (2022)
33. Lourenço, C., Tjepkema-Cloostermans, M.C., Teixeira, L.F. and van Putten, M.J., 2020. Deep learning for interictal epileptiform discharge detection from scalp EEG recordings. In Proceedings of MEDICON 2019, Coimbra, Portugal (pp. 1984–1997). Springer International Publishing
34. Najafi T., Jaafar R., Remli R., Wan Zaidi W.A. A Classification Model of EEG Signals Based on RNN-LSTM for Diagnosing Focal and Generalized Epilepsy. Sensors. 22(19):7269. https://doi.org/10.3390/s22197269 (2022)
35. Geng, D., Alkhachroum, A., Bicchi, M.A.M., Jagid, J.R., Cajigas, I. and Chen, Z.S. Deep learning for robust detection of interictal epileptiform discharges. Journal of neural engineering, 18(5), p.056015 (2021)
36. Tan, K., Huang, W., Liu, X., Hu, J. and Dong, S. A multi-modal fusion framework based on multi-task correlation learning for cancer prognosis prediction. Artificial Intelligence in Medicine, 126, p.102260 (2022)
37. Thung, K. H., Wee, C. Y., Yap, P. T., Shen, D., & Alzheimer's Disease Neuroimaging Initiative. Neurodegenerative disease diagnosis using incomplete multi-modality data via matrix shrinkage and completion. NeuroImage, 91, 386–400 (2014)
38. Broti, N. M., Sawada, M., Takayama, Y., Iijima, K., Iwasaki, M., & Ono, Y. Automated Detection of Interictal High-frequency Oscillations for Epileptogenic Zone Localization. Advanced Biomedical Engineering, 13, 100-107 (2024)
39. Zelmann, R., Mari, F., Jacobs, J., Zijlmans, M., Dubeau, F., & Gotman, J. A comparison between detectors of high frequency oscillations. Clinical Neurophysiology, 123(1), 106-116 (2012)
40. Gardner, A. B., Worrell, G. A., Marsh, E., Dlugos, D., & Litt, B. Human and automated detection of high-frequency oscillations in clinical intracranial EEG recordings. Clinical neurophysiology, 118(5), 1134-1143 (2007)
41. Hosseini, M. P., Tran, T. X., Pompili, D., Elisevich, K., & Soltanian-Zadeh, H. Multimodal data analysis of epileptic EEG and rs-fMRI via deep learning and edge computing. Artificial Intelligence in Medicine, 104, 101813 (2020)
42. Kuroda, N., Sonoda, M., Miyakoshi, M., Nariai, H., Jeong, J.W., Motoi, H., Luat, A.F., Sood, S. and Asano, E. Objective interictal electrophysiology biomarkers optimize prediction of epilepsy surgery outcome. Brain communications, 3(2), p.fcab042 (2021)

43. He, X., Zhao, K., and Chu, X. AutoML: A survey of the state-of-the-art. Knowledge-based systems, 212, 106622 (2021)

Learning Neural Networks for Multi-label Medical Image Retrieval Using Hamming Distance Fabricated with Jaccard Similarity Coefficient

Asim Manna(✉) and Debdoot Sheet

Indian Institute of Technology Kharagpur, Kharagpur 721302, West Bengal, India
asimmanna17@kgpian.iitkgp.ac.in, debdoot@ee.iitkgp.ac.in

Abstract. Deep neural hashing (DNH) has demonstrated its effectiveness in content-based medical image retrieval (CBMIR) for efficient nearest-neighbor search in large image datasets. It learns a hash function to generate hash codes from the images. Conventional pairwise DNH methods are inadequate for multi-label CBMIR as they do not incorporate between the Hamming distance (HD) of hash codes and the Jaccard similarity coefficient (JSC) of label sets for an image pair. This work introduces a JSC-based loss function called adaptive HD loss (AHDL) for learning HD between hash pairs using a deep neural network to retrieve multi-label medical images. AHDL helps the model assign an appropriate HD between a pair of hash codes based on their image similarity level. We also adopt pairwise multi-label classification loss to generate unique features for each class combination. Experiments are demonstrated on the publicly available NIH chest X-ray dataset. Our method achieves 3.98% higher normalized discounted cumulative gain compared to the state-of-the-art method for a top-100 image retrieval task.

Keywords: Content-based medical image retrieval · Deep neural hashing network · Jaccard coefficient · Hamming distance · Pairwise similarity

1 Introduction

Content-based medical image retrieval (CBMIR) offers clinicians valuable evidence for assessing cases with similar symptoms or pathological representations [3,22]. Feature extraction and ranking of images based on the similarity of a query image are two critical steps in CBMIR. With the rapid growth in medical imaging, efficient and accurate retrieval of relevant information from large databases remains a challenge in routine radiology workflows [8,18]. The interpretation of medical images often relies on the expertise of professionals. The utilization of case reports limited to medical images as diagnostic benchmarks has been shown to affect expert interpretation significantly. CBMIR is capable of retrieving similar cases of medical images for supplementary analysis to bridge variations across diagnoses by experts [8,27].

The task of multi-label medical image retrieval is challenging because it entails assigning multiple labels or categories to a given medical image [11,25]. Consequently, there is a growing focus on multi-label CBMIR tasks due to their practicality in real-life scenarios. As an instance, an organ in the human body is affected by multiple pathologies simultaneously. The similarity associated with various pathology in medical images could have been very subtle, so it is necessary to use advanced feature extraction approaches for more comprehensive analysis [7]. Challenges in multi-label medical image retrieval include the ability to accurately capture complex relationships represented by different labels, address class imbalance issues, and developing efficient retrieval algorithms capable of handling large-scale datasets [11,13]. These challenges create difficulties for traditional retrieval methods, thereby necessitating the development of advanced algorithms to enhance the retrieval performance.

Deep neural hashing (DNH) methods [21,23] have emerged as promising solutions for medical image retrieval. These methods involve encoding medical images into compact binary codes while preserving similarity relationships in the Hamming space [4]. It learns advanced hash functions that can generate hash codes from images. Hashing is used for feature extractors by transforming the images into binary representations. These hash codes facilitate efficient nearest-neighbor searches within extensive image datasets. After generating hash codes from images, the Hamming distance (HD) is often used to measure the semantic similarity between images. HD between a pair of hash codes is finite and inversely proportional to the similarity of the images they represent. Therefore, incorporating HD effectively into the ranking algorithm ensures that similar images are prioritized higher in the retrieval list, enhancing the overall performance of the retrieval system.

In the context of multi-label medical images, a pair of images may exhibit comorbid pathologies as well as distinct pathologies. So, HD between generated hash codes of a multi-label image pair should depend on the proportion of shared labels out of the total possible labels for the pair. An example regarding this is illustrated in Figure 1. There are three possible labels: Atelectasis, Effusion, and Infiltration. Consider three images $\mathbf{x}_i, \mathbf{x}_j,$ and $\mathbf{x}_k$ with their label sets denoted as $\mathbf{y}_i = \{1,0,0\}, \mathbf{y}_j = \{1,1,0\},$ and $\mathbf{y}_k = \{0,1,1\}$ respectively. The value 1 in a label set indicates that the pathology is associated with this image, while 0 indicates otherwise. The number of all possible labels, shared labels, and Jaccard similarity coefficient (JSC) [2] between an image pair $\mathbf{x}_i, \mathbf{x}_j$ are denoted as $n_{ij}^{(1)}(= |\mathbf{y}_i \cup \mathbf{y}_j|), n_{ij}^{(2)} (= |\mathbf{y}_i \cap \mathbf{y}_j|)$, and $\frac{n_{ij}^{(2)}}{n_{ij}^{(1)}}$ respectively. In this example, $n_{ij}^{(1)} = 2, n_{ij}^{(2)} = 1, n_{jk}^{(1)} = 3, n_{jk}^{(2)} = 1$. Since, $\frac{n_{ij}^{(2)}}{n_{ij}^{(1)}} > \frac{n_{jk}^{(2)}}{n_{jk}^{(1)}}$, the preferable scenario is that the HD between hash codes for $\mathbf{x}_j, \mathbf{x}_k$ is higher than the HD between hash codes for $\mathbf{x}_i, \mathbf{x}_j$. In the other words, for each unique combination of $n_{ij}^{(1)}$ and $n_{ij}^{(2)}$, the HD should vary. Indeed, as the JSC between the label sets of an image pair increases, indicating a higher degree of similarity in their pathology manifestations, the HD between the corresponding hash pair should decrease. This implies that the JSC-based similarity measurement has the capability to delineate nuanced multi-level semantic similarities. So, there is a need for an advanced learning hash function capable of establishing a one-to-one relationship between the HD of hash pairs and the JSC. Such a function would effectively capture the intricate relationships between label similarities and HD, facilitating more accurate representation and retrieval of multi-label medical images.

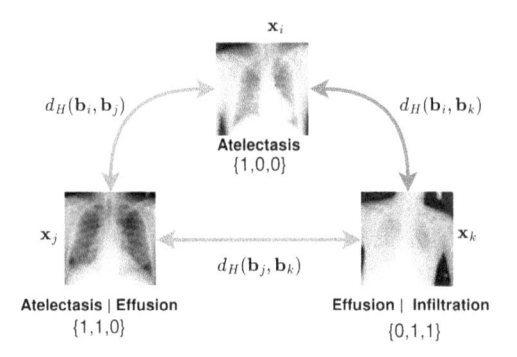

Fig. 1. An example overview of our objective using three images $\mathbf{x}_i$ (Atelectasis), $\mathbf{x}_j$ (Atelectasis, Effusion), $\mathbf{x}_k$ (Effusion, Infiltration). Here $\frac{n_{ij}^{(2)}}{n_{ij}^{(1)}} > \frac{n_{jk}^{(2)}}{n_{jk}^{(1)}} > \frac{n_{ik}^{(2)}}{n_{ik}^{(1)}} = 0$, where each ratio is calculated from corresponding image label sets. In this scenario, $d_H(\mathbf{b}_i, \mathbf{b}_j) < d_H(\mathbf{b}_j, \mathbf{b}_k) < d_H(\mathbf{b}_i, \mathbf{b}_k)$ will be followed, where $\mathbf{b}_i, \mathbf{b}_j, \mathbf{b}_k$ are the hash codes corresponding to each of the images.

In this work, we propose a loss function to foster similarity learning between a pair of images, utilizing the JSC as a metric. Adaptive HD loss (AHDL) is employed to assign suitable Hamming distance (HD) between a pair of hash codes according to their image similarity level based on the value of the JSC. Besides HD learning, semantic classification is another significant learning objective to learn hash representation from images. We adopt pairwise multi-label classification loss to generate unique features for each different label combination. The main contributions of this work are summarized as follows:

- Method of learning hash codes using a neural network in order to retrieve images contextually sensitive to their semantic similarity of multiple pathologies imaged in an organ.
- To the best of our knowledge, no existing work learns hash codes that simultaneously consider both the HD and the JSC for multi-label CBMIR. In this work, we develop an advanced learning hash function that establishes a one-to-one relationship between the HD of hash pairs and the JSC of the label set of the corresponding image pairs.

- A loss function designed to generate appropriate hash codes so that accurate HD is based on the similarity levels between a pair of images.
- The commonly used metrics, normalized discounted cumulative gain (nDCG), average cumulative gains (ACG), and wighted mean average precision (wMAP) are utilized in order to measure the retrieval performance of the proposed method.

The paper is organized as follows. The prior art of DNH for image retrieval is presented in Section 2. The proposed method is introduced in Section 3. Experimental details are discussed in Section 4. Results and discussions are presented in Section 5. This work is concluded in Section 6.

2 Prior Art

In this section, we will primarily discuss some related works in this domain, including methods related to medical image retrieval and DNH for multi-label image retrieval. The methods for image retrieval utilizing hashing can be broadly categorized into two categories: data-independent hashing (DIH) and data-dependent hashing (DDH).

DIH refers to a hashing technique where the hash functions are generated without relying on the specific characteristics or content of the data being hashed. This technique does not utilize any labeled data or information for the hashing process [24]. Local sensitive hashing (LSH) is a widely used DIH technique that employs randomized projections or permutations to design different hash functions, aiming to return the identical codes for similar data items with high probability [14,24].

The rapid expansion of data coupled with the advancement of deep neural networks (DNN) [20] is reshaping the landscape of various fields, including image retrieval. DNH has gained significant attention in recent times due to the advancements in deep neural networks (DNN) [20] for image representation. These methods incorporate DNN into the process of constructing binary codes for images. By combining the strengths of deep learning with the computational efficiency and storage benefits of hashing techniques, DNH-based methods effectively map the image representation space learned by deep models into a binary space. DNH techniques are suitable for large datasets, which is a common requirement in the medical field where medical images are continually being generated and stored. The concept of DNH [19,24,26] is introduced with the help of DNN to build a hash function that effectively leverages data distributions and incorporates information regarding class labels present in a dataset. Its objective is to ensure that the nearest neighbor of any pattern in the space of hash codes closely resembles the neighboring patterns in the original space [24]. Preserving data similarity in the Hamming space is the primary objective of majority of learning-based data-dependent hashing techniques [1,9,28]. Several studies have previously concentrated on supervised hashing-based image retrieval using the similarity matrix and quantization loss [29]. HashNet proposes a method to learn non-smooth binary activation in order to generate binary hash codes from imbalanced similarity data [5]. Deep Cauchy hashing (DCH) model utilizes a pairwise cross-entropy loss based on the Cauchy distribution to generate binary hash codes [4]. OrthoHash is based on one loss, eliminating the need for balancing coefficient tuning in various losses [12]. OrthHash generates center hash codes [15] using Bernoulli distributions. Attention-based triplet hashing (ATH) network [10] is an end-to-end system designed to learn low-dimensional hash codes that preserve the categorization, region of interest, and small-sample information. Multi-scale triplet hashing [6] and deep semantic ranking hashing based on self-attention (DSHA) [28] offer an effective and scalable solution by leveraging multi-scale information and triplet loss to achieve accurate and efficient retrieval of medical images. The previous learning-based hashing methods [31] are able to only take care of $n_{ij}^{(1)}$ but do not properly incorporate the information between $n_{ij}^{(1)}, n_{ij}^{(2)}$, and the HD between hash codes. This affects the ranking in the image retrieval list and, therefore, retrieval performance. Images with the same similarity level can be more finely differentiated using the JSC. However, a learning method that generates hash codes considering both the HD and the JSC has not been properly developed yet. In this work, we introduce a pairwse learning approach to generate hash codes from image pairs while accounting for the JSC between the label sets of these multi-label medical image pairs. Our method fulfills the following requirements. (i) generate unique features for different combinations of pathology present in images. (ii) HD between a pair of hash codes of multi-label images depends on the number of match pathologies. The overview of our approach and CBMIR using DNH is illustrated in Figure 2.

3 Proposed Methodology

3.1 Problem Statement

Consider a training set of images represented as $\mathbf{X}_{\mathfrak{T}} = \{\mathbf{x}_1^{\mathfrak{T}}, \mathbf{x}_2^{\mathfrak{T}}, \ldots, \mathbf{x}_i^{\mathfrak{T}}, \ldots, \mathbf{x}_{U_1}^{\mathfrak{T}}\}$. $\mathbf{y}_i^{\mathfrak{T}} \in \{0,1\}^L$ represents the label set of image $\mathbf{x}_i^{\mathfrak{T}} \in \mathbb{R}^{M \times N}$, where L denotes the number of possible labels in the dataset. Consider a non-linear hash function $F : \mathbb{R}^{M \times N} \mapsto \{-1,1\}^K$ such that each $\mathbf{x}_i^{\mathfrak{T}} \in \mathbb{R}^{M \times N}$ is an image to be hashed into a K-length binary hash code $\mathbf{b}_i^{\mathfrak{T}} \in \{-1,1\}^K$.

Let, the number of all possible labels and shared labels between an image pair $\mathbf{x}_i^{\mathfrak{T}}, \mathbf{x}_j^{\mathfrak{T}}$ are denoted as $n_{ij}^{(1)} (= |\mathbf{y}_i^{\mathfrak{T}} \cup \mathbf{y}_j^{\mathfrak{T}}| \neq 0), n_{ij}^{(2)} (= |\mathbf{y}_i^{\mathfrak{T}} \cap \mathbf{y}_j^{\mathfrak{T}}|)$ respectively. Our aims to learn $F(\cdot)$ following a supervised learning approach such that $d_H(\mathbf{b}_i^{\mathfrak{T}}, \mathbf{b}_j^{\mathfrak{T}}) \leq d_H(\mathbf{b}_i^{\mathfrak{T}}, \mathbf{b}_k^{\mathfrak{T}})$ if and only if $\frac{n_{ij}^{(2)}}{n_{ij}^{(1)}} \geq \frac{n_{ik}^{(2)}}{n_{ik}^{(1)}}$, where $\mathbf{b}_i^{\mathfrak{T}} = F(\mathbf{x}_i^{\mathfrak{T}})$ and $d_H(\cdot)$ represents the HD between two hash codes of length K. The idea is that if the number of common labels between a pair of images is more, then the HD between a pair of generated hash codes should be less.

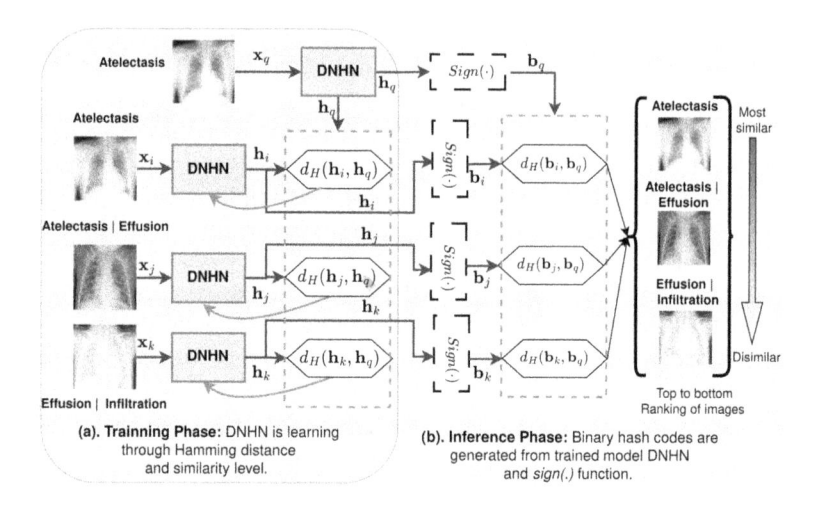

Fig. 2. The figure on the left illustrates the training process of a deep neural hashing network (DNHN), where the network learns by the HD between real-valued hash codes $(\mathbf{h}_q, \mathbf{h}_i, \mathbf{h}_j, \mathbf{h}_k)$. We aim to learn correct order of $d_H(\mathbf{h}_i, \mathbf{h}_q), d_H(\mathbf{h}_j, \mathbf{h}_q), d_H(\mathbf{h}_k, \mathbf{h}_q)$ through the Jaccard coefficient, where $d_H(\cdot)$ represents HD between two hash codes. Conversely, the figure on the right demonstrates the generation of binary hash codes $(\mathbf{b}_q, \mathbf{b}_i, \mathbf{b}_j, \mathbf{b}_k)$ achieved by applying the $sign(\cdot)$ function. The order of the HD is $d_H(\mathbf{b}_i, \mathbf{b}_q) < d_H(\mathbf{b}_j, \mathbf{b}_q) < d_H(\mathbf{b}_k, \mathbf{b}_q)$ and ideal image ranking with respect to $\mathbf{x}_q$.

3.2 Hash Code Generation

The hash code generation process from images can be expressed through $F(\cdot)$ using the following equations,

$$\mathbf{b}_i^{\mathfrak{T}} = sign(\mathbf{h}_i) = F(\mathbf{x}_i^{\mathfrak{T}}) \tag{1}$$

$$\mathbf{h}_i = Tanh(\mathtt{fc_h}(\mathbf{z}_i)) \tag{2}$$

$$\mathbf{z}_i = \mathtt{net_e}(\mathbf{x}_i^{\mathfrak{T}}) \tag{3}$$

where, $\mathbf{h}_i \in [-1, 1]^K$ and $\mathbf{b}_i^{\mathfrak{T}} \in \{-1, 1\}^K$ represent real valued and binary hash codes respectively. The $sign(\cdot)$ function [4] is employed to convert the real valued hash to a binary hash representation. $\mathtt{net_e}(\cdot)$ represents the CNN-based feature encoding function is given by,

$$\begin{aligned}\mathtt{net_e}(\cdot) \mapsto\ &\mathtt{Conv2D:64c11w4s2p} \to \mathtt{ReLU} \to \mathtt{MaxPool2D:3w2s} \\ \to\ &\mathtt{Conv2D:192c5w1s2p} \to \mathtt{ReLU} \to \mathtt{MaxPool2D:3w2s} \\ \to\ &\mathtt{Conv2D:384c3w1s1p} \to \mathtt{ReLU} \to \mathtt{Conv2D:256c3w1s1p} \to \mathtt{ReLU} \\ \to\ &\mathtt{Conv2D:256c3w1s1p} \to \mathtt{MaxPool2D:3w2s} \to \mathtt{Flatten}\end{aligned} \tag{4}$$

$\mathtt{fc_h}(\cdot)$ is the real valued hash generating fully connected layers.

$$\mathtt{fc_h}(\cdot) \mapsto \mathtt{Linear:4096} \to \mathtt{ReLU} \to \mathtt{Linear:K} \tag{5}$$

(2) and (3) are used during training to avoid the vanishing gradient [4] challenge faced in (1) on account of the $sign(\cdot)$ function. Thus, we utilize $\mathbf{h}_i$ and $\mathbf{b}_i^{\mathfrak{T}}$ during training and inference respectively.

During the training process, we utilize two distinct loss functions: adaptive HD loss (AHDL) and pairwise multi-label classification loss (PMCL). The purpose of the PMCL is to create distinctive features for various combinations of pathologies present in the images. Meanwhile, AHDL is applied to generate hash codes, ensuring that the Hamming distance between these codes appropriately reflects the similarity levels between images. Our method is trained in an end-to-end manner, in which image feature learning and HD learning via hash codes from image pairs are performed simultaneously.

3.3 Adaptive Hamming Distance Loss (AHDL)

The idea of designing adaptive HD loss for multi-label image retrieval is that HD between hash codes of a image pair should be depended on the number of total possible pathology $(n_{ij}^{(1)} = |\mathbf{y}_i^{\mathfrak{T}} \cup \mathbf{y}_j^{\mathfrak{T}}|)$ and the number of common pathology $(n_{ij}^{(2)}) = |\mathbf{y}_i^{\mathfrak{T}} \cap \mathbf{y}_j^{\mathfrak{T}}|$ between $\mathbf{x}_i^{\mathfrak{T}}$ and $\mathbf{x}_j^{\mathfrak{T}}$. When $\frac{n_{ij}^{(2)}}{n_{ij}^{(1)}}$ is increased HD should be less and vice-versa. Let, $\mathbf{h}_i, \mathbf{h}_j$ be the real valued hash codes of this image pair $\mathbf{x}_i^{\mathfrak{T}}$ and $\mathbf{x}_j^{\mathfrak{T}}$ respectively. Here are some constraints specified for this image pair,

1. $0 \leq n_{ij}^{(2)} \leq n_{ij}^{(1)} \leq L$ and $n_{ij}^{(1)} \neq 0 \; \forall i, j$.
2. $n_{ij}^{(1)} = 1$ implies $\mathbf{y}_i^{\mathfrak{T}} = \mathbf{y}_j^{\mathfrak{T}}$. In this scenario, $n_{ij}^{(2)} = n_{ij}^{(1)} = 1 \; \forall i, j$.
3. $0 \leq d_H(\mathbf{h}_i, \mathbf{h}_j) \leq K, \; \forall i, j$.

Table 1. An example of computing HD based on similarity level of an image pair. Here $L = 3$ and $K = 16$. $L_{HD}^{(n_{ij}^{(1)})}(\mathbf{h}_i, \mathbf{h}_j)$ is the list of HD for given value of $n_{ij}^{(1)}$. $D_H^{(n_{ij}^{(1)}, n_{ij}^{(2)})}(\mathbf{h}_i, \mathbf{h}_j)$ is the HD, collected from the list $L_{HD}^{(n_{ij}^{(1)})}(\mathbf{h}_i, \mathbf{h}_j)$ for given specific value of $n_{ij}^{(2)}$.

$\mathbf{y}_i^{\mathfrak{T}}$	$\mathbf{y}_j^{\mathfrak{T}}$	$n_{ij}^{(1)}$	$n_{ij}^{(2)}$	$L_{HD}^{(n_{ij}^{(1)})}(\mathbf{h}_i, \mathbf{h}_j)$	$D_H^{(n_{ij}^{(1)}, n_{ij}^{(2)})}(\mathbf{h}_i, \mathbf{h}_j)$
{1,0,1}	{0,1,0}	3	0	[16, 10, 5, 0]	16
{1,1,1}	{1,0,0}		1		10
{1,1,1}	{1,1,0}		2		5
{1,1,1}	{1,1,1}		3		0
{1,0,0}	{0,1,0}	2	0	[16, 8, 0]	16
{1,1,0}	{1,0,0}		1		8
{1,1,0}	{1,1,0}		2		0
{1,0,0}	{1,0,0}	1	1	[16, 0]	0

The adaptive HD between $\mathbf{h}_i, \mathbf{h}_j$ is based on the value of $n_{ij}^{(1)}, n_{ij}^{(2)}$ and defined by,

$$D_H^{(n_{ij}^{(1)}, n_{ij}^{(2)})}(\mathbf{h}_i, \mathbf{h}_j) = L_{HD}^{(n_{ij}^{(1)})}(\mathbf{h}_i, \mathbf{h}_j)[n_{ij}^{(2)}] \tag{6}$$

$$L_{HD}^{(n_{ij}^{(1)})}(\mathbf{h}_i, \mathbf{h}_j) = \left[K, \left\lfloor \frac{(n_{ij}^{(1)} - 1)K}{n_{ij}^{(1)}} \right\rfloor, \left\lfloor \frac{(n_{ij}^{(1)} - 2)K}{n_{ij}^{(1)}} \right\rfloor, \ldots, 0 \right] \tag{7}$$

The HD between any hash pair ranges from 0 to K. Given that $0 \leq n_{ij}^{(2)} \leq n_{ij}^{(1)}$, the number of possible values for $n_{ij}^{(2)}$ is $(n_{ij}^{(1)} + 1)$. $(n_{ij}^{(1)} + 1)$ approximately equidistant points are selected from the interval $[0, K]$ using the floor function. We then store these HD in a descending order list, denoted as $L_{HD}^{(n_{ij}^{(1)})}(\mathbf{h}_i, \mathbf{h}_j)$ in (7). $L_{HD}^{(n_{ij}^{(1)})}(\mathbf{h}_i, \mathbf{h}_j)$ is a descending order list of HD between $\mathbf{h}_i, \mathbf{h}_j$ based on the value of $n_{ij}^{(1)}$. From (6), we can observe that as the number of shared levels increases, the value of $D_H^{(n_{ij}^{(1)}, n_{ij}^{(2)})}(\mathbf{h}_i, \mathbf{h}_j)$ decreases. When $n_{ij}^{(2)} = n_{ij}^{(1)} = 1$, $L_{HD}^{(1)}(\mathbf{h}_i, \mathbf{h}_j) = [K, 0]$ implies $D_H^{(1,1)}(\mathbf{h}_i, \mathbf{h}_j) = 0$. An illustrative example is depicted in Table 1. When $n_{ij}^{(1)} = 3$, the possible values of $n_{ij}^{(2)}$ are $0, 1, 2, 3$. Then we get, $L_{HD}^{(3)}(\mathbf{h}_i, \mathbf{h}_j) = [16, 10, 5, 0]$ from (7). If there is no shared label i.e, $n_{ij}^{(2)} = 0$ then $D_H^{(3,0)}(\mathbf{h}_i, \mathbf{h}_j) = 16$. Similarly, $D_H^{(3,1)}(\mathbf{h}_i, \mathbf{h}_j) = 10$, $D_H^{(3,2)}(\mathbf{h}_i, \mathbf{h}_j) = 5$, $D_H^{(3,3)}(\mathbf{h}_i, \mathbf{h}_j) = 0$. The value of $D_H^{(n_{ij}^{(1)}, n_{ij}^{(2)})}(\mathbf{h}_i, \mathbf{h}_j)$ is distinct for each unique combination of $n_{ij}^{(1)}$ and $n_{ij}^{(2)}$. The AHDL for image pair on $\mathbf{X}_{\mathfrak{T}}$ is computed as,

$$J_1 = \sum_{\mathbf{x}_i^{\mathfrak{T}}, \mathbf{x}_j^{\mathfrak{T}} \in \mathbf{X}_{\mathfrak{T}}} \log\left(\cosh\left(\frac{D_H^{(n_{ij}^{(1)}, n_{ij}^{(2)})}(\mathbf{h}_i, \mathbf{h}_j) - d_H(\mathbf{h}_i, \mathbf{h}_j)}{K}\right)\right) \tag{8}$$

where the predicted HD $d_H(\mathbf{h}_i, \mathbf{h}_j)$ is defined by,

$$d_H((\mathbf{h}_i, \mathbf{h}_j) = \frac{K}{2}(1 - \cos(\mathbf{h}_i, \mathbf{h}_j)) \tag{9}$$

The above formula describes the relationship between cosine similarity and normalized Euclidean distance for hash codes $\mathbf{h}_i$ and $\mathbf{h}_j$ of length K, where $\cos(\mathbf{h}_i, \mathbf{h}_j)$ represents the cosine similarity between the two hash codes. The above loss in (8) is deduced from absolute value $|\cdot|$. Given a real number x, we can write,

$$|x| \approx \log(\cosh(x)) \tag{10}$$

$D_H^{(n_{ij}^{(1)}, n_{ij}^{(2)})}(\mathbf{h}_i, \mathbf{h}_j)$ and $d_H(\mathbf{h}_i, \mathbf{h}_j)$ respectively can be considered as ground truth and predicted HD for this image pair. The idea is based on the value of $n_{ij}^{(1)}$ and $n_{ij}^{(2)}$; this loss forces the predicted HD to closely trail the ground truth HD.

3.4 Pairwise Multi-label Classification Loss (PMCL)

The PMCL on a train set $\mathbf{X}_\mathfrak{T}$ is defined by,

$$J_2 = -\sum_{\mathbf{x}_i^\mathfrak{T},\mathbf{x}_j^\mathfrak{T} \in \mathbf{X}_\mathfrak{T}} \left\{ \sum_{l=1}^{L} \left(\mathbf{y}_{il}^\mathfrak{T} \log(\sigma(\hat{\mathbf{y}}_{il}^\mathfrak{T})) + (1-\mathbf{y}_{il}^\mathfrak{T})\log(1-\sigma(\hat{\mathbf{y}}_{il}^\mathfrak{T})) \right) \right. \\ \left. + \left(\mathbf{y}_{jl}^\mathfrak{T} \log(\sigma(\hat{\mathbf{y}}_{jl}^\mathfrak{T})) + (1-\mathbf{y}_{jl}^\mathfrak{T})\log(1-\sigma(\hat{\mathbf{y}}_{jl}^\mathfrak{T})) \right) \right\} \quad (11)$$

where ground truth labels $\mathbf{y}_{il}^\mathfrak{T}, \mathbf{y}_{jl}^\mathfrak{T} \in \{0,1\}$ indicates whether the l-th label is present in samples $\mathbf{x}_i^\mathfrak{T}$ and $\mathbf{x}_j^\mathfrak{T}$. $\sigma(\cdot)$ is sigmoid function. $\hat{\mathbf{y}}_i^\mathfrak{T}, \hat{\mathbf{y}}_j^\mathfrak{T}$ are the predicted classes of $\mathbf{x}_i^\mathfrak{T}, \mathbf{x}_j^\mathfrak{T}$ respectively. These are obtained from the classification network $\text{net}_c(\cdot)$. $\text{net}_c(\cdot)$ is defined as,

$$\text{net}_c(\cdot) \mapsto \text{Linear}: 4096 \to \text{ReLU} \to \text{Linear}: L \quad (12)$$

Since multi-label classification serves as a fundamental loss function to generate distinctive features for various combinations of pathologies present in the images, obtaining accurate feature vectors $\mathbf{z}_i$ and $\mathbf{z}_j$ is crucial for deriving accurate representations $\mathbf{h}_i$ and $\mathbf{h}_j$ for an image pair. For this purpose, we utilize $\text{net}_c(\cdot)$, applying the PMCL loss to these feature vectors, which are the output of $\text{net}_e(\cdot)$.

3.5 Overall Loss

The overall loss is computed as,

$$J = \lambda_1 J_1 + \lambda_2 J_2 \quad (13)$$

where, λ_1 and λ_2 is scale hyperparameters. Minimizing J, thereby updating the parameters of $\text{net}_e(\cdot)$, $\text{net}_c(\cdot)$, and $\text{fc}_h(\cdot)$, which enable us to achieve our objectives. The overall training procedure is illustrated in Figure 3.

4 Experiments

4.1 Experimental Setup

We have implemented a modified AlexNet architecture [17] to build the $\text{net}_e(\cdot)$ and $\text{fc}_h(\cdot)$. $\text{net}_c(\cdot)$ comprises of two linear layers inherited from $\text{net}_e(\cdot)$. Adam optimizer [16] is used to learn the parameters of the above three networks. The weight decay parameter and batch size are set to 5×10^{-3} and 512, respectively. The training was initialized with a learning rate of 1×10^{-4}, and then the learning rate scheduler was used with patient 40 and factor 0.4. The values of hyperparameters $\lambda_1 = 1$ and $\lambda_2 = 1.5$ in (13). λ_1 and λ_2 are chosen through hyperparameter tuning using random search within the range $[0, 5]$ with a step size of 0.5. The experiments are conducted on a server equipped with 2× Intel Xeon 4110 CPUs, 12 × 8 GB DDR4 ECC Reg. RAM, 2 × 4 TB HDD, 4× Nvidia GTX 1080Ti GPUs, each with 11 GB DDR5 RAM, and Ubuntu 20.04 LTS operating system. The algorithms are implemented using Python 3.9 with PyTorch 1.11 and CUDA 11.2.

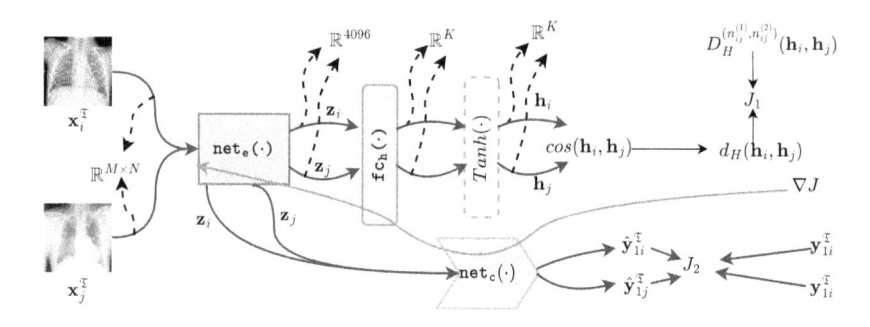

Fig. 3. An overview of training procedure of our method. Given an image pair $\mathbf{x}_i^{\mathfrak{T}}, \mathbf{x}_j^{\mathfrak{T}}$ along with their corresponding pair label sets $\mathbf{y}_i^{\mathfrak{T}}, \mathbf{y}_j^{\mathfrak{T}}$, we first compute the ground truth HD $D_H^{(n_{ij}^{(1)}, n_{ij}^{(2)})}(\mathbf{h}_i, \mathbf{h}_j)$. Next, we pass the generated feature vectors $\mathbf{z}_i$ and $\mathbf{z}_j$ from the network $\texttt{net}_\texttt{e}(\cdot)$ through both the networks $\texttt{fc}_\texttt{h}(\cdot)$. The real-valued hash vectors $\mathbf{h}_i$ and $\mathbf{h}_j$ are then obtained after applying $Tanh(\cdot)$ on the outputs of $\texttt{fc}_\texttt{h}(\cdot)$. Subsequently, utilizing (8), we calculate the loss terms J_1. J_2 is computed using (11) based on the predicted class by $\texttt{net}_\texttt{c}(\cdot)$. Minimizing the total loss J allows for the optimization of the weights of $\texttt{net}_\texttt{e}(\cdot)$, $\texttt{fc}_\texttt{h}(\cdot)$, and $\texttt{net}_\texttt{c}(\cdot)$.

4.2 Dataset

We have utilized the publicly available NIH Chest X-ray database [1], which consists of 112,120 frontal-view X-ray images from 30,805 unique patients. Each image is associated with one or more of the 14 common thoracic pathologies identified from the accompanying radiological reports. We have selected 51,480 images depicting the 13 most frequent pathologies, including Atelectasis, Consolidation, Infiltration, Pneumothorax, Edema, Emphysema, Fibrosis, Effusion, Pneumonia, Pleural thickening, Cardiomegaly, Nodule, and Mass. These images are divided into three non-overlapping image sets: training set ($\mathbf{X}_{\mathfrak{T}}$), gallery set ($\mathbf{X}_G$), and query set ($\mathbf{X}_Q$), where $|\mathbf{X}_{\mathfrak{T}}| = 38,610$, $|\mathbf{X}_G| = 10,296$ and $|\mathbf{X}_Q| = 2,574$.

4.3 Evaluation metrics

The metrics most commonly used by multi-label CBMIR methods are: normalized discounted cumulative gain (nDCG), average cumulative gain (ACG), and weighted mean average precision (wMAP) [23]. These metrics provide a comprehensive evaluation by assessing ranking quality, overall relevance, and precision-recall balance respectively. During training, we use the normalized relevance score (i.e., JSC). During evaluation, we use only the relevance score, adhering to the same strategy as existing methods on multi-label image retrieval [31].

---

[1] https://www.kaggle.com/datasets/nih-chest-xrays/data

Normalized discounted cumulative gain (nDCG) In order to compute $nDCG@p$ for top-p retrieval, first we need to calculate $DCG@p$ for top-p retrieval. The mathematical formulation for $DCG_q@p$ of query image $\mathbf{x}_q^Q$ is given by,

$$DCG_q@p = \sum_{r=1}^{p} \frac{2^{R_q(r)} - 1}{log_2(r+1)} \qquad (14)$$

where relevance score $R_q(r) (= |\mathbf{y}_q^Q \cap \mathbf{y}_r^G|)$ represents the number of pathologies between $\mathbf{x}_q^Q \in \mathbf{X}_Q$ and $\mathbf{x}_r^G \in \mathbf{X}_G$ are matched.

We normalize this by dividing it with the maximally achievable value or Ideal DCG (iDCG). Finally to obtain,

$$nDCG_q@p = \frac{DCG_q@p}{iDCG_q@p} \qquad (15)$$

where $iDCG_q@p = DCG_q@p$ of ideal ranking or best possible ranking.

Finally,

$$nDCG@p = \frac{1}{|\mathbf{X}_Q|} \sum_{\mathbf{x}_q^Q \in \mathbf{X}_Q} nDCG_q@p \qquad (16)$$

Average cumulative gains (ACG) The $ACG@p$ metric quantifies the cumulative similarity between a query image and the $top-p$ retrieved images. It is calculated by summing the similarities between the query image and each of the $top-p$ retrieve in the retrieval list. $ACG@p$ can be formulated as,

$$ACG@p = \frac{1}{|\mathbf{X}_Q|} \sum_{\mathbf{x}_q^Q \in \mathbf{X}_Q} \sum_{r=1}^{p} \frac{R_q(r)}{p} \qquad (17)$$

Weighted mean average precision $(wMAP)$ The $wMAP$ can be formulated by,

$$wMAP = \frac{1}{|\mathbf{X}_Q|} \sum_{\mathbf{x}_q^Q \in \mathbf{X}_Q} \left(\frac{\sum_{r=1}^{p} \delta(R_q(r) > 0) ACG@r}{\sum_{r=1}^{p} (\delta(R_q(r) > 0))} \right) \qquad (18)$$

where $\delta(\cdot)$ is the indicator function.

5 Results and Discussions

5.1 Comparison with existing methods

Since our method is based on pairwise learning, we compare it with four recent state-of-the-art (SOTA) pairwise multi-label DNH methods IDHN [31], DCH [4], OrthoHash [12], CSQ [30], HSDH [1]. The same network i.e., AlexNet, is used for a fair comparison. The parameters of SOTA models are selected

Table 2. Comparison of nDCG and ACG with pairwise deep hashing methods for different hash code lengths. - indicates that it is not evaluated since it is not applicable for the specific hash code length.

Method	nDCG@100				ACG@100			
	16	32	48	64	16	32	48	64
IDHN [31]	0.5900	0.5931	0.5797	0.5955	0.3283	0.3620	0.3323	0.3158
DCH [4]	0.5916	0.6139	0.6005	0.6084	0.3330	0.3550	0.3363	0.3383
OrthoHash [12]	0.5905	0.5947	0.6244	0.5916	0.3294	0.3387	0.3463	0.3256
CSQ [30]	0.5905	0.6215	-	0.6194	0.3308	0.3780	-	0.3624
HSDH [1]	0.5920	0.5917	0.6059	0.5910	0.3330	0.3326	0.3416	0.3309
Ours	**0.6318**	**0.6363**	**0.6426**	**0.6362**	**0.3874**	**0.3869**	**0.4028**	**0.3930**

Table 3. Comparison of $wMAP$ with the SOTA for different hash code lengths.

Method	wMAP			
	16	32	48	64
IDHN [31]	0.3619	0.4016	0.3776	0.3705
DCH [4]	0.3671	0.4174	0.3903	0.3990
OrthoHash [12]	0.3626	0.3713	0.4318	0.3644
CSQ [30]	0.3642	0.4377	-	0.4319
HSDH [1]	0.3674	0.3660	0.3895	0.3773
Ours	**0.4572**	**0.4600**	**0.4767**	**0.4664**

according to those specified in their respective original publications. We evaluate four variants of our proposed methods with four different hash code lengths $K = \{16, 32, 48, 64\}$. The comparison results for nDCG and ACG are presented in Table 2, while those for $wMAP$ are presented in Table 3. These demonstrate the substantial superiority of our proposed method over SOTA methods. CSQ achieves the best results for $K = 32$ hash code length across all SOTA compared. Our method shows an improvements over CSQ, with an approximately improvement of $4.13\%, 1.48\%, 1.68\%$ for hash code lengths $K = \{16, 32, 64\}$ respectively in terms of $nDCG@100$. Furthermore, our method demonstrates relative increases of $5.66\%, 0.89\%, 3.06\%$ in $ACG@100$, and $9.30\%, 3.23\%, 3.45\%$ in $wMAP$ for hash code lengths $K = \{16, 32, 64\}$ respectively over CSQ. CSQ is not applicable for $K = 48$. Notably, our method achieves $1.82\%, 5.65\%, 4.49\%$ higher $nDCG@100, ACG@100$, and $wMAP$ than OrthoHash for $K = 48$ respectively. Our method surpasses HSDH with improvements of $3.98\%, 5.44\%,$ and 8.98% in $nDCG@100, ACG@100$, and $wMAP$, respectively, for $K = 16$. Our method demonstrates comparable or even superior performance with shorter binary codes. For instance, our method with a 16-bit binary code significantly outperforms all other hashing methods by large margins for each metric. These

Jaccard Coefficient to Define Hamming Distance for Multi-label CBMIR 263

results highlight the effectiveness of our proposed method for retrieving multi-label medical images.

(a) $\mathbf{x}_1^{\mathfrak{T}}(\mathbf{h}_1)$ (b) $\mathbf{x}_2^{\mathfrak{T}}(\mathbf{h}_2)$ (c) $\mathbf{x}_3^{\mathfrak{T}}(\mathbf{h}_3)$ (d) $\mathbf{x}_4^{\mathfrak{T}}(\mathbf{h}_4)$ (e) $\mathbf{x}_5^{\mathfrak{T}}(\mathbf{h}_5)$

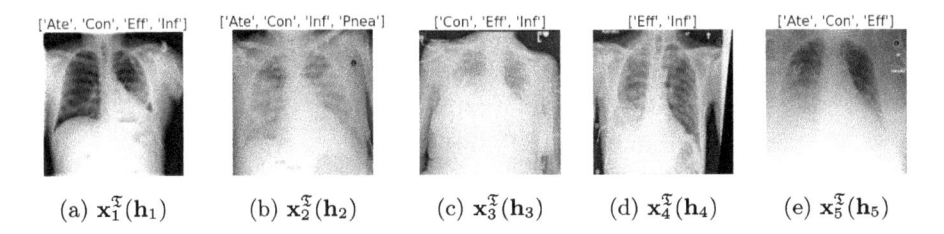

Fig. 4. The above figures depict five random images selected from the training set and generated hash codes, with their respective labels annotated. Atelectasis ('Ate'), Pneumonia ('Pnea'), Consolidation ('Con'), Effusion ('Eff'), Infiltration ('Inf').

5.2 Learning of Hamming Distances During Training

We consider five randomly picked training images $\mathbf{x}_1^{\mathfrak{T}}, \mathbf{x}_2^{\mathfrak{T}}, \mathbf{x}_3^{\mathfrak{T}}, \mathbf{x}_4^{\mathfrak{T}}, \mathbf{x}_5^{\mathfrak{T}}$ along with their hash codes $\mathbf{h}_1, \mathbf{h}_2, \mathbf{h}_3, \mathbf{h}_4, \mathbf{h}_5$. These images are illustrated in Figure 4 with their labels. Here $\frac{n_{12}^{(2)}}{n_{12}^{(1)}} = \frac{3}{5}, \frac{n_{13}^{(2)}}{n_{13}^{(1)}} = \frac{3}{4}, \frac{n_{14}^{(2)}}{n_{14}^{(1)}} = \frac{2}{4} = \frac{1}{2}, \frac{n_{15}^{(2)}}{n_{15}^{(1)}} = \frac{3}{4}$. The ground truth Hamming distances are calculated using (6) and (7), while the predicted Hamming distances are calculated using (9). The detailed calculations are presented in Table 4. It can be observed that the predicted Hamming distance ($d_H(\cdot)$) closely approximates the ground truth Hamming distance ($D_H(\cdot)$) between all pairs of hash codes for $K = \{16, 32\}$. For $K = 48$, the distances are close except for the hash pair $(\mathbf{h}_1, \mathbf{h}_4)$, and for $K = 64$, except for $(\mathbf{h}_1, \mathbf{h}_2)$, and $(\mathbf{h}_1, \mathbf{h}_4)$. The mean and standard deviation of the ground truth HD and predicted HD are presented in Table 4. These results show that smaller hash codes exhibit less disparity between these HD values. These findings confirm that our approach is particularly effective when employing shorter hash code lengths for these five images that are analyzed.

5.3 Top Retrieved Images

In Figure 5, we present retrieved images obtained using our approach. Images exhibiting more similarity in pathologies with the query image are prioritized to appear at the top of the ranking. This verifies the capability of our method to maintain multi-level similarity, thereby retrieving images that offer a higher level of similarity for enhanced assessment support.

Table 4. The comparison between ground truth HD ($D_H(\cdot)$) and predicted HD ($d_H(\cdot)$) for five randomly selected images (depicted in Figure 4) across different hash code lengths. The last column indicates the mean and standard deviation (SD) between these HD.

	Hash code pairs	$(\mathbf{h}_1, \mathbf{h}_2)$	$(\mathbf{h}_1, \mathbf{h}_3)$	$(\mathbf{h}_1, \mathbf{h}_4)$	$(\mathbf{h}_1, \mathbf{h}_5)$	mean ± SD
$K = 16$	ground truth HD	6	4	8	4	5.5 ± 1.65
	predicted HD	5.16	3.74	7.78	4.04	5.18 ± 1.59
$K = 32$	ground truth HD	12	8	16	8	11 ± 3.31
	predicted HD	13.29	9.71	13.67	7.55	11.05 ± 2.54
$K = 48$	ground truth HD	19	12	24	12	16.75 ± 5.06
	predicted HD	17.47	15.78	21.24	25.33	19.95 ± 3.67
$K = 64$	ground truth HD	25	16	32	16	22.25 ± 6.72
	predicted HD	17.27	14.45	17.10	13.23	15.51 ± 1.72

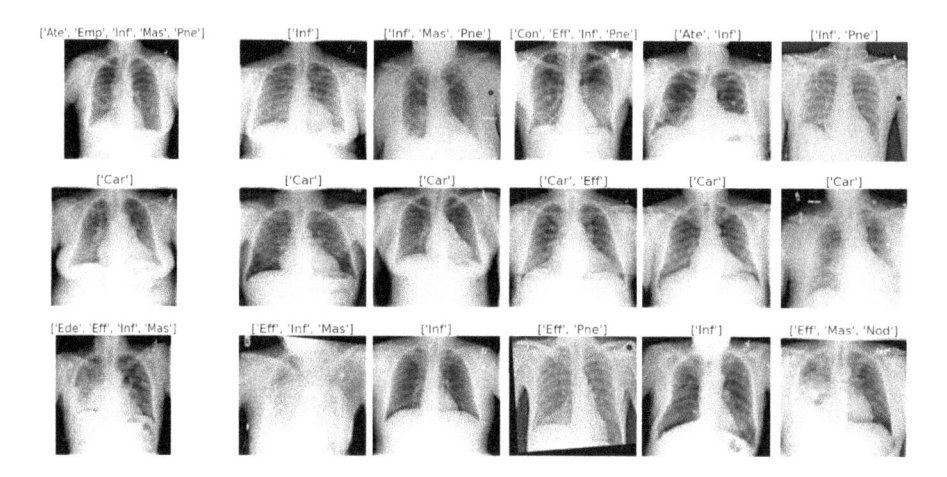

Fig. 5. Qualitative results for our method. The images on the left side represent the query images, while those on the right side depict the top-5 retrieved images. Atelectasis ('Ate'), Pneumonia ('Pnea'), Consolidation ('Con'), Effusion ('Eff'), Infiltration ('Inf'), Pneumothorax ('Pne'), Mass ('Mas'), Cardiomegaly ('Car'), Edema ('Ede').

6 Conclusion

In this work, we have developed an effective CBMIR system tailored for large-scale multi-label CBMIR. Our approach leverages DNH, which learns HD between hash codes to generate image-specific hash codes. We design a loss function for effective HD learning using the JSC between image labels. Through extensive experiments conducted with a publicly available multi-label medical image datasets, our proposed method demonstrated superior performance compared to existing methods. It effectively learns features and hash codes, enhancing the performance of multi-label CBMIR. This work provides insight into the advantageous nature of leveraging the complementarity between image labels for hash learning and their HD.

References

1. Alizadeh, S.M., Helfroush, M.S., Müller, H.: A novel siamese deep hashing model for histopathology image retrieval. Expert Syst. Appl. **225**, 120169 (2023)
2. Bag, S., Kumar, S.K., Tiwari, M.K.: An efficient recommendation generation using relevant jaccard similarity. Inf. Sciences **483**, 53–64 (2019)
3. Cai, T.W., Kim, J., Feng, D.D.: Content-based medical image retrieval. In: Biomed. Inf. Technol., pp. 83–113. Elsevier (2008)
4. Cao, Y., Long, M., Liu, B., Wang, J.: Deep cauchy hashing for hamming space retrieval. In: Proc. Conf. Comp. Vision Pattern Recognit. pp. 1229–1237 (2018)
5. Cao, Z., Long, M., Wang, J., Yu, P.S.: Hashnet: deep learning to hash by continuation. In: Proc. Conf. Comp. Vision Pattern Recognit. pp. 5608–5617 (2017)
6. Chen, Y., Tang, Y., Huang, J., Xiong, S.: Multi-scale triplet hashing for medical image retrieval. Computers Biol. Medicine **155**, 106633 (2023)
7. Chen, Z., Cai, R., Lu, J., Feng, J., Zhou, J.: Order-sensitive deep hashing for multimorbidity medical image retrieval. In: Proc. Med. Image Comput. Comput. Assisted Intervention. pp. 620–628. Springer (2018)
8. Das, P., Neelima, A.: An overview of approaches for content-based medical image retrieval. Int. J. Multimedia Inf. Retrieval **6**(4), 271–280 (2017)
9. Doan, K.D., Yang, P., Li, P.: One loss for quantization: Deep hashing with discrete wasserstein distributional matching. In: Proc. Conf. Comp. Vision Recognit. pp. 9447–9457 (2022)
10. Fang, J., Fu, H., Liu, J.: Deep triplet hashing network for case-based medical image retrieval. Med. Image Anal. **69**, 101981 (2021)
11. Guo, X., Duan, J., Gichoya, J., Trivedi, H., Purkayastha, S., Sharma, A., Banerjee, I.: Multi-label medical image retrieval via learning multi-class similarity. Available at SSRN 4149616 (2022)
12. Hoe, J.T., Ng, K.W., Zhang, T., Chan, C.S., Song, Y.Z., Xiang, T.: One loss for all: Deep hashing with a single cosine similarity based learning objective. Advances Neural Inf. Process. Syst. **34**, 24286–24298 (2021)
13. Hou, D., Zhao, Z., Hu, S.: Multi-label learning with visual-semantic embedded knowledge graph for diagnosis of radiology imaging. IEEE Access **9**, 15720–15730 (2021)
14. Indyk, P., Motwani, R.: Approximate nearest neighbors: towards removing the curse of dimensionality. In: Proc. ACM Symp. Theory Cmput. pp. 604–613 (1998)
15. Jose, A., Filbert, D., Rohlfing, C., Ohm, J.R.: Deep hashing with hash center update for efficient image retrieval. In: IEEE Int. Conf. Acoust. Speech and Signal Proc. pp. 4773–4777. IEEE (2022)
16. Kingma, D., Ba, J.: Adam: A method for stochastic optimization. In: Int. Conf. Learn. Representations. San Diega, CA, USA (2015)
17. Krizhevsky, A., Sutskever, I., Hinton, G.E.: Imagenet classification with deep convolutional neural networks. Advances Neural Inf. Process. Syst. **25** (2012)
18. Kumar, A., Kim, J., Cai, W., Fulham, M., Feng, D.: Content-based medical image retrieval: a survey of applications to multidimensional and multimodality data. J. Digit. Imag. **26**, 1025–1039 (2013)

19. Liu, H., Wang, R., Shan, S., Chen, X.: Deep supervised hashing for fast image retrieval. In: Proc. Conf. Comp. Vision Pattern Recognit. pp. 2064–2072 (2016)
20. Liu, W., Wang, Z., Liu, X., Zeng, N., Liu, Y., Alsaadi, F.E.: A survey of deep neural network architectures and their applications. Neurocomputing **234**, 11–26 (2017)
21. Luo, X., Wang, H., Wu, D., Chen, C., Deng, M., Huang, J., Hua, X.S.: A survey on deep hashing methods. ACM Trans. Knowl. Discovery Data **17**(1), 1–50 (2023)
22. Müller, H., Deserno, T.M.: Content-based medical image retrieval. In: Biomed. Image Process., pp. 471–494. Springer (2010)
23. Rodrigues, J., Cristo, M., Colonna, J.G.: Deep hashing for multi-label image retrieval: a survey. Artif. Intell. Rev. **53**(7), 5261–5307 (2020)
24. Singh, A., Gupta, S.: Learning to hash: a comprehensive survey of deep learning-based hashing methods. Knowl. Inf. Syst. **64**(10), 2565–2597 (2022)
25. Sorower, M.S.: A literature survey on algorithms for multi-label learning. Oregon State University, Corvallis **18**(1), 25 (2010)
26. Su, S., Zhang, C., Han, K., Tian, Y.: Greedy hash: Towards fast optimization for accurate hash coding in cnn. Advances Neural Inf. Process. Syst. **31** (2018)
27. Tagare, H.D., Jaffe, C.C., Duncan, J.: Medical image databases: A content-based retrieval approach. J. Amer. Med. Inform. Assoc. **4**(3), 184–198 (1997)
28. Tang, Y., Chen, Y., Xiong, S.: Deep semantic ranking hashing based on self-attention for medical image retrieval. In: Int. Conf. Pattern Rec. pp. 4960–4966. IEEE (2022)
29. Wang, J., Zhang, T., Sebe, N., Shen, H.T., et al.: A survey on learning to hash. IEEE Trans. Pattern Anal. Mach. Intell. **40**(4), 769–790 (2017)
30. Yuan, L., Wang, T., Zhang, X., Tay, F.E., Jie, Z., Liu, W., Feng, J.: Central similarity quantization for efficient image and video retrieval. In: Proc. Conf. Comp. Vision Pattern Recognit. pp. 3083–3092 (2020)
31. Zhang, Z., Zou, Q., Lin, Y., Chen, L., Wang, S.: Improved deep hashing with soft pairwise similarity for multi-label image retrieval. IEEE Trans. Multimedia **22**(2), 540–553 (2019)

Adaptive Cross-Modal Representation Learning for Heterogeneous Data Types in Alzheimer Disease Progression Prediction with Missing Time Point and Modalities

S. P. Dhivyaa[1], Duy-Phuong Dao[1], Hyung-Jeong Yang[1(✉)], and Jahae Kim[1,2]

[1] Department of AI Convergence, Chonnam National University, Gwangju 61186, South Korea
hjyang@jnu.ac.kr
[2] Department of Nuclear Medicine, Chonnam National University Hospital, Gwangju 61469, South Korea

Abstract. Alzheimer's treatment requires early detection; yet, predicting progression is challenging due to significant missing information in medical data for biomarkers and neuroimages. Recent studies tackled the missing data issue in biomarker, by introducing an imputation module to handle the missing values. However, for neuroimaging modalities such as Magnetic Resonance Imaging (MRI) and Positron Emission Tomography (PET), we still need a reliable system to handle this major issue. To overcome this, we propose an end-to-end hybrid model that is capable of handling both missing biomarker data as well as neuroimages. The proposed model employs a two-fold approach: first, it uses an attention-based multimodal variational autoencoder to impute missing neuroimages and a mask imputation strategy for biomarker data. Second, it leverages a recurrent neural network (RNN) to predict AD progression in future years, effectively handling missing modalities by reconstructing the missing data before making predictions. We performed our experiments on 1369 patients from the Alzheimer's Disease Neuroimaging Initiative (ADNI) dataset and our model achieved 0.6059 ± 0.0151, 0.6074 ± 0.0163, 0.6166 ± 0.0203, 0.7749 ± 0.0135 in terms of accuracy, precision, recall, and mAUC, respectively. The results confirm that our proposed model can be useful for handling missing data and by utilizing both biomarker and neuroimaging data simultaneously, we can precisely predict the progression of Alzheimer's disease in clinical settings.

Keywords: Disease Progression Prediction · Multimodal Variational Autoencoders · Alzheimer's Disease

1 Introduction

Alzheimer's disease (AD), a progressive neurodegenerative disorder, is the most common cause of dementia among older adults, characterized by cognitive decline and loss of brain function [1]. Early and accurate prediction of AD progression is crucial for effective clinical management, patient care, and the development of therapeutic strategies

---
S. P. Dhivyaa and D.-P. Dao—Equal contribution.

aimed at altering the disease course. Recent advances in neuroimaging and biomarker research have provided invaluable insights into the pathophysiological foundations of AD, offering potential predictors for disease progression. However, one of the significant challenges in utilizing this wealth of data is the frequent occurrence of missing information, which can significantly hinder the analysis and interpretation of longitudinal datasets.

Neuroimaging techniques like magnetic resonance imaging (MRI) and positron emission tomography (PET) capture the structural and functional details of the human brain. Recently, predictive models have evolved from unimodal to multi-modal approaches, as they can provide complementary information of the brain thus improving the disease characterization and diagnosis. Deep learning models which directly learns the features from neuroimage, take advantage of the complementary information provided by multimodal data. However, adopting a multi-modal approach has introduced greater complexities in data management, particularly the heightened risk of dealing with incomplete datasets. Lately, the generative adversarial networks have been employed to fill in missing modality information by integrating multimodal features [2]. However, it should be noted that these models are developed for classifying AD diseases rather than predicting its progression. [3] proposed a multimodal deep-learning framework combining a 3-dimensional convolutional neural network (CNN) to capture intra-slice features from MRI volumes and a bi-directional recurrent neural network (BRNN) to identify inter-sequence patterns for Alzheimer's Disease progression prediction. This model utilizes MRI, biomarkers, and demographics to enhance prediction accuracy. Missing data points were handled using forward and backward filling techniques, which can introduce bias and reduce model accuracy by not accounting for underlying data distributions. Hence, there is a need for more appropriate imputation methods that can better capture the underlying data patterns to improve the robustness and accuracy of the disease progression predictions.

On the other hand, for biomarker data, early efforts relied on linear statistical models, but the complexity of AD pathology soon necessitated more sophisticated, non-linear computational techniques [4]. In response to this challenge, a progressive shift has occurred towards leveraging deep learning models, which excel in handling vast and incomplete datasets, thus marking a new era in the prediction of disease progression. The incorporation of machine learning, and more specifically recurrent neural networks (RNNs), has revolutionized predictive modeling in AD by capturing the temporal dependencies and patterns within complex longitudinal data. RNNs, and their more advanced variants such as long short-term memory (LSTM) networks, have been increasingly applied in the field, demonstrating significant improvements in the prediction of disease progression by effectively handling time-series data, including irregular time intervals, varying sampling rates, and missing data [5][6][7]. However, it is important to emphasize that these models are limited to imputing only biomarker data.

In this paper we propose Alzheimer's Disease Progression Prediction using X Modalities (ADxPro) with missing of neuroimaging modalities as well as biomarker data. Uniquely, our model addresses the dual challenge of imputing missing data in longitudinal studies and provides a better prediction result for the AD progression in the follow-up years. The key contributions are outlined as follows:

- Our proposed model, ADxPro conducts multi-tasking such as missing neuro-image reconstruction, imputation of biomarkers and diseases clinical status prediction across multiple time points over a 5-year courses prediction, offering significant development in understanding and managing AD.
- We propose a joint representation learning module for neuroimaging data that consist of an attention-based multimodal variational autoencoder for unified low-dimensional representation of diverse neuroimaging data such as MRI and PET, enhancing learning and imputation, especially for missing modalities. Unlike traditional models, ADxPro excels in processing incomplete datasets, handling neuroimaging and biomarker data efficiently.

The rest of this paper is arranged as follow. Section 2 gives a brief description of the related studied on variational autoencoder models and AD progression prediction models. Section 3 provides the details of our proposed ADxPro model. The experimental setup and result analysis are highlighted in Sect. 4. Lastly, Sect. 5 concludes our study along with limitations and future work.

2 Related Works

In this section, we review previous works on AD progression prediction models related to data imputation. Recently, deep generative models have demonstrated remarkable flexibility and expressiveness as unsupervised methods, capable of uncovering the latent structures within complex, high-dimensional datasets. Variational autoencoders (VAEs) are being recognized for generating and imputing missing data across various domains, including healthcare. [8] introduced an approximation of complex data distributions, offering an unsupervised learning approach in complex datasets. [9] captures biologically relevant features within its latent space for cancer stratification and predict gene expression patterns influenced by genetic variations or treatments. [10] proposed the imputation in mixed-type datasets, employing VAEs to handle both numerical and categorical data. Multi-modal VAE (MVAE) introduced in [11, 12] delivers a principled probabilistic formulation to handle missing data. [13] employed a MVAE for normative modeling, enabling the precise identification of abnormal brain volume deviations in Alzheimer's disease by integrating and analyzing multimodal neuroimaging data. These studies demonstrate the broad applicability of VAEs in dealing with missing data across various types of datasets and modalities, highlighting their potential in healthcare.

In literature, the AD progression prediction can be classified into three categories: classification, estimation and modeling. Under classification category, researchers try to identify whether the disease is stable or progressive over a period of time. Deep InfoMax (DIM) was developed by [14] to classify the stable and progressive patients in Cognitive Normal (CN), Mild Cognitive Impairment (MCI) and Alzheimer's Disease (AD). In estimation category, cognitive scores such as mini-mental state examination and Alzheimer's disease assessment scale-cognitive subscale are estimated by a Multisource Multitask Learner [15]. These estimates help the doctors to identify the complexity and progression rate of AD. For modelling category, [16] proposed a probabilistic approach, Diseases Progression Modeling (DPM), which quantifies the uncertainty in diagnosing the severity of individual diseases, with respect to missing data, biomarkers, and follow-up information.

To tackle the missing information problem, [17] utilized two representations of informative missingness patterns: masking and time interval. They proposed a deep recurrent network know as gated recurrent unit with decay (GRU-D) which efficiently use missing patterns to identify long-term temporal correlations in time series data. MinimalRNN [6] is employed in predicting patient diagnoses, ventricular volumes, and cognitive scores. [18] proposed a RNN model called BiPro, which utilizes bidirectional approaches to integrate past and future data for imputation. However, it is essential to highlight that these models are solely focused on the imputation of biomarker data.

3 Alzheimer's Disease Progression Prediction Using X Modalities (ADxPro)

In this section, we propose an end-to-end framework called ADxPro, a hybrid model designed to address the challenges of analyzing complex, multimodal medical data, including neuroimaging and clinical data. The comprehensive framework of the proposed approach is shown in Fig. 1.

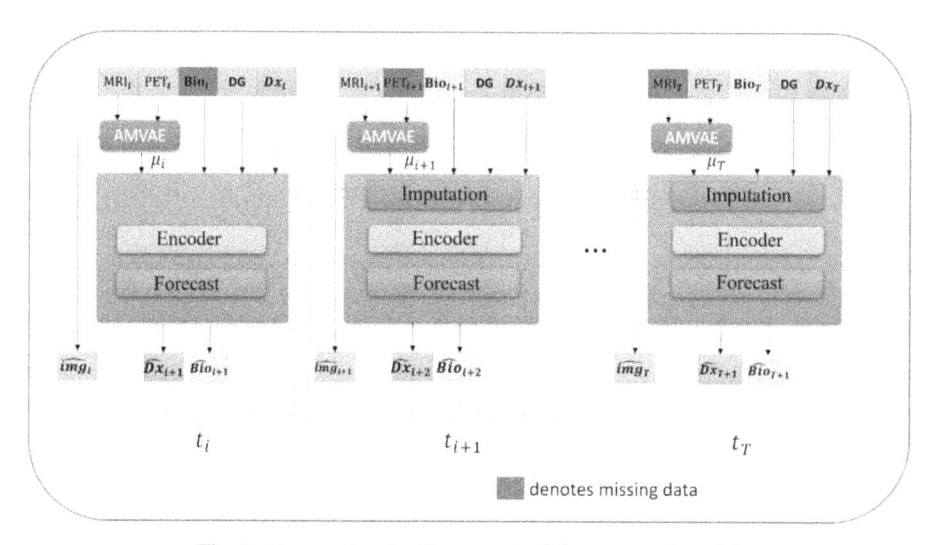

Fig. 1. Comprehensive framework of the proposed model

Our proposed model consists of four main modules: Joint Representation Learning for Neuroimaging Data, Imputation, Encoder, and Forecast. The joint representation learning module consists of an attention-based multimodal variational autoencoder-based architecture that facilitates the integration of different neuroimaging modalities into a lower-dimensional latent space. The imputation module fills in missing values for follow-up visits, the encoder captures the data's temporal dynamics, and the forecast module uses this information to predict diagnoses and biomarker levels in future visits.

3.1 Joint Representation Learning for Neuroimaging Data

In the proposed model, we hypothesis that $X_t = \{\dot{x}_1, \dot{x}_2, ..., \dot{x}_T\}$ represents an input sequence of T discrete temporal observations. At a given times step t, the input is represented as a multi-dimensional array $\dot{x}_t$, which encapsulates various data modalities. Let img_t, bio_t and dx_t denote the neuro imaging modalities MRI and PET, biomarker and diagnosis data, respectively, at time step t. dg denotes the demographic data as well as genetic data, which has constant features such as gender, education, APOE4 allele except for the age which is incremented yearly. Therefore, the input array at time step t can be expressed as $\dot{x}_t = [\, img_t, bio_t, dg, dx_t]$, integrating both dynamic and static data sources into the model's framework.

We consider Δ_t as time elapsed since the most recent observed visit prior to the tth visit. To represent the observation status of elements at the tth visit, we introduce a binary masking vector $m_t \in \{0, 1\}$. The tth component of the masking vector $m_{t,k}^{type}$, is assigned a value according to the presence of the kth element at time t, such that:

$$m_{t,k}^{type} = \begin{cases} 1, \text{if } \dot{x}_{t,k} \text{ is present} \\ 0, \text{otherwise} \end{cases} \quad (1)$$

where *type* represents the data modality and $\dot{x}_{t,k}$ indicates the actual observed status of the kth element at time t.

The time interval Δ_t is computed based on the sequential visit and observation records, and is defined as:

$$\Delta_{t,i} = \begin{cases} 0 \text{ if } t = 1 \\ s_t - s_{t-1} \text{ if } t > 1, m_{type,t-1,k} \text{ is present} \\ s_t - s_{t-1} + \Delta_{t-1,i} \text{ if } t > 1, m_{type,t-1,k} \text{ is missed} \end{cases} \quad (2)$$

Here, s_t denotes the time stamp for the visit, and the term $\Delta_{t-1,k}$ carries the accumulated time delay from the previous visit if the kth element was not observed.

The proposed an attention-based multimodal variational auto-encoder (AMVAE) encompasses of a modality specific encoder-decoder system specifically designed for the separate yet simultaneous processing of MRI and PET scans shown in Fig. 2. It captures the distinct attributes of each scan type while merging their learned features. The encoders highlight crucial aspects within the scans, tackling the challenge of differing relevance across areas. In addition to that, the encoders are incorporated with an attention mechanism that selectively emphasizes the salient features from each modality, improving the relevance and quality of the learned representations. Meanwhile, decoders are carefully built using convolutional layers, instance normalization, and activations to reconstruct the original images from compressed latent space.

We assume M neuroimaging modalities $x_1, x_2, \ldots, x_M$, which are conditionally independent over a shared latent space, z. The generative model is assumed to be in the form $p_\theta(x_1, x_2, \ldots, x_M, z) = p(z) p_\theta(x_1|z) p_\theta(x_2|z) \ldots p_\theta(x_M|z)$ where $p(z)$ is a prior, typically a spherical gaussian distribution. The decoders, denoted as $p_\theta(x_1|z) p_\theta(x_2|z) \ldots p_\theta(x_M|z)$, is composed of a deep neural network with parameters θ, which is coupled with a straightforward likelihood model, such as Bernoulli or Gaussian. This factorization allows for the exclusion of unobserved modalities during the evaluation of marginal likelihood. When a data point is defined as the collection of existing modalities, that is $X = \{ x_i | i^{th} modality\, is\, observed \}$, then the evidence lower bound (ELBO) is defined as:

$$\text{ELBO}(X) \triangleq E_{q_\phi(z|X)}\left[\sum_{x_i \in X} \lambda_i \log p_\theta(x_i|z)\right] - \beta \text{KL}\left[q_\phi(z|X), p(z)\right] \qquad (3)$$

where $q_\phi(z|X)$ is the inference network for optimizing ELBO, KL $[p,q]$ denotes the Kullback-Leibler divergence, which is a degree of the dissimilarity between two probability distributions, p and q. The parameters λ and β serve as weighting coefficients.

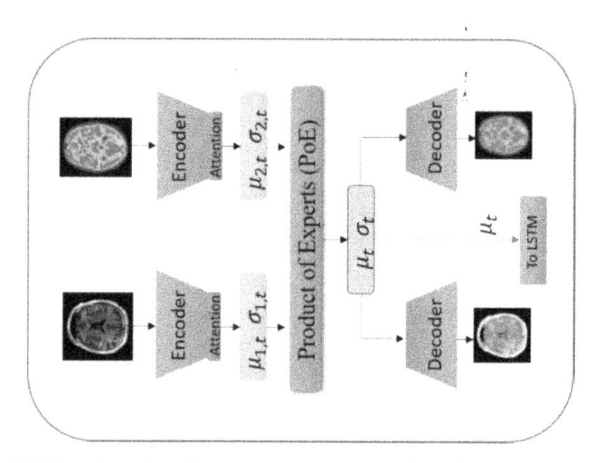

Fig. 2. AMVAE module for joint representation learning of neuroimaging modalities

Contrast to the conventional MVAE, for AD progress prediction we need to consider a time-series dataset consisting of multiple modalities. Let us denote $X^M = \{x_1^M, x_2^M, \ldots, x_T^M\}$ as the series of data points for modality M, observed at times $t = 1, \ldots, T$. The Evidence Lower BOund (ELBO) for a time-series with multiple modalities $\{M_1, M_2, \ldots, M_N\}$ along with modality specific masking, m_t^i, is given as:

$$L_{ELBO(X^{M_1},X^{M_2},\ldots X^{M_N})} \triangleq E_{q_\phi(z|X^{M_1},X^{M_2},\ldots X^{M_N})}\left[\sum_{t=1}^{T}\sum_{i=1}^{N} \lambda_t^i \log p_\theta(x_t^i|z)\right] \odot m_t^i - \\ \beta\, \text{KL}\left[q_\phi(z|X^{M_1}, X^{M_2}, \ldots X^{M_N}), p(z)\right] \odot m_t^i \qquad (4)$$

where X^{M_i}, x_t^i and λ_t^i are time-series data for modality i, the data point at time t for modality i and a weight factor for the importance of modality i at time t. $\odot$ denotes element-wise multiplication.

Unlike MVAEs, which require all modalities for training and inference, AMVAE adapts to incomplete data scenarios by incorporating product-of-experts (PoE) approach to effectively combine information from available modalities. This is particularly advantageous in real-world clinical settings where missing data is common. While training AMVAE, we must specify $2^M - 1$ inference networks, $q_\phi(z|X)$, corresponding to each possible subset of modalities $X \subseteq x_1, x_2, \ldots, x_M$. To approximate joint posterior, we follow the factorization approach (PoE), that uses a mix of unimodal variational posteriors $q_\phi(z|X)$ rather than relying on a single variational network which demands the presence of all images simultaneously. The PoE is defined as:

$$q_\phi(z|X) = p(z) \prod_{x_i \in X} q_\phi(z|x_i) \qquad (5)$$

where $p(z) = N(z; 0, I)$ is a uniform gaussian prior distribution and $q_\phi(z|x_i)$ is a gaussian distribution with diagonal covariance, defined through CNNs.

The distributions derived from above-mentioned product operations generally cannot be resolved into closed-form expressions. However, since $p(z)$ and $q_\phi(z|x_i)$ are both gaussian distribution, the product of these two distributions will also be gaussian. The gaussian experts are defined as:

$$\mu_t = \left(\sum_i \mu_{i,t} T_{i,t} m_{i,t}\right) \left(\sum_i T_{i,t} m_{i,t}\right)^{-1} \qquad (6)$$

$$V_t = \left(\sum_i T_{i,t} m_{i,t}\right)^{-1} \qquad (7)$$

where $\mu_{i,t}$ and $T_{i,t}$ are the mean, precision (which is the inverse of the variance) of the ith gaussian expert, respectively and $T_{i,t} = V_t^{-1}$ is the precision. $m_{i,t}$ represents the mask for the ith expert. When $m_{i,t}$ is equal to 1, the ith expert is included in the evaluation; when it is equal to 0, the ith expert is excluded. This allows for selective integration of experts based on the presence or absence of data. μ_t and V_t are the joint mean and variance at time point t that will be used for imputing the missing modality via separate decoder units. μ_t will be used in downstream units for predicting the progression of Alzheimer's Diseases.

3.2 Imputation module

It is commonly noticed that AD patients check the conditions irregularly and different data are collected at each visit. Therefore, it is natural there are missing data in AD data set. The imputation unit is responsible for imputing the missing values in neuroimages, biomarker, diagnosis and age feature in demographic data. It should be noted that for the imputation of values at the initial timepoint, we employ a global mean approach. For imputing the follow-up timepoints, we use the predicted values from the previous visit, $\widehat{v}_t$ and the masking M_t, which indicates the observed data modalities. Let X_t and Y_t be the observed data modalities and diagnosis at time point t, respectively. $\check{v}_t$ represents the combined form of X_t and Y_t at time point t.

$$X_t = [\mu_t \| bio_t \| dg] \& Y_t = [dx_t] \qquad (8)$$

$$\check{v}_t = [X_t \| Y_t] \qquad (9)$$

$$M_t = [m_t^{img} \| m_t^{bio} \| m_t^{dg} \| m_t^{dx}] \qquad (10)$$

$$\widehat{v}_t = [\widehat{X_{t-1}} \| \widehat{Y_{t-1}}] \qquad (11)$$

$$v_t = M_t \odot \breve{v}_t + (1 - M_t) \odot \widehat{v}_t \tag{12}$$

where v_t represents the imputed data at time step t, $\|$ is the concatenation operator and $\odot$ is element-wise multiplication.

3.3 Encoding and Forecasting

The encoding unit is responsible for capturing the temporal information from one time step to the subsequent one within the network. We employ a LSTM unit for temporal feature learning in time-series data with the masking vector. We initially used a hidden decay weight, ψh_t, derived from the interval time Δ_t, to predict the new hidden state, $\widehat{h}_t$. This characteristic enables the model to selectively retain information from past visits by adjusting the value of ψh_t based on Δ_t. The $\widehat{h}_t$ is defined as:

$$\psi h_t = e^{-(\max(0, W_{\psi h} \Delta_t + b_{\psi h}))} \tag{13}$$

$$\widehat{h}_t = W_{\psi h} \odot \psi h_t \tag{14}$$

where $W_{\psi h}$ and $b_{\psi h}$ are the weight and bias parameters, respectively. LSTM along with the masking vector m_t is given by:

$$f_t = \sigma\left(W_{vf} v_t + W_{hf} \widehat{h}_t + W_{mf} m_t + b_f\right) \tag{15}$$

$$\widehat{c}_t = \tanh\left(W_{vc} v_t + W_{hc} \widehat{h}_t + W_{mc} m_t + b_c\right) \tag{16}$$

$$o_t = \sigma\left(W_{vo} v_t + W_{ho} \widehat{h}_t + W_{mo} m_t + b_o\right) \tag{17}$$

The updated cell state, c_t and hidden state, h_t are defined as:

$$c_t = f_t \odot c_{t-1} + \odot \widehat{c}_t \tag{18}$$

$$h_t = o_t \odot t(1 - f_t) anh(c_t) \tag{19}$$

The forecasting unit leverages the current hidden state h_t to predict future values for both biomarkers and diagnosis data at time step $t + 1$. For biomarker data, the prediction is done by combining the hidden state h_t with the input v_t. On the other hand, for diagnosis data, the prediction is done solely based on the hidden state h_t utilizing a SoftMax function for multi-class classification. These functions are defined as:

$$\widehat{x}_{bio,t+1} = W_v h_t + v_t \tag{20}$$

$$\widehat{y}_{dx,t+1} = softmax(W_y h_t + b_y) \tag{21}$$

Our proposed model consists of multiple tasks such as AD diagnosis progression prediction, imputation of clinical biomarker data and neuroimage joint representation learning. The loss function for diagnosis prediction employs the cross-entropy loss between the model's predictions and the ground truth labels, specifically focusing on portions of the data indicated by a mask. For imputation loss, the difference between the model's imputations and actual observed values are calculated. And finally for joint representation learning, we employ Kullback-Leibler Divergence and reconstruction loss for optimizing variational autoencoders by balancing the trade-off between enhancing latent space regularization and improving input data reconstruction quality.

The loss functions for each task are defined as follows:

$$L_{dx} = -\sum_{t=2}^{T}(y_t \log \hat{y}_t) \qquad (22)$$

$$L_{impute} = \sum_{t=2}^{T} |\hat{x}_t^{bio} - \dot{x}_t^{bio}| \qquad (23)$$

In addressing multi-task learning challenges, we adopt the uncertainty-based loss weighting methodology as proposed by [19], which leverages the concept of homoscedastic uncertainty to dynamically balance the loss contributions of individual tasks. By adopting this, our overall loss function is defined as:

$$L_{total} = \frac{1}{2\sigma_1^2} L_{dx}(W_1) + \frac{1}{2\sigma_2^2} L_{impute}(W_2) + \frac{1}{2\sigma_3^2} L_{ELBO}(W_3) + \log \sigma_1 \sigma_2 \sigma_3 \qquad (24)$$

where W_1, W_2, W_3 are weights and $\sigma_1, \sigma_2, \sigma_3$ are the variance of loss terms L_{dx}, L_{impute} and L_{image}, respectively. The logarithmic term at the end acts as the regularization term in the variational inference setting.

4 Experiment setup and result analysis

4.1 Data set and preprocessing

In our study, we employed the ADNIMERGE data obtained from Alzheimer's Disease Neuroimaging Initiative (ADNI) database[1], which plays a crucial role in the research of Alzheimer's disease. Our dataset includes MRI and PET imaging data, biomarkers from T1-weighted MRI scans that offer precise volumetric measurements of brain regions, including the Ventricles, Hippocampus, Whole Brain, Entorhinal, Fusiform, and MidTemp, alongside genetic profiles such as APOE4 allele, known for its correlation with heightened Alzheimer's disease risk [20]. Our study integrates neuroimaging data and six MRI-derived volumetric variables to capture Alzheimer's Disease's complexity, leveraging MRI and PET images to reveal pathological changes and a broader spectrum of biomarkers not evident through volumetric analysis alone. In addition to that, we incorporated demographic information such as age, education, and gender, which are essential for assessing disease risk and progression. The selected patient's class-wise data distribution is shown in Fig. 3(a).

---

[1] Alzheimer's Disease Neuroimaging Initiative, is a global research project that collects and analyses clinical, imaging, genomic, and biomarker data to study the progression of Alzheimer's disease. For additional details, please visit www.adni-info.org

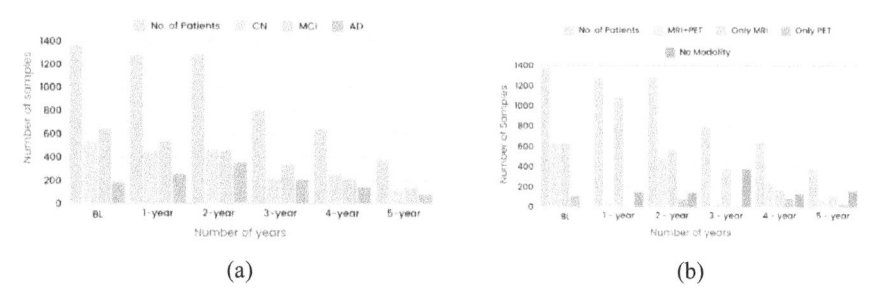

Fig. 3. (a) Selected set of longitudinal patient distribution from the ADNI study over 5 years. BL denotes Baseline visit, CN denotes Cognitive Normal, MCI denotes Mild Cognitive Impairment and AD denotes Alzheimer's Disease. (b) Availability of neuroimaging modalities across patient visits over 5 years. MRI + PET denotes patients with both MRI and PET scans. Only MRI denotes patients who only have MRI scans. Only PET denotes patients who only have PET scans. No Modality denotes patients without any MRI or PET scans recorded.

The ADNIMERGE dataset combines longitudinal data from 2,430 participants spanning 16,340 visits across various phases (ADNI-GO, ADNI-1, ADNI-2, and ADNI-3), recording up to 17 years of follow-up through 36 distinct visits. In our study, subject selection was based on specific criteria: firstly, the prediction of yearly progression was narrowed to baseline visits extending through five years of follow-up (i.e., M0, M12, M24, M36, M48, and M60) [21]. This timeframe is selected based on the progression rate of Alzheimer's disease, ensuring that the model provides meaningful predictions for early intervention while maintaining manageable computational complexity. Secondly, only subjects with a minimum of two visits, including the baseline visit, accompanied by a clinical diagnosis, were chosen. Thirdly, participants with reversible diagnoses, particularly those transitioning from MCI to CN or from AD to either MCI or CN, were excluded. Therefore, our dataset comprised 1,369 patients across 5,768 visits, which included 971 stable and 398 progressive patients. The comparison of available neuroimaging data over 5 years is shown in Fig. 3(b).

Following the approach by [22], adjustments were made for variations in brain size among subjects by normalizing the volumetric measurements from MRI biomarkers using each individual's intracranial volume. Subsequently, MRI features were transformed to fit a standard normal distribution through standardization with respect to their mean and standard deviation, facilitating a uniform evaluation metric. Subject ages, initially recorded at baseline visits, are progressively incremented throughout the follow-up period. Age and education, being numerical variables, were standardized using z-score normalization, while gender was encoded into a one-hot encoding vector for analytical compatibility. The APOE4 allele is transformed into a one-hot encoding vector for analysis. Following the neuroimaging pre-processing steps outlined in [2], our study incorporates image registration, cropping, and normalization as foundational preprocessing techniques. Initially, MRI and PET images are registered to align with the MNI152 standard brain template, utilizing SPM12 in Matlab, resulting in dimensions of 182 × 218 × 182 voxels. Given the large size, peripheral regions of the images are removed to both decrease dimensions and focus on the relevant brain structure information, leading to a refined size of 128 × 160 × 128 for both MRI and PET images. Next, voxel intensities are adjusted to a scale between 0 and 1 through min-max normalization, optimizing them for analysis.

4.2 Result Analysis

ADxPro model was implemented on a computational platform with the following specifications: The hardware included an NVIDIA A100 Tensor Core graphics processing unit featuring Multi-Instance GPU with 7 GB memory. The software environment consisted of the PyTorch library, utilizing the Adam optimizer with a learning rate set to 0.0002 for training the neural networks. The model's weights were initialized using the Xavier uniform initialization method to promote convergence during training. For model validation, a 5-fold cross-validation approach was employed to rigorously evaluate the model's generalizability across different subsets of the data. The robustness of the model was further ensured by setting up an annealing schedule over a predefined number of epochs, allowing for a gradual adjustment of the learning rate during the training process.

Table 1. Results for Alzheimer's Disease Progression prediction task up to 5 years (*marked as statistically significant with a p-value < 0.05)

Data Type	Model	ACC (↑)	PRE (↑)	REC (↑)	mAUC (↑)
Bio + Dg + Gen	GRU-D [17]	0.5750 ± 0.0122*	0.5722 ± 0.0069*	0.5934 ± 0.0128	0.7632 ± 0.0095
	MinimalRNN [6]	0.5680 ± 0.0203*	0.5661 ± 0.0211*	0.5885 ± 0.0185*	0.7549 ± 0.0091*
	BiPro [18]	0.5847 ± 0.0166	0.5867 ± 0.0219	0.5876 ± 0.0113*	0.7650 ± 0.0087
MRI + Bio + Dg + Gen	CNN-BRNN [3]	0.5947 ± 0.0209	0.5955 ± 0.0211	0.6000 ± 0.0161*	0.7654 ± 0.0178
MRI + PET + Bio + Dg + Gen	ADxPro	**0.6059 ± 0.0151**	**0.6074 ± 0.0163**	**0.6166 ± 0.0203**	**0.7749 ± 0.0132**

For the AD diagnosis prediction task, commonly utilized metrics such as accuracy (ACC), precision (PRE), recall (REC), and mean area under the curve (mAUC) were employed to gauge the model's diagnostic capability, underscoring the importance of these measures in assessing the performance of predictive models within clinical settings. For biomarker data imputation, mean absolute error (MAE) and mean relative error (MRE) served as the benchmarks, reflecting the precision of our model in handling missing data. Importantly, we applied the same settings mentioned above to all competing models to ensure a fair comparison.

Table 1 shows the efficacy of several predictive models in the task of clinical status prediction for Alzheimer's Disease over a span of five years. The proposed model, which imputes neuroimage, biomarkers, along with demographic data, demonstrates a superior performance across all the measured indices—accuracy, precision, recall, and mean area under the curve with 0.6059 ± 0.0151, 0.6074 ± 0.0163, 0.6166 ± 0.0203, and 0.7749 ± 0.013, respectively. This shows that our model's performance marginally exceeds the competing models thus illustrating a more consistent ability to identify true positive instances of the disease's progression. Incorporating multi modal data provides a holistic view of the disease biomarkers, which in turn enables a more distinct and accurate prediction of disease progression. Furthermore, a t-test was conducted to determine the statistical significance of our model's performance against that of the competing models, with findings detailed in Table 1.

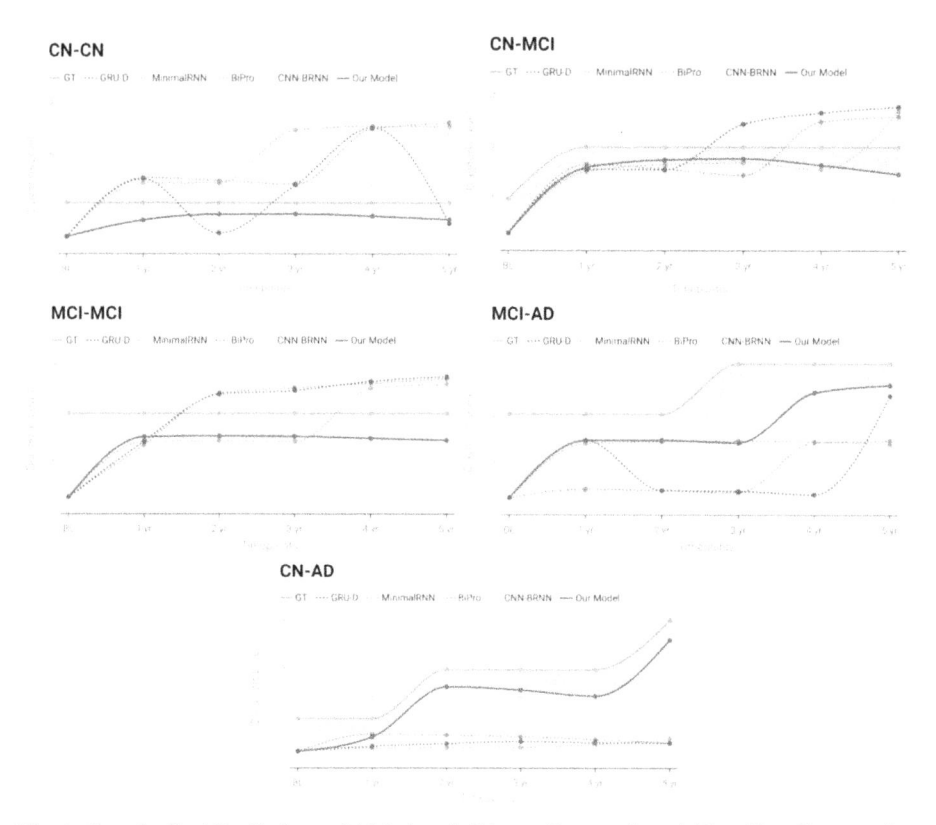

Fig. 4. Longitudinal Predictions of Alzheimer's Disease Progression: A Five-Year Comparative Analysis of Diagnostic Models. GT denotes the ground truth. Level 1,2 and 3 denotes CN, MCI and AD, respectively.

Figure 4 depicts a five-year longitudinal study comparing the effectiveness of various diagnostic models in predicting Alzheimer's Disease progression, marked by transitions between clinical states: CN to CN, CN to MCI, MCI to MCI, MCI to AD, and CN to AD. Our model highlights superior year-wise predictions compared to GRU-D, Minimal RNN, BiPro and CNN-BRNN, consistently aligning closely with the ground truth over five years. In contrast, the competing models show less stability and consistency, with fluctuations not as closely aligned to the ground truth.

The stable and accurate year-wise predictions are essential for effective clinical decision-making and long-term patient management. By precisely predicting AD progression patterns, our model can assist the clinicians in understanding the disease progression and plan interventions accordingly. The statistical significance and alignment with the ground truth confirm the effectiveness of our method, demonstrating that integrating multimodal data improves Alzheimer's Disease progression prediction.

Table 2. Results for Imputation task

Data Type	Model	MAE (ml) (↓)	MRE (%) (↓)
Bio + Dg + Gen	GRU-D [17]	5.5422 ± 0.3055	7.3789 ± 0.2946
	MinimalRNN [6]	3.9373 ± 0.3163	5.7233 ± 0.2599
	BiPro [18]	4.0700 ± 0.4840	5.9741 ± 0.3026
MRI + Bio + Dg + Gen	CNN-BRNN [3]	5.6713 ± 0.3005	7.5433 ± 0.2819
MRI + PET + Bio + Dg + Gen	ADxPro	4.0363 ± 0.3373	5.8039 ± 0.4032

In evaluating the effectiveness of various models for the biomarker imputation task, our model stands out as the optimal choice, despite a challenging multi-task environment. Our model's slight decrease in performance compared to MinimalRNN is offset by its ability to analyze a wider range of data—images, diagnostics, and biomarkers—making it a more versatile and comprehensive tool. Table 2 presents the outcomes of the imputation task.

Table 3. Results for Alzheimer's Disease Progression prediction with and without diagnosis data (Dx) at baseline

Data Type	Model	ACC (↑)	PRE (↑)	REC (↑)	mAUC (↑)
MRI + PET + Bio + Dg + Gen	Without Dx	0.6059 ± 0.0151	0.6074 ± 0.0163	0.6166 ± 0.0203	0.7749 ± 0.0132
	With Dx	0.8507 ± 0.0111	0.8333 ± 0.0102	0.8184 ± 0.0097	0.9396 ± 0.0084

Table 3 compares the predictive accuracy of our proposed model for AD progression using two sets of input data: one including diagnosis data (Dx) at the baseline visit and one without it. Our model's performance is markedly higher when the diagnosis data is included at the baseline, indicating the significant impact of initial diagnostic information on the model's predictive capabilities.

Figure 5 provides a comparison of the diagnostic performance across different neuroimaging modalities using unimodal and multimodal VAEs. The experimental setup for

Fig. 5 follows a similar procedure as our proposed model. The primary difference lies in the calculation of unimodal diagnostic performance, where we employed a vanilla variational autoencoder combined with a recurrent neural network for disease progression prediction. In this setup, each unimodal neuroimaging modality (MRI and PET) was evaluated individually but combined with other data types, including biomarkers (Bio), diagnostic data (Dx), genetic (Gen) and demographic information (Dg). Separate models were trained for each modality using a 5-fold cross-validation approach to ensure generalizability. The data suggests that the combined use of MRI and PET imaging modalities yields the highest performance metrics, indicating that a multi-modal imaging approach is more effective for the accurate diagnosis and monitoring of disease progression in clinical settings.

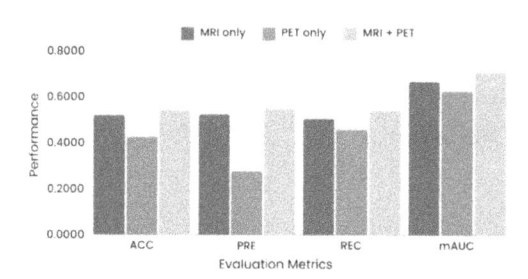

Fig. 5. Performance comparison of neuroimaging modalities: MRI, PET, and Combined MRI + PET.

5 Conclusion and Future Work

In this paper, we proposed ADxPro, a hybrid model for predicting longitudinal disease progression consisting of an attention-based multimodal variational autoencoder combined with RNN architecture. Our model performs three tasks simultaneously: missing neuroimage reconstruction, imputation of biomarkers and diseases clinical status prediction across multiple timepoints over a 5-year course. Overall, our research highlighted the significance of multimodal data integration in the accurate prediction of Alzheimer's Disease progression with missing data. Additionally, our findings have shown that the inclusion of diagnosis data at baseline substantially enhances predictive accuracy.

However, our model still has certain limitation. Given that our model primarily aims at predicting progression, we placed greater emphasis on optimizing this task over others, such as image reconstruction. Hence the quality of the reconstructed images is not good enough. The necessity to improve the quality of reconstructed MRI and PET images is important because of their role in monitoring disease progression and optimizing treatment. Additionally, enhanced images can reduce costs for patients by eliminating the need for expensive PET scans at every timepoint. In future work, we plan to employ Versatile Diffusion technique to enhance the quality of the generated images. This approach is expected to overcome the above limitation by using multimodal data and advanced diffusion techniques for higher quality image reconstructions.

Acknowledgement. This work was supported by the National Research Foundation of Korea (NRF) grant (MSIT)(RS-2023–00208397) and by Institute of Information & communications Technology Planning & Evaluation (IITP) under the Artificial Intelligence Convergence Innovation Human Resources Development (IITP-2023-RS-2023–00256629) grant funded by the Korea government (MSIT). This research was also supported by the MSIT (Ministry of Science and ICT), Korea, under the ITRC (Information Technology Research Center) support program (IITP-2024-RS-2024–00437718) supervised by the IITP (Institute for Information & Communications Technology Planning & Evaluation).

References

1. Alzheimer's Association: Alzheimer's disease facts and figures. Alzheimer's & Dementia 16(3), 391-460 (2020)
2. Gao, X., Shi, F., Shen, D., Liu, M.: Task-induced pyramid and attention GAN for multimodal brain image imputation and classification in Alzheimer's disease. IEEE Journal of Biomedical and Health Informatics 26(1), 36-43 (2021)
3. Rahim, N., El-Sappagh, S., Ali, S., Muhammad, K., Del Ser, J., & Abuhmed, T. Prediction of Alzheimer's progression based on multimodal deep-learning-based fusion and visual explainability of time-series data. Information Fusion, 92, 363-388 (2023)
4. Huang, Y., Mucke, L.: Alzheimer mechanisms and therapeutic strategies. Cell 148(6), 1204-1222 (2012)
5. Lipton, Z.C., Kale, D.C., Elkan, C., Wetzel, R.: Learning to diagnose with LSTM recurrent neural networks. arXiv preprint arXiv:1511.03677 (2015)
6. Nguyen, M., He, T., An, L., Alexander, D.C., Feng, J., Yeo, B.T., Alzheimer's Disease Neuroimaging Initiative: Predicting Alzheimer's disease progression using deep recurrent neural networks. NeuroImage 222, 117203 (2020)
7. Dao, D. P., Ho, N. H., Kim, J., & Yang, H. J.: Improving Recurrent Gate Mechanism for Time-to-Conversion Prediction of Alzheimer's Disease. In: The 9th International Conference on Smart Media and Applications, pp. 66–71 (2020)
8. Kingma, D.P., Welling, M.: Auto-encoding variational bayes. arXiv preprint arXiv:1312.6114 (2013)
9. Way, G.P., Greene, C.S.: Extracting a biologically relevant latent space from cancer transcriptomes with variational autoencoders. In: Pacific Symposium on Biocomputing 2018: Proceedings of the Pacific Symposium, pp. 80–91 (2018)
10. Mattei, P.A., Frellsen, J.: MIWAE: Deep generative modelling and imputation of incomplete data sets. In: International Conference on Machine Learning, pp. 4413–4423. PMLR (2019)
11. Shi, Y., Paige, B., Torr, P.: Variational mixture-of-experts autoencoders for multi-modal deep generative models. Advances in Neural Information Processing Systems 32 (2019)
12. Wu, M., Goodman, N.: Multimodal generative models for scalable weakly-supervised learning. Advances in Neural Information Processing Systems 31 (2018)
13. Kumar, S., Payne, P.R., Sotiras, A.: Normative modeling using multi-modal variational autoencoders to identify abnormal brain volume deviations in Alzheimer's disease. In: Medical Imaging 2023: Computer-Aided Diagnosis, Vol. 12465, p. 1246503. SPIE (2023)
14. Fedorov, A., Hjelm, R.D., Abrol, A., Fu, Z., Du, Y., Plis, S., Calhoun, V.D.: Prediction of progression to Alzheimer's disease with deep infomax. In: 2019 IEEE EMBS International Conference on Biomedical & Health Informatics (BHI), pp. 1–5. IEEE (2019)
15. Nie, L., Zhang, L., Meng, L., Song, X., Chang, X., Li, X.: Modeling disease progression via multisource multitask learners: A case study with Alzheimer's disease. IEEE Transactions on Neural Networks and Learning Systems 28(7), 1508-1519 (2016)

16. Lorenzi, M., Filippone, M., Frisoni, G.B., Alexander, D.C., Ourselin, S., Alzheimer's Disease Neuroimaging Initiative: Probabilistic disease progression modeling to characterize diagnostic uncertainty: application to staging and prediction in Alzheimer's disease. NeuroImage 190, 56-68 (2019)
17. Che, Z., Purushotham, S., Cho, K., Sontag, D., Liu, Y.: Recurrent neural networks for multivariate time series with missing values. Scientific Reports 8(1), 6085 (2018)
18. Ho, N.H., Yang, H.J., Kim, J., Dao, D.P., Park, H.R., Pant, S.: Predicting progression of Alzheimer's disease using forward-to-backward bi-directional network with integrative imputation. Neural Networks 150, 422-439 (2022)
19. Kendall, A., Gal, Y., Cipolla, R.: Multi-task learning using uncertainty to weigh losses for scene geometry and semantics. In: Proceedings of the IEEE Conference on Computer Vision and Pattern Recognition, pp. 7482–7491 (2018).
20. Bryant, E.: Study reveals how APOE4 gene may increase risk for dementia. National Institute on Aging. Available at: https://www.nia.nih.gov/news/study-reveals-how-apoe4-gene-may-increase-risk-dementia. Last accessed 2024/03/06
21. Zandifar, A., Fonov, V.S., Ducharme, S., Belleville, S., Collins, D.L., Alzheimer's Disease Neuroimaging Initiative: MRI and cognitive scores complement each other to accurately predict Alzheimer's dementia 2 to 7 years before clinical onset. NeuroImage: Clinical 25, 1021 (2020)
22. Ghazi, M.M., Nielsen, M., Pai, A., Cardoso, M.J., Modat, M., Ourselin, S., Sorensen, L., Alzheimer's Disease Neuroimaging Initiative: Training recurrent neural networks robust to incomplete data: Application to Alzheimer's disease progression modeling. Medical Image Analysis 53, 39-46 (2019)

Enhancing Medical Image Analysis with MA-DTNet: A Dual Task Network Guided by Morphological Attention

Susmita Ghosh[✉] and Swagatam Das

Electronics and Communication Sciences Unit, Indian Statistical Institute, Kolkata, India
ghoshsusmitasen@gmail.com, swagatam.das@isical.ac.in

Abstract. Accurate breast tumor segmentation and malignancy detection are crucial for early cancer diagnosis. In this context, we propose a novel lightweight multi-task learning framework, MA-DTNet, designed to perform both tasks simultaneously in an encoder-shared scenario. This approach leverages shared representations and contextual information, enabling mutual enhancement of the tasks. Unlike existing methods that require a large number of trainable parameters, MA-DTNet integrates a Spatial Morphological Attention (SMA) module alongside a Channel Attention (CA) mechanism to strategically enhance crucial morphological features and emphasize informative channels within the extracted representations. The SMA mechanism combines traditional morphological operations with trainable, adaptive structuring elements, effectively highlighting critical morphological attributes of regions of interest (ROIs) of various shapes and sizes within medical images. This targeted emphasis on morphological features translates to improved performance in both segmentation and classification tasks. Notably, MA-DTNet demonstrates superior performance compared to state-of-the-art multi-task learning (MTL) and single-task models on two publicly available breast ultrasound datasets. Specifically, on the UDIAT dataset, our approach achieves a 3.28% and 1.05% enhancement in dice score (segmentation) and F1 score (classification), respectively. Similarly, for the BUSI dataset, MA-DTNet exhibits a 1.62% and 4.96% improvement in dice score and accuracy, respectively. Significantly, MA-DTNet achieves these performance gains with significantly fewer trainable parameters than existing methods, underscoring its efficiency and potential for real-world applications. The method's generalization ability is further tested on two additional multi-task learning tasks: segmenting and classifying glands in histology images and segmenting and classifying skin lesions in dermoscopic images.

Supplementary Information The online version contains supplementary material available at https://doi.org/10.1007/978-3-031-78198-8_19.

Keywords: Multitask Learning · Medical Image Segmentation · Medical Image Classification · Spatial Morphological Attention Module · Channel Attention

1 Introduction

Medical image analysis underpins various healthcare applications, including diagnosis, treatment planning, and prognosis. Accurate segmentation and classification of anatomical structures and abnormalities within these images are essential for practical analysis. These tasks have been traditionally addressed using separate models trained for each specific task. However, this approach can be inefficient and resource-intensive, mainly when dealing with limited datasets or complex tasks requiring extensive training data. The emergence of multi-task learning (MTL) has revolutionized the field of medical image analysis, offering a compelling alternative to single-task learning approaches. MTL leverages the inherent correlations and shared information between related tasks within a single model, leading to several potential benefits like improved data efficiency[33], reduced overfitting [20], enhanced feature learning[20], among others.

Successful applications of MTL in medical image analysis include glioma segmentation and isocitrate dehydrogenase genotyping from brain MRI [6], skin lesion segmentation and classification from dermoscopic images [27], and kidney segmentation and domain translation from urographic images [31]. Building upon this growing body of research, we focus on the tasks of breast tumor segmentation and malignancy detection, as documented in recent studies highlighting the effectiveness of MTL for these specific tasks [7,28,29,32]. These MTL networks comprise trainable parameters in the range of 90-110 M. However, MTL models with fewer parameters are highly desirable for practical implementation in real-world settings. We design a lighter MTL architecture of 22.95 M parameters to address this challenge.

Ultrasound imaging typically identifies cancerous tumors as hypoechoic regions with poorly defined borders [10]. Consequently, more accurate tumor segmentation improves tumor-type diagnosis. Through experimentation, we found that the encoder-shared multi-task learning (MTL) model outperforms individual models designed separately for segmentation and classification. However, there is potential to further enhance segmentation performance and, by extension, malignancy detection. To address this, we propose a Channel and Spatial Morphological Attention Module (C-SMA), which integrates channel attention (CA) to prioritize crucial feature maps and spatial morphological attention (SMA) to focus on morphological attributes within the feature maps. While the successful integration of morphological operations in deep learning has been reported for tasks like image de-raining and image restoration [16], this work is the first, to our knowledge, to incorporate a trainable, multi-scale morphological operation in the form of morphological attention within an MTL network. Our MTL architecture integrates the C-SMA block into the skip connections between the encoder and decoder at various stages. Although we do not directly use the segmentation outcome to detect tumor malignancy, the encoder shared by both segmentation and classification branches is optimized to improve segmentation results, thereby enhancing malignancy detection. The contributions of this study are as follows:

- We propose a novel C-SMA module that integrates channel attention with morphological operations. The spatial morphological operations with trainable, multi-scale structuring elements effectively highlight the morphological attributes of feature maps, allowing for the identification of regions of interest (ROIs) with variable shapes and sizes.
- We propose a lightweight multi-task learning network for breast tumor segmentation and malignancy detection from ultrasound images by incorporating the C-SMA module within the multi-task framework.
- To evaluate efficiency, we employed the proposed method on public datasets (UDIAT and BUSI), outperforming both single-task and multi-task baselines and SOTA methods. We achieve 3.28% and 1.62% improvement in dice score in tumor segmentation tasks for UDIAT and BUSI datasets, respectively. In the classification task, an enhancement of 1.05% in the F1 score and 4.96% in accuracy is observed for the UDIAT and BUSI datasets, respectively.
- The generalization ability of our method is extended to two other multi-tasking scenarios: segmenting and classifying skin lesions in dermoscopic images and segmenting and predicting malignancy of glands in histology images.

The rest of the paper is organized as follows. Section 2 describes the existing literature concerning relevant MTL and their limitations. Section 3 elaborates on the proposed MA-DTNet and its components. Section 4 presents the experimental result, description of the datasets, and training protocol followed in this study. Lastly, section 5 concludes the study.

2 Related Studies

In this section, we delve into the existing literature concerning pertinent segmentation, classification, and multi-task learning for breast tumor segmentation and malignancy detection from ultrasound images, aiming to identify the gaps and limitations in current methodologies that our research seeks to address.

In 2021, Zhang *et al.* [32] devised an integrated segmentation and classification network, incorporating attention gates to utilize information from lesion regions effectively. However, this approach fails to address the computational cost of the model, which is a critical factor for real-world applications, especially in clinical settings where resources may be limited. This highlights the need to develop more computationally efficient models.

In 2022, Xu *et al.* [28] introduced an MTL framework for segmenting breast ultrasound tumors and predicting their malignancy. This approach leverages segmentation outcomes as prior knowledge to enhance contextual relationships. Later, in 2023, Xu *et al.* [29] introduced a regional-attentive multi-task learning framework by integrating a regional attention (RA) module. This incorporation enhances representation, improving performance in segmentation and classification tasks for each breast ultrasound image. Despite these advancements, the models used by Xu *et al.* employed a self-attention mechanism with a large number of trainable parameters (109 M and 93 M). This makes them less suitable for real-time applications due to their computational demands. Existing research shows a key challenge: developing lightweight and efficient MTL models that maintain performance for real-world clinical use. To address this, we propose a more computationally efficient MTL network with fewer parameters, making it suitable for real-time applications.

Moreover, while existing methods have integrated various attention mechanisms within segmentation and MTL networks, none have explored the use of trainable morphological operations with adaptive structuring elements. To our knowledge, this is the first attempt to integrate such operations within MTL networks. Our novel C-SMA mechanism aims to enhance both segmentation and classification tasks by focusing on important channels and spatial locations in the feature maps based on their morphological attributes while maintaining a lightweight model architecture.

By addressing the computational inefficiencies and introducing a novel attention mechanism, our work fills a critical gap in the existing literature, contributing to the development of more practical and effective MTL models for breast tumor segmentation and malignancy detection from ultrasound images.

3 Methods

Spatial Morphological Attention (SMA): We introduce a Spatial Morphological Attention (SMA) module designed to emphasize the spatial pixel position in the feature maps according to the morphological characteristics. This module performs two types of morphological operations - dilation (D) and erosion(E) on the inputs feature map X, described by the equation 2 and 1 respectively.

$$\mathbf{E}_{(i,j,k)}(X) = \overset{C}{\underset{c=1}{\|}} \min_{m,n=0,...K} X(i-m, j-n, c) + \text{SE}(m, n, c), \quad (1)$$

$$\mathbf{D}_{(i,j,k)}(X) = \overset{C}{\underset{c=1}{\|}} \max_{m,n=0,...K} X(i-m, j-n, c) + \text{SE}(m, n, c), \quad (2)$$

where $i = 1, 2, 3, \ldots, H$ and $j = 1, 2, 3, \ldots, W$. Here, X represents the input feature map, and SE denotes the structuring element of size $m \times n$ that characterizes the pattern of interest within the provided feature map X. The traditional morphological operations have two limitations. Firstly, structuring elements of a particular size is insufficient to capture the morphological characteristics of the ROIs of different sizes. We have considered structuring elements of three different sizes to overcome this drawback. This is achieved by incorporating dilation rates within the structuring elements to achieve a larger receptive field with the

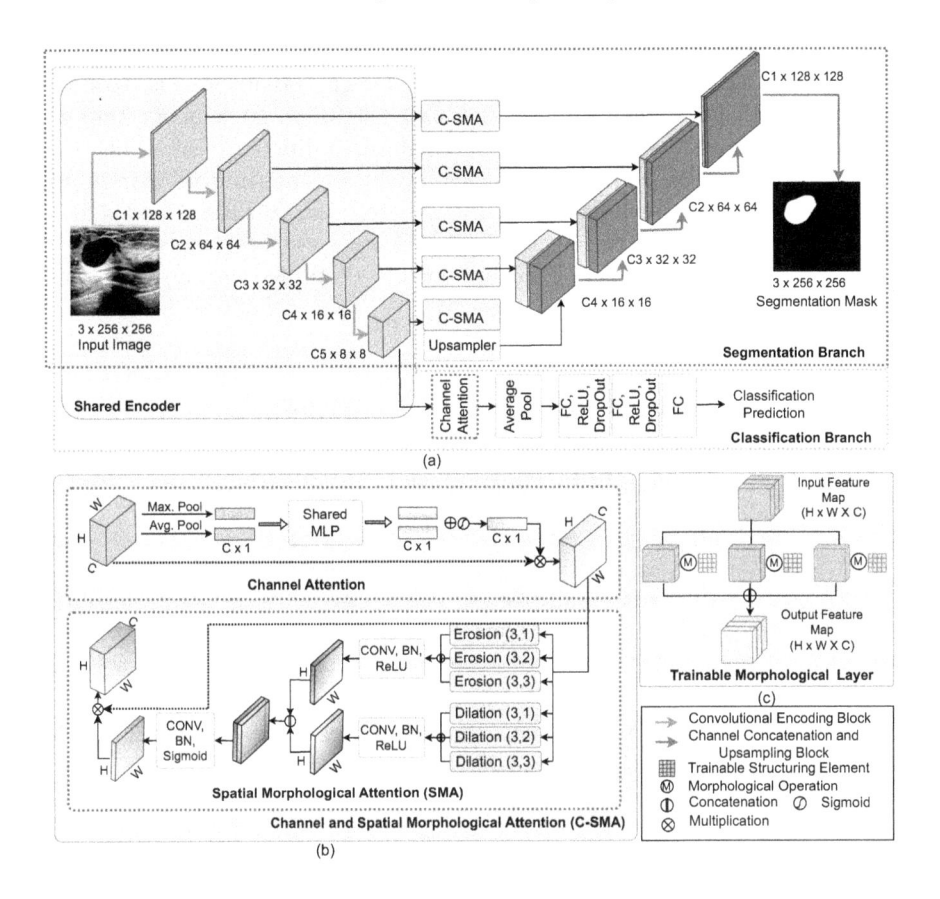

Fig. 1. (a) The proposed multitask learning architecture for medical image segmentation and classification, (b) The channel and spatial morphological attention module (C-SMA).

same computational memory. The dilated erosion and dilation operation can be defined by the equation 4 and 3, respectively.

$$\mathbf{DE}^r_{(i,j,k)}(X) = \underset{c=1}{\overset{C}{\|}} \min_{m,n=0,...K} X(i-rm, j-rn, c) + \mathrm{SE}(m,n,c), \quad (3)$$

$$\mathbf{DD}^r_{(i,j,k)}(X) = \underset{c=1}{\overset{C}{\|}} \max_{m,n=0,...K} X(i-rm, j-rn, c) + \mathrm{SE}(m,n,c), \quad (4)$$

where r is the dilation rate. Secondly, a predefined structuring element is necessary to perform traditional morphological operations, which are unsuitable for precisely segmenting tumours of various shapes. Unlike traditional morphological operations where the structuring elements are predefined, we learn the structuring elements via backpropagation while training [19].

For a particular feature map, each of two morphological operations, i.e., dilated erosion and dilated dilation operations, are carried out with three sizes of structuring elements implemented using dilated structuring elements, thus consequently capturing morphological features at three different scales. The three resulting feature maps from morphological operations are merged using a convolutional block that includes convolution, batch normalization, and a ReLU activation function. These two resulting feature maps, each corresponding to one of the two morphological operations, are combined and fed into another convolutional block, generating a weight map (W_{SMA}) that estimates the weights for every pixel position.

$$W_{SMA} = \sigma(BN(Conv(\|(X_{DE}, X_{DD})))), \tag{5}$$

where,

$$X_{DE} = ReLU(BN(Conv(\|_{r=1,2,3} DE^r))), \tag{6}$$

$$\text{and } X_{DD} = ReLU(BN(Conv(\|_{r=1,2,3} DD^r))). \tag{7}$$

The original feature map multiplied by the weight maps results in the output of the proposed SMAM module. Thus, spatial attention is computed by the morphological characteristics of the feature map, which is learned via training.

Channel Attention(CA): Given an input feature map X with dimensions $C \times H \times W$, where C is the number of channels and H and W are the spatial dimensions, the channel attention mechanism computes attention weights W_{CA} as follows [25].

$$W_{CA} = \sigma(\mathbf{MLP}(X_{avg}) + \mathbf{MLP}(X_{max})), \tag{8}$$

where X_{avg} and X_{max} are feature descriptors obtained by performing average-pooling and max-pooling operations on input X. **MLP** comprises two fully connected layers with ReLU activation function and are shared by both parallel paths. MLP squeezes pooled feature descriptor of dimension $C \times 1 \times 1$ to $C/r \times 1 \times 1$, with a reduction ratio of r and again excites it back to $C \times 1 \times 1$. The sigmoid function (σ) normalizes the attention scores across channels, allowing the model to selectively amplify or suppress specific channels based on their importance for the task. Finally, these attention weights are applied element-wise to the input feature map X to obtain the attended feature map X'.

$$X' = W_{CA} \odot X, \tag{9}$$

where $\odot$ denotes element-wise multiplication.

Channel and Spatial Morphological Attention (C-SMA): In line with the inspiration drawn from the Convolutional Block Attention Module (CBAM)[25], we systematically integrate a channel attention module with the Spatial Morphological Attention Module to enhance overall performance as channel attention leverages the association among the channels of the feature map. We denote this combined attention as channel and spatial morphological attention (C-SMA).

3.1 MA-DTNet

The proposed network, MA-DTNet architecture for multi-task learning, is rooted in the encoder-decoder structure of the UNet[18] architecture. There are two paths, one for segmentation and the other for classification. These two paths share the encoder part of the UNet. This encoder generates feature maps at different scales, which are later combined with the decoder's output at their respective levels. Instead of passing the outputs from the encoder directly to the decoder, we route the feature maps through our proposed C-SMA to enrich the morphological features. This concludes the segmentation network pathway. The prediction for the classification outcome is generated by the classification head, which is attached to the encoder's final output via the channel attention module. The detailed diagram of the proposed architecture is shown in Figure 1. We used a popular CNN network, ResNet34, as the backbone of MA-DTNet.

4 Experimental Results

4.1 Datasets

In this study, we utilize two publicly accessible datasets, namely UDIAT [30] and BUSI [1], to assess the efficacy of the proposed multitask learning approach. The UDIAT dataset comprises 163 ultrasound images, of which 110 pertain to benign tumors and the remaining depict malignant tumors, while the BUSI dataset encompasses 647 ultrasound images of breast tumors, with 437 depicting benign tumors and the remainder of malignancies. The primary objective for both datasets involves tumor segmentation, followed by the secondary task of malignancy detection. To standardize the datasets for analysis, we resize both the images and corresponding masks to dimensions of 256 × 256. Employing a 5-fold cross-validation methodology, we compute the mean and standard deviation of the performance metrics. To enhance the diversity of the training dataset and improve model robustness, we incorporate data augmentation techniques such as horizontal and vertical flips and rotations.

4.2 Training Protocol

The overall loss of the proposed network is computed by a weighted combination of the segmentation loss and classification loss given by the following equation.

$$Loss = \lambda * loss_{seg} + (1 - \lambda) * loss_{cls}. \qquad (10)$$

The value of λ lies between 0 and 1 and is determined to be 0.8 for optimal performance (please refer 4.6 for detailed experiment). $loss_{seg}$ consists of binary cross entropy loss and Dice loss between the original segmentation mask and predicted segmentation mask, whereas $loss_{cls}$ is computed by the binary cross entropy between the actual class label and the predicted class label. The proposed multitask network is trained up to 1000 epochs by optimizing the total loss using Adam optimizer with an initial learning rate of 0.0001. The PyTorch implementation is available at https://github.com/SusmitaSenGhosh/MA-DTNET.

We have used pixel similarity measuring metrics — dice score (DS), intersection over union (IoU), sensitivity, specificity and precision metric to quantify the segmentation performance. We additionally included the 95th percentile Hausdorff distance (HD95) to assess shape similarity. Unlike pixel-based metrics, a lower HD95 indicates greater shape resemblance. To evaluate the malignancy classification performance, we considered threshold-dependent metrics like accuracy and F1 score as well as the receiver operating characteristic curve (AUROC). This threshold-independent metric reflects the overall discrimination ability of the model. Detailed mathematical definitions of the evaluation metrics are available in the supplementary material.

4.3 Comparison with SOTA

In this section, we compare the performance of the proposed model with state-of-the-art methods for classification, segmentation, and MTL. Our evaluation includes state-of-the-art generalized segmentation models like UNet [18], UNet++ [35], DeepLabV3+ [5], utilizing ResNet34, a popular CNN backbone, as well as classification models like ResNet34. Furthermore, we take into account the contemporary segmentation, classification, and multitask models [2,8,9,13,26,32] that are tailored to the breast tumor segmentation as well as malignancy detection task. We have considered Dice score, IOU, sensitivity, and specificity as segmentation performance accessing metrics and accuracy, F1 score, and AUROC as classification metrics following [29].

Result of the above experiments for UDIAT and BUSI dataset are reported in Table 1 and 2 respectively, where the top three performances are highlighted in red, blue, and green, corresponding to the first, second, and third best results. From both the tables, it is evident that our proposed method has MA-DTNet outperformed others in both segmentation and classification tasks, achieving the highest scores in terms of almost all the metrics considered here. Compared to the state-of-the-art MTL network, our proposed approach demonstrates a 3.28% increase in dice score and a 3.14% improvement in IoU metrics. Additionally, we observe a 1.05% enhancement in the F1 score for the UDIAT dataset. Furthermore, our proposed method exhibits a better balance between sensitivity and specificity compared to other MTL methods. A similar trend is also observed in the case of the BUSI dataset (Table 2). The segmentation performance experienced an improvement of 1.62% and 1.52% in terms of dice-score and IoU metrics, respectively, while 4.96% improvement is observed in accuracy metrics for the malignancy detection task. While the segmentation specificity of the proposed model may not surpass that of the other MTL models, it does exhibit a superior balance between sensitivity and specificity. High sensitivity ensures that the model does not miss any tumor regions, which is critical for accurate diagnosis and treatment planning. High specificity helps minimize false alarms or misclassifications of non-tumor regions as tumors, which can reduce unnecessary medical interventions. In tumor segmentation, striking the right balance between

Table 1. A comparison of the performance of the proposed method with a related state-of-the-art method for the UDIAT dataset. The first, second, and third-best performances are highlighted in red, blue, and green, respectively.

Task	Method	Params.	Segmentation Performance				Classification Performance		
			DS↑	IoU↑	Sen.↑	Spec.↑	Acc.↑	F1 score↑	AUROC↑
Seg.	UNet[18]	22.72 M	87.79 ± 2.55	79.86 ± 3.01	91.10 ± 3.01	99.44 ± 0.11	-	-	-
	UNet++[35]	24.42 M	87.36 ± 2.58	79.50 ± 3.20	89.33 ± 2.33	99.52 ± 0.10	-	-	-
	DeepLabV3+[5]	22.44 M	87.45 ± 2.89	79.34 ± 3.20	90.32 ± 1.97	99.48 ± 0.17	-	-	-
	AAU-net[2]†	-	78.14 ± 2.41	69.10 ± 2.98	82.22 ± 3.84	98.82 ± 0.35	-	-	-
	ESKNet[4]†	44.57 M	78.71 ±2.37	70.20 ±2.28	82.41 ±2.84	97.47 ±0.35	-	-	-
	NU-net[3]†	77.05 M	80.80 ±0.57	72.03 ±0.82	84.13 ±1.73	98.96 ±0.17	-	-	-
	SMU-Net[17]†	-	87.03 ±1.25	78.49 ±1.49	88.85 ±1.72	-	-	-	-
Clas.	ResNet34[11]	21.35 M	-	-	-	-	94.53 ± 3.94	91.21 ± 6.57	0.937 ± 0.066
	HoVer-Trans[15]†	79.38 M	-	-	-	-	77.40 ±6.10	61.90 ±9.90	0.781 ±0.118
	EPTM[21]†	192 M	-	-	-	-	90.90	93.20	0.932
MT	MTL-Net[29]†	93.50 M	80.95 ± 5.00	72.73 ± 5.49	84.28 ± 5.05	99.25 ± 0.14	87.08 ± 2.79	90.51 ± 2.29	0.936 ± 0.046
	MTL-COSA[28]†	109.24 M	84.07 ± 3.25	76.05 ± 3.71	86.97 ± 2.76	99.27 ± 0.25	91.44 ± 3.90	93.85 ± 2.58	0.946 ± 0.034
	RMTL-Net[29]†	93.51 M	85.69 ± 2.00	77.84 ± 2.45	89.51 ± 0.91	99.25 ± 0.19	95.74 ± 3.45	92.84 ± 5.98	0.935 ± .081
	MA-DTNet (ours)	22.95 M	88.97 ± 2.50	80.98 ± 3.17	91.55 ± 1.98	99.45 ± 0.23	96.35 ± 3.95	93.89 ± 6.81	0.954 ± 0.062

Seg. : Segmentation, Clas. : Classification, MT : Multitask, DS : Dice score, IoU : Intersection over union, Sen. : Sensitivity, Spec. : Specificity, Acc. : Accuracy Results reported that the methods marked by † are taken from respective studies. The proposed method outperformed all the multitask learning models for segmentation and classification tasks with significantly fewer trainable parameters.

sensitivity and specificity is essential to ensure the model's effectiveness in accurately identifying tumor regions while minimizing false positives and negatives. Moreover, MA-DTNet has at least four times fewer parameters than other MTL models, indicating an efficient model design that outperforms the other MTL methods in both tumor segmentation and malignancy detection tasks.

Table 2. A comparison of the performance of the proposed method with a related state-of-the-art method for the BUSI dataset. The first, second, and third-best performances are highlighted in red, blue, and green, respectively.

Task	Method	Params.	Segmentation Performance DS↑	IoU↑	Sen.↑	Spec.↑	Classification Performance Acc.↑	F1 score↑	AUROC↑
Seg.	UNet[18]	22.72 M	81.10 ± 1.60	73.32 ± 1.64	83.33 ± 3.20	97.83 ± 0.55	-	-	-
	UNet++[35]	24.42 M	81.09 ± 1.36	72.92 ± 1.47	83.51 ± 2.47	97.91 ± 0.34	-	-	-
	DeepLabV3+[5]	22.44 M	80.88 ± 1.56	72.69 ± 1.67	82.68 ± 2.23	97.80 ± 0.26	-	-	-
	AAU-net[2]†	-	77.51 ± 0.68	68.82 ± 0.44	80.10 ± 0.52	97.57 ± 0.24	-	-	-
	ESKNet[4]†	44.57 M	79.92 ± 2.21	71.65 ± 2.39	82.66 ± 1.40	99.01 ± 0.35	-	-	-
	AMS-PAN[14]†	-	80.71	68.53	79.30	98.54	-	-	-
	NU-net[3]†	77.05 M	78.62 ± 1.38	70.35 ± 1.54	82.46 ± 1.02	97.48 ± 0.49	-	-	-
Clas.	ResNet34	21.35 M	-	-	-	-	95.21 ± 1.74	92.49 ± 2.71	0.968 ± 0.019
	HoVer-Trans[15]†	79.38 M	-	-	-	-	85.50 ± 5.00	87.20 ± 8.00	0.865 ± 0.066
	MS-GOF[34]†	105.69M	-	-	-	-	76.48 ± 5.93	74.90 ± 7.53	0.790 ± 0.064
MT	MTL-Net[29]†	93.50 M	77.76 ± 3.11	69.33 ± 2.89	78.91 ± 2.22	98.30 ± 0.25	90.18 ± 3.25	93.07 ± 2.41	0.962 ± 0.021
	MTL-COSA[28]†	109.24 M	78.90 ± 2.03	70.65 ± 2.01	79.31 ± 2.48	98.31 ± 0.11	91.49 ± 3.02	93.66 ± 2.36	0.968 ± 0.016
	RMTL-Net[29]†	93.51 M	80.04 ± 2.47	71.93 ± 2.15	82.54 ± 2.31	98.00 ± 0.3	91.02 ± 3.42	93.32 ± 3.35	0.967 ± 0.015
	MA-DTNet (ours)	22.95 M	81.66 ± 1.56	73.45 ± 1.55	83.61 ± 2.51	97.85 ± 0.11	95.98 ± 1.48	93.71 ± 2.29	0.978 ± 0.010

Seg. : Segmentation, Clas. : Classification, MT : Multitask, DS : Dice score, IoU : Intersection over union, Sen. : Sensitivity, Spec. : Specificity, Acc. : Accuracy Results reported the methods that are marked by † are taken from respective studies. Proposed method outperformed all the multitask learning model for both segmentation and classification tasks with significantly lower number of trainable parameters.

4.4 Comparison among various attention mechanisms

The effectiveness of the proposed C-SMA is compared with the performance of the other attention mechanisms for segmentation tasks. For this experiment, we have considered CBAM[25] and multi-head self-attention (MHSA)[24]. CBAM is a spatial and channel attention mechanism designed to enhance CNN's representational power, while self-attention is a mechanism commonly used in transformer architectures to capture dependencies within input sequences. Along with vanilla UNet, we have considered three types of attention modules — CBAM,

Table 3. Performance comparison of different attention mechanism integrated within UNet architecture for segmentation task.

Dataset	Method	Params.	Dice Score↑	IoU↑	Precision↑	Recall↑	HD95↓
UDIAT	UNet	1.81 M	79.98 ± 2.82	70.48 ± 3.78	80.20 ± 2.97	83.14 ± 2.81	24.72 ± 7.35
	UNet-CBAM	1.82 M	82.55 ± 1.35	73.10 ± 1.93	82.26 ± 1.35	86.13 ± 1.21	19.68 ± 4.87
	UNet-MHSA	3.31 M	84.97 ± 3.53	75.87 ± 4.12	83.13 ± 3.01	**88.80** ± 3.38	16.91 ± 6.81
	UNet-CA	1.95 M	83.48 ± 1.79	74.31 ± 2.30	82.49 ± 2.51	86.43 ± 2.37	18.41 ± 3.51
	UNet-SMA	1.98 M	83.50 ± 2.30	74.49 ± 2.74	83.98 ± 2.55	87.74 ± 2.63	19.43 ± 5.96
	UNet-C-SMA (Ours)	1.99 M	**85.14** ± 0.94*	**75.95** ± 1.27*	**85.35** ± 2.07*	87.34 ± 1.26	**15.38** ± 1.40
BUSI	UNet	1.81 M	74.79 ± 1.53	65.92 ± 1.57	78.86 ± 1.28	76.65 ± 2.19	36.00 ± 4.77
	UNet-CBAM	1.82 M	78.63 ± 1.52	69.88 ± 1.48	81.42 ± 1.30	80.32 ± 1.45	27.17 ± 2.83
	UNet-MHSA	3.31 M	79.61 ± 1.70	70.72 ± 1.83	80.58 ± 2.08	**83.01** ± 2.21	27.90 ± 2.95
	UNet-CA	1.95 M	78.88 ± 1.27	70.02 ± 1.06	80.19 ± 1.24	82.02 ± 1.55	29.45 ± 1.37
	UNet-SMA	1.98 M	76.87 ± 1.79	67.97 ± 1.65	79.54 ± 2.14	79.52 ± 1.24	31.96 ± 3.98
	UNet-C-SMA (Ours)	1.99 M	**79.73** ± 1.61*	**71.00** ± 1.37*	**81.92** ± 1.75*	81.49 ± 1.20*	**26.29** ± 1.98

UNet-C-SMA achieved statistically significant performance improvements over UNet-CBAM, as evidenced by Wilcoxon signed-rank test results (indicated by *). UNet-C-SMA exhibited greater efficiency compared to UNet-MHSA in terms of dice score, intersection over union (IoU), and Hausdorff Distance (HD95) metrics, while also requiring fewer parameters. Notably, UNet-C-SMA maintained a superior balance between precision and recall.

MHSA and C-SMA that are fused with UNet architecture. In UNet with CBAM and C-SMA, attention computation is conducted within each skip connection connecting the encoder to the decoder. Conversely, in UNet with MHSA, MHSA is exclusively applied to three skip connections containing smaller feature maps. This selective application is attributed to the heightened memory usage and computational cost associated with MHSA, rendering it impractical to employ across all skip connections.

Table 3 summarizes the results of the experiment. All three attention mechanisms (MHSA, CBAM, and C-SMA) yielded improvements in segmentation metrics compared to the baseline model without attention. Notably, MHSA led to significant improvement, but at the cost of nearly doubling the trainable parameters. Conversely, CBAM and C-SMA increased the parameter count by a smaller margin. Focusing on the comparison between UNet-CBAM and UNet-C-SMA, the latter achieved statistically significant (Wilcoxon signed-rank test) improvements of 2.59% and 1.15% on the UDIAT and BUSI datasets, respectively. This improvement came at the expense of only 0.17 million additional trainable parameters.

To assess the individual contributions of CA and SMA components within the proposed C-SMA module, we evaluate the segmentation performance of the UNet network integrated with each component separately. These results, presented alongside the performance with the full C-SMA module, are crucial for understanding the efficacy of each component. As expected, the inclusion of either CA or SMA independently improves segmentation performance compared to the baseline UNet. Notably, integrating the combined C-SMA module

(CA+SMA) further enhances segmentation accuracy, demonstrating the synergistic effect of these attention mechanisms.

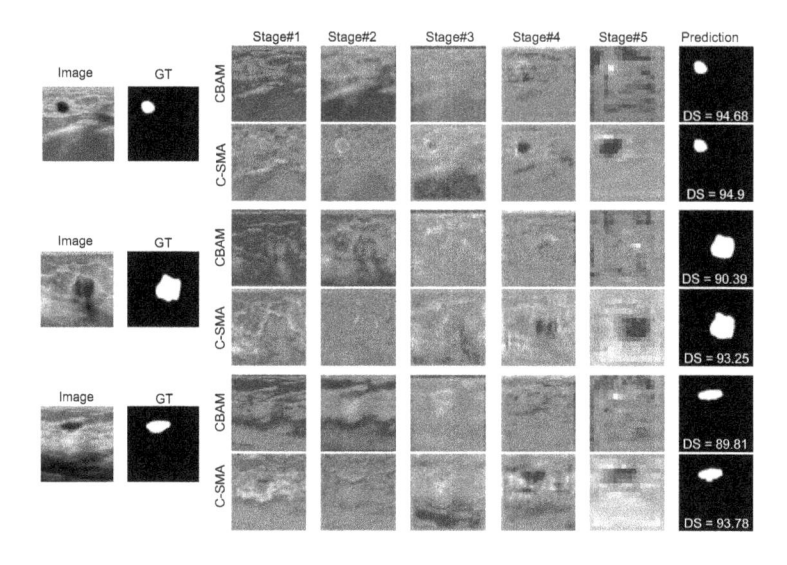

Fig. 2. The qualitative comparison of the feature maps obtained at different stages of encoder after the application of CBAM and proposed C-SMA.

4.5 Ablation study

The qualitative analysis of CBAM and C-SMA module is also conducted by analysing the feature maps obtained at different stages of encoder post application of respective attention mechanism. It is evident from Figure 2 that C-SMA is capable to capture morphological attributes such as shape and size by of the tumor in each of the feature map more accurately by distinctly enhancing the boundary pixels.

MA-DTNet tackles multi-task learning by sharing an encoder for improved efficiency. It seamlessly integrates C-SMA for enhanced segmentation while incorporating CA within the classification branch to boost performance. We performed an ablation study to validate each component of the proposed network. This involved training a series of models with each component progressively removed. The performance of these ablated models was then compared to the full model's performance on the combined task. Additionally, to evaluate the effectiveness of the multi-task learning approach, we compared the performance of individual tasks (segmentation and classification) trained in isolation to their performance within the MA-DTNet framework. The segmentation task leverages a U-shaped network architecture, serving as the foundation for the proposed multi-task learning (MTL) network. For the classification task, the

Table 4. Performance comparison of the ablation study on each component of the proposed MA-DTNet for UDIAT and BUSI dataset.

Dataset	Task	Method	Segmentation Performance					Classification Performance		
			DS↑	IoU↑	Prec.↑	Rec.↑	HD95↓	Acc.↑	F1 score↑	AUROC↑
UDIAT	Seg.	UNet	79.98 ± 2.82	70.48 ± 3.78	80.20 ± 2.97	83.14 ± 2.81	24.72 ± 7.35	-	-	-
	Clas.	UNet Encoder	-	-	-	-	-	90.21 ± 5.79	84.62 ± 8.88	0.8886 ± 0.0811
	MT	ES-MTL	80.88 ± 4.31	71.34 ± 4.87	81.56 ± 5.38	84.37 ± 3.37	24.21 ± 5.75	90.81 ± 3.67	84.23 ± 6.94	0.8831 ± 0.0756
		ES-MTL +C-SMA	84.99 ± 1.79	75.79 ± 2.06	**86.98** ± 2.44	86.41 ± 2.56	16.98 ± 2.85	93.26 ± 3.30	89.27 ± 4.94	0.9175 ± 0.0416
		ES-MTL +C-SMA+CA	**85.33** ± 2.41	**76.13** ± 2.91	84.54 ± 3.00	**88.66** ± 1.09	**16.39** ± 5.53	**93.26** ± 2.57	**89.63** ± 4.66	**0.9445** ± 0.0457
BUSI	Seg.	UNet	74.79 ± 1.53	65.92 ± 1.57	78.86 ± 1.28	76.65 ± 2.19	36.00 ± 4.77	-	-	-
	Clas.	UNet Encoder	-	-	-	-	-	89.18 ± 1.79	81.86 ± 3.92	0.9205 ± 0.0266
	MT	ES-MTL	75.65 ± 1.44	66.67 ± 1.50	78.39 ± 2.00	79.04 ± 1.84	36.20 ± 3.00	91.97 ± 1.28	87.46 ± 1.92	0.9429 ± 0.0155
		ES-MTL +C-SMA	**79.56** ± 1.42	70.73 ± 1.31	81.72 ± 1.20	81.42 ± 1.24	27.15 ± 0.74	92.12 ± 1.83	87.54 ± 2.91	0.9441 ± 0.0254
		ES-MTL +C-SMA+CA	79.42 ± 1.69	**70.78** ± 1.44	**82.35** ± 1.38	80.67 ± 2.63	**25.51** ± 2.16	**93.20** ± 1.58	**89.14** ± 3.11	**0.9521** ± 0.0116

Seg. : Segmentation, Clas. : Classification, MT : Multitask
ES-MTL: Encoder-shared multitask learning model, ES-MTL+C-SMA: Encoder-shared multitask learning model with channel and spatial morphological attention (C-SMA) in encoder-decoder skip connection path, ES-MTL+MA+CA: Encoder-shared multitask learning model with channel and spatial morphological attention (C-SMA) in encoder-decoder skip connection path and channel attention (CA) in encoder-classifier path.

encoder portion of the U-Net is directly integrated with a classification head. The result of the ablation study is reported in table 4 for both datasets.

It can be observed that the encoder-shared vanilla MTL network surpassed both individual segmentation and classification performance for both UDIAT and BUSI datasets, suggesting the efficiency of an encoder-shared MTL (referred to as ES-MTL). The Encoder-shared MTL with C-SMAs integrated within the skip connections of the encoder and decoder) referred to as ES-MTL+C-SMA) further improves the segmentation performance in terms of all the segmentation metrics by a significant margin. Notably, this enhancement in segmentation performance was accompanied by a corresponding improvement in classification performance, particularly for the UDIAT dataset and marginally for BUSI. This suggests that accurate tumor segmentation plays a crucial role in achieving better malignancy detection, thus supporting the effectiveness of the proposed C-SMA for the MTL framework. Finally, the inclusion of class activation (CA) within the classification path yielded a further improvement in classification performance, underlining its importance in this context.

4.6 Hyperparameter Optimization

The optimal value of the hyperparameter λ in equation 10 was determined by conducting a grid search across λ values ranging from 0 to 1 in increments of 0.1. As shown in figure 3, λ value of 0.8 yields the best balance between segmentation performance and malignancy detection performance.

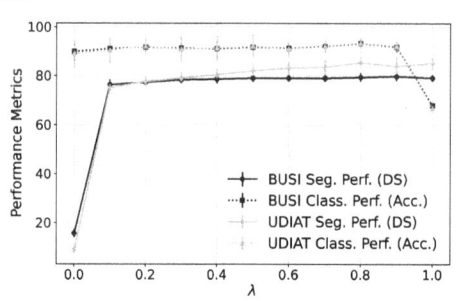

Fig. 3. The segmentation (dice score) and classification performances (accuracy) of ES-MTL+C-SMA+CA for different λ

The size of the structuring element (SE) of morphological operation within the C-SMA module is another hyperparameter that requires optimization. It as also determined empirically. As shown in table 5, increasing the SE size from 3 to 5 leads to a growth in the number of trainable parameters. However, this increase did not translate to significant performance improvements. Therefore, we opted for a 3×3 SE as the optimal configuration for the C-SMA module, balancing model complexity with performance.

4.7 Generalization Ability

The efficacy of the proposed methodology is demonstrated in the context of segmenting breast tumors and detecting malignancies in ultrasound images. This section delves into examining the generalization capacity of the proposed method across two additional multitask learning scenarios pertinent to medical imaging.

Table 5. Performance comparison for different size of SE in C-SMA

Dataset	SE Size	Param.	DS↑	HD95↓
BUSI	3	1.95 M	79.73 ± 1.44	26.29 ± 1.77
	5	2.04 M	79.37 ± 1.33	25.94 ± 1.47
UDIAT	3	1.95 M	85.14 ± 0.84	15.38 ± 1.25
	5	2.04 M	83.66 ± 0.75	15.70 ± 3.28

The first scenario involves segmenting skin lesions and classifying diseases using dermoscopic images, while the second entails segmenting glands and predicting malignancies from histological images. To evaluate these tasks, we utilized the HAM10000 [23] and GlaS [22] datasets (for detailed description refer to the supplementary material). Comparative analyses of the proposed method's performance against state-of-the-art multitask learning approaches relevant to these tasks are presented in Table 6.

Regarding the HAM10000 dataset, the proposed method has exhibited superior performance compared to CTAN [12], surpassing it by 1.84% in dice score and 27.98% in accuracy metrics for skin lesion segmentation and classification tasks, respectively. The assessment of segmentation performance on the GlaS dataset entails an evaluation of two distinct subsets of the test data, GlaS A

Table 6. Comparison of the performance of the proposed method on HAM10000 and GlaS dataset with related state-of-the-art methods.

Skin lesion segmentation and classification (HAM10000)			Gland segmentation and malignancy detection (GlaS)				
	Seg. perf.	Clas. Perf.		Seg. perf. (GlaS A/ GlaS B)		Clas. perf.	
Method	DS↑	Acc.↑	Method	Object DS↑	HD↓	F1 score↑	Acc.↑
CTAN[12]	92.91	57.85	MTUNet[9]	95.60/90.90	23.17/71.53	97.77	97.50
MA-DTNet (Ours)	94.75 ±0.17	85.83 ±0.47	MA-DTNet (Ours)	92.99/91.51	19.89/21.08	100.00	100.00

Seg. perf. : Segmentation performance, Clas. perf. : Classification performance

For GlaS, the original train test split was used, whereas for HAM10000, 5-fold cross-validation is adapted for the proposed method.

and GlaS B. Both the object dice score[9] and Hausdorff Distance (HD) metrics exhibit enhancements on the GlaS B dataset when compared to MTUNet [9]. Conversely, for GlaS A, an improvement over MTUNet is observed solely in the Hausdorff distance metric. Despite these segmentation disparities, our proposed method achieves perfect accuracy in predicting gland malignancy. Overall, beyond its success in segmenting and classifying breast tumors in ultrasound images, the method performs well in the other two multitask settings, suggesting its generalizability for medical image analysis.

5 Conclusion

This study introduces a novel Multi-Task Learning (MTL) model specifically designed for breast tumor segmentation and malignancy detection from ultrasound images. The model consists of a shared UNet encoder and UNet decoder, along with novel Channel-Spatial Morphological Attention Modules (C-SMAs) integrated at multiple resolution stages for semantic segmentation. Additionally, it includes a classification head with a channel attention module for disease grade prediction. The channel attention module highlights significant channels within a multitude, while the C-SMA simultaneously focuses on important channels and spatial locations in the feature maps based on their morphological attributes.

The proposed approach demonstrates effectiveness through experiments on ultrasound datasets for breast tumor segmentation and malignancy detection. To further test the versatility of the proposed method, it has been generalized to other multitask scenarios such as microscopic gland segmentation and detection, as well as skin lesion segmentation and disease detection. Further research must explore its adaptability to a wider range of medical imaging modalities beyond those investigated here.

Furthermore, the C-SMA module currently utilizes fundamental morphological operations. Future research could potentially investigate the application of more complex morphological operations like opening, closing, and gradients to improve model performance.

References

1. Al-Dhabyani, W., Gomaa, M., Khaled, H., Fahmy, A.: Dataset of breast ultrasound images. Data Brief **28**, 104863 (2020)
2. Chen, G., Li, L., Dai, Y., Zhang, J., Yap, M.H.: AAU-net: an adaptive attention u-net for breast lesions segmentation in ultrasound images. IEEE Trans. Med, Imaging (2022)
3. Chen, G., Li, L., Zhang, J., Dai, Y.: Rethinking the unpretentious u-net for medical ultrasound image segmentation. Pattern Recogn. **142**, 109728 (2023)
4. Chen, G., Zhou, L., Zhang, J., Yin, X., Cui, L., Dai, Y.: Esknet: An enhanced adaptive selection kernel convolution for ultrasound breast tumors segmentation. Expert Syst. Appl. **246**, 123265 (2024)
5. Chen, L.C., Zhu, Y., Papandreou, G., Schroff, F., Adam, H.: Encoder-decoder with atrous separable convolution for semantic image segmentation. In: Proceedings of the ECCV. pp. 801–818 (2018)
6. Cheng, J., Liu, J., Kuang, H., Wang, J.: A fully automated multimodal MRI-based multi-task learning for glioma segmentation and IDH genotyping. IEEE Trans. Med. Imaging **41**(6), 1520–1532 (2022)
7. Chowdary, J., Yogarajah, P., Chaurasia, P., Guruviah, V.: A multi-task learning framework for automated segmentation and classification of breast tumors from ultrasound images. Ultrason. Imaging **44**(1), 3–12 (2022)
8. Cui, W., Peng, Y., Yuan, G., et al.: FMRNet: A fused network of multiple tumoral regions for breast tumor classification with ultrasound images. Med. Phys. **49**(1), 144–157 (2022)
9. Dabass, M., Vashisth, ., Vig, R.: MTU: A multi-tasking u-net with hybrid convolutional learning and attention modules for cancer classification and gland segmentation in colon histopathological images. Comput. Biol. Med. **150**, 106095 (2022)
10. Gokhale, S.: Ultrasound characterization of breast masses. Indian Journal of Radiology and Imaging **19**(03), 242–247 (2009)
11. He, K., Zhang, X., Ren, S., Sun, J.: Deep residual learning for image recognition. In: Proceedings of the IEEE conference on computer vision and pattern recognition. pp. 770–778 (2016)
12. Kim, S., Purdie, T.G., McIntosh, C.: Cross-task attention network: Improving multi-task learning for medical imaging applications. In: International Conference on Medical Image Computing and Computer-Assisted Intervention. pp. 119–128. Springer (2023)
13. Luo, H., Changdong, Y., Selvan, R.: Hybrid ladder transformers with efficient parallel-cross attention for medical image segmentation. In: International Conference on Medical Imaging with Deep Learning. pp. 808–819. PMLR (2022)
14. Lyu, Y., Xu, Y., Jiang, X., Liu, J., Zhao, X., Zhu, X.: Ams-pan: Breast ultrasound image segmentation model combining attention mechanism and multi-scale features. Biomed. Signal Process. Control **81**, 104425 (2023)
15. Mo, Y., Han, C., Liu, Y., Liu, M., Shi, Z., Lin, J., Zhao, B., Huang, C., Qiu, B., Cui, Y., et al.: Hover-trans: Anatomy-aware hover-transformer for roi-free breast cancer diagnosis in ultrasound images. IEEE Transactions on Medical Imaging (2023)

16. Mondal, R., Purkait, P., Santra, S., Chanda, B.: Morphological networks for image de-raining. In: International Conference on Discrete Geometry for Computer Imagery. pp. 262–275. Springer (2019)
17. Ning, Z., Zhong, S., Feng, Q., Chen, W., Zhang, Y.: Smu-net: Saliency-guided morphology-aware u-net for breast lesion segmentation in ultrasound image. IEEE Trans. Med. Imaging **41**(2), 476–490 (2021)
18. Ronneberger, O., Fischer, P., Brox, T.: U-Net: Convolutional Networks for Biomedical Image Segmentation. In: Navab, N., Hornegger, J., Wells, W.M., Frangi, A.F. (eds.) MICCAI 2015. LNCS, vol. 9351, pp. 234–241. Springer, Cham (2015). https://doi.org/10.1007/978-3-319-24574-4_28
19. Roy, S.K., Mondal, R., Paoletti, M.E., Haut, J.M., Plaza, A.: Morphological convolutional neural networks for hyperspectral image classification. IEEE J. Sel. Top. Appl. Earth Obs. Remote Sens. **14**, 8689–8702 (2021)
20. Ruder, S.: An overview of multi-task learning in deep neural networks. arXiv preprint arXiv:1706.05098 (2017)
21. Singh, V.K., Mohamed, E.M., Abdel-Nasser, M.: Aggregating efficient transformer and cnn networks using learnable fuzzy measure for breast tumor malignancy prediction in ultrasound images. Neural Computing and Applications pp. 1–17 (2024)
22. Sirinukunwattana, K., Pluim, J.P., Chen, H., et al.: Gland segmentation in colon histology images: The glas challenge contest. Med. Image Anal. **35**, 489–502 (2017)
23. Tschandl, P., Rosendahl, C., Kittler, H.: The ham10000 dataset, a large collection of multi-source dermatoscopic images of common pigmented skin lesions. Scientific data **5**(1), 1–9 (2018)
24. Vaswani, A., Shazeer, N., Parmar, N., Uszkoreit, J., Jones, L., Gomez, A.N., Kaiser, Ł., Polosukhin, I.: Attention is all you need. Advances in neural information processing systems **30** (2017)
25. Woo, S., Park, J., Lee, J.Y., Kweon, I.S.: CBAM: Convolutional block attention module. In: Proceedings of the ECCV. pp. 3–19 (2018)
26. Wu, H., Huang, X., Guo, X., Wen, Z., Qin, J.: Cross-image dependency modelling for breast ultrasound segmentation. IEEE Trans. Med, Imaging (2023)
27. Xie, Y., Zhang, J., Xia, Y., Shen, C.: A mutual bootstrapping model for automated skin lesion segmentation and classification. IEEE Trans. Med. Imaging **39**(7), 2482–2493 (2020)
28. Xu, M., Huang, K., Qi, X.: Multi-task learning with context-oriented self-attention for breast ultrasound image classification and segmentation. In: 2022 IEEE 19th ISBI. pp. 1–5. IEEE (2022)
29. Xu, M., Huang, K., Qi, X.: A regional-attentive multi-task learning framework for breast ultrasound image segmentation and classification. IEEE Access **11**, 5377–5392 (2023)
30. Yap, M.H., Pons, G., Marti, J., Ganau, S., Sentis, M., Zwiggelaar, R., Davison, A.K., Marti, R.: Automated breast ultrasound lesions detection using convolutional neural networks. IEEE J. Biomed. Health Inform. **22**(4), 1218–1226 (2017)
31. Zeng, W., Fan, W., Chen, R., et al.: Accurate 3d kidney segmentation using unsupervised domain translation and adversarial networks. In: 2021 IEEE 18th ISBI. pp. 598–602. IEEE (2021)
32. Zhang, G., Zhao, K., Hong, Y., Qiu, X., Zhang, K., Wei, B.: SHA-MTL: soft and hard attention multi-task learning for automated breast cancer ultrasound image segmentation and classification. Int. J. Comput. Assist. Radiol. Surg. **16**(10), 1719–1725 (2021). https://doi.org/10.1007/s11548-021-02445-7
33. Zhang, Y., Yang, Q.: A survey on multi-task learning. IEEE Trans. Knowl. Data Eng. **34**(12), 5586–5609 (2021)

34. Zhong, S., Tu, C., Dong, X., Feng, Q., Chen, W., Zhang, Y.: MsGoF: breast lesion classification on ultrasound images by multi-scale gradational-order fusion framework. Comput. Methods Programs Biomed. **230**, 107346 (2023)
35. Zhou, Z., Rahman S., M.M., Tajbakhsh, N., et al.: Unet++: A nested u-net architecture for medical image segmentation. In: 4th International Workshop, DLMIA 2018, and 8th International Workshop, ML-CDS 2018, Held in Conjunction with MICCAI 2018, Granada, Spain, September 20, 2018, Proceedings 4. pp. 3–11. Springer (2018)

TransNetOCT: An Efficient Transformer-Based Model for 3D-OCT Segmentation Using Prior Shape

Mohamed Elsharkawy[1], Ibrahim Abdelhalim[1], Mohammed Ghazal[2], Mohammad Z. Haq[3], Rayan Haq[1], Ali Mahmoud[1], Harpal S. Sandhu[1], Aristomenis Thanos[4], and Ayman El-Baz[1](✉)

[1] Department of Bioengineering, University of Louisville, Louisville, KY, USA
aselba01@louisville.edu
[2] Electrical and Computer Engineering Department, Abu Dhabi University, Abu Dhabi, UAE
[3] School of Medicine, University of Louisville, Louisville, KY, USA
[4] Legacy Devers Eye Institute, Portland, OR, USA

Abstract. Segmentation of retinal layers in three-dimensional optical coherence tomography (3D-OCT) images plays a pivotal role in disease identification and prognosis. For instance, analyzing variations in layer curvature within 3D-OCT images offers crucial insights into age-related macular degeneration (AMD) progression. This paper presents TransNetOCT, a novel approach employing a transformer architecture for 3D-OCT volume segmentation. The method delineates 3D-OCT scans into twelve surfaces, outlining the background and eleven layers. Initially, we segment the central macular B-scan, identifiable by the fovea, using a joint Markov Gibbs random field model. This model integrates shape, intensity, and spatial characteristics across the twelve retinal surfaces. Subsequently, a probability prior shape algorithm is applied to the adjacent slices, utilizing the middle segmented B-scan image as a reference. TransNetOCT is then trained on the central slice along with the extracted probability prior shapes of adjacent slices. An evaluation of 85 patients, including those with normal, early, and intermediate AMD OCT scans, demonstrates the method's effectiveness. It achieves an Absolute Sum of Surface Distance (ASSD) of 1.3149 and Mean Absolute Surface Distance (MASD) of 1.7091, outperforming well-known models like the UNet and Feature Pyramid Network (FPN). Specifically, it surpasses UNet and FPN with a 78% reduction for ASSD and a 76% reduction for MASD, respectively, and a 48% reduction for both ASSD and MASD, respectively. These results highlight its superior capability in segmenting a broader spectrum of retinal layers.

Keywords: 3D-OCT · TransNetOCT · OCT Segmentation · Prior Shape

1 Introduction

Age-related macular degeneration (AMD) remains the essential reason for central vision impairment in individuals aged 50 and older in North America and other developed regions [2]. AMD can be categorized into early and late stages through various classification schemes like the Age-Related Eye Diseases Study (AREDS) severity scale [5] and basic clinical classification [1]. The clinical classification of AMD is based on different stages, which are determined by the presence and size of drusen and pigment irregularities: (i) No signs of drusen or pigment abnormalities in the eye; (ii) Only small drusen ($\leq 63\mu$m) are present, without any pigment irregularities; (iii) Medium-sized drusen ($> 63\mu$m and $\leq 125\mu$m) are present, but no pigment irregularities; (iv) Large drusen ($> 125\mu$m) or any pigment irregularities (abnormal pigmentation) are present; (v) Advanced AMD is characterized by either "wet" AMD (new abnormal blood vessels growing under the retina) or "dry" AMD (geographic atrophy, which is the death of retinal cells). In patients with AMD, early diagnosis can facilitate timely intervention which is instrumental in preventing vision loss. Due to the pathophysiology of AMD, the diagnostic imaging utilized relies heavily on identifying small structural alterations in the layers of the retina. Such alterations include thickening of the retinal pigment epithelium (RPE) and Bruch's membrane, and abnormalities in pigmentation. The formation of drusen, which are deposits of cellular debris within Bruch's membrane, is a characteristic hallmark of AMD.

Optical Coherence Tomography builds three-dimensional (3D-OCT) images of the retina by combining multiple segments known as A-scans onto each other. A-scans detect the ways that light bounces off of the retinal tissue to perceive depth, functioning much like a radar. Scans are taken linearly across the retina to provide a cross-sectional view, also known as a B-scan. The three-dimensional OCT volume is then formed by stacking multiple B-scans together [20]. Different layers formed in the image can be used as early markers of various retinal diseases such as AMD [4,11] through examining differences in the RPE [18]. The precise delineation and separation of retinal tissue layers in OCT images is essential for the prompt detection and diagnosis of AMD. When determining the borders of different layers, manual OCT image interpretation is laborious and prone to error. Because of this, throughout the past few decades, a significant amount of research has focused on automating the segmentation of retinal layers in OCT images. The primary objective, shared by many studies including ours, is to achieve smooth and continuous surfaces of retinal layers in their proper anatomical order, as depicted in Fig.1 [3,9,10,14,15,17,21,22].

1.1 Related Works

Thus, further development in segmentation methodologies has become an innovation of interest. For example, a two-step approach for extracting five surfaces from the layer estimates was introduced by Pekala et al. [15]. They integrated DenseNet-based semantic segmentation with Gaussian process regression-based post-processing in their approach. Kumar et al. [17] introduced a machine

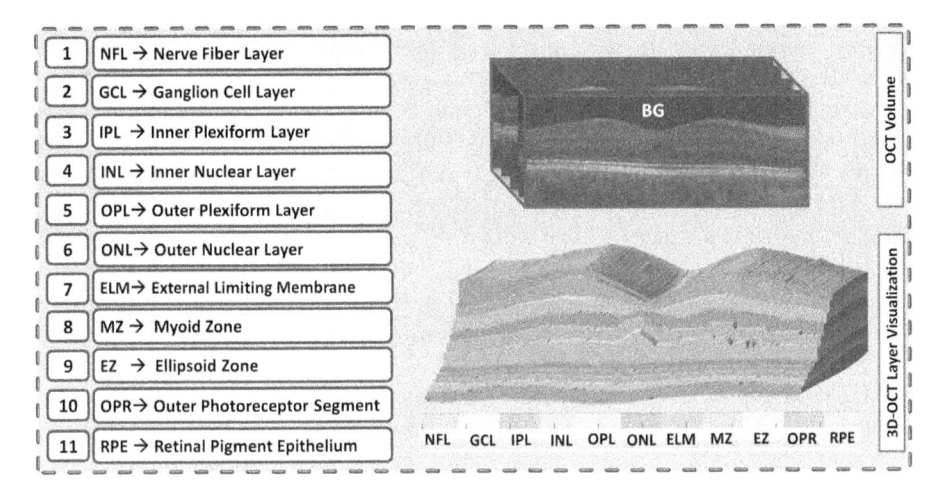

Fig. 1. An illustrative figure of the retinal 3D-OCT volume visualization, providing background (BG) and delineation of eleven layers.

learning-based approach to segment retinal layers in OCT images. The proposed model structure comprises two stacks, each composed of an encoder and a decoder. Stack 1 utilized an encoder-decoder pair with a Modified Attention U-Net decoder, focusing on local context. In contrast, Stack 2 employed a similar architecture with a decoder optimized for a broader context, functioning as a denoiser. This segmentation network predicts eight retinal layer maps from OCT scans. The study introduced by [10] annotated ten different labels on OCT datasets. Their segmentation framework comprises three components: an optic disc detection network, a retinal layer segmentation network, and a fusion module. Initially, the optic disc detection network processes OCT images to generate a mask and feature map indicating the optic disc's location. The mask is then applied to create a disc-free image fed into the retinal layer segmentation network, which produces segmented outputs. These outputs are generated by concatenating feature maps from both optic disc detection and retinal layer segmentation stages, trained end-to-end with two defined loss functions penalizing intermediate disc detection and final segmentation. Mukherjee et al. [14] introduced a method employing a 3D deep neural network to directly predict surface locations, aimed at ensuring precision, and smoothness in delineating retinal layers. The network comprises two interconnected subnetworks: (i) a 3D UNet for multi-class voxel labeling of retinal layer surfaces, and (ii) a 3D convolutional-autoencoder, which refines the UNet's output to estimate smooth contours delineating the boundaries between retinal layers. The approach proposed by Viedma et al. [21] with Mask R-CNN for retinal and choroidal layer instance segmentation raises certain concerns. While it introduces a novel methodology, relying solely on a large dataset of OCT images from healthy participants might limit its applicability to more diverse clinical scenarios. Xie et al. [22] developed a deep learning approach for the segmentation of retinal OCTs that combined a U-Net framework with differentiable dynamic programming (DDP) for feature extraction and feature smoothness, respectively. Their proposed model segmented 9 surfaces. Additional image gradient channels were utilized in the U-Net to improve boundary discrimination. The fully differentiable DDP module minimized total surface area cost while respecting smoothness constraints described by pixel-wise cost associations and parameterization of smoothness constraints. Finally, a multiple-surface cross-entropy loss and L1 loss function were utilized. Cao G et al. [3] proposed a model that combined encoder-decoder architecture with self-attention mechanisms for retinal segmentation that yielded 9 layers. These steps aimed to effectively capture both local and global features. First, fine details were extracted from the scan by the encoder utilizing convolutional layers. Then, a one-dimensional transformer block computing only self-attention in the vertical direction captured global semantic information, thus limiting computational complexity. Finally, an attention gate block and channel attention were incorporated into the decoder to assign weights to feature channels. Also, this approach [19] started by creating a 2D shape prior, which was subsequently adjusted using a first-order appearance model combined with a second-order spatial interaction MGRF model. After successfully segmenting the central macula, known as the "foveal area," the layer labels and appearances from this segmentation were then applied to segment adjacent slices. This final step was repeated recursively until the full 3D OCT scan of the patient was segmented into 12 layers. This approach relied solely on the MGRF model, overlooking the importance of incorporating additional features or methods that could potentially improve the segmentation accuracy. In summary, upon reviewing the existing literature, we identified several critical limitations. Firstly, most works are confined to segmenting only a subset of retinal layers, not all twelve, thereby failing to convey the complete picture of the intricate details of the retina. Furthermore, previous works heavily rely on CNN-based models, neglecting the potential of transformer-based models which have achieved state-of-the-art results in numerous computer vision tasks [7].

In response to these limitations, this paper introduces TransNetOCT, a novel method that employs a transformer architecture to segment 3D-OCT volumes into twelve distinct retinal surfaces. Our approach begins by segmenting the central macular B-scan using a joint Markov Gibbs Random Field (MGRF) model, integrating shape, intensity, and spatial characteristics across the layers. To enhance the segmentation accuracy of the remaining slices within the OCT volume, we introduce a novel propagation criterion based on the middle segmented B-scan image and extract shape probability priors for adjacent slices. The training of TransNetOCT is conducted using the middle slice, along with the slices on its immediate left and right, from which we also extract probability priors. This method results in an enhancement of the segmentation performance. Importantly, this approach offers critical insights into diseases such as AMD.

2 Method

Our proposed TransNetOCT method, depicted in Fig. 2, involves three primary stages for the precise segmentation of 3D-OCT volumes into twelve unique reti-

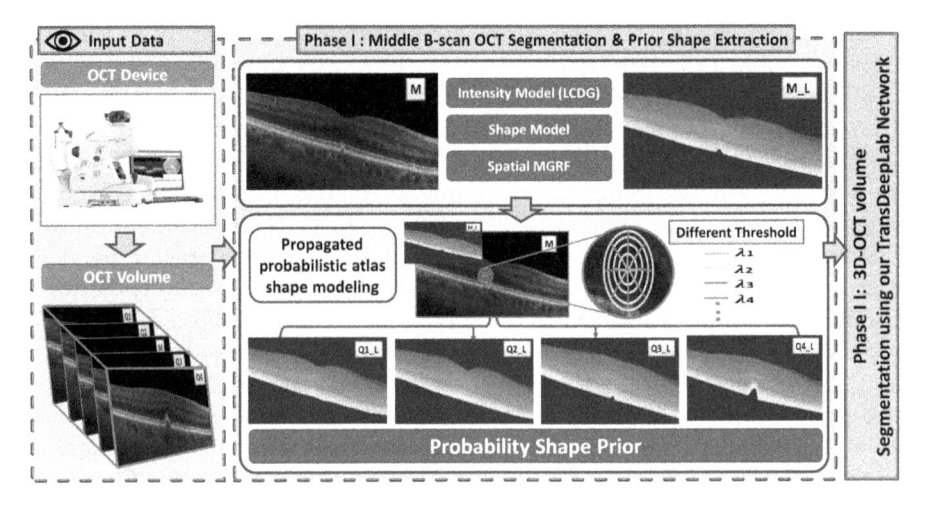

Fig. 2. Proposed system framework for segmentation of 3D-OCT images using two-phase segmentation approach: (I) Mid-slice segmentation and probability prior shape propagation. (II) 3D-OCT volume segmentation utilizing our TransNetOCT network.

nal surfaces. Initially, we segment the central macular B-scan, where the fovea is clearly visible, utilizing an MGRF model. This model integrates shape, intensity, and spatial attributes across the retinal surfaces, capitalizing on the unique features of the central B-scan. Following this, we implement a probability prior shape model for the residual slices within the OCT volume. This model applies a propagation standard based on the segmented middle B-scan image, facilitating the extraction of shape priors for neighboring slices. The segmentation of the 3D-OCT volume is accomplished by leveraging the central slice, the slices to its left and right, and the extracted probability prior shapes for these slices. The specifics of our proposed method are expanded in the subsequent sections.

2.1 Propagated Probabilistic Atlas shape modeling

To enhance the robustness of OCT volume segmentation and make it fast, we have adopted a novel approach: probabilistic prior shape modeling for the 3D-OCT volume, with a focus on individual 2D images (OCT B-scans). First, to model the appearance of an OCT B-scan as a MGRF, first consider the image as a field of reflectivity measurements $r(x)$ over a two-dimensional discrete space X. Formally, $r : X \rightarrow G$, where co-domain G is the set of gray levels (i.e., pixel values) in the encoded image. There is a corresponding segmentation $s : X \rightarrow L$, i.e., a map of region labels (L) over the same domain. To establish the shape prior, we utilize a collection of training OCT scans in our experiments. These scans were carefully selected to represent the variability present in the entire dataset. Expert retina specialists supervised the delineation of ground truth region maps from these scans. These maps provide a manual segmentation of the retina, highlighting its 12 distinct surfaces with 11 different layers.

Using a high-quality reference scan that's straight and perfectly focused on the fovea, we make all other training images match it by adjusting them through Thin-Plate Spline (TPS) deformations. We make the same adjustments to their corresponding maps that show the correct labels. Then, we combine these aligned maps pixel by pixel to get the prior shape probabilities. The image is aligned with a shape database using TPS and multi-resolution edge tracking for control point identification. The à trous algorithm decomposes the scan via wavelet transform, revealing distinct retinal bands. Gradient maxima tracing provides initial boundary estimates. The foveal peak, positioned strategically between vitreous/ Nerve Fiber Layer (NFL) and Myoid Zone (MZ)/ Ellipsoid Zone (EZ) contours, serves as the reference point for control points. These guide the optimized TPS alignment of the input image with the shape database, facilitating shape prior utilization in image segmentation. The input image's intensity histogram is normalized, creating the empirical marginal probability distribution. It's approximated using Linear Combination of sign-alternate Discrete Gaussians (LCDG) and divided into components, each representing a dominant mode in distinct regions. To achieve enhanced spatial homogeneity in segmentation and to account for noise with greater precision, a second-order MGRF model, which captures spatial interactions between region labels, is integrated with the aforementioned shape prior and intensity model. Combining shape prior, intensity model and MGRF enhances robust and accurate retinal layer delineation for the middle slice image. For more details about this approach to segment single OCT B-scan, see our previous work in [6].

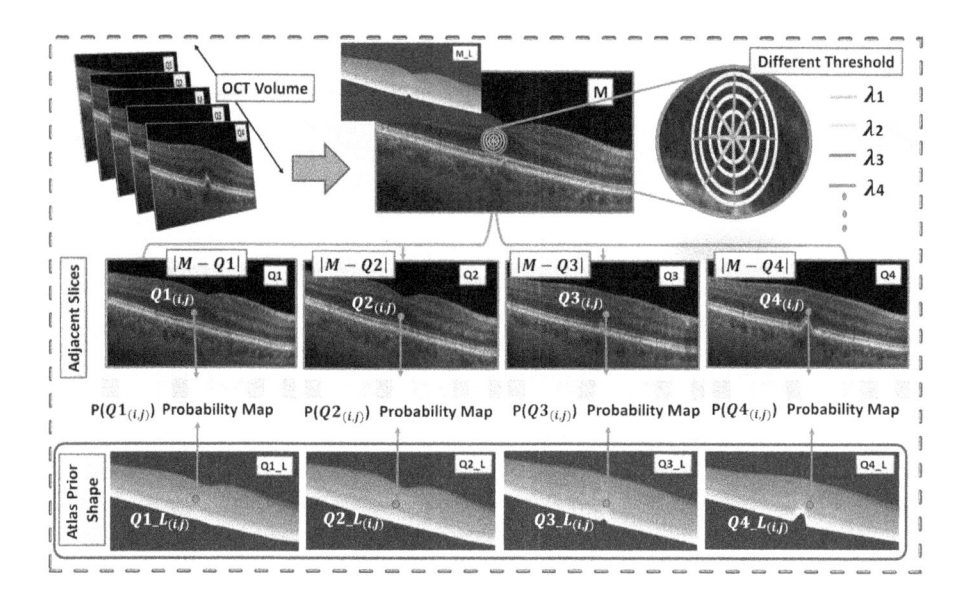

Fig. 3. Illustration of probability shape prior extraction for subsequent OCT slices from the central B-scan OCT.

After segmenting the middle OCT B-scan, we propagated prior shape modeling for adjacent slices. The algorithm outlined in Algorithm 1 explains the propagated appearance shape modeling, designed to segment and label the left and right adjacent slices of a 3D-OCT volume using information from the labeled middle slice. The algorithm takes as input the middle slice image (M), the labeled middle slice image (M_L), and the four adjacent slices (Q_1, Q_2, Q_3, Q_4) where Q_1 and Q_2 are to the left of the middle slice, and Q_3 and Q_4 are to the right of the middle slice. The goal is to obtain the labeled shape prior images (Q_1^L, Q_2^L, Q_3^L, Q_4^L) for these adjacent slices. The algorithm starts by calculating the absolute difference (Δ_i) between the M and each of the $Q_i, i = 1, 2, 3, 4$. It then determines a range of thresholds (λ) to be considered for calculating the prior shape. For each pixel p in the difference images (Δ_i), the algorithm computes a probability vector $v(p)$ that represents the likelihood of each pixel being associated with a specific layer. Hence, $v(p)$ is a vector of size 1×12 representing the twelve surfaces. Suppose the pixel p falls within the range of thresholds λ. In that case, the algorithm updates the probability vector $v(p)$ by setting the value corresponding to M_L at that pixel to 1 and all other values to 0. Otherwise, the algorithm defines a window W_p around the pixel p and computes the rotation-invariant absolute difference between p in the Q_i and the neighboring pixels of p (β_k^p) in the M. Based on the values of β_k^p and the thresholds λ, the algorithm adaptively updates the probability vector $v(p)$. If pixels of β_k^p falls within the range of λ, the corresponding value in $v(p)$ is set to the label of the M_L at that pixel. If β_k^p is outside the range of λ, the window size W_p is adaptively increased. If W_p exceeds a certain limit (Ω), the corresponding value in $v(p)$ is set to 0. Finally, the algorithm assigns the index of the maximum probability of $v(p)$ as the label to the corresponding pixel in Q_i. The labeled shape prior images ($Q_1^L, Q_2^L, Q_3^L, Q_4^L$) are then returned as the output. For visual explanation, Fig. 3 demonstrates a guided probabilistic atlas shape model for subsequent OCT slices derived from the central B-scan OCT.

2.2 TransNetOCT

Given a volume $x \in \mathbb{R}^{W \times H \times 9}$, as discussed in Section 2.1, it is encoded using ResNet-50 [8], where W and H denote the width and height of the image, respectively. It yields feature maps F_1 and F_4, which are the outputs from layer 1, low-level features, and layer 4, high-level features, of ResNet-50, respectively. Given that the initial convolution layer, Conv1, of ResNet-50 is designed to receive input with 3 channels, we adapted it to accept 9 channels by replacing it with a modified version and duplicating the pretrained ImageNet weights of Conv1 for this new layer. In our task, we aim to segment the 12 layers of the retina, which share visual similarities. We hypothesize that modeling global features is inevitable. Given that transformers excel at this task, we decided to model these features with a transformer encoder. Specifically, $F_4 \in \mathbb{R}^{N \times C \times P \times P}$, where P represents the patch size, N is the number of patches, and C is the number of channels received by the transformer encoder. These feature maps are reshaped and linearly projected to a sequence of patch embeddings

Algorithm 1. Propagated Appearance Shape Modeling

Require: M (Middle Slice Image), M_L (Labeled Middle image), (Q_1, Q_2, Q_3, Q_4) (left and right Adjacent B-scan Slices)
Ensure: $(Q_1^L, Q_2^L, Q_3^L, Q_4^L)$ (Labeled Shape Prior Images for (Q_1, Q_2, Q_3, Q_4))
1: Let $Q = \{Q_1, \ldots, Q_4\}$ and $Q_L = \{Q_1^L, \ldots, Q_4^L\}$ denote sets of gray levels q and region labels L, respectively.
2: Calculate the absolute difference between the middle slice and other four images:
$$\Delta_i = |M - Q_i|, \quad i = 1, 2, 3, 4$$
3: Determine a range of thresholds $\lambda = \{\lambda_1, \lambda_2, \ldots, \lambda_n\}$ to be considered for calculation of prior shape
4: **for** each pixel p in Δ_i **do**
5:     Let $\mathbf{v}(p) = (v_1(p), \ldots, v_{12}(p))$ be the probability descriptors for p, where $v_j(p)$ expresses the degree of belonging to each layer j.
6:     **if** the pixel $p \in$ a range of thresholds λ_1^n **then**
7:         Update probability vector $\mathbf{v}(p)$ to set $v_j(p) = j$ for the layer j corresponding to the label of M at this pixel, and $v_k(p) = 0$ for all other layers $k \neq j$.
8:     **else**
9:         Define a window of size W_p around p of M^p
10:        Compute the rotation-invariant absolute difference between p of Q_i^p and the neighboring pixels of p (N_k^p) in M: ,   k = 1, 2, ..., $W_p - 1$
$$\beta_k^p = |Q_i^p - N_k^p|, \quad i = 1, 2, 3, 4$$
11:        Update the probability vector $\mathbf{v}(p)$ based on β_k^p and λ:
12:        **if** $\beta_k^p \in \lambda$ **then**
13:            $v_j(p) = M_L(p)$
14:        **else if** $\beta_k^p \notin \lambda$ **then**
15:            Adaptively increase W_p to some size Ω
16:        **else if** $W_p > \Omega$ **then**
17:            $v_j(p) = 0$
18:        **end if**
19:     **end if**
20: **end for**
21: Assign the label $Q_i^L(p)$ to each pixel p in the adjacent slices Q_i based on the maximum value in $\mathbf{v}(p)$:
$$Q_i^L(p) = \arg\max_j v_j(p)$$
22: **return** $(Q_1^L, Q_2^L, Q_3^L, Q_4^L)$

$f = [e_{x_1}, \ldots, e_{x_N}] \in \mathbb{R}^{N \times P^2 \times D}$. To retain positional information, learnable position embeddings $p = [p_1, \ldots, p_N] \in \mathbb{R}^{P^2 \times D}$ are added to the patch sequence, resulting in the input sequence of tokens $\alpha = f + p$. The transformer encoder, composed of M layers, processes the input sequence α. Each layer comprises a multi-headed self-attention (MSA) [13] block followed by a point-wise MLP block, with layer normalization (LN) applied before and residual connections

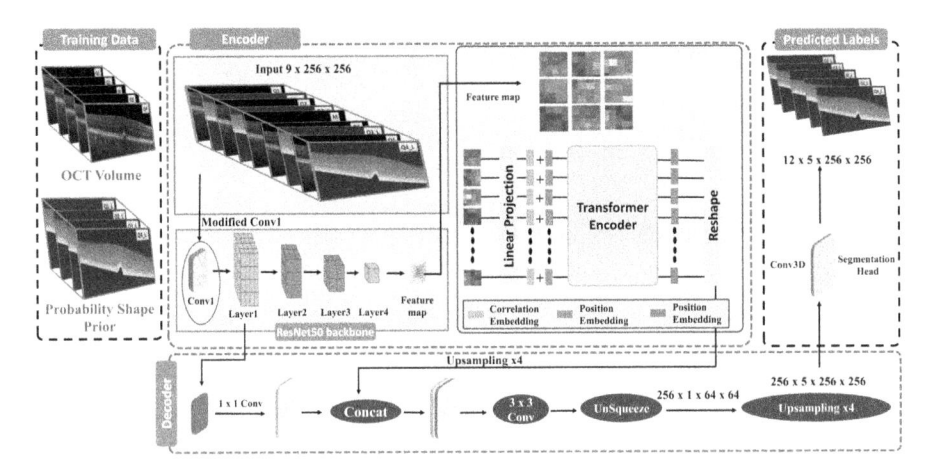

Fig. 4. An illustrative figure showcasing our network, TransNetOCT, which effectively segments the entire 3D-OCT volume into twelve distinct surfaces (eleven OCT layers).

added after each block.

$$a_{i-1} = \text{MSA}(\text{LN}(\alpha_{i-1})) + \alpha_{i-1},$$
$$\alpha_i = \text{MLP}(\text{LN}(a_{i-1})) + a_{i-1},$$

where $i \in 1, ..., M$. The self-attention mechanism computes queries $\mathbf{Q} \in \mathbb{R}^{N \times d}$, keys $\mathbf{K} \in \mathbb{R}^{N \times d}$, and values $\mathbf{V} \in \mathbb{R}^{N \times d}$ via three point-wise linear layers, followed by self-attention calculation:

$$\text{MSA}(\mathbf{Q}, \mathbf{K}, \mathbf{V}) = \text{softmax}\left(\frac{\mathbf{Q}\mathbf{K}^T}{\sqrt{d}}\right)\mathbf{V}.$$

The transformer encoder maps the input sequences to a contextualized encoding sequence containing rich, salient information $\alpha_M = [\alpha_{M,1}, ..., \alpha_{M,N}]$. Then, the output of the transformer encoder is reshaped back to $\alpha \in \mathbb{R}^{N \times C \times P \times P}$. Next, the low-level features and the contextualized features are input into a decoder. Finally, the decoder's output is reshaped and passed through a 3D convolution layer to produce the final 3D mask (i.e., a $12 \times 5 \times 256 \times 256$ mask). For a more detailed explanation of our network's architecture, please refer to Figure 4.

3 Experimental results

In this paper, the proposed approach is assessed on a cohort of 85 patients. Among these, 45 were individuals diagnosed with early or intermediate AMD, while the remaining data pertained to individuals with normal retinas. The data were collected through a retrospective study involving patients at the University of Louisville and Legacy Devers Eye Institute. Approval for the study was obtained from the institutional review boards (IRB) of both institutions (IRB #07.0296 and IRB #1971, respectively), and it was conducted in compliance with the principles outlined in the Declaration of Helsinki. OCT imaging was performed using two distinct devices: Zeiss Cirrus OCT models 4000 and 5000, in addition to the Heidelberg device. The macular B-scans within each volume were acquired with varying pixel dimensions, covering a depth of $2mm$ and a width of $6mm$. Along the anterior-posterior axis, the pixel spacing remained consistently at 1.955 μm, while along the naso-temporal axis, it ranged from 5.865 μm to 11.74 μm.

Table 1. Segmentation accuracy comparison between proposed approach using prior (ours) and without prior

Metric	DSC		ASSD		MASD		Sensitivity	
Method	With Prior	Without prior	With Prior	Without prior	With Prior	Without prior	With Prior	Without prior
Background	0.9935	0.9918	0.4830	0.6212	0.4777	0.6079	0.9947	0.9926
NFL	0.5759	0.5314	1.7433	2.0805	2.4047	2.7970	0.5518	0.5106
GCL	0.6921	0.6430	1.1403	1.3673	1.1611	1.3443	0.6979	0.6569
IPL	0.7057	0.6540	1.0314	1.2697	1.0328	1.2695	0.6986	0.6493
INL	0.7352	0.6784	0.9683	1.1864	0.9550	1.1290	0.7423	0.6902
OPL	0.6373	0.5728	1.0097	1.3632	1.0627	1.5847	0.6244	0.5597
ONL	0.8376	0.8003	1.0872	1.3376	1.0697	1.2742	0.8639	0.8415
ELM	0.3616	0.2586	4.5690	7.2680	8.2979	13.4460	0.3044	0.2034
MZ	0.5645	0.4917	0.8298	1.3233	0.8188	1.5102	0.5982	0.5358
EZ	0.4698	0.3883	1.1841	2.0168	1.5166	0.9980	0.4362	0.3514
OPR	0.5572	0.4890	0.8431	1.2665	0.8399	2.9265	0.5673	0.4934
RPE	0.7431	0.6863	0.8891	1.1384	0.8729	1.4071	0.7512	0.7083
Average	**0.6561**	**0.5988**	**1.3149**	**1.8532**	**1.7091**	**2.6633**	**0.6526**	**0.5994**

The system proposed in this study underwent training utilizing the AdamW optimizer, configured with a learning rate of 0.0001, and a cosine annealing scheduler. Moreover, a batch size of 8 was employed. Both training and testing phases adopted a leave-one-out cross-validation strategy, utilizing the ResNet-50 encoder with pretrained ImageNet weights. Additionally, cross-entropy loss was employed for optimization. The implementation was carried out using PyTorch, leveraging a single NVIDIA RTX A2000 GPU with a memory capacity of 12 GB. It's important to note that from each volume of 3D-OCT data, we selected the 5 most significant slices where the AMD was clearly visible.

In this study, we assess the performance of our proposed method for 3D-OCT segmentation using four distinct sets of metrics. These metrics are computed layer-wise, allowing for a detailed analysis of the model's performance across various retinal structures. Initially, we employ the Dice Similarity Coefficient (DSC) to gauge the overlap between predicted and ground truth masks for each retinal layer. Additionally, we utilize sensitivity to evaluate the model's ability to accurately detect the presence of specific retinal layers. Finally, we integrate the Mean Absolute Surface Distance (MASD) and Absolute Sum of Surface Distance (ASSD) metrics to quantify the average absolute distance and the sum of absolute distances between the predicted and ground truth boundaries for each retinal layer, respectively.

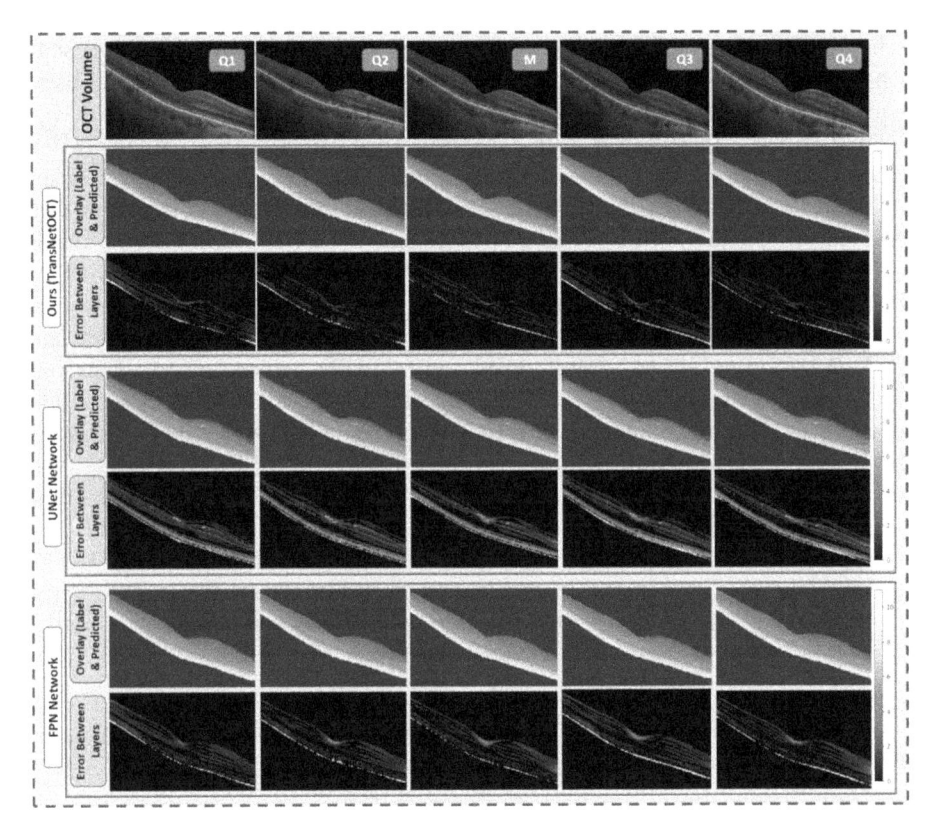

Fig. 5. An illustrative figure showcasing a comparison of retinal layer segmentation performance among TransNetOCT, UNet, and FPN using OCT scans of a patient with early AMD.

Table 1 presents a comprehensive comparison of segmentation performance between the proposed approach using our novel prior model and the approach without the prior. The metrics evaluated include the DSC, ASSD, MASD, and sensitivity for each retinal layer in the OCT volume. The results highlight the significant improvement achieved by incorporating prior knowledge into the segmentation model. On average, the proposed approach with prior knowledge outperforms the approach without prior knowledge across all metrics. Specifically, the average DSC improves from 0.5988 to 0.6561, indicating a higher overlap between the predicted and ground truth segmentation masks. The average Sensitivity also increases from 0.5994 to 0.6526, demonstrating an enhanced ability to detect the presence of retinal layers accurately. Based on the MASD, it is evident that the proposed approach with the prior shape model performs significantly better than the approach without the prior model. The average MASD with the prior model is 1.7091, while the average MASD without the prior model is 2.6633, which is approximately 55% higher. Moreover, our approach consistently outperforms the one without prior knowledge across all retinal layers. Notably, in the segmentation of the RPE layer, the ASSD decreases from 1.1384 to 0.8891 with prior information, demonstrating the effectiveness of adding a prior shape to our model. Specifically, this RPE layer is important for the diagnosis of AMD.

Furthermore, TransNetOCT is compared with other popular segmentation models, including UNet and Feature Pyramid Network (FPN). The setup for UNet and FPN is the same as our proposed method, ensuring a fair comparison. According to Table 2, our approach outperforms both UNet and FPN in terms of DSC, achieving a DSC of 0.6561, compared to 0.4178 for UNet and 0.5512 for FPN. Additionally, our approach also achieves lower ASSD and MASD values of 1.3149 and 1.7091, respectively, in contrast to UNet's 6.0690 and 7.0618, and FPN's 2.5183 and 3.2558. Lower ASSD and MASD values indicate that our segmentation boundaries are closer to the ground truth, making our method more precise. Notably, our method achieves a DSC of 0.7431 for the RPE layer, significantly higher than 0.5323 of UNet and 0.6501 of FPN, which is a crucial layer for diagnosing and monitoring AMD. Additionally, Figure 5 shows that the proposed TransNetOCT system outperforms UNet and FPN in accurately segmenting retinal layers from OCT scans of a patient with early AMD. The narrower rainbow-colored regions in the TransNetOCT row indicate lower error between the segmented layers and the true layer boundaries, compared to the wider misalignments observed in UNet and FPN.

Table 2. Segmentation accuracy comparison between proposed approach (TransNetOCT), UNet [16], and FPN [12] using prior shape

Metric	DSC			ASSD			MASD		
Method	Ours	UNet [16]	FPN [12]	Ours	UNet [16]	FPN [12]	Ours	UNet [16]	FPN [12]
Background	0.9935	0.9773	0.9886	0.4830	1.8867	0.8586	0.4777	1.4915	0.8717
NFL	0.5759	0.2639	0.4603	1.7433	5.7292	3.3211	2.4047	7.0180	4.7451
GCL	0.6921	0.4934	0.5909	1.1403	3.6967	1.9825	1.1611	2.4212	1.9784
IPL	0.7057	0.4867	0.5849	1.0314	3.0128	1.9214	1.0328	2.7536	1.9291
INL	0.7352	0.5158	0.6059	0.9683	2.8586	1.8263	0.9550	2.0353	1.8124
OPL	0.6373	0.3404	0.5102	1.0097	5.7695	2.0291	1.0627	7.4806	2.2008
ONL	0.8376	0.6503	0.7514	1.0872	3.4969	1.8880	1.0697	2.2680	1.8130
ELM	0.3616	0.0750	0.2260	4.5690	13.2581	6.8553	8.2979	18.5708	12.3897
MZ	0.5645	0.2515	0.4488	0.8298	10.1028	2.2235	0.8188	15.5480	2.5081
EZ	0.4698	0.1588	0.3419	1.1841	12.2631	3.0230	1.5166	14.3538	4.2505
OPR	0.5572	0.2689	0.4550	0.8431	7.4679	2.3210	0.8399	9.2824	2.6376
RPE	0.7431	0.5323	0.6501	0.8891	3.2860	1.9704	0.8729	1.5185	1.9334
Average	0.6561	0.4178	0.5512	1.3149	6.0690	2.5183	1.7091	7.0618	3.2558

4 Conclusions and Future work

Accurate segmentation of retinal layers within 3D-OCT volumes is crucial for disease prediction, notably in understanding AMD progression. This study introduces TransNetOCT, a transformer-based method designed for segmenting 3D-OCT volumes into twelve surfaces, delineating the background and eleven retinal layers. Initially, we applied a joint MGRF model to segment the central macular B-scan, incorporating shape, intensity, and spatial characteristics across retinal surfaces. Subsequently, a probability prior shape model was implemented for the remaining slices, utilizing a propagation criterion based on the segmented middle B-scan image. TransNetOCT was trained to utilize the middle slice, adjacent left and right slices, and extracted probability prior shapes for accurate 3D-OCT segmentation. Evaluation of an 85-patient cohort, including those with normal, early, and intermediate AMD, showcased the potential of TransNetOCT. It outperformed well-known models like UNet and FPN by approximately 57% and 19%, respectively, as measured by the Dice coefficient. Additionally, TransNetOCT achieved an ASSD of 1.3149, representing a 78% reduction compared to UNet (6.0690) and a 48% reduction compared to FPN (2.5183). Similarly, the MASD of TransNetOCT (1.7091) was 76% lower than UNet (7.0618) and 48% lower than FPN (3.2558). These notable enhancements in DSC, ASSD, and MASD metrics underscore the superior performance of our approach in precisely segmenting retinal layers from OCT images. In future work, we will collect more data to further explore TransNetOCT's potential, investigate additional slices, and aim to make TransNetOCT a purely 3D model.

Disclosure of Interests. The authors have no competing interests.

References

1. Bird, A.C., Bressler, N.M., Bressler, S.B., Chisholm, I.H., Coscas, G., Davis, M.D., de Jong, P.T., Klaver, C., Klein, B., Klein, R., et al.: An international classification and grading system for age-related maculopathy and age-related macular degeneration. Surv. Ophthalmol. **39**(5), 367–374 (1995)
2. Bressler, N.M.: Age-related macular degeneration is the leading cause of blindness... Jama **291**(15), 1900–1901 (2004)
3. Cao, G., Wu, Y., Peng, Z., Zhou, Z., Dai, C.: Self-attention cnn for retinal layer segmentation in oct. Biomed. Opt. Express **15**(3), 1605–1617 (2024)
4. Chiu, S.J., Izatt, J.A., O'Connell, R.V., Winter, K.P., Toth, C.A., Farsiu, S.: Validated automatic segmentation of amd pathology including drusen and geographic atrophy in sd-oct images. Investigative ophthalmology & visual science **53**(1), 53–61 (2012)
5. Davis, M.D., Gangnon, R.E., Lee, L.Y., Hubbard, L.D., Klein, B., Klein, R., Ferris, F.L., Bressler, S.B., Milton, R.C., et al.: The age-related eye disease study severity scale for age-related macular degeneration: Areds report no. 17. Archives of ophthalmology (Chicago, Ill.: 1960) **123**(11), 1484–1498 (2005)

6. ElTanboly, A., Ismail, M., Shalaby, A., Switala, A., El-Baz, A., Schaal, S., Gimel'farb, G., El-Azab, M.: A computer-aided diagnostic system for detecting diabetic retinopathy in optical coherence tomography images. Medical physics **44**(3), 914–923 (2017)
7. Han, K., Wang, Y., Chen, H., Chen, X., Guo, J., Liu, Z., Tang, Y., Xiao, A., Xu, C., Xu, Y., et al.: A survey on vision transformer. IEEE Trans. Pattern Anal. Mach. Intell. **45**(1), 87–110 (2022)
8. He, K., Zhang, X., Ren, S., Sun, J.: Deep residual learning for image recognition. In: Proceedings of the IEEE conference on computer vision and pattern recognition. pp. 770–778 (2016)
9. Hussain, M.A., Bhuiyan, A., Ramamohanarao, K.: Disc segmentation and bmo-mrw measurement from sd-oct image using graph search and tracing of three bench mark reference layers of retina. In: 2015 IEEE International Conference on Image Processing (ICIP). pp. 4087–4091. IEEE (2015)
10. Li, J., Jin, P., Zhu, J., Zou, H., Xu, X., Tang, M., Zhou, M., Gan, Y., He, J., Ling, Y., et al.: Multi-scale gcn-assisted two-stage network for joint segmentation of retinal layers and discs in peripapillary oct images. Biomed. Opt. Express **12**(4), 2204–2220 (2021)
11. Lim, L.S., Mitchell, P., Seddon, J.M., Holz, F.G., Wong, T.Y.: Age-related macular degeneration. The Lancet **379**(9827), 1728–1738 (2012)
12. Lin, T.Y., Dollár, P., Girshick, R., He, K., Hariharan, B., Belongie, S.: Feature pyramid networks for object detection. In: Proceedings of the IEEE conference on computer vision and pattern recognition. pp. 2117–2125 (2017)
13. Mao, X., Qi, G., Chen, Y., Li, X., Duan, R., Ye, S., He, Y., Xue, H.: Towards robust vision transformer. In: Proceedings of the IEEE/CVF conference on Computer Vision and Pattern Recognition. pp. 12042–12051 (2022)
14. Mukherjee, S., De Silva, T., Grisso, P., Wiley, H., Tiarnan, D.K., Thavikulwat, A.T., Chew, E., Cukras, C.: Retinal layer segmentation in optical coherence tomography (oct) using a 3d deep-convolutional regression network for patients with age-related macular degeneration. Biomed. Opt. Express **13**(6), 3195–3210 (2022)
15. Pekala, M., Joshi, N., Liu, T.A., Bressler, N.M., DeBuc, D.C., Burlina, P.: Deep learning based retinal oct segmentation. Comput. Biol. Med. **114**, 103445 (2019)
16. Ronneberger, O., Fischer, P., Brox, T.: U-net: Convolutional networks for biomedical image segmentation. In: Medical image computing and computer-assisted intervention–MICCAI 2015: 18th international conference, Munich, Germany, October 5-9, 2015, proceedings, part III 18. pp. 234–241. Springer (2015)
17. Sampath Kumar, A., Schlosser, T., Langner, H., Ritter, M., Kowerko, D.: Improving oct image segmentation of retinal layers by utilizing a machine learning based multistage system of stacked multiscale encoders and decoders. Bioengineering **10**(10), 1177 (2023)
18. Schütze, C., Wedl, M., Baumann, B., Pircher, M., Hitzenberger, C.K., Schmidt-Erfurth, U.: Progression of retinal pigment epithelial atrophy in antiangiogenic therapy of neovascular age-related macular degeneration. Am. J. Ophthalmol. **159**(6), 1100–1114 (2015)
19. Sleman, A.A., Soliman, A., Elsharkawy, M., Giridharan, G., Ghazal, M., Sandhu, H., Schaal, S., Keynton, R., Elmaghraby, A., El-Baz, A.: A novel 3d segmentation approach for extracting retinal layers from optical coherence tomography images. Med. Phys. **48**(4), 1584–1595 (2021)
20. Stankiewicz, A., Marciniak, T., Dąbrowski, A., Stopa, M., Marciniak, E., Michalski, A.: Matching 3d oct retina images into super-resolution dataset. In: 2016 Signal

Processing: Algorithms, Architectures, Arrangements, and Applications (SPA). pp. 130–137 (Sept 2016)
21. Viedma, I.A., Alonso-Caneiro, D., Read, S.A., Collins, M.J.: Oct retinal and choroidal layer instance segmentation using mask r-cnn. Sensors **22**(5), 2016 (2022)
22. Xie, H., Xu, W., Wang, Y.X., Wu, X.: Deep learning network with differentiable dynamic programming for retina oct surface segmentation. Biomed. Opt. Express **14**(7), 3190–3202 (2023)

On the Importance of Local and Global Feature Learning for Automated Measurable Residual Disease Detection in Flow Cytometry Data

Lisa Weijler[1](✉), Michael Reiter[1,2], Pedro Hermosilla[1], Margarita Maurer-Granofszky[2], and Michael Dworzak[2]

[1] TU Wien, Vienna, Austria
lweijler@cvl.tuwien.ac.at
[2] St.Anna CCRI, Vienna, Austria

Abstract. This paper evaluates various deep learning methods for measurable residual disease (MRD) detection in flow cytometry (FCM) data, addressing questions regarding the benefits of modeling long-range dependencies, methods of obtaining global information, and the importance of learning local features. We propose two adaptations to the current state-of-the-art (SOTA) model based on our findings. Our contributions include an enhanced SOTA model, demonstrating superior performance on publicly available datasets and improved generalization across laboratories, as well as valuable insights for the FCM community, guiding future DL architecture designs for FCM data analysis. The code is available at https://github.com/lisaweijler/flowNetworks.

Keywords: Flow Cytometry · Automated MRD Detection · Deep Learning · Self-Attention · Graph Neural Networks

1 Introduction

The detection and monitoring of measurable residual disease (MRD) in pediatric acute leukemia represent a critical aspect of patient care and treatment evaluation [5]. MRD defined as the proportion of residual cancer cells in patients after therapy, serves as a prognostic indicator for disease relapse and guides therapeutic decisions towards achieving better clinical outcomes [8,26]. Flow cyotmetry (FCM), with its ability to analyze cellular characteristics at a single-cell level, has emerged as a cornerstone technique for MRD assessment due to its sensitivity and specificity [18].

However, the accurate identification and quantification of MRD amidst heterogeneous cell populations remain challenging, often necessitating complex data analysis methodologies and training of medical experts. In recent years, the advent of deep learning (DL) approaches has revolutionized the landscape of biomedical data analysis [7], offering promising solutions to address the inherent complexities of MRD detection in FCM data for pediatric acute leukemia.

Given the unstructured characteristic of single cell FCM data, it does not fit in well-researched modalities such as text or images, and hence, applying DL methods is not straightforward. While traditional machine learning approaches have become standard practice for the analysis of FCM data [6,10,25], there have only been a handful of approaches applying DL to FCM data directly, primarily relying on convolutional neural networks (CNN) for e.g. imaging FCM or attention-based networks that are able to process unstructured data for single cell FCM data.

FCM data samples are essentially sets of F-dim feature vectors (events) corresponding to single cells, in a feature space $\mathbb{R}^F$ comprised of the properties measured by FCM, where F is usually between 10 and 15 and can vary between different samples. Similar cell types share similar feature vectors and tend to form clusters, representing a composition of different cell populations.

Since events within one sample are not i.i.d., previous works suggest that the relative position of cell populations within one sample contains crucial information for successful MRD detection, meaning it is beneficial to process full samples at once rather than considering single events detached from their origin-sample as input to DL models [17,23,24]. In other words, global feature extraction is suggested to be beneficial for single-cell classification as in the task of MRD detection. However, to the best of our knowledge, there exists no extensive evaluation of this assumption except for minor baseline testing with simple MLPs. Further, there are several ways of learning global features and infusing them with single-cell features for classification, which have not been evaluated. Dominating methods in the literature rely on gaussian mixture models (GMM) or self-attention, while the latter holds the current state-of-the-art (SOTA) for automated MRD in pediatric b-cell acute lymphoblastic leukemia (b-ALL) [33]. Self-attention allows for modeling long-range dependencies, yet does not explicitly learn local features, i.e., introduce an inductive bias of spatial locality, which is a crucial component of common successful architectures using convolutions and local feature aggregation, especially for tasks requiring fine semantic perception [9].

In this work, we provide an extensive evaluation of several methods guided by asking the following questions. "*Does automated MRD detection benefit from modelling long-range dependencies?*", "*Does it matter how the global information is obtained?*" and "*Is it beneficial to explicitly learn local features?*" Based on the findings of the analysis we propose two adaptations of the current SOTA that give a performance increase and better generalization abilities. First, the current SOTA is based on the set transformer (ST) [14], which uses learned query vectors, called inducing points, to mitigate the quadratic complexity issue of self-attention; instead of using learned query vectors we propose to use feature vectors from the FCM sample directly sampled by farthest-point-sampling (FPS). Second, we introduce explicit learning of local features by infusing the self-attention layers with graph neural network (GNN) layers.

In summary, our main contributions are,

1. an enhanced version of the current SOTA model that leads to a new SOTA performance on publicly available datasets as well as better generalization abilities between datasets of different laboratories,
2. providing an extensive evaluation of several DL methods for FCM data with valuable insights for the FCM community, on which future designs of DL architectures can be based.

2 Related Work

Given the wide range of applications of FCM data, the developed approaches for automated analysis are highly task-specific. In this work, we focus on single-cell classification for rare cell populations and give an overview of techniques related to this task.

The most direct approach to automatically predicting the class label of each event of a patient sample is to pool events from the training set of different samples together and train a classifier using pairs of single events and corresponding labels. Abdelaal et al. Authors in [1] propose a linear discriminant analysis classifier (LDA) for this task, leading to interpretable results due to the simplicity of LDA. However, the assumption of equal covariance matrices of classes is not valid for rare cell population detection as in MRD quantification. In [19], authors suggest training one support vector machine model per patient, implicitly incorporating prior knowledge of the patient's specific phenotype. However, this requires the availability of labeled training data for each patient.

Other methods are based on neural networks [17]. However, methods using single events as input are restricted to learning fixed decision regions. One way to circumvent this is to register samples by transforming them into a common feature space [16,30] or by creating landmarks based on prior biological knowledge that guides classification [13]. Another way is to process full samples at once, yet those methods need to be equivariant to the ordering and handle the volume of events present in a sample. Representing a sample based on its statistical parameters by, e.g., Gaussian Mixture Models [23] is an option. Others propose sample-wise clustering-based approaches [4,29,32].

Methods based on DL that process full samples at once are scarce, given the characteristics of FCM data. One line of research is to transform FCM samples to images and apply convolutional neural networks as in [2]. More recently, methods based on the attention mechanism [27] have been proposed [12,31,33]. Attention-based models are a way for event-level classification that learns and incorporates the relevance of other cell populations in a sample for the specific task. A limiting factor is the complexity of the standard self-attention operation that increases quadratically with the input sequence length $\mathcal{O}(n^2)$; this is infeasible for FCM data and efficient variants such as the ST [14] have to be used, as authors in [33] do.

In the context of FCM data, to the best of our knowledge, graph-based methods have only been applied for unsupervised clustering, where the clustering algorithm itself is graph-based [3,15], but not for targeted cell classification and modeling local spatial structure. Additionally, as far as we are aware this work is the first to introduce the benefits of GNN to implicitly make use of sample- or patient-specific features.

3 Methods

In this section, the problem setup is described in detail (Section 3.1), all methods analyzed are introduced (Section 3.2), and the experimental setup is outlined (Section 3.3).

3.1 Preleminaries

We treat the problem of MRD detection as a binary classification of single events into healthy or cancerous cells. We use the following definitions of FCM data sets throughout the paper.

Definition 1. *A FCM data set* $\mathcal{X} = \{X_1, \ldots, X_N\}$ *contains N samples $X_i \in \mathbb{R}^{n_i \times F_i}, i = 1 \ldots N$, where F_i is the feature space dimension of sample X_i and $n_i = |X_i|$ the number of events $x_{i_j}, j = 1, \ldots, n_i$ per sample.*

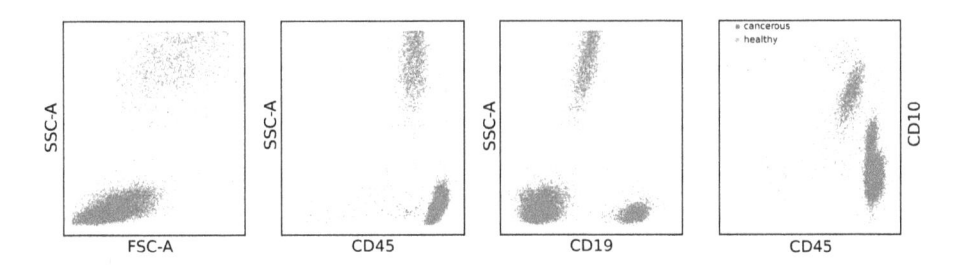

Fig. 1. This figure shows 2D projections of an FCM sample on pairs of features, where each dot represents the feature vector of a cell (event). Healthy cells are denoted in grey, and cancerous cells in red. FSC, SSC stands for forward-, side-scatter, and CD for cluster of differentiation.

In general, the proportion of leukemic cells varies between samples; it can be as low as 0.01%, while the number of measured cells is up to $n_i = 10^6$. Fig. 1 shows 2D projections of an FCM sample.

When applying GNN to FCM data, each sample X_i is converted into a graph $\mathcal{G}_i$ by the k-NN algorithm utilizing its full individual feature space $\mathbb{R}^{F_i}$. The graph is constructed using all events $x_{i_j} \in X_i$; every node of the graph thus represents a single event.

3.2 Methods assessed

In the following we introduce the architectures analysed to answer the posed questions. The range of models used is not exhaustive, yet yields a good representation of various domains ranging from simple multi-layer perceptron (MLP) over GNN to variations of attention-based networks.

We distinguish between *no-context* models, where cells are classified based on the information of the single cell only, *global-context* models, where the sample as a whole is included in the prediction of single cells, and *local-context* models, where the local context of similar cells in the sample is used for classification. Finally, we present our proposed architecture, *local-global-context*, where explicit extraction of local features is combined with long-range dependency modeling.

All architectures are comprised of 4 layers of the specific network layers and a linear layer as prediction-head using a hidden dimensionality of 32 unless stated otherwise. If multi-head self-attention is part of the architecture, 4 heads are used. For methods that use local feature learning, a k-NN graph is constructed with $k = 10$, unless stated otherwise. As non-linearity, the GELU activation function is employed.

No-context model A baseline model for comparing to single cell processing with neural networks, which translates into fixed decision boundaries independent of in-sample-context.

`MLP:` A simple MLP with batch normalizations after the non-linearity.

Global-context models Baseline models and proposed FCM-specific adaptations to compare different versions of incorporating global information i.e. taking the whole input sample into consideration for single cell predictions.

`MLP-mean:` Same as `MLP` but a global feature vector obtained with mean-aggregation is infused by concatenation to the single-event feature vectors before passing through the last linear layer

`MLP-max:` Same as `MLP-mean` but using max-aggregation.

`MLP-pma:` Same as `MLP-mean` but using a learnt query vector for aggregation. This aggregation is equivalent to the pooled multi-head attention (PMA) as proposed in [14] using one seed vector and one attention head, where a learnt query vector is cross-attended to all single-event feature vectors.

`PointNet:` The PointNet architecture introduced in [20] for 3D point cloud classification and segmentation.

`PointNet-adapted:` Same as `PointNet` but to reweigh the focus on the global information to be equally distributed among event-wise and global vectors, we increase the dimensionality of the event-wise feature vectors to match the dimensionality of the global vectors, namely 1024.

ST: The ST [14] which holds the current SOTA for automated MRD detection in b-ALL [33]. It uses learned query vectors, called *inducing* points to circumvent the quadratic complexity of self-attention. The induced set attention block (ISAB) is defined as

$$ISAB_m(X) = MAB(X, H) \in \mathbb{R}^{n \times d},$$
$$\text{where } H = MAB(I, X) \in \mathbb{R}^{m \times d}, \tag{1}$$

with $I \in \mathbb{R}^{m \times d}$ being the m d-dimensional inducing points, $X \in \mathbb{R}^{n \times d}$ the set to be processed with cardinality n and MAB a multi-head attention block. Such an ISAB reduces the complexity from $\mathcal{O}(n^2)$ to $\mathcal{O}(mn)$ with $m << n$. Intuitively, ISAB summarizes the sample in the learned queries and induces the information back with cross-attention. $m = 16$ as proposed in [33].

$\mathit{reluFormer}$: An adaptation of the cosFormer proposed in [21] that uses ReLU instead of softmax to remove the non-linearity in the attention calculation in order to be able to rearrange the matrix multiplications to get linear complexity of self-attention while using the full input sequence length. Authors in [21] propose a cosine-based distance re-weighting scheme instead of the softmax function that concentrates the focus of the attention values on neighboring tokens. Since we do not have an ordered sequence as input we omit the re-weighting scheme in this work.

$\mathit{ST\text{-}FPS}$: Inspired by the intuition of ISAB we propose to select event feature vectors instead of learning queries to compose the matrix as inducing points. We use FPS, a widely used sampling method, on the input FCM sample to select the indices of events used to create the I matrix. We use a sampling ratio of $r = 0.0005$, which results on average in ≈ 150 events.

Local-context models Baseline models that introduce an inductive bias based on prior knowledge of homophily (biologically similar events share similar feature measurements) by explicitly learning local features.

GCN: The graph convolution network (GCN) layers as proposed in [11].

GAT: The graph attention network (GAT) layers as proposed in [28].

GIN: The graph isomorphism network (GIN) layers as proposed in [35].

$\mathit{GAT\text{-}AsAP}$: The adaptive structure aware pooling (ASAP) layers as proposed in [22] combined with GAT layers. The architecture used is similar to [22] with two GAT blocks, constituted of 2 GAT layers each and two ASAP layers pooling to 100 and 50 nodes, respectively.

$\mathit{GIN\text{-}AsAP}$: Same as GAT-ASAP but with a GIN layer instead of GAT.

Local- and global-context model The proposed architecture combines explicit local feature learning with long-range dependency modeling.

`GAT-ST-FPS`: The adapted SOTA architecture based on the findings in Section 4, GAT infused ST with FPS self-attention. We use one GAT layer for local feature extraction and concatenate those with the input feature vectors before inserting them into the 3 `ST-FPS` layer.

`GIN-ST-FPS`: GIN infused ST with FPS self-attention. Same as `GAT-ST-FPS` but with a GIN layer instead of GAT.

3.3 Experimental setup

Datasets. We conduct our experiments on publicly available[1] data sets of bone marrow samples from pediatric patients with b-ALL. In our main experiments, we use the biggest dataset *Vie*; for our inter-laboratory experiments to assess the architectures' ability to generalize to different laboratories we use *Bln* and *Bue* for testing.

- *Vie* contains 519 samples collected between 2009 and 2020 at the St. Anna Children's Cancer Research Institute (CCRI) with an LSR II flow cytometer (Becton Dickinson, San Jose, CA) and FACSDiva v6.2. The samples collected between 2009 and 2014 were stained using a conventional seven-colour drop-in panel ("B7") consisting of the liquid fluorescent reagents: CD20-FITC/ CD10-PE/ CD45-PerCP/ CD34-PE-Cy7/ CD19-APC/ CD38-AlexaFluor700 and SYTO 41. The samples collected between 2016 and 2020 were stained using dried format tubes (DuraClone^{TM}, "ReALB") consisting of the fluorochrome-conjugated antibodies CD58-FITC/ CD34-ECD/ CD10-PC5.5/ CD19-PC7/ CD38-APC-Alexa700/ CD20-APC-Alexa750/ CD45-Krome Orange plus drop-in SYTO 41.
- *Bue* contains 65 samples collected between 2016 and 2017 at the Garrahan Hospital in Buenos Aires. The samples were recorded with FACSCanto II flow cytometer with FACSDiva v8.0.1 and stained with the following panel: CD58, FITC/CD10, PE/CD34, PerCPCy5.5/CD19, PC7/CD38, APC/CD20, APC-Alexa750/CD45, Krome-Orange plus drop-in SYTO 41.
- *Bln* consists of 72 samples collected in 2016 at the Charité Berlin. The samples were collected with a Navios flow cytometer and stained with the same panel as *Bue*.

All data were collected on day 15 after induction therapy. Sampling and research were approved by local Ethics Committees, and informed consent was obtained from patients or patients' parents or legal guardians according to the Declaration of Helsinki. Ground truth was obtained using manual gating by at least two experts. For every sample, the resulting labels from different experts are then combined into a final gating for each sample to obtain reliable ground truth data.

Table 1 provides a tabular overview of the data sets.

---

[1] flowrepository.org

Table 1. Description of the FCM data sets.

Name	City	Years	Samples
Vie	Vienna	2009-2020	519
Bln	Berlin	2016	72
Bue	Buenos Aires	2016-2017	65

Metrics. We use precision p, recall r, and F_1-score for evaluation, with correctly identified cancer cells as true positives. The metrics are computed per FCM sample and then averaged to obtain the final score. We report the mean and standard deviation of at least 5 runs for each experiment.

Training details. For comparability we keep the same training setup for each experiment. We use a batch size of 4 with $5*10^4$ randomly sampled events per sample. For augmentation, we employ random jitter with a scale parameter of 0.01 and label smoothing with $eps = 0.1$. We use a train, validation, and test split of 50%, 25%, and 25%, respectively. Each model is trained for 150 epochs using AdamW optimization with a learning rate of 0.001 and cosine annealing with a starting value of 0.001, a minimum value of 0.0002, and a maximum of 10 iterations as learning rate scheduler. To mitigate overfitting in GAT layers we employ dropout and weight decay with a rate of 0.2. Training for all experiments is conducted on an NVIDIA GeForce RTX 3090. The model that performed best on the validation split based on mean F_1-score is used for testing.

4 Experiments

In this section we analyse the results of the methods described above and thereby aim to find answers to the questions posed. First, in Section 4.1, the importance of global information for the success of the current SOTA ST [33] is assessed. Second, in Section 4.2, different ways of obtaining global features are compared. Third, in Section 4.3, methods that explicitly learn local features are evaluated. And finally, in Section 4.4 we combine the findings into the proposed enhanced architecture.

4.1 Does automated MRD detection benefit from modelling long-range dependencies?

To answer this question, we take the current SOTA method ST and compare it to its identical architecture but with removed self-attention, which is substituted by summed query, key, and value vectors. This way, we can test for the impact of modeling long-range dependencies without architecture noise. Tabel 2 shows that modeling long-range dependencies brings a clear benefit, and the architecture itself has no contribution but rather impairs results for event-wise processing when looking at the simple MLP for comparison. This question can thus be clearly answered, with an *yes*. However, there are several ways of obtaining global features, which are discussed in the next section.

Table 2. Results for removing the ability to model long-range dependencies without an architecture change from the SOTA model.

Method	p	r	avg F_1	med F_1
ST-No-Att	0.7463	0.8766	0.771 ±0.0038	0.8925 ±0.0036
MLP	0.7912	0.8701	0.8032 ±0.0073	0.9224 ±0.0084
ST	0.8251	0.8601	**0.8284** ±0.0117	0.9405 ±0.0085

4.2 Does it matter how the global information is obtained?

The most simple way to inject global features in the network is by single-cell feature vector aggregation and combining the obtained global vector with each single-cell feature vector for classification. We evaluate max and mean aggregation (MLP-max, MLP-mean) as well as using a learned query vector (MLP-pma). Further, we evaluate PointNet, being a pioneering architecture for 3D point cloud classification and segmentation. Table 3 shows that there is no real difference in performance using mean or max aggregation and only a minor performance increase when using a learned query vector. PointNet, however, performs better, yet the hidden dimensions are much higher (128 and 1024 for single-event and global feature vectors, respectively) than 32 as used throughout our experiments. When increasing the feature dimension for the single-event vectors to 1024 as well in PointNet-adapted to remove the increased focus on global information from the standard PointNet architecture, the model slightly outperforms the current SOTA ST relying on self-attention. This is interesting given the simplicity of PointNet. However, note that the feature dimension in PointNet-adapted is 32 times higher than in ST.

Further, self-attention can essentially be seen as another way of directly obtaining global information since no local spatial structure is imposed. Table 3 shows a clear benefit using feature vectors sampled directly from the FCM sample (ST-FPS) rather than learned query vectors as inducing points. One explanation is that relying more on the sample at hand helps to generalize between patient- or sample-specific shifts and variations (Table 6 supports this interpretation). Further, to compare with a full-range self-attention method, we look at the results obtained with the reluFormer. Although this approach solely relies on the sample at hand and has all sample-specific information, i.e., the full sequence length, the performance in comparison to ST only improves from $F_1 = 0.8284$ to $F_1 = 0.8313$. The reluFormer is missing the non-linearity of the softmax function in the self-attention operation, which could be an explanation for this result. To rule out that the performance increase of ST-FPS solely comes from using more inducing points, ≈ 150 compared to 16, we train the ST with 150 inducing points and denote this experiment as ST-150I. We can see however, that this worsens the results, meaning that the performance increase of ST-FPS does stain from sampling features vectors of the sample as inducing points. An explanation here could be that sampling the inducing points directly from the data reduces the parameters to be learned, which is beneficial in low data regimes, often the case in MRD detection of pediatric leukemia.

Finally, we can answer this question with *yes*, it matters how the global information is obtained, where self-attention that relies solely on the sample at hand and uses a non-linearity to calculate the attention matrices performs best.

Table 3. Results for the methods learning global features in a direct manner.

	Method	p	r	avg F_1	med F_1
single-cell feature vec. agg.	MLP-max	0.7855	0.8749	0.8015 ±0.0044	0.9179 ±0.0098
	MLP-mean	0.7887	0.8748	0.8041 ±0.0071	0.9203 ±0.0087
	MLP-pma	0.7912	0.8755	0.8058 ±0.0026	0.9228 ±0.0061
	PointNet	0.8027	0.8686	0.8117 ±0.0045	0.9155 ±0.0073
	PointNet-adapted	0.8191	0.8792	**0.83** ±0.0037	0.9437 ±0.006
self-attention	ST-150I	0.817	0.8597	0.8211 ±0.012	0.9346 ±0.008
	ST	0.8251	0.8601	0.8284 ±0.0117	0.9405 ±0.0085
	reluFormer	0.8298	0.867	0.8313 ±0.0059	0.9466 ±0.0029
	ST-FPS	0.8332	0.8636	**0.8369**±0.0076	0.9454 ±0.0063

4.3 Is it beneficial to explicitly learn local features?

To answer this question we look at local neighborhood aggregation methods and find that GNNs are a good fit for FCM samples. The graph for each sample, e.g., a sample's local neighborhoods, can be constructed by using the full sample-specific features (marker panels). Since marker panels can vary from sample to sample this is an easy way of incorporating sample-specific information through the structure of the spatial relations, which has to be completely dismissed in the other methods since all models assessed expect the same set of input features for each sample (see Section 4.4 and Table 7).

We look at three main GNN types, GCN, GAT, and GIN, where the latter outperforms GCN and GAT. The performance of GAT and GCN are similar with incrementally better median F_1 score of GAT. Hence we conduct all following experiments involving GNN with those two. All reach similar results to our baseline ST, with GIN even slightly outperforming. Although those methods do not directly learn global features, we assume that this is because with $k = 10$, we have connections between every cluster present in the FCM sample, and using 4 layers means that the receptive field of each node (cell) is 400. The local neighborhoods are thus highly overlapping and information can flow on a global scale. Results for GAT-ASAP, GIN-ASAP and only using $k = 3$ in the k-NN graph construction (GAT-3, GIN-3) support this assumption: adding explicit aggregation with ASAP has some benefit, yet not significantly, but capping global information flow by using a k-NN graph with $k = 3$, which results in disconnected clusters within the sample, impairs the results. Although the model cannot learn full-range dependencies, the results are competitive or outperform our baseline ST indicating that local feature learning is beneficial and that GNNs are suitable methods for FCM data.

Based on those results, an initial answer to the question stated is that at least compared to the performance of the current SOTA, explicitly modeling local spatial relationships leads to competitive or slightly better results in terms of F_1 score and hence modeling long-range dependencies as with self-attention is not strictly necessary to reach those results (Table 4).

Table 4. Results for methods explicitly learning local features based on GNNs.

Method	p	r	avg F_1	med F_1
GCN	0.7827	0.8598	0.8288 ±0.007	0.9405±0.0015
GAT	0.7941	0.8339	0.8255 ±0.0063	0.9458 ±0.0036
GIN	0.7902	0.8486	**0.8317** ±0.0082	0.9415 ±0.0045
GIN-3	0.8018	0.843	0.813 ±0.0149	0.9268 ±0.0138
GAT-3	0.8104	0.8504	0.8147 ±0.0068	0.9383 ±0.005
GAT-ASAP	0.7877	0.8481	0.8274 ±0.0086	0.9443 ±0.0017
GIN-ASAP	0.7994	0.8458	**0.8378** ±0.0217	0.9457 ±0.0050

4.4 GNN infused ST with FPS self-attention

Based on the findings above using methods based on either purely learning global or local features, we ask if we can benefit from combining those two strains, similarly to work in [34]. Table 5 shows that by using FPS based ISABs and explicitly learning local features by replacing one layer by a GNN layer, GAT or GIN, we reach new SOTA performance on the *Vie* dataset. Further, Table 6 shows that by introducing the above-mentioned adaptations, the inter-laboratory generalization ability improves significantly due to relying more on the sample at hand by using FPS and the sample's specific spatial local structures. In Figure 2 the network features of the last layer before the prediction head, which is a linear layer, are plotted using PCA with 2 components of each feature space. The features of ST-FPS, GIN, and the combined GIN-ST-FPS are compared to get insights into how the models might complement each other and create the performance increase. We can see that GIN compresses the cell types to tight clusters, yet struggles with cancer cells that closely adhere to healthy ones. We postulate that this can be traced back to the k-NN graph, where healthy cells might be also strongly connected to cancerous cells on the edge of the different cell population clusters. The ST-FPS model seems to have a smoother but yet rather blurry transition. GIN-ST-FPS, can make use of the spatial locality stored in the graph structure while also being able to balance it out by incorporating the context of other cell populations regardless of spacial proximity.

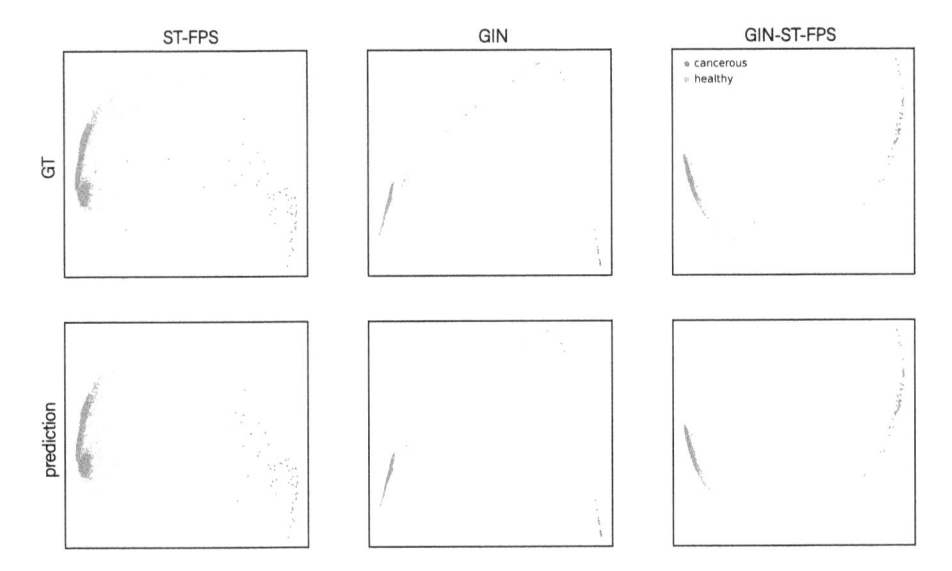

Fig. 2. Network features of the last layer of each model before the prediction head (a linear layer) are plotted using PCA with 2 components. Each model had the same FCM sample as input. Healthy cells are denoted in grey, and cancerous cells in red.

As stated in Section 4.3, in the case of varying marker panels between samples, modeling a FCM sample as a graph can be beneficial since the information of those sample-specific features can be indirectly incorporated via the graph structure. To analyze if the network is actually able to make use of this information, we remove 3 of the most important features for b-ALL detection, CD10, CD19, and CD45, from the input features but keep them for graph construction. Table 6 shows that while the performance drops for all methods, the GAT-ST-FPS and GIN-ST-FPS give the best results, confirming our assumption.

Table 5. Results for our proposed architectural changes.

Method	p	r	avg F_1	med F_1
ST	0.8251	0.8601	0.8284 ±0.0117	0.9405 ±0.0085
ST-FPS	0.8332	0.8636	0.8369 ±0.0076	0.9454 ±0.0063
GAT-ST-FPS	0.8242	0.8829	0.8465 ±0.0094	0.9529 ±0.0043
GIN-ST-FPS	0.8335	0.8775	**0.8665** ±0.0083	0.9561 ±0.0044

Table 6. Results testing generalization ability to datasets of other laboratories.

	Bln		Bue	
Method	avg F_1	med F_1	avg F_1	med F_1
ST	0.6089 ±0.0577	0.7225 ±0.0972	0.7274 ±0.0214	0.9246 ±0.0122
ST-FPS	**0.7265** ±0.0275	0.9052 ±0.0060	0.7939 ±0.0164	0.9445 ±0.0095
GAT-ST-FPS	0.7136 ±0.0587	0.8952 ±0.0488	0.7802 ±0.0375	0.9519 ±0.0156
GIN-ST-FPS	0.7046 ±0.0501	0.9064 ±0.0406	**0.8356** ±0.0111	0.9628 ±0.0025

Table 7. Results showing the model's capability of making use of implicit information stored in the graph structure. The 3 most important features for b-ALL detection, CD10, CD19, and CD45, were removed from the input features (node features) but kept for graph construction.

Method	avg F_1	med F_1
ST	0.3462 ±0.0170	0.1452 ±0.0565
ST-FPS	0.3668 ±0.0142	0.2023 ±0.0554
GAT-ST-FPS	0.4343 ±0.0086	0.3943 ±0.0285
GIN-ST-FPS	**0.4545** ±0.0113	0.4647 ±0.0303

Revisiting the question addressed in Section 4.3, we can now give a clear *yes* as an answer; explicitly learning local features with GNN layers, is beneficial, especially when combined with the long-range dependency modeling capabilities of self-attention.

5 Conclusion

This paper presents an evaluation of different DL methods for FCM data processing in the problem setting of automated MRD detection. Several methods divided into global and local feature learning methods are evaluated, and based on the findings, two adaptations to the current SOTA model are proposed. The evaluation shows that modeling long-range dependencies is indeed important for automated MRD detection, where self-attention based on FPS performs best. Further, methods based solely on local feature learning can reach similar performance and, in some cases, even outperform self-attention-based methods given overlapping local receptive fields. Using feature vectors sampled from the sample at hand by FPS instead of learned query vectors combined with introducing a local feature learning layer complement each other and result in a new SOTA performance and better inter-laboratory generalization abilities tested on publicly available datasets. While the DL methods evaluated for automated FCM processing in this paper cover a wide range of model types, it is by far not an exhaustive evaluation of possible architectures; extending this work by e.g. drawing inspiration from 3D point cloud processing for semantic segmentation is thus an interesting topic for future work.

References

1. Abdelaal, T., van Unen, V., Höllt, T., Koning, F., Reinders, M.J., Mahfouz, A.: Predicting cell populations in single cell mass cytometry data. Cytometry A **95**(7), 769–781 (2019)
2. Arvaniti, E., Claassen, M.: Sensitive detection of rare disease-associated cell subsets via representation learning. Nat. Commun. **8**(14825), 2041–1723 (2017)
3. Becht, E., McInnes, L., Healy, J., Dutertre, C.A., Kwok, I.W., Ng, L.G., Ginhoux, F., Newell, E.W.: Dimensionality reduction for visualizing single-cell data using umap. Nat. Biotechnol. **37**(1), 38–44 (2019)
4. Bruggner, R.V., Bodenmiller, B., Dill, D.L., Tibshirani, R.J., Nolan, G.P.: Automated identification of stratifying signatures in cellular subpopulations. Proc. Natl. Acad. Sci. **111**(26), E2770–E2777 (2014)
5. Campana, D.: Minimal residual disease in acute lymphoblastic leukemia. Hematology 2010, the American Society of Hematology Education Program Book **2010**(1), 7–12 (2010)
6. Cheung, M., Campbell, J.J., Whitby, L., Thomas, R.J., Braybrook, J., Petzing, J.: Current trends in flow cytometry automated data analysis software. Cytometry Part A pp. 1–15 (2021)
7. Dash, S., Acharya, B.R., Mittal, M., Abraham, A., Kelemen, A.: Deep learning techniques for biomedical and health informatics. Springer (2020)
8. Dworzak, M.N., Gaipa, G., Ratei, R., Veltroni, M., Schumich, A., Maglia, O., Karawajew, L., Benetello, A., Pötschger, U., Husak, Z., et al.: Standardization of flow cytometric minimal residual disease evaluation in acute lymphoblastic leukemia: Multicentric assessment is feasible. Cytometry Part B: Clinical Cytometry: The Journal of the International Society for Analytical Cytology **74**(6), 331–340 (2008)
9. Guo, Y., Wang, H., Hu, Q., Liu, H., Liu, L., Bennamoun, M.: Deep learning for 3d point clouds: A survey. IEEE Trans. Pattern Anal. Mach. Intell. **43**(12), 4338–4364 (2020)
10. Hu, Z., Bhattacharya, S., Butte, A.J.: Application of machine learning for cytometry data. Front. Immunol. **12**, 787574 (2022)

11. Kipf, T.N., Welling, M.: Semi-supervised classification with graph convolutional networks. arXiv preprint arXiv:1609.02907 (2016)
12. Kowarsch, F., Weijler, L., Wödlinger, M., Reiter, M., Maurer-Granofszky, M., Schumich, A., Sajaroff, E.O., Groeneveld-Krentz, S., Rossi, J.G., Karawajew, L., Ratei, R., Dworzak, M.N.: Towards self-explainable transformers for cell classification in flow cytometry data. In: Interpretability of Machine Intelligence in Medical Image Computing: 5th International Workshop, IMIMIC 2022, Held in Conjunction with MICCAI 2022, Singapore, Singapore, September 22, 2022, Proceedings. p. 22–32. Springer-Verlag, Berlin, Heidelberg (2022)
13. Lee, H.C., Kosoy, R., Becker, C., Kidd, B.: Automated cell type discovery and classification through knowledge transfer. Bioinformatics (Oxford, England) **33** (01 2017)
14. Lee, J., Lee, Y., Kim, J., Kosiorek, A., Choi, S., Teh, Y.W.: Set transformer: A framework for attention-based permutation-invariant neural networks. In: International Conference on Machine Learning. pp. 3744–3753. PMLR (2019)
15. Levine, J.H., Simonds, E.F., Bendall, S.C., Davis, K.L., Amir, E.a.D., Tadmor, M.D., Litvin, O., Fienberg, H.G., Jager, A., Zunder, E.R., Finck, R., Gedman, A.L., Radtke, I., Downing, J.R., Pe'er, D., Nolan, G.P.: Data-driven phenotypic dissection of aml reveals progenitor-like cells that correlate with prognosis. Cell **162**(1), 184—-197 (2015)
16. Li, H., Shaham, U., Stanton, K.P., Yao, Y., Montgomery, R.R., Kluger, Y.: Gating mass cytometry data by deep learning. Bioinformatics **33**(21), 3423–3430 (2017)
17. Licandro, R., Schlegl, T., Reiter, M., Diem, M., Dworzak, M., Schumich, A., Langs, G., Kampel, M.: Wgan latent space embeddings for blast identification in childhood acute myeloid leukaemia. In: 2018 24th International Conference on Pattern Recognition (ICPR). pp. 3868–3873. IEEE (2018)
18. McKinnon, K.M.: Flow cytometry: an overview. Curr. Protoc. Immunol. **120**(1), 5–1 (2018)
19. Ni, W., Hu, B., Zheng, C., Tong, Y., Wang, L., Li, Q.q., Tong, X., Han, Y.: Automated analysis of acute myeloid leukemia minimal residual disease using a support vector machine. Oncotarget **7**(44), 71915–71921 (2016)
20. Qi, C.R., Su, H., Mo, K., Guibas, L.J.: Pointnet: Deep learning on point sets for 3d classification and segmentation. In: Proceedings of the IEEE conference on computer vision and pattern recognition. pp. 652–660 (2017)
21. Qin, Z., Sun, W., Deng, H., Li, D., Wei, Y., Lv, B., Yan, J., Kong, L., Zhong, Y.: cosformer: Rethinking softmax in attention. arXiv preprint arXiv:2202.08791 (2022)
22. Ranjan, E., Sanyal, S., Talukdar, P.: Asap: Adaptive structure aware pooling for learning hierarchical graph representations. In: Proceedings of the AAAI conference on artificial intelligence. vol. 34, pp. 5470–5477 (2020)
23. Reiter, M., Diem, M., Schumich, A., Maurer-Granofszky, M., Karawajew, L., Rossi, J.G., Ratei, R., Groeneveld-Krentz, S., Sajaroff, E.O., Suhendra, S., et al.: Automated flow cytometric mrd assessment in childhood acute b-lymphoblastic leukemia using supervised machine learning. Cytometry A **95**(9), 966–975 (2019)
24. Rota, P., Kleber, F., Reiter, M., Groeneveld-Krentz, S., Kampel, M.: The role of machine learning in medical data analysis. a case study: Flow cytometry. In: VISIGRAPP (3: VISAPP). pp. 305–312 (2016)
25. Suffian, M., Montagna, S., Bogliolo, A., Ortolani, C., Papa, S., D'Atri, M., et al.: Machine learning for automated gating of flow cytometry data. In: CEUR WORKSHOP PROCEEDINGS. vol. 3307, pp. 47–56. Sun SITE Central Europe, RWTH Aachen University (2022)

26. Testi, A.M., Attarbaschi, A., Valsecchi, M.G., Möricke, A., Cario, G., Niggli, F., Silvestri, D., Bader, P., Kuhlen, M., Parasole, R., et al.: Outcome of adolescent patients with acute lymphoblastic leukaemia aged 10–14 years as compared with those aged 15–17 years: Long-term results of 1094 patients of the aieop-bfm all 2000 study. Eur. J. Cancer **122**, 61–71 (2019)
27. Vaswani, A., Shazeer, N., Parmar, N., Uszkoreit, J., Jones, L., Gomez, A.N., Kaiser, Ł., Polosukhin, I.: Attention is all you need. In: Advances in neural information processing systems. pp. 5998–6008 (2017)
28. Veličković, P., Cucurull, G., Casanova, A., Romero, A., Liò, P., Bengio, Y.: Graph attention networks. In: International Conference on Learning Representations (2018)
29. Weber, L.M., Nowicka, M., Soneson, C., Robinson, M.D.: diffcyt: Differential discovery in high-dimensional cytometry via high-resolution clustering. Communications Biology **2**(183), 2399–3642 (2019)
30. Weijler, L., Diem, M., Reiter, M., Maurer-Granofszky, M.: Detecting rare cell populations in flow cytometry data using umap. In: 2020 25th International Conference on Pattern Recognition (ICPR). pp. 4903–4909 (2021)
31. Weijler, L., Kowarsch, F., Reiter, M., Hermosilla, P., Maurer-Granofszky, M., Dworzak, M.: Fate: Feature-agnostic transformer-based encoder for learning generalized embedding spaces in flow cytometry data. In: Proceedings of the IEEE/CVF Winter Conference on Applications of Computer Vision. pp. 7956–7964 (2024)
32. Weijler, L., Kowarsch, F., Wödlinger, M., Reiter, M., Maurer-Granofszky, M., Schumich, A., Dworzak, M.N.: Umap based anomaly detection for minimal residual disease quantification within acute myeloid leukemia. Cancers **14**(4) (2022)
33. Wodlinger, M., Reiter, M., Weijler, L., Maurer-Granofszky, M., Schumich, A., Groeneveld-Krentz, S., Ratei, R., Karawajew, L., Sajaroff, E., Rossi, J., Dworzak, M.N.: Automated identification of cell populations in flow cytometry data with transformers. Computers in Biology and Medicine p. 105314 (2022)
34. Wu, Z., Jain, P., Wright, M., Mirhoseini, A., Gonzalez, J.E., Stoica, I.: Representing long-range context for graph neural networks with global attention. Adv. Neural. Inf. Process. Syst. **34**, 13266–13279 (2021)
35. Xu, K., Hu, W., Leskovec, J., Jegelka, S.: How powerful are graph neural networks? In: International Conference on Learning Representations (2019)

Self-supervised Siamese Network Using Vision Transformer for Depth Estimation in Endoscopic Surgeries

Snigdha Agarwal[1(✉)] and Neelam Sinha[2]

[1] International Institute of Information Technology, Bangalore, Bengaluru, India
snigdha.agarwal@iiitb.ac.in
[2] Indian Institute of Science, Bangalore, Bengaluru, India
neelamsinha@iisc.ac.in

Abstract. Depth estimation in real-world videos has been extensively researched, however, surgical videos pose unique challenges, such as specular reflections, multifaceted occlusions from tissues, fluid and surgical instruments. Accurately estimating depth from a 2D perspective amidst these conditions demands expertise, experience and cognitive effort. Implementing real-time depth estimation especially in endoscopic surgeries, could assist surgeons, leading to a decrease in post-operative complications. In this paper, we present a novel methodology for self-supervised monocular depth estimation using stereo pair images. The contributions of this study are, 1) a modified Siamese(Twin) network encoder-decoder architecture with a Gated Fusion DiffNet using Vision Transformers(DVTwin), 2) utility of a combination loss, fusing Structural Similarity Index Measure and L1 Loss, 3) utility of a random-masking technique. The ViT encoder leverages self-attention and captures global spatial information. The Gated Fusion DiffNet in the encoder calculates disparity at each stage of the network. The combination loss captures structural information from the stereo images while preserving sharp discontinuities at the edges. Training the model with random masking teaches it to learn depth from missing information enabling it to function efficiently with occlusions. We train and evaluate our model on 12 videos of the publicly available Hamlyn dataset. The videos comprise of challenging intra-corporeal scenes from endoscopic surgeries. On the holdout set, we report an Absolute Relative Error(AbsRel) of 0.084 and RMSE of 8.352, a 3% improvement from SOTA. To test for generalizability, we evaluate our model on the test set of SCARED dataset. We achieve an AbsRel of 0.067 and RMSE of 5.953, on par with SOTA, illustrating our model's generalizability. We fine-tune the model on the train set of SCARED dataset and evaluate on the test set to achieve an improvement of 16% in AbsRel and 18% in RMSE.

Keywords: monocular depth estimation · endoscopic surgery · self-supervised learning · vision transformer · siamese network · stereo endoscopy

1 Introduction

Depth estimation in endoscopic surgeries plays a critical role in the realm of minimally invasive procedures. These surgeries utilize small incisions and specialized instruments like endoscopes. An endoscope, equipped with a camera, provides surgeons with a visual guide inside the patient's body. This guide is inherently two-dimensional, lacking depth perception, which is a crucial element for precise surgical maneuvers [18]. Depth perception enables surgeons to accurately gauge the distance and spatial relationships between various anatomical structures avoiding damage to vital organs and ensuring surgical precision. Currently, surgeons rely heavily on their experience and tactile feedback to "fill in" the missing depth information. However, this approach has limitations and is subject to human error [12].

With the advancements in deep learning [13] and availability of large datasets of real-world images, we can develop systems capable of providing real-time, accurate depth information in everyday tasks. Convolutional Neural Networks (CNNs), Optical Flow [27], Vision Transformers (ViTs) [5] have emerged as powerful techniques that recognize visual cues indicative of depth in images. However, depth estimation in surgical videos presents unique challenges often featuring homogenous textures, limited color contrasts, variable lighting conditions, presence of fluids, occlusion due to surgical tools and specular reflections. Additionally, unavailability of large datasets with ground truth depth, the camera's limited field of view and the need for real-time processing further complicate the task.

In our work, we present a novel approach to mitigate these challenges of depth estimation in surgical videos using two large stereo endoscopy datasets, [24], [1]. We use a self-supervised stereo matching approach, wherein the disparity labels for each image are derived from the disparity between the captures of the left and right cameras. Our system is light-weight, enabling real-time depth estimation during an endoscopic procedure. At the inference stage, the camera feed provides the left image, while a subtly sheared version of this left image serves as the right image, facilitating the estimation of depth from a single-lens setup. To this end, our main contributions are as follows,

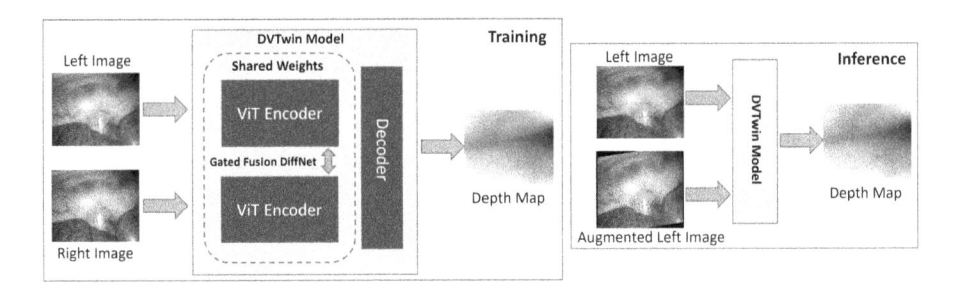

Fig. 1. Overall framework during training and inferencing

- a novel siamese network [3] based architecture using Vision Transformer (DVTwin) as encoder. The encoder extracts meaningful features capturing global context and long-range dependencies in understanding complex scenes. We design a Gated Fusion DiffNet to encode contributions from feature maps of the left and right images of stereo pairs to generate disparity maps.
- a meticulously selected combination loss optimizing for structural consistency and maintaining sharp transition at edges to train our model.
- we run thorough ablation experiments to evaluate our model and demonstrate the effectiveness of our proposed methods

The overall framework of our approach is presented in Fig. 1.

2 Related Work

Previous studies in dense depth estimation can be categorized into three distinct approaches: supervised, self-supervised, and unsupervised.

2.1 Supervised Depth Estimation

The initial research on depth estimation from monocular images, conducted in a supervised manner, was pioneered by [21], [22]. The authors introduce a model that uses a discriminatively-trained Markov Random Field (MRF). This model incorporates multi-scale local and global image features. In [6], the authors propose regression to estimate depth with multi-scale networks and a scale-invariant loss. In [14] and [9], the authors leverage deep convolutional neural networks with conditional random fields. Since, availability of large datasets with ground-truth depth is challenging due to requirement of specialized hardware, authors in [4] create synthetic colon data from CT scans and use adversarially-trained convolutional neural network for depth prediction.

2.2 Unsupervised Depth Estimation

One of the pioneering works in unsupervised depth estimation was proposed by [7]. The authors use stereo image pairs to train a convolutional encoder and use an inverse warp to reconstruct the source image from the target image. In [28], [17], [8], the authors use single-view depth and multi-view pose estimation networks trained on unlabeled video sequences. They introduce loss function enforcing consistency between the left-right image pairs. In [25], the authors propose an optical flow based CNN method using extracted warped features of the right image and features of the left image.

2.3 Self-supervised Depth Estimation

Self-supervised methods have achieved remarkable accuracy by formulating the problem as an image reconstruction task. In [26], the authors use a CNN based auto-encoder approach to predict disparity maps using a linear interpolation based spatial transformer. In [16] and [19] the authors propose a U-net based depth network and an auto-encoder based pose network to predict camera angle between frames. The authors also propose appearance based photometric loss function from which we take inspiration for our work. In [23], the authors propose an appearance flow based method to tackle brightness inconsistencies between images. The authors also show the capability of generalizing their model across domains. In [15], the authors propose an extension to [23] and predict camera intrinsic parameters using CNN. In [10], the authors propose a depth estimator utilizing the concealed 3D geometric structural information embedded within stereo pairs.

In this paper, we discuss our methodology that builds upon the self-supervised paradigm, taking advantage of the availability of stereo image pairs in endoscopic surgeries.

3 Methods

3.1 Problem Definition

For monocular depth estimation using a single RGB image of size $h \times w \times 3$, $I(x, y)$, our goal is to estimate a disparity map $d(x, y)$ of size $h \times w$ such that $d(x, y) = f(I(x, y))$ for each pixel (x, y) in the image. During training, we learn the function $f()$ from a pair of stereo images $(I_l, I_r) \in \mathbb{R}^{h \times w \times 3}$ and calculate a depth map $d \in \mathbb{R}^{h \times w}$.

3.2 Architecture

Our architecture is an expansion of the well-studied encoder-decoder framework, exemplified by models like the U-Net [20]. We propose a vision transformer based encoder-decoder architecture using siamese networks (DVTwin). We leverage the twin network system in the encoder of our architecture with weight sharing.

Encoder We introduce a patchify stem to convert the RGB stereo pair images into patches. The input to each branch of the encoder is a sequence of n patch embeddings, $x_e \in \mathbb{R}^{n \times p^2}$, of resolution (p, p) generated from the left and right stereo images, I_l and I_r, of size $x \in \mathbb{R}^{h \times w \times 3}$ respectively. We directly incorporate 4 Transformer Blocks as initially presented in [27]. The input x_e is linearly transformed to queries Q, keys K, and values V using different linear projections: $Q = x_e W^Q, K = x_e W^K, V = x_e W^V$, where $W^Q, W^K, W^V \in \mathbb{R}^{p \times p_k}$. This enables Multi-Headed Self Attention(MHSA) allowing the model to simultaneously attend to different parts of the input capturing a more diverse range of information. The output is then added to the original input and Layer normalized before being fed into a Feed-Forward Network. The output from each transformer block is subjected to down-sampling through a Maxpooling layer, defined as $F_l^i = Maxpool_{2 \times 2}(F_l^{i-1})$. This results in feature maps, F_l^i and F_r^i

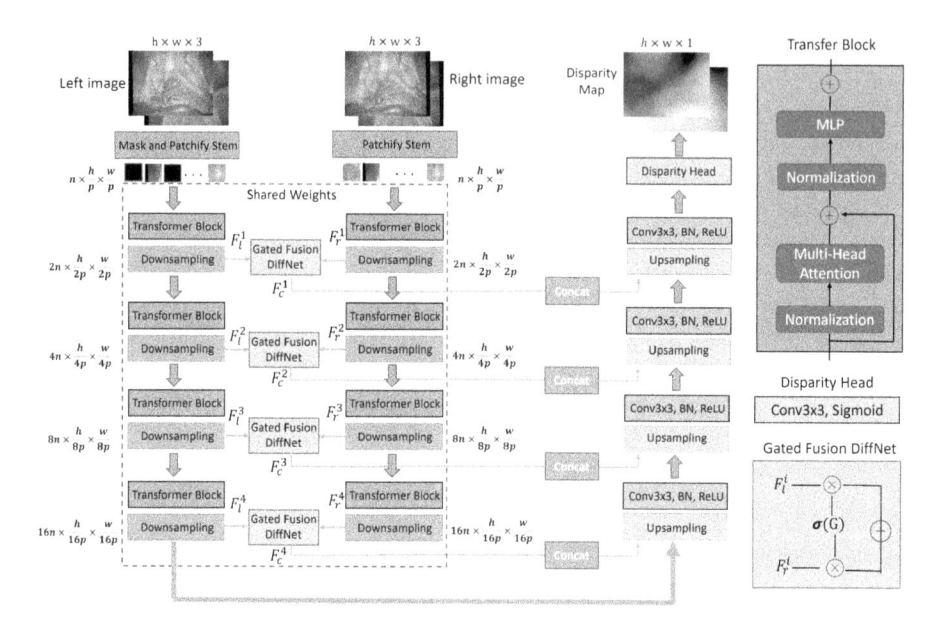

Fig. 2. Our proposed DVTwin - Siamese network using Vision transformer encoder-decoder architecture

for each respective branch, with i representing each stage. To encode disparity between the stereo image pairs, we introduce a Gated Fusion DiffNet between the feature maps from each branch at respective stage. This is defined as,

$$G_l = \sigma(W_l \cdot F_l + b_l)$$
$$G_r = \sigma(W_r \cdot F_r + b_r)$$
$$F'_l = G_l \odot F_l$$
$$F'_r = G_r \odot F_r$$
$$F_c = \text{Concat}(F'_l, F'_r)$$

This allows the network to adaptively learn how much influence each stereo image's features should have in the final feature map based on the content of the images. The output F_c is concatenated to the decoder at each stage to maintain information flow of disparity estimation from the encoder to the decoder preserving information that may be lost during down-sampling.

Decoder In the decoder, the down-sampled image from the left branch of the encoder is fed into a bilinear interpolation up-sampling layer, This is followed by a Convolution layer with a 3x3 kernel, Batch Normalization and ReLU activation function. Each stage of the encoder can be defined as,

$$z_l = Concat(F_l^i, F_c^i)$$
$$z_{l+1} = BilinearUpsample_{2\times 2}(z_l)$$
$$y = Conv_{3\times 3}(RLU\{(BN(z_l))\}) + z_{l+1}$$

Here, RLU is the ReLU activation function defined as, $RLU_{x_e} = max(0, x_e)$. BN is Batch Normalization as in [11]. Finally, a disparity head is added after the final stage to get the predicted depth maps, defined as $d_{pred} = \sigma(Conv_{3\times 3})$. The proposed architecture is presented in Fig. 2.

3.3 Loss Function

Our proposed architecture is trained using a fusion of loss functions, related to [8]. The total loss is calculated as,

$$y_L = y_{SCL} + y_{ESL} \tag{1}$$

To maintain structural integrity and feature similarity between the stereo pair images, we employ a combination of the Structural Similarity Index Measure (SSIM) and L1 Loss. This combination is referred to as the Structural Consistency Loss (SCL), defined as,

$$y_{SCL} = \alpha \frac{(2\mu_x\mu_y + c_1)(2\sigma_{xy} + c_2)}{(\mu_x^2 + \mu_y^2 + c_1)(\sigma_x^2 + \sigma_y^2 + c_2)} + \beta \frac{1}{N} \sum |x - y| \tag{2}$$

where, μ_x and μ_y are the average values of x and y, σ_x and σ_y are the variances and σ_{xy} is the covariance of x and y. The constants c_1 and c_2 are used to stabilize the division with weak denominator. α and β are hyperparameters.

To preserve edge information while encouraging smooth transitions in other regions of the depth map we incorporate Edge-aware Smoothness Loss (ESL). This is defined as,

$$L_{smooth} = \sum_{pixel} |\nabla d_{pixel}| e^{-\alpha |\nabla I_{pixel}|} \tag{3}$$

where, ∇d_{pixel} is the gradient of the depth map at a given pixel and ∇I_{pixel} is the gradient of the image intensity at the same pixel. The exponential term serves as a weighting function, reducing the smoothing effect at image edges where intensity gradients are high.

3.4 Evaluation Metrics

Consistent with the evaluation metrics employed in prior studies [8], [26], we adopt the same criteria to assess our model, facilitating comparative analysis. We use Absolute Relative Error (Abs Rel) defined as,

$$AbsoluteRelativeError = \frac{1}{N} \sum_{i=1}^{N} \left| \frac{d_{gt} - d_{pred}}{d_{gt}} \right| \tag{4}$$

and Root Mean Square Error (RMSE) defined as,

$$RMSE = \sqrt{\frac{1}{N}\sum_{i=1}^{N}(d_{gt} - d_{pred})^2} \qquad (5)$$

where, N is the total number of pixels, d_{gt} is groundtruth depth and d_{pred} is predicted depth value. For both evaluation metrics, lower values indicate better performance.

4 Experiments

4.1 Dataset

We train and evaluate our model on the Hamlyn Dataset [24] consisting of in vivo endoscopic videos from different surgical procedures with their ground-truth point cloud information. The images from this dataset are very challenging for depth estimation and vary in terms of lighting conditions, reflections, occlusions due to surgical tools. We use 12 videos from this dataset. 10 videos comprising of 65535 frames were used for training the model and 2 videos comprising of 6345 frames were used for evaluation. We report our performance metrics on these 2 videos.

We also use the Stereo Correspondence and Reconstruction of Endoscopic Data Challenge (SCARED) [1] dataset as a second validation dataset collected from the abdominal anatomy of porcine cadavers. This dataset consists of 35 endoscopic videos from 9 different subjects with their ground-truth point cloud and ego-motion information. 27 videos from 7 subjects comprise the train set with a total of 22232 images in the challenge and 8 videos from 2 subjects are used as the test set with a total of 5907 frames. We first report performance metrics on the test set from the model trained on the Hamlyn dataset using Transfer Learning. We, then, finetune this model on the train set of the SCARED dataset and report metrics with the finetuned model.

Table 1. Comparison with previous works on Hamlyn Dataset

Architecture	Abs Rel	RMSE
Endodepth [19]	0.119	10.590
DVTwin (Ours)	**0.084**	**8.352**

4.2 Pre-processing and Post-processing

The image size in the Hamlyn dataset is 640 × 480 pixels and 1280 × 1024 pixels in the SCARED dataset. In the Hamlyn dataset, we observe artifacts

Table 2. Comparison with previous works on SCARED Dataset

Architecture	Abs Rel	RMSE
AF-SfMLearner [15]	0.065	5.416
END-Flow [25]	0.071	7.550
M3 Depth [10]	0.116	9.274
DVTwin - Transfer Learned (Ours)	0.067	5.953
DVTwin - Fine-tuned (Ours)	**0.056**	**4.879**

Table 3. Comparison with SOTA architectures

	Hamlyn Dataset		SCARED Dataset	
Architecture	Abs Rel	RMSE	Abs Rel	RMSE
SfMLearner [28]	0.096	10.876	0.082	6.894
Monodepth [8]	0.113	9.130	0.079	6.520
Monodepth2 [16]	0.101	8.946	0.071	5.606
Ye et. al [26]	0.124	11.082	0.098	8.137
DVTwin (Ours)	**0.084**	**8.352**	**0.056**	**4.879**

on the left and right side of the images for the stereo pair images. We remove these artifacts using cropping. We resize all the input images to 224 × 224 to be fed into the model. Taking inspiration from the approach outlined in [2], we integrate random masking into our input image. This technique aids the model in accurately predicting depth in areas with occlusions, arising from both information gaps between stereo images and obstructions caused by surgical instruments. We use percentage of masking as a hyperparameter during model training. As is common methodology, we use hole filling as a post-processing step in the generated depth maps.

4.3 Training Strategy

We train our models on the p3.8xlarge GPU instances on Amazon Web Services using our proposed architecture and minimizing on our loss function. We leverage Hyperparameter Tuning Optimization Jobs with a maximum of 50 jobs and running 5 jobs in parallel. We run hyperparameter search for batch size (8, 12, 16), learning rate ranging from 0.000001 to 0.001, percentage of masking ranging from 10 to 50 with increments of 10, using a bayesian search strategy, α and β ranging from 0.1 to 0.99 using random search and We run each model for 60 epochs using AdamW optimizer with dropping learning rate by a factor of 10 every 5 epochs. We select the best performing model with the minimum loss and Absolute Relative Error by comparing performances of all the 50 models on our test dataset. Our best model uses a batch size of 8, with a learning rate of 0.000195 and random masking percentage of 20%. To enable monocular depth estimation during inference, we use synthetically generated right image from the input image. We use shearing transformation as a data augmentation strategy with a small shearing angle ranging from -10 degrees to +10 degrees. We report all our results using this methodology.

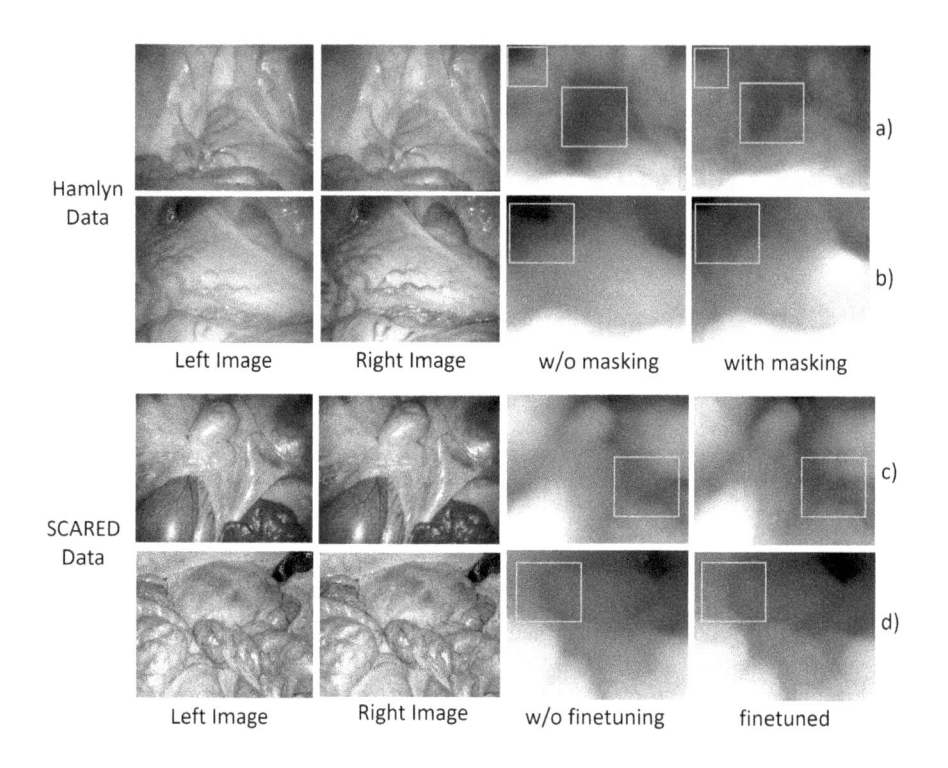

Fig. 3. Sample left and right images from the stereo pair images of the Hamlyn and SCARED dataset. We also show the outputs for Hamlyn dataset using random masking and without masking. For the SCARED dataset, we show the outputs with finetuning and without finetuning.

5 Results and Discussion

Sample outputs from our model are presented in Fig. 3. Right image in both the datasets is included as a reference and is not used during model inference. We evaluate the performance of our model using Abs Rel and RMSE. We compare the results of our approach with previous works on the Hamlyn dataset in Table 1. Given the continuous updates to the Hamlyn dataset, there is a scarcity of studies that serve as reliable benchmarks. To the best of our knowledge, the research presented in [19] is the most comparable to our training and test dataset, especially in terms of scene complexity and lighting variations.

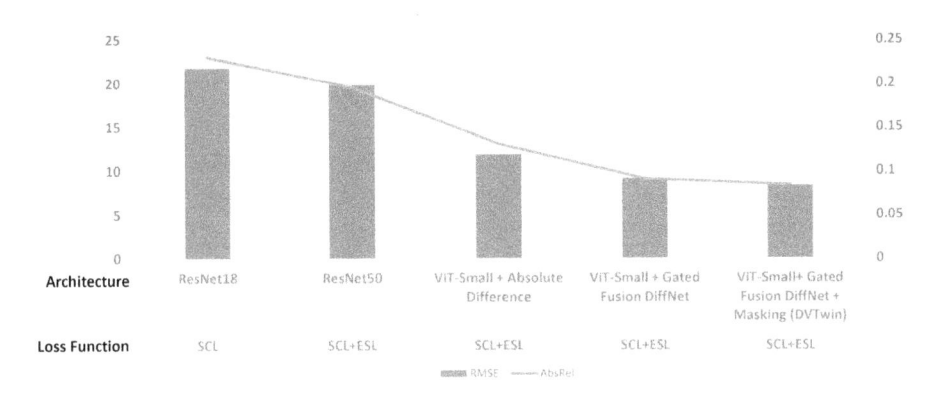

Fig. 4. Results of the Ablation Study. The primary axis corresponds to RMSE and the secondary axis corresponds to AbsRel. Both the metrics, if lower are better. We see that our incremental modifications significantly improve results.

We also compare the results of our model with the previous works on SCARED dataset in Table 2. We see that our methodology achieves improved results by a margin of 1% to 3%. We further compare our results with other state-of-the-art depth estimation architectures on both Hamlyn and SCARED datasets in Table 3. All the architectures for this analysis, are used off-the-shelf from authors' repositories without any modifications or are public implementations.

We run multiple ablation studies to illustrate the impact of our proposed methodology. As ResNet proved to be the backbone of choice in the previous works [8], [16], [23], we begin by employing the ResNet architecture in our encoder. We minimize for Stereo Consistency Loss and progressively introduce modifications. Should the evaluation metrics on the test dataset show a decrease, we proceed with these adjustments. To improve performance, we further introduce the Edge-aware smoothness loss as changes in tissue textures are very subtle. This loss reduces the risk of over-smoothing and missing out on edge details by allowing for sharp transitions at edges while promoting smoothness in less detailed regions. To improve feature extraction from the images, we introduce ViT in model encoder. ViTs have proven to be strong features extractors enabling global context awareness and maintaining long range dependencies between the different parts of the images. The MHSA enables parallelization with the multiple heads focusing on different features such as edges, textures, object sizes and maintaining context with the distant parts of the image for wholistic learning. To learn depth-related information from stereo disparity during training, we introduce absolute difference between the feature maps of both the branches of the encoder. We see that introduction of ViT and the differencing methodology significantly improves the metrics. Further, to adaptively control the contribution of each feature map of the stereo images through learned weights during model training, we introduce Gated Fusion DiffNet. This effectively reduces noise and surgical artifacts present in the images.

As shown in Fig. 3 a), we see that our model is not able to accurately estimate depth on the edges of the input image owing to occlusion from the missing information in the left and right images. In Fig. 3 b), we see that occlusion due to the presence of surgical tool also hinders accurate depth calculation. To solve for this, we introduce random masking while training to make the model better at inferring missing information. The results of our ablation study are present in Fig. 4. We use both Transfer learning and fine-tuning to test our model on the SCARED dataset.

6 Conclusion

We conclude with this study that our proposed method yields best results on the task of depth estimation. We attribute the promising performance of our approach to the following contributions : a) a novel siamese network approach incorporating the advantages of Vision Transformers, b) Gated Fusion DiffNet to improve the contribution in disparity calculation from both images, c) a carefully selected loss function to improve learning, d) random masking strategy to improve performance on occluded regions. Our evaluation on the SCARED dataset confirms generalizability of our model. However, as shown in Fig. 3 c), it is noticeable that the depth estimation along edges lacks sharpness and exhibits blurring. Fig. 3 d) shows that our model can achieve sharp edge transitions when fine-tuned on the SCARED dataset. The architecture of our model, being a ViT-small, comprises of 22M parameters. This is on par with the ResNet architecture, thus enabling real-time depth estimation for surgeons during minimally invasive endoscopic surgeries.

Acknowledgements. No funding was received for conducting this study. The authors have no relevant financial or non-financial interests to disclose.

References

1. Allan, M., Mcleod, J., Wang, C., Rosenthal, J.C., Hu, Z., Gard, N., Eisert, P., Fu, K.X., Zeffiro, T., Xia, W., et al.: Stereo correspondence and reconstruction of endoscopic data challenge. arXiv preprint arXiv:2101.01133 (2021)
2. Assran, M., Caron, M., Misra, I., Bojanowski, P., Bordes, F., Vincent, P., Joulin, A., Rabbat, M., Ballas, N.: Masked siamese networks for label-efficient learning. In: European Conference on Computer Vision. pp. 456–473. Springer (2022)
3. Bromley, J., Guyon, I., LeCun, Y., Säckinger, E., Shah, R.: Signature verification using a" siamese" time delay neural network. Advances in neural information processing systems **6** (1993)
4. Chen, R.J., Bobrow, T.L., Athey, T., Mahmood, F., Durr, N.J.: Slam endoscopy enhanced by adversarial depth prediction. arXiv preprint arXiv:1907.00283 (2019)

5. Dosovitskiy, A., Beyer, L., Kolesnikov, A., Weissenborn, D., Zhai, X., Unterthiner, T., Dehghani, M., Minderer, M., Heigold, G., Gelly, S., et al.: An image is worth 16x16 words: Transformers for image recognition at scale. arXiv preprint arXiv:2010.11929 (2020)
6. Eigen, D., Puhrsch, C., Fergus, R.: Depth map prediction from a single image using a multi-scale deep network. Advances in neural information processing systems **27** (2014)
7. Garg, R., B.G., V.K., Carneiro, G., Reid, I.: Unsupervised CNN for Single View Depth Estimation: Geometry to the Rescue. In: Leibe, B., Matas, J., Sebe, N., Welling, M. (eds.) ECCV 2016. LNCS, vol. 9912, pp. 740–756. Springer, Cham (2016). https://doi.org/10.1007/978-3-319-46484-8_45
8. Godard, C., Mac Aodha, O., Brostow, G.J.: Unsupervised monocular depth estimation with left-right consistency. In: Proceedings of the IEEE conference on computer vision and pattern recognition. pp. 270–279 (2017)
9. He, L., Wang, G., Hu, Z.: Learning depth from single images with deep neural network embedding focal length. IEEE Trans. Image Process. **27**(9), 4676–4689 (2018)
10. Huang, B., Zheng, J.Q., Nguyen, A., Xu, C., Gkouzionis, I., Vyas, K., Tuch, D., Giannarou, S., Elson, D.S.: Self-supervised depth estimation in laparoscopic image using 3d geometric consistency. In: International Conference on Medical Image Computing and Computer-Assisted Intervention. pp. 13–22. Springer (2022)
11. Ioffe, S., Szegedy, C.: Batch normalization: Accelerating deep network training by reducing internal covariate shift. In: International conference on machine learning. pp. 448–456. pmlr (2015)
12. Kavic, S.M., Basson, M.D.: Complications of endoscopy. Am. J. Surg. **181**(4), 319–332 (2001)
13. LeCun, Y., Bengio, Y., Hinton, G.: Deep learning. nature **521**(7553), 436–444 (2015)
14. Liu, F., Shen, C., Lin, G., Reid, I.: Learning depth from single monocular images using deep convolutional neural fields. IEEE Trans. Pattern Anal. Mach. Intell. **38**(10), 2024–2039 (2015)
15. Lou, A., Noble, J.: Ws-sfmlearner: Self-supervised monocular depth and ego-motion estimation on surgical videos with unknown camera parameters. arXiv preprint arXiv:2308.11776 (2023)
16. Mac Aodha, O., Firman, M., Brostow, G.J., et al.: Digging into self-supervised monocular depth estimation. In: 2019 IEEE/CVF International Conference on Computer Vision (ICCV)(2019). pp. 3827–3837 (2019)
17. Ozyoruk, K.B., Gokceler, G.I., Bobrow, T.L., Coskun, G., Incetan, K., Almalioglu, Y., Mahmood, F., Curto, E., Perdigoto, L., Oliveira, M., et al.: Endoslam dataset and an unsupervised monocular visual odometry and depth estimation approach for endoscopic videos. Med. Image Anal. **71**, 102058 (2021)
18. P. Breedveld, H. G. Stassen, D.W.M., Stassen, L.P.S.: Theoretical background and conceptual solution for depth perception and eye-hand coordination problems in laparoscopic surgery. Minimally Invasive Therapy & Allied Technologies **8**(4), 227–234 (1999). https://doi.org/10.3109/13645709909153166, https://doi.org/10.3109/13645709909153166
19. Recasens, D., Lamarca, J., Fácil, J.M., Montiel, J., Civera, J.: Endo-depth-and-motion: Reconstruction and tracking in endoscopic videos using depth networks and photometric constraints. IEEE Robotics and Automation Letters **6**(4), 7225–7232 (2021)

20. Ronneberger, O., Fischer, P., Brox, T.: U-net: Convolutional networks for biomedical image segmentation. In: Medical image computing and computer-assisted intervention–MICCAI 2015: 18th international conference, Munich, Germany, October 5-9, 2015, proceedings, part III 18. pp. 234–241. Springer (2015)
21. Saxena, A., Chung, S., Ng, A.: Learning depth from single monocular images. In: Weiss, Y., Schölkopf, B., Platt, J. (eds.) Advances in Neural Information Processing Systems. vol. 18. MIT Press (2005)
22. Saxena, A., Sun, M., Ng, A.Y.: Make3d: Learning 3d scene structure from a single still image. IEEE Trans. Pattern Anal. Mach. Intell. **31**(5), 824–840 (2008)
23. Shao, S., Pei, Z., Chen, W., Zhu, W., Wu, X., Sun, D., Zhang, B.: Self-supervised monocular depth and ego-motion estimation in endoscopy: Appearance flow to the rescue. Med. Image Anal. **77**, 102338 (2022)
24. Stoyanov, D., Scarzanella, M.V., Pratt, P., Yang, G.Z.: Real-time stereo reconstruction in robotically assisted minimally invasive surgery. Med Image Comput Comput Assist Interv **13**(Pt 1), 275–282 (2010)
25. Yang, Z., Simon, R., Li, Y., Linte, C.A.: Dense Depth Estimation from Stereo Endoscopy Videos Using Unsupervised Optical Flow Methods. In: Papież, B.W., Yaqub, M., Jiao, J., Namburete, A.I.L., Noble, J.A. (eds.) MIUA 2021. LNCS, vol. 12722, pp. 337–349. Springer, Cham (2021). https://doi.org/10.1007/978-3-030-80432-9_26
26. Ye, M., Johns, E., Handa, A., Zhang, L., Pratt, P., Yang, G.Z.: Self-supervised siamese learning on stereo image pairs for depth estimation in robotic surgery. arXiv preprint arXiv:1705.08260 (2017)
27. Zhai, M., Xiang, X., Lv, N., Kong, X.: Optical flow and scene flow estimation: A survey. Pattern Recognition **114**, 107861 (2021). https://doi.org/10.1016/j.patcog.2021.107861, https://www.sciencedirect.com/science/article/pii/S0031320321000480
28. Zhou, T., Brown, M., Snavely, N., Lowe, D.G.: Unsupervised learning of depth and ego-motion from video. In: Proceedings of the IEEE conference on computer vision and pattern recognition. pp. 1851–1858 (2017)

Enhanced 3D Dense U-Net with Two Independent Teachers for Infant Brain Image Segmentation

Afifa Khaled[1(✉)] and Ahmed Elazab[2]

[1] School of Computer Science and Technology, Huazhong University of Science and Technology, Wuhan 430074, China
I202055095@hust.edu.cn

[2] School of Biomedical Engineering, Shenzhen University Medical School, Shenzhen University, Shenzhen 518060, China

Abstract. Accurate segmentation of infant brain images from magnetic resonance imaging (MRI) scans is crucial for studying brain development. Existing deep learning methods often rely on encoder-decoder structures with local operators, limiting their ability to efficiently capture long-range information. Moreover, these models struggle to integrate diverse tissue properties from different MRI sequences, leading to computational and memory challenges during inference. To address these limitations, we propose a novel model, 3D-DenseUNet, which incorporates adaptable global aggregation blocks to mitigate spatial information loss during down-sampling. We enhance the model with a self-attention module that integrates feature maps across spatial and channel dimensions, improving representation potential and discrimination ability. Furthermore, we introduce a novel learning scheme, termed "two independent teachers," which leverages model weights instead of label predictions. In this scheme, each teacher model is trained on a specific MRI sequence (T1 and T2) to capture diverse tissue properties. A fusion model is then employed to enhance test accuracy while reducing computational overhead. Empirical evaluations on two distinct datasets demonstrate the effectiveness of our approach. Our source code is publicly available at https://github.com/AfifaKhaled/Two-Independent-Teachers-are-Better-Role-Model.

Keywords: Infant brain segmentation · U-Net · Self-attention mechanism · Teacher model · Fusion model

1 Introduction

Brain tissue segmentation from magnetic resonance images (MRI) is a crucial task in clinical practice for visualizing anatomical structures, understanding brain development, and monitoring disease progression, particularly in the infant brain. Typically, brain tissues are segmented into white matter (WM), gray matter (GM), and cerebrospinal fluid (CSF). However, manual segmentation of these tissues is subjective, labor-intensive, and time-consuming. Therefore, automating this task is highly imperative. Recently, deep learning techniques have revolutionized research in various tasks related to medical image analysis and have achieved state-of-the-art performance, including in brain tissue segmentation.

Various methods in the literature have leveraged convolutional neural networks (CNNs) [12], [19], [1] for brain segmentation, achieving notable performance. For instance, Dolz et al. [6] proposed HyperDenseNet, a 3D CNN model for multi-modal segmentation tasks. Bao and Chung [2] introduced a multi-scale structured CNN to capture discriminative features for each brain sub-cortical structure. Similarly, the widely used U-Net [14] architecture and its variants have demonstrated significant success in segmenting diverse medical images. To enhance the aggregation of global information, Wang et al. [18] proposed a non-local U-Net with an aggregate global block. Qamar et al. [13] introduced a new variant of U-Net designed to extract volumetric contextual information from MRI data. Hoang et al. [9] presented a dilated attention mechanism and attention loss network, incorporating skip block layers and atrous block convolution. Zheng et al. [20] introduced SCU-Net, an enhanced U-Net model with a serial encoding-decoding structure and hybrid dilated convolution. Furthermore, generative adversarial networks (GANs) [7] have been employed to address the limited availability of medical image datasets by synthesizing realistic images. For example, Arnab et al. [11] utilized GANs to augment the segmentation task with limited training samples. Similarly, Delannoy et al. [4] adopted GANs and introduced the SegSRGAN model, which incorporates a ResNet network to synthesize high-resolution images and improve segmentation accuracy.

Conversely, teacher models aim to leverage historical data to enhance segmentation accuracy. For instance, Cui et al. [3] proposed a semi-supervised learning approach utilizing the mean teacher model and a consistency loss function to ensure segmentation coherence within a self-ensembling framework. This method involves creating both student and teacher models with identical CNN architectures. Similarly, Wang et al. [16] introduced a novel approach involving foreground and background reconstruction tasks and a signed distance field prediction task to capture semantic information and shape constraints, respectively. They used a mean teacher architecture to explore the synergistic effect between these auxiliary tasks and segmentation tasks. To address potential bias in the teacher model due to annotation scarcity, they developed a tripled-uncertainty guided framework to improve the reliability of knowledge transfer to the student model.

Despite the remarkable performance of previous methods, some models face limitations in capturing global contextual information, particularly when dealing with subtle anatomical features, such as those in neonatal brain imaging. Moreover, the downsampling and upsampling operations in the contracting and expansive pathways, particularly in U-Net-based methods, can lead to resolution loss and reduced spatial detail in segmented images. Another issue arises from variations in textural information across different MRI sequences (e.g., T1 and T2), potentially impacting segmentation accuracy. In this work, we address the aforementioned limitations and challenges by introducing a novel 3D-DenseUNet model. This model aims to mitigate information loss and improve the utilization of overall structural relationships using multi-head attention mechanisms. Furthermore, we incorporate two independent teacher models for deep supervision, each focusing on a specific type of brain MRI data (T1 and T2). This enables us to leverage the unique characteristics of each MRI sequence to enhance segmentation accuracy. Additionally, we introduce a fusion module to integrate the outputs of the two teacher models, reducing the number of learning parameters throughout the network and ultimately enhancing segmentation performance. The main contributions of this paper are as follows:

– We propose a 3D-DenseUNet model with multi-head attention to effectively encode wide context information into local features and explore interdependencies between channel maps. Skip-connections between modules are also utilized to aggregate contextual information, improving performance compared to traditional concatenation-based approaches in the U-Net model.
– We establish two independent teacher models (2IT), each operating on a distinct MRI sequence. This approach enables efficient learning of deep feature information for each modality, reducing information uncertainty and boundary effects, and facilitating weight extraction.
– We construct a fusion model with weights updated from the two teacher modules, which facilitates in-depth analysis of both brain data (T1 and T2) while minimizing computational and memory requirements during inference and training.
– To avoid oscillation throughout the training stage of the fusion model, as well as overfitting and noise issues resulting from the summation of the teacher models' weights, we propose a function to calculate the fusing coefficient α value based on metric values of the model and the current epochs of some parameters.

2 Methodology

In this paper, we propose a 3D-DenseUNet model for segmenting 3D MRI images of infants' brains. The overall architecture of our model is shown in Fig. 1. Our model consists of a downsampling encoder, which contains three blocks that form a fully residual network within each block, and an upsampling decoder, which also contains three blocks forming a fully residual network, with skip connections between them serving as long-term residual connections. Additionally, we introduce a global attention block to fully utilize multi-scale contextual features. Moreover, our framework incorporates two independent teacher models and a fusion model, all utilizing the 3D-DenseUNet architecture. The details of each model are presented in the following subsections.

2.1 The Framework of 3D-DenseUNet Model

Down-Sampling and Up-Sampling In the proposed 3D-DenseUNet, the downsampling module contains three blocks, each forming a fully residual network. Each residual network employs three different convolutional layers, with batch normalization and ReLU6 activation functions applied before each convolutional layer. Specifically, the first block serves as the input layer with a $3 \times 3 \times 3$ convolution and a stride of 1, while the output block is similarly structured, followed by a $1 \times 1 \times 1$ convolution with a stride of 1. To address the first issue mentioned above, features are extracted at multiple

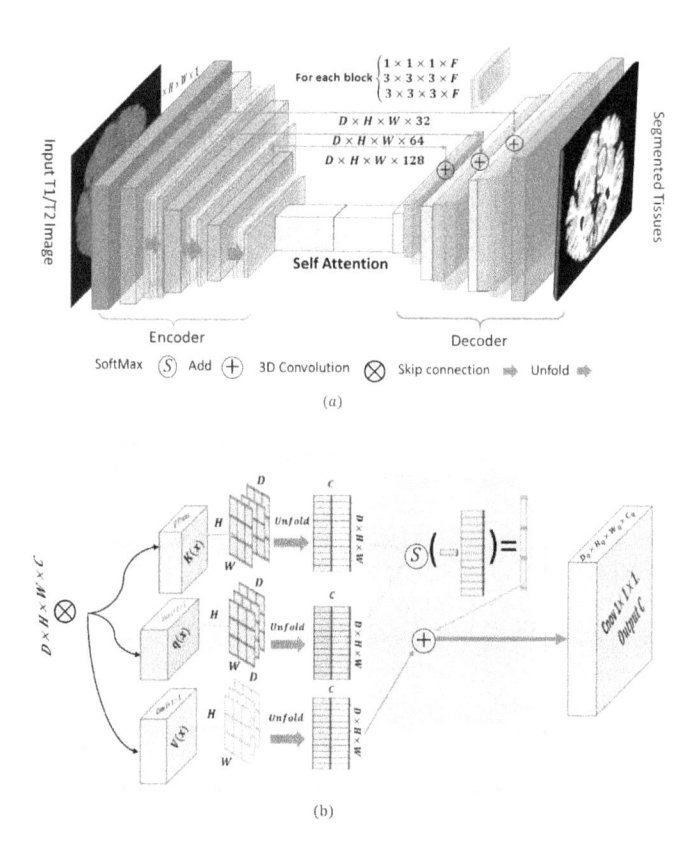

Fig. 1. Framework of the proposed model with (a) the 3D-DenseUNet architecture and (b) the global attention block featuring 3D layers based on multi-head attention as the self-attention.

scales using the first residual block after the sum of skip connections, thereby avoiding the loss of spatial information. In our architecture, we utilize summation instead of concatenation, as seen in the standard U-Net model, for several reasons. Firstly, summation effectively manages the increase in feature maps, thereby reducing the number of trainable parameters in subsequent layers. Secondly, employing summation with skip connections can be likened to long-term residual connections, consequently reducing the time required for model training. We incorporate three different convolutions to construct three-level features, which are then combined to form the final feature for each block in the downsampling process. These diverse features ensure that downsampling preserves various characteristics for the next level. Finally, an end block aggregates global information, leading to the encoder output. In our model's upsampling process, the $Q_{T}transform_{C}(\cdot)$ in the global attention block is formed by a $3 \times 3 \times 3$ deconvolution with a stride of 2 to avoid information loss and retain more accurate details. Finally, the output of the decoder block is the segmentation probability map, which produces probabilities for each segmentation class using a single $\arg\max$ operation.

Multi-head Self-attention Mechanism To fully utilize multi-scale contextual features, we gather global information from low-level layers and combine it with the high-level features from deeper layers to form the final feature representations in the end block of the encoder. However, a one-step sequence correlation might overlook valuable details during the large-scale upsampling process, leading to significant information loss. To address this, we propose an attention mechanism based on the multi-head self-attention function from the Transformer [15]. This mechanism, depicted in Fig. 1 (b), is applied after the encoder stage to incorporate global information from feature maps of any size. It facilitates the adaptive integration of information on local features and image edges, effectively treating the end block of the up/down-sampling blocks as a unified entity.

Given an image $X^{H \times W \times D \times C}$, where H, W, and D denote the image dimensions, and C represents the number of image channels, the end block of the downsampling stage accepts the latent cube representation X as input and computes it using multi-head attention with $1 \times 1 \times 1$ convolution and a stride of 1. Initially, the attention mechanism generates query (Q), key (K), and value (V) matrices using:

$$Q = Q\_Transform_{C_K}(X), K = Conv_{C_K}(X), V = Conv_{C_V}(X), \qquad (1)$$

where the operation $Q\_Transform_{C_K}(\cdot)$ produces C_K feature maps, and C_K and C_V are hyperparameters representing the dimensions of keys and values.

These matrices are decomposed into multiple heads in the second stage through parallel and independent computations. This allows simpler stacking of multiple transformer blocks as well as identity skip connections. Therefore, we unfold the $D \times H \times W \times C$ tensor into a $(D \times H \times W) \times C$ matrix using $Unfold(\cdot)$. Consequently, the V and K dimensions become $(H \times W \times D) \times C_V$ and $(H \times W \times D) \times C_K$, respectively.

Next, a scaled dot-product operation with SoftMax normalization between Q and the transposed version of K is conducted to generate the matrix of the contextual attention map A with dimensions $(D_Q \times H_Q \times W_Q) \times (D \times H \times W)$, which defines the similarities of the given features from Q concerning the global elements of K. To calculate the aggregation of values weighted by attention weights, A is multiplied by V, producing the output matrix O as follows:

$$O = Softmax\left(\frac{QK^T}{\sqrt{c_K}}\right)V, \qquad (2)$$

where $\sqrt{c_K}$ is the dimension of the query Q and K the key-value sequence. Finally, dropout is used to avoid overfitting, and Y reshapes the optimized feature maps to obtain the final output:

$$Y = Conv_{C_O}(Fold(O)), \qquad (3)$$

where $Fold(\cdot)$ is the reverse operation of $Unfold(\cdot)$ and C_O represents the output dimension. Ultimately, the size of the output Y is $D_Q \times H_Q \times W_Q \times C_O$.

At the end of an upsampling residual block, we use a $3 \times 3 \times 3$ deconvolution with a stride of 2 instead of an identity residual connection. This choice is made to incorporate global information, ensuring that the upsampling block can recover more accurate details.

2.2 Two-Independent-Teacher Model

We aim to develop two independent teacher models, referred to as TM1 and TM2, each specialized for different types of data (TM1 for T1-Weighted and TM2 for T2-Weighted images). To ensure their effectiveness, several adjustments are necessary. In the initial training step for each model, we employ the strategy proposed in [8] for weight initialization to facilitate the rapid convergence of the model structure based on the specific characteristics of the data.

2.3 The Fusion Model

We introduce a fusion model that leverages the weights of model layers from the teacher models throughout the training process. Instead of directly utilizing the final weights, we employ these weights during training to construct a more accurate model, as depicted in Eq. (4). The fusion model updates weights by combining the weights from the two independent teacher models at each iteration. Through empirical observation, we have identified two advantages of using multiple independent teachers. Firstly, the summation increases the feature maps of both T1 and T2, enhancing the model's capacity. Secondly, integrating weights from multiple models provides additional information, which has been shown to improve model performance [21]. Given the difficulty in acquiring labeled data, merging weights from multiple models offers a promising approach to enhancing segmentation performance without requiring additional labeled data. The key idea behind our proposed model is to update each parameter's weight in the fusion model by computing the summation of weights from the corresponding parameters in the independent teacher models, adjusted by a factor α to reduce noise and prevent overfitting.

$$W^L = \alpha W^L + (1-\alpha) \sum (W^{L_1}, W^{L_2}), \qquad (4)$$

where W^{L_1} is the weight from the first teacher model and W^{L_2} is the weight from the second teacher model. The parameter α, as defined in Eq. (5), is utilized to fine-tune the model weights at each iteration of the 3D-DenseUNet.

$$\alpha = \frac{(P+1)}{((1/L)+N)}, \qquad (5)$$

where P is the accuracy, L is the loss, and N is the batch size for each iteration. The α value, dynamically determined based on accuracy, loss values, and batch size, is crucial for performance. In the early training phase, when model accuracy and loss have not yet converged, α is set to small values to retain global information from the teacher models. Conversely, if α is too large, the model relies more on current weights, resulting in slow convergence and potential overfitting. Therefore, α should gradually approach 1. Experimentation shows that starting with $\alpha = 0.01$ and gradually increasing it during training yields better performance. Finally, the cross-entropy function serves as the cost function, quantifying the average discrepancy between the predicted output P and the ground truth T across the entire input domain $W \times H \times D \times C$. Denoting the network parameters as θ (including convolution weights, biases, and parameters from parametric rectifier units), and Y_s^v as the label of voxel V in the S-th image segment, we optimize the cross-entropy equation as follows:

$$J(\theta) = -\frac{1}{S \cdot V} \sum_{s=1}^{S} \sum_{v=1}^{V} \sum_{c=1}^{C} \delta(Y_s^v = C) \cdot \log P_c^v(X_S), \qquad (6)$$

where $P_c^v(X_s)$ is the SoftMax output of the network for voxel V and class C, and the input segment is X_s. The overall fusion process is presented in Algorithm 1.

Algorithm 1: The Fusion model algorithm.

---
1   **Input:** Initial model parameters P_0, dataset X, number of epochs N, data batch $D^i = \{x_i, y_i\}$, initial weights w.
2   **Output:** Prediction for segmenting MRI into a number of tissues GM, WM, and CSF.
3   **for** $n_i \in \mathcal{N}$ **do**
       $x_i \leftarrow X$
       if *train model is TM1* **then**
           for $j \in \mathcal{W}^i$ **do**
               $w^{L_1} \leftarrow w^{mij}$
           end
       end
       if *train model is TM1* **then**
           for $j \in \mathcal{W}^i$ **do**
               $w^{L_2} \leftarrow w^{ij}$
           end
       end
       if *train model is Fuse* **then**
           for $j \in \mathcal{W}^i$ **do**
               $Calculate\ \alpha$ by Eq. 5
               $Update\ W^{ij}$ by Eq. 4
           end
       end
       $loss = \mathcal{L}_{c\varepsilon}(x_i, y_i)$
4   **end**

Datasets To evaluate the performance of our proposed model, we conduct experiments on the $MICCAI\ iSEG$ dataset [17]. Upon analyzing this dataset, one can observe significant differences in the characteristics of the image data. The $MICCAI\ iSEG$ dataset contains a total of 10 training images, labeled *T1-1* through *T1-10*, and *T2-1* through *T2-10*, along with corresponding ground truth labels. The test set includes 13 images, labeled *T-11* through *T-23*. Fig. 2 shows an example from the $MICCAI\ iSEG$ dataset. The parameters used to create the $T1$ and $T2$ images are listed in Table 1.

The $MRBrainS$ [1] dataset is an adult dataset containing only 20 subjects. In this paper, $T1$- and $T2$-fluid-attenuated inversion recovery ($FLAIR$) images are used for segmentation. In $MRBrainS$, the number of training images is 5 (i.e., 2 male and 3 female), while the number of testing images is 15. The dataset includes segmentation

---
[1] https://mrbrains13.isi.uu.nl/

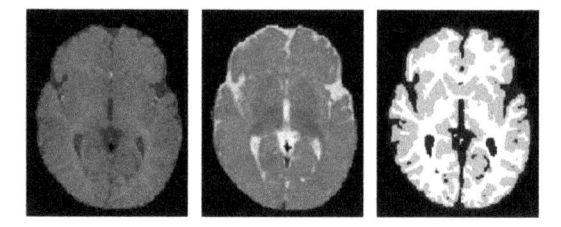

Fig. 2. An example of the $MICCAI\ iSEG$ dataset. Left: $T1$, middle: $T2$, right: manual reference contour.

Table 1. Parameters used to generate $T1$ and $T2$.

Data	TR/TE	Flip angle	Resolution
$T1$	1,900/4.38 ms	7	$1\times1\times1$
$T2$	7,380/119 ms	150	$1.25\times1.25\times1.25$

of the following 8 tissue types: (a) peripheral cerebrospinal fluid, (b) basal ganglia, (c) cerebellum, (d) white matter lesions, (e) brainstem, (f) lateral ventricles, (g) white matter, and (h) cortical gray matter.

2.4 Experimental Results

The proposed model is extensively assessed on the $3D$ multi-modal isointense infant brain tissues in $T1$- and $T2$-weighted brain MRI scans task. The task involves automatically segmenting MRI images into the cerebrospinal fluid (CSF), gray matter (GM), and white matter (WM) regions. In our experiment, the $MICCAI\ iSEG$ dataset is used. For each tissue type, we first replicate the model as independent teacher models. Each model was trained on a specific data type (the first teacher model trained on $T1$ and the second on $T2$), with one sample as a validation set and the remaining nine as training sets.

At each iteration, 2000 patches of $32\times32\times32$ are randomly selected as the training and validation datasets and processed in a batch size of 8. In the fusion model process, we use patches of the same size as those employed in both teacher models, taken from the original image with a fixed overlapping step size. The size of the overlapping step should not be greater than the patch size. Thus, the fusion model, which benefits from the teacher models' weights, accurately segments tissues and refines tissue probability maps for every voxel in the original image for GM, WM, and CSF. The proposed model was trained for 5000 epochs on the training and validation datasets.

Fig. 3 plots the proposed model's accuracy and loss during the validation stage at different iterations. This figure demonstrates that the teacher models exhibited poor accuracy and high loss in the early training phase. In contrast, the fusion model produced better results in the early training phase. As training and validation continued, any improvement in the teacher models corresponded to an improvement in the fusion model.

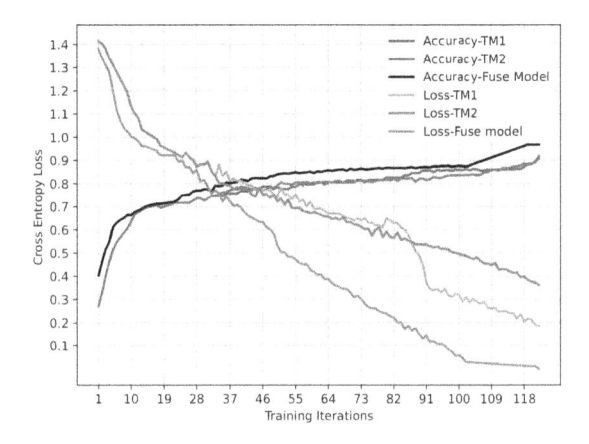

Fig. 3. The accuracy and loss validation of the proposed model on $MICCAI$ $iSEG$ dataset.

The Dice Coefficient (DC) is used as the evaluation metric to assess the accuracy of the fusion model for the validation subject. The DC measures the overlap level between the segmentation space and the ground truth label. As shown in Table 4, the results of the proposed model were compared to advanced state-of-the-art deep learning models in terms of metrics for the segmentation of CSF, WM, and GM brain tissues. Higher DC values indicate a greater overlap between manual and automatic segmentation boundaries. As a result, our model has values equivalent to some models in certain cases and outperforms others. Particularly, our model yields the best DC values for CSF and GM and ranks second for WM.

Similarly, the 13 unlabeled images were used as the test set during the testing phase. The proposed model outperforms state-of-the-art models in segmenting GM, WM, and CSF, as shown in Table 2. The results demonstrate excellent performance compared to other methods. Specifically, the proposed model produces the best DC values for GM, equal values for CSF, and slightly lower values for WM brain tissues.

Furthermore, Fig. 4 illustrates the accuracy metric values of the proposed model for the 13 subjects in the test set. As the figure shows, the DC values exhibit some variance across different subjects. This can be attributed to the fact that providing additional training information through fused weights from the two teacher models enhances the segmentation results, significantly improving the distinction between the edges of white and gray matter.

In addition, we test our model on the $MRBrainS$ dataset to further verify its efficiency. Table 3 shows the performance of our proposed model on the test set compared with state-of-the-art models in terms of the DC metric for the segmentation of CSF, WM, and GM brain tissues. Our model produces strong DC values compared to other models, achieving the top results for CSF and GM and slightly lower results for WM.

To accurately evaluate efficiency and effectiveness, training time was taken as a metric, as shown in Table 5, which presents the average training time (in hours) and standard deviation (SD). To evaluate this, we compare our model with related works

Table 2. Segmentation performance in Dice Coefficient (DC) obtained on the $MICCAI$ dataset for the 13 unlabeled images used as the test set. The best performance for each tissue class is highlighted in **bold**.

Model	Dice Coefficient (DC)
	CSF GM WM
Wang et al. [18]	0.95 0.92 **0.91**
Hoang et al. [9]	0.95 0.91 **0.91**
Dolz et al. [6]	**0.96** 0.92 0.90
Qamar et al. [13]	**0.96** 0.92 **0.91**
First Teacher Model	0.82 0.80 0.89
Second Teacher Model	0.88 0.90 0.81
Fusion Model	**0.96 0.93** 0.90

Table 3. Segmentation performance (Dice Coefficient, DC) on the $MRBrainS$ dataset. The best performance for each tissue class is highlighted in **bold**.

Model	Dice Coefficient (DC)
	CSF GM WM
Mahbod et al. [10]	0.83 0.85 0.88
Dolz et al. [6]	0.83 0.86 **0.89**
First Teacher Model	0.39 0.40 0.50
Second Teacher Model	0.43 0.60 0.44
Fusion Model	**0.86 0.87** 0.87

that reported training time. As seen in the table, the average execution time of the proposed framework, which includes the summation of the two teacher models and the fusion model, is lower than that of HyperDenseNet [6]. Conversely, it is higher than the model in [13], despite our framework being composed of three independent models. This suggests that our proposed framework's architecture has fewer learned parameters.

Accordingly, we investigate the number of parameters used in our model. As shown in Table 6, our model has 20% fewer parameters than the state-of-the-art models. The architecture of our proposed model is much deeper than all state-of-the-art models because it is composed of three independent models, which consist of 82 layers with 2.12 million learned parameters. Therefore, the proposed architecture is deeper compared to other existing approaches and achieves 96% accuracy. In terms of the number of parameters, the model in [18] is comparable to ours.

Moreover, Fig 5 shows the visualization results of the proposed model on the image used as the validation set. We can observe that the results achieved by the proposed model are fairly close to the ground truth. Specifically, the segmentation accuracy is high at the boundaries.

Overall, our results match those of some models and surpass others in terms of image segmentation accuracy. Despite our model's architecture being much deeper

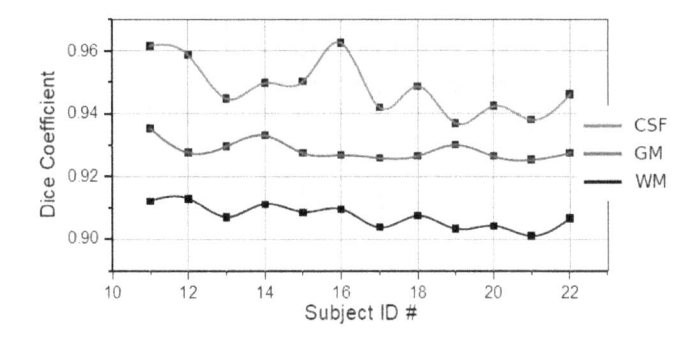

Fig. 4. Performance of the proposed model on 13 different subjects from the $MICCAI\ iSEG$ dataset as the test set.

Table 4. Segmentation performance in Dice Coefficient (DC) obtained on the $MICCAI$ dataset for the image used as the validation set. The best performance for each tissue class is highlighted in **bold**.

Model	Dice Coefficient (DC)
	CSF GM WM
Wang et al. [18]	0.95 **0.92 0.91**
Hoang et al. [9]	0.95 0.91 **0.91**
Dolz et al. [5]	**0.96** 0.92 0.90
Qamar et al. [13]	**0.96 0.92 0.91**
First Teacher Model	0.80 0.79 0.90
Second Teacher Model	0.89 0.82 0.81
Fusion Model	**0.96 0.92** 0.90

than state-of-the-art approaches, it contains fewer parameters and requires less execution time, leading to improved performance comparable to that of the state-of-the-art approaches.

2.5 Ablation Study

To evaluate the effectiveness of the component blocks used in our proposed model with the fusion method, we conduct ablation studies on the $MICCAI$ training and testing sets. We apply different α values to determine what best suits our model. Two models, one with α from Eq. 5 and the other with constant values for α, are constructed to establish a suitable procedure. These models are trained for 5000 epochs with a batch size of 32 and are supervised by the proposed combination loss function. Additionally, the same number of down-sampling and up-sampling blocks are used. The lowest value of α starts at 0.1, with multiple runs adjusting α to 0.2, 0.3, and so on, up to 0.9.

We can observe from the results that the model improves in some areas, while it fluctuates in terms of loss and accuracy values in others, particularly at the minimum

Table 5. Average execution time (in hours) and standard deviation on the $MICCAI$ dataset.

Model	Time (h)
Dolz et al. [6]	105.67 ± 14.7
Saqib et al. [13]	38.00 ± n/a
First Teacher Model	**36.45 ± 0.12**
Second Teacher Model	**36.45 ± 0.12**
Fusion Model	**36.45 ±0.12**

Table 6. Comparison of the number of parameters.

Model	# Parameters
Wang et al. [18]	2,534,276
Dolz et al. [6]	10,349,450
First Teacher Model	**2,331,160**
Second Teacher Model	**2,331,160**
Fusion Model	**2,331,160**

and maximum values. This is why we use Eq. 5, in that Eq. 5 as it leads to better performance. Table 7 and Fig. 6 present comparisons from the ablation studies on the validation set, focusing on accuracy and loss metrics. In the early stages of training, the model's accuracy and loss did not fit well, and α had very small values. In contrast, when α is much larger, it results in slow convergence and tends to overfit, leading to low accuracy. Therefore, α values are determined according to Eq. 5 to mitigate fluctuations in the early stages and to address the overfitting issue throughout the training process.

Additionally, from the perspective of the Dice metric in Table 7, it can be observed that adopting the fusion method with α from Eq. 5 is an effective approach to improving the performance of brain MRI segmentation. These experimental results demonstrate that the α hyper-parameter enhances the performance of the proposed model components for brain MRI segmentation.

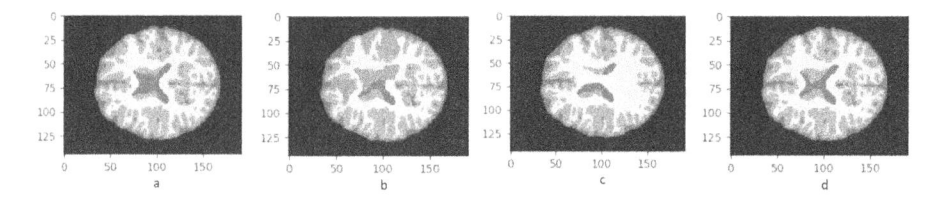

Fig. 5. Comparison of the results on the 150th slice in the $MICCAI\ iSEG$ dataset with (a) Ground Truth, (b) First Teacher Model, (c) Second Teacher Model, and (d) Fusion Model.

Table 7. Segmentation performance in Dice Coefficient (DC) for the $MICCAI\ iSEG$ dataset. The best performance for each tissue class is highlighted in **bold**.

Fusion model with different α	Dice Coefficient (DC) Accuracy CSF GM WM
0.1	0.78 0.74 0.70
0.2	0.79 0.75 0.71
0.3	0.79 0.75 0.71
0.4	0.83 0.79 0.76
0.5	0.83 0.79 0.76
0.6	0.87 0.83 0.80
0.7	0.80 0.75 0.71
0.8	0.81 0.77 0.70
0.9	0.79 0.74 0.70
Defined in Eq. 5	**0.96 0.92 0.90**

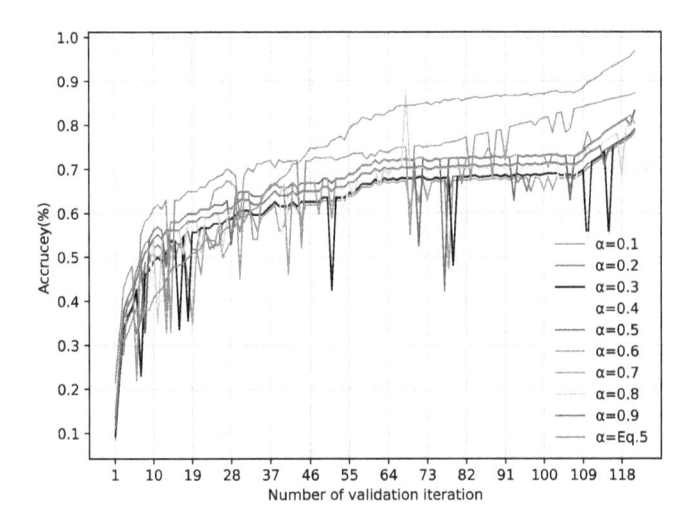

Fig. 6. The accuracy of the validation dataset on $MICCAI\ iSEG$ with various settings of α.

3 Conclusion

In this paper, we introduced 3D-DenseUNet, a novel deep-learning model designed to address spatial information loss during down-sampling. By incorporating adaptable global aggregation blocks and a self-attention module, we enhanced the model's ability to integrate feature maps across spatial and channel dimensions, thereby improving representation and discrimination. Additionally, we developed two independent teacher models trained on different sequences of infant brain MRI scans and integrated a fusion model to enhance test accuracy while reducing the number of parameters and labels.

Our empirical results demonstrate the effectiveness of the proposed model in brain image segmentation, achieving comparable performance with fewer parameters and shorter implementation time. Future work could include an ablation study on the summation operation that replaces the concatenation operation. Moreover, we will investigate the impact of deactivating the multi-head self-attention mechanism to assess the specific improvements brought by this component compared to the rest of the architecture.

4 Declarations

4.1 Competing interests

The authors declare that they have no competing interests.

4.2 Availability of data and materials

The data that supports the findings of this study is available at MICCAI Grand challenge on 6-month infant brain MRI seg-mentation (http://iseg2017.web.unc.edu) and MRBrains (https://mrbrains13.isi.uu.nl/results.php) and are both publicly available.

4.3 Ethics approval and consent to participate

Not applicable

4.4 Consent for publication

Not applicable

4.5 Ethics Approval and Consent to Participate

Huazhong University of Science and Technology the ethics committee that approved our study and the committee's reference number is 430074.

References

1. Bansal, P., Singh, S.P., Gopal, K.: Human brain mri segmentation approaches and challenges: A review. In: Jain, S., Marriwala, N., Tripathi, C.C., Kumar, D. (eds.) Emergent Converging Technologies and Biomedical Systems, pp. 1–8. Springer Nature Singapore, Singapore (2023)
2. Bao, S., Chung, A.C.S.: Multi-scale structured cnn with label consistency for brain mr image segmentation. Computer Methods in Biomechanics and Biomedical Engineering: Imaging & Visualization **6**, 113–117 (2018)
3. Cui, W., Liu, Y., Li, Y., Guo, M., Li, Y., Li, X., Wang, T., Zeng, X., Ye, C.: Semi-supervised brain lesion segmentation with an adapted mean teacher model. In: Information Processing in Medical Imaging: 26th International Conference, IPMI 2019, Hong Kong, China, June 2–7, 2019, Proceedings 26. pp. 554–565 (2019)

4. Delannoy, Q., Pham, C.H., Cazorla, C., Tor-Díez, C., Dollé, G., Meunier, H., Bednarek, N., Fablet, R., Passat, N., Rousseau, F.: Segsrgan: Super-resolution and segmentation using generative adversarial networks - application to neonatal brain mri. Comput. Biol. Med. **120**, 103755 (2020)
5. Dolz, J., Desrosiers, C., Wang, L., Yuan, J., Shen, D., Ayed, I.B.: Deep CNN ensembles and suggestive annotations for infant brain MRI segmentation. Comput. Med. Imaging Graph. **79**, 101660 (2020)
6. Dolz, J., Gopinath, K., Yuan, J., Lombaert, H., Desrosiers, C., ben Ayed, I.: Hyperdense-ne: A hyper-densely connected CNN for multi-modal image segmentation. IEEE Transactions on Medical Imaging **38**, 1116–1126 (2018)
7. Goodfellow, I.J., Pouget-Abadie, J., Mirza, M., Xu, B., Warde-Farley, D., Ozair, S., Courville, A., Bengio, Y.: Generative adversarial networks (2014)
8. He, K., Zhang, X., Ren, S., Sun, J.: Delving deep into rectifiers: Surpassing human-level performance on ImageNet classification (2015)
9. Hoang, D.H., Diep, G.H., Tran, M.T., Le, N.T.H.: DAM-AL. In: Proceedings of the 37th ACM/SIGAPP Symposium on Applied Computing (April 2022)
10. Mahbod, A., Chowdhury, M., orjan Smedby, Wang, C.: Automatic brain segmentation using artificial neural networks with shape context. Pattern Recognition Letters **101**, 74–79 (2018)
11. Mondal, A.K., Dolz, J., Desrosiers, C.: Few-shot 3d multi-modal medical image segmentation using generative adversarial learning (2018)
12. O'Shea, K., Nash, R.: An introduction to convolutional neural networks (2015)
13. Qamar, S., Jin, H., Zheng, R., Ahmad, P., Usama, M.: A variant form of 3D-UNet for infant brain segmentation. Futur. Gener. Comput. Syst. **108**, 613–623 (2020)
14. Ronneberger, O., Fischer, P., Brox, T.: U-Net: Convolutional networks for biomedical image segmentation. In: Navab, N., Hornegger, J., Wells, W.M., Frangi, A.F. (eds.) Medical Image Computing and Computer-Assisted Intervention – MICCAI 2015. pp. 234–241 (2015)
15. Vaswani, A., Shazeer, N., Parmar, N., Uszkoreit, J., Jones, L., Gomez, A.N., Kaiser, Ł., Polosukhin, I.: Attention is all you need. Advances in Neural Information Processing Systems **30** (2017)
16. Wang, K., Zhan, B., Zu, C., Wu, X., Zhou, J., Zhou, L., Wang, Y.: Semi-supervised medical image segmentation via a tripled-uncertainty guided mean teacher model with contrastive learning. Med. Image Anal. **79**, 102447 (2022)
17. Wang, L., Nie, D., Li, G., Puybareau, É., Dolz, J., Zhang, Q., Wang, F., Xia, J., Wu, Z., Chen, J.W., et al.: Benchmark on automatic six-month-old infant brain segmentation algorithms: the iseg-2017 challenge. IEEE Trans. Med. Imaging **38**(9), 2219–2230 (2019)
18. Wang, Z., Zou, N., Shen, D., Ji, S.: Non-local U-nets for biomedical image segmentation. In: Proceedings of the AAAI Conference on Artificial Intelligence. vol. 34, pp. 6315–6322 (2020)
19. Yang, S., Li, X., Mei, J., Chen, J., Xie, C., Zhou, Y.: 3d-transunet for brain metastases segmentation in the brats2023 challenge (2024)
20. Zheng, P., Zhu, X., Guo, W.: Brain tumour segmentation based on an improved u-net. BMC Med. Imaging **22**(1), 1–9 (2022)
21. Zhou, Y., Wang, Y., Tang, P., Bai, S., Shen, W., Fishman, E., Yuille, A.: Semi-supervised 3D abdominal multi-organ segmentation via deep multi-planar co-training. In: 2019 IEEE Winter Conference on Applications of Computer Vision (WACV). pp. 121–140 (2019)

IDQCE: Instance Discrimination Learning Through Quantized Contextual Embeddings for Medical Images

Azad Singh[✉] and Deepak Mishra

Indian Institute of Technology, Jodhpur, Jodhpur, India
{singh.63,dmishra}@iitj.ac.in

Abstract. Self-supervised pre-training is effective in learning discriminative features from unlabeled medical images. However, typical self-supervised models lead to sub-optimal representations due to negligence of high anatomical similarity present in the medical images. This affects the negative and positive pairs in discriminative self-supervised models to learn view-invariant representations. Various methods are proposed to address this issue. However, many of them either concentrate on preserving pixel-level details or offer solutions for specific modalities. In this context, we propose a generalized solution to leverage the anatomical similarities while relaxing the requirements of complex pixel-preservation learning. Specifically, we introduce IDQCE: Instance Discrimination Learning through Quantized Contextual Embeddings. The proposed approach leverages the sparse discrete contextual information to guide the self-supervised framework to learn more informative representations for medical images. We evaluate the representations learned by IDQCE through comprehensive experiments and observe more than 3% performance gain under linear evaluation protocol over other SOTA approaches in multiple downstream tasks.

Keywords: Self-supervised Learning · Codebook · VQ-VAE · X-ray · Fundus

1 Introduction

Recent advancements in self-supervised learning (SSL) alleviate the need for manual annotations. By exposing the models to diverse data augmentations and self-imposed tasks, SSL encourages learning generalized transferable features [12]. Despite the advantages, there are limitations, mainly when using instance discriminative approaches such as contrastive and non-contrastive methods. These methods require the creation of positive (related instances) and negative (unrelated instances) pairs from augmented versions of the input images, which pose significant challenges in the context of medical imaging. Though methods like SimCLR [3], MoCo [10], BYOL [8], and VICReg [2] have shown success in medical images (e.g., Chest Radiographs and Fundus images), however, they learn sub-optimal visual representations as due to oversight of significant anatomical similarities. The inherent anatomical similarities can lead to uncertainty in defining genuinely dissimilar pairs, which forces an incorrect constraint on capturing instance-invariant features [7,26]. Moreover, these methods rely solely on augmented versions of the same image for positive pairs and ignore the semantically similar information in other images.

Recently, several approaches [26,29,37,38] addressed this by incorporating local contextual features within contrastive learning or leveraging metadata to create more informed positive pairs. However, these methods overlook the inherent anatomical similarities and focus on pixel-level restoration in their learning objectives. More recent techniques such as AWCL [7] and Alice [13] explicitly consider domain-specific anatomical features in 3D volumes and ultrasound scans to create positive and negative pairs and pixel-level restoration. Nevertheless, these methods showed improvements; however, they require the domain knowledge of a specific modality.

In response to the mentioned constraints, we introduce the SSL framework, denoted as **IDQCE**: Instance Discrimination Learning through Quantized Contextual Embeddings. We employ a codebook-based vector quantization process to represent the data in a discrete, quantized format. The quantization implicitly capitalizes on anatomical overlaps within intra and inter-medical images, so eliminates the need for modality-specific knowledge. In medical images like chest X-rays, the quantization process results in discrete codebook vectors, which carry features related to essential anatomical structures in chest X-rays like lungs, ribs, heart, and any present abnormalities or conditions. Since different images often share these anatomical features and pathological information, the codebook vectors for various X-ray images may be similar or closely related, leading to shared features across different images.

By grouping analogous visual features through vector quantization, the codebook implicitly ensures the preservation of essential anatomical information in a concise and structured manner. The objective is to utilize this compact structured information to further boost the capacity of the SSL framework to acquire more informative and context-aware visual features. To learn the codebook, we use vector quantization proposed in VQ-VAE [28]. This preserves most of the spatial information of the input data at the discrete latent space [23]. To demonstrate the efficacy of the proposed method, we conduct multiple experiments and in-depth ablation studies using chest X-rays and fundus images.

2 Related works

2.1 Discriminative SSL for Medical Images

Discriminative SSL methods capture transformations invariant representations by making a pixel-level comparison of the positive and negative pairs [17,21]. Methods like SimCLR [3], MoCo [10], PIRL [19], and others [24,31] based on contrastive learning, minimize the distance between positive pairs and maximize the distance between negative pairs. Non-contrastive methods like BYOL [8],

SimSiam [4], Barlow-Twins [35] and VICReg [2] omit the requirement of negative pairs by opting architectural constraints like asymmetric siamese network or by deploying variance-covariance regularization.

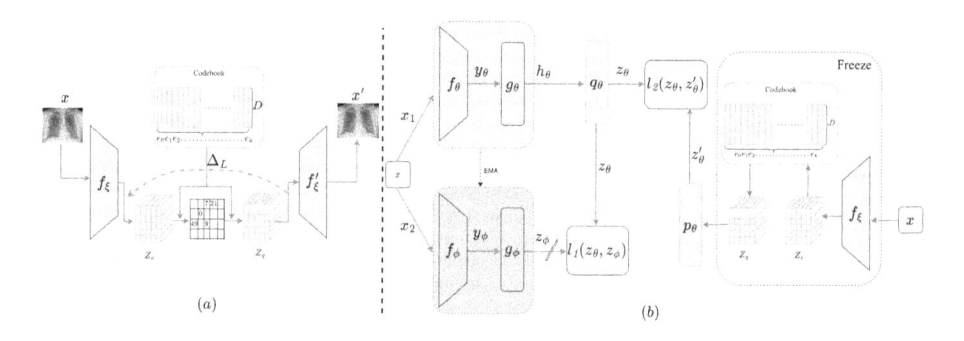

Fig. 1. The proposed IDQCE framework. (a) Represents the VQ-VAE. f_ξ processes input X-ray image x and and f'_ξ serves as the decoder. Δ_L signifies the gradient from the decoder to the encoder. In (b), the IDQCE includes two augmented views, x_1, and x_2 process by f_θ and f_ϕ, respectively. g_θ and g_ϕ are MLP projection heads in each branch while q_θ and p_θ are the trainable projection heads. f_ξ and codebook are frozen during the pre-training. // is the stop-gradient operator.

Contrastive SSL methods mainly predominate in the field of medical image analysis. For instance, in [39] Zhou et al. introduced C2L by applying contrastive loss on 2D chest radiographs, emphasizing feature-level contrast to learn general image representations. Zhou et al. proposed preservational contrastive representation learning (PCRL) for 2D and 3D medical images by incorporating diverse image reconstruction methods and cross-model mixup to encode more information in the representations [38]. Vu et al. in MedAug [29] discussed the creation of positive pairs using patient metadata such as study number and laterality to identify pairs of images sharing common pathological features. Azizi et al. explored similar ideas but in slightly different ways [1]. Taher et al. highlighted the limitations of standard discriminative SSL methods in medical image analysis and proposed enhancement by adding a pixel-level restoration branch in the SSL framework. DiRA [9] proposed by Haghighi et al. integrates discriminative, restorative, and adversarial learning components to leverage complementary visual information. In [37] Zhou et al. introduced a unified SSL framework for preserving pixel-level information, invariant semantics, and multi-scale representations in visual data, emphasizing the importance of such preservation. Fu et al. introduced anatomy-aware contrastive learning (AWCL) [7] for fetal ultrasound imaging tasks, where domain-specific anatomy information is leveraged to improve the quality of visual representations. Yan et al. proposed a self-supervised anatomical embedding (SAM) [32] contrastive learning framework to learn universal anatomical embeddings from unlabeled radiological images, enabling applications like landmark detection and lesion matching. Most existing methods highlighted the limitations of standard discriminative SSL methods.

While some methods addressed it through pixel-level restoration and domain-specific solutions, we present an innovative approach that implicitly harnesses the anatomical similarities, offering a straightforward yet effective solution for SSL in the context of medical imaging.

2.2 Vector Quantization in Medical Image Analysis

Vector quantization divides the data points into smaller discrete groups called codebook vectors and reduces the data dimensionality while preserving essential information [33]. Recently, VQ-VAE [28] successfully managed to learn discrete features of images by using a codebook. Multiple studies [5,11,18,33,36] demonstrated the effectiveness of learning discrete representations in image retrieval, generation, recognition, and compression. While vector quantization finds broader applications in many vision domains, its utilization for medical images is limited, especially for SSL. Tudosiu et al. presented a 3D VQ-VAE that reconstructs high-fidelity, full-resolution, and neuro-morphologically accurate brain images for potential Alzheimer's disease detection [27]. Patel et al. in [22] introduced self-supervised anomaly detection framework based upon the VQ-VAE - transformers pipeline. Kobayashi et al. in [16] proposed a feature decomposing network that explicitly separates semantic features of medical images into normal and abnormal anatomy codes, enabling a novel Content-Based Image Retrieval (CBIR) system for glaucoma. Existing approaches often combine VQ-VAE with autoregressive models for anomaly detection or segmentation; our work stands out for leveraging the VQ-VAE's codebook for learning richer and more generalized representations.

3 Method

In this section, we first discuss the preliminary VQ-VAE module and subsequently provide the details of the proposed IDQCE framework. Figure 1 provides the architectural overview of the proposed approach, which is divided into two parts: the left side (Figure 1(a)) presents the VQ-VAE while the right side (Figure 1(b)) shows the proposed approach.

3.1 VQ-VAE

VQ-VAE [28] successfully maps the input data to discrete, quantized codes, which are then used to reconstruct the data. This discrete representation enables improved data compression, generation, and transfer in image and audio processing applications. As shown in Figure 1(a), VQ-VAE consists of a CNN encoder f_ξ, a decoder f'_ξ and a codebook. The codebook consists of randomly initialized K vectors, denoted as e_k, each of size $\mathcal{D}$. The f_ξ maps the input x to continuous latent space $z_e \in \mathbb{R}^{\mathcal{C} \times \mathcal{H} \times \mathcal{W}}$, where $\mathcal{C}$ denotes the number of channels and $\mathcal{H}$, $\mathcal{W}$ represents height and width of the feature maps respectively. The $\mathcal{H}$ and $\mathcal{W}$ are reshaped to $\mathcal{D}$ to enable the quantization in the channel space. The latent space z_e is then quantized by selecting the nearest codebook vector from the predefined codebook to output a quantized feature map $z_q \in \mathbb{R}^{C \times \mathcal{H} \times \mathcal{W}}$, which serves as a compact representation of the input x.

The decoder f'_ξ reconstructs the input data from discrete feature map z_q into final output x', with a distribution $p(x|z_q)$ aiming to minimize the VQ-VAE loss. The VQ-VAE loss function consists of three terms: reconstruction loss to optimize f_ξ and f'_ξ, codebook loss to move the codebook vectors e_k towards z_e and the commitment loss to ensure that the f_ξ sticks to specific embeddings, preventing uncontrolled expansion of the latent space. The loss function is defined in equation (1).

$$\mathcal{L}_{\mathcal{VQ}} = -\log(p(x|z_q)) + ||SG[z_e] - e_k||_2^2 + \lambda ||z_e - SG[e_k]||_2^2 \qquad (1)$$

where the first component corresponds to the reconstruction loss, the second component is codebook loss, and the third component is the commitment loss. SG denotes stop-gradient, and λ controls the weightage of the commitment loss.

3.2 IDQCE Framework

IDQCE follows a two-stage training process, wherein in the first stage, we train the VQ-VAE network according to the description provided in sub-section 3.1. Here, f_ξ and the codebook are trained to learn discrete latent representations z_q of the input x. Subsequently, in the second stage, the discrete representations z_q are utilized to enhance the discriminative feature representations for the proposed method during SSL pre-training. IDQCE follows the non-contrastive paradigm of the SSL and is inspired by the self-distillation SSL method BYOL. However, it is also compatible with other SSL frameworks.

Figure 1(b) provides an overview of the proposed framework. The IDQCE framework includes two weight-shared CNN encoders, f_θ, and f_ϕ, each followed by an MLP projection head, g_θ and g_ϕ, respectively. Additionally, it incorporates the VQ-VAE's encoder, f_ξ, and the codebook, which are trained during the first stage. The framework also features two additional MLP networks, q_θ and p_θ. Here, q_θ serves as the prediction head, responsible for predicting the target embeddings, while p_θ acts as the projection head, mapping the quantized features into a lower-dimensional representational space. The proposed framework creates two augmented views, x_1 and x_2, of input x by applying a random set of augmentations sampled from $\mathcal{T}$. x_1 and x_2 are processed by the respective encoders f_θ and f_ϕ to output representations, y_θ and y_ϕ respectively. The y_θ and y_ϕ are then projected onto a lower-dimensional space using g_θ and g_ϕ in their respective branches, resulting in outputs denoted as h_θ and z_ϕ.

To enhance the representations learned by f_θ, we incorporate the quantized embeddings z_q obtained from the pre-trained VQ-VAE encoder f_ξ and its associated codebook. The rationale behind this step is that the z_q is inherently sparse and contains distinctive anatomical characteristics in a condensed form. During VQ-VAE training, the quantization process leads to similar anatomical features associated with the specific codebook vectors, and thus, z_q represents the clusters of similar anatomical features. Particularly, each codebook vector in z_q corresponds to a cluster center, summarizing a group of similar latent representations. This clustering ensures that z_q not only compresses complex information but also enhances the interpretability of features by aligning them with clinically relevant anatomical priors. By integrating z_q into the SSL pre-training, f_θ can leverage these structured embeddings to extract more discriminative features from input data. This integration allows the model to exploit prior knowledge encoded in z_q, which enhances the quality of the learned features, ultimately leading to improved performance in discriminative SSL. To maintain the consistency of the anatomical insights embedded in the quantized embeddings, we keep f_ξ and the associated codebook in a fixed state during IDQCE pre-training. This serves not only to preserve consistency in z_q but also provides regularization benefits that enhance model robustness in medical image analysis tasks by ensuring that the learned features remain grounded in clinically relevant anatomical characteristics. Further, to effectively integrate the quantized embeddings z_q the projection head p_θ projects z_q into z'_θ, where the dimensions of z'_θ match those of z_θ. By doing this, p_θ facilitates the utilization of anatomical insights encoded in z_q across subsequent stages of feature extraction and learning. This process enhances the model's ability to leverage the rich anatomical information captured by f_ξ during the initial training phase, thereby improving the discriminative power and generalization capability of the IDQCE model in medical image analysis.

Loss function: θ are the trainable parameters while the parameters of f_ϕ and g_ϕ are are exponential moving average of the parameters of the f_θ and g_θ as $\phi \leftarrow \mu\phi + (1-\mu)\theta$, where $\mu \in [0,1]$ is the decay rate. To optimize the parameters θ, we calculate the similarity scores between z_θ and z_ϕ, as well as between z_θ and z'_θ using the loss function defined in equation (2).

$$\mathcal{L}_1 = \frac{\langle z_\theta, z_\phi \rangle}{||z_\theta||_2 \cdot ||z_\phi||_2}, \mathcal{L}_2 = \frac{\langle z_\theta, z'_\theta \rangle}{||z_\theta||_2 \cdot ||z'_\theta||_2} \qquad (2)$$

The final loss of IDQCE is given as $L_\theta = \mathcal{L}_1 + \mathcal{L}_2$. Further, we use the symmetric form of the loss L_θ by interchangeably feeding the views x_1 and x_2 to f_θ and f_ϕ.

4 Experiments and Results

This section overviews the experimental setup, including dataset descriptions, implementation details, baseline methods, and results.

4.1 Description of Datasets

For pre-training, we utilize two chest X-ray datasets: MIMIC-CXR v2.0.0 [14] and NIH-Chest X-ray 14 [30] and a diabetic retinopathy dataset EyePACS [6] contains 35,126 images. MIMIC-CXR comprises 201,083 samples for pre-training, while we reserve 42,164 samples for downstream evaluations. The NIH dataset comprises 112,120 X-ray images from 30,805 distinct patients, and we use the official train test split for pre-training and downstream evaluations. For downstream evaluation, we employ three X-ray datasets, MIMIC, NIH, and VinBig-CXR [20], and two retinopathy datasets, MuReD [25], and ODIR [15,40]. The VinBig-CXR dataset has 18,000 chest X-ray images, each annotated with 14 different pathologies. The ODIR has 7,000 retinal images with 8 labels, while MuReD has 2,208 samples with 20 classes.

4.2 Implementation Details

The f_θ and f_ϕ are ResNet-50 models that output the representations y_θ and y_ϕ from the global average pooling layer. The projection heads g_θ and g_ϕ are implemented as two-layer MLP networks with an output size of 256 and a hidden layer size of 4096. Additionally, the prediction head q_θ and p_θ have the exact dimensions, resulting in a final embedding size of 256. We update parameters θ using LARS [34] optimization technique with a base learning rate of 0.02 and cosine decay without restarts as the learning rate scheduler. In the case of the MIMIC-CXR dataset, we pre-train the IDQCE framework with a batch size of 64 for 100 epochs, while in the case of the EyePACS dataset, the batch size is 256 for 300 epochs. With the NIH-Chest X-ray dataset, we match the settings of PCRLv2 and CAiD to keep a batch size of 256 and pre-train the model for 500 epochs. We use Adam as the optimizer during downstream evaluations, with 1e-4 as the learning rate. We also pre-train the VQ-VAE model on the respective IDQCE pre-training datasets (MIMIC, NIH, and EyePACS). For VQ-VAE, we employ a shallow CNN network as encoders and decoders, with three different sets of codebook vectors, having sizes of 512, 1024, and 4096. The dimension of codebook vectors is 256. We utilize the Adam optimizer with a batch size of 16 and a learning rate of 1e-3 for a total of 100 epochs.

4.3 Baselines

For comparison, we evaluate our method against both supervised and SOTA SSL methods. For supervised approaches, we consider models initialized with random weights and ImageNet pre-trained weights. Our SSL comparisons include contrastive methods like SimCLR [3] and MoCo [10], as well as non-contrastive methods such as BYOL [8], SimSiam [4], Barlow Twins (BT)[35], and VICReg[2]. Additionally, we assess the performance of IDQCE against recently proposed medical-specific SSL methods, including PCRLv2 [37] and CAiD [26]. To ensure a fair and meaningful comparison, we pre-train all baseline methods using their original implementations on the same datasets as IDQCE. We maintain consistent batch sizes, the number of pre-training epochs, and sample distribution across all methods. This uniformity ensures that our evaluations are reliable and directly comparable, highlighting the efficacy of IDQCE in various contexts.

4.4 Results

To assess the learned representations from IDQCE pre-training, we follow the standard evaluation protocol commonly used by SSL methods like SimCLR [3]

Table 1. Experimental results from test set samples in terms of AUC for linear classification on the MIMIC and NIH dataset with frozen representations. The pre-training is done on MIMIC-CXR images.

Methods	MIMIC→MIMIC 1%	10%	all	MIMIC→NIH 1%	10%	all
Ran. Init.	0.579	0.589	0.703	0.514	0.551	0.597
Img. Init.	0.639	0.694	0.766	0.594	0.658	0.697
SimCLR [3]	0.738	0.762	0.799	0.619	0.676	0.715
MoCo [10]	0.730	0.754	0.789	0.598	0.664	0.706
BYOL [8]	0.741	0.768	0.813	0.629	0.703	0.727
SimSiam [4]	0.733	0.757	0.791	0.613	0.693	0.716
BT [35]	0.734	0.766	0.806	0.618	0.677	0.710
VICReg [2]	0.733	0.761	0.792	0.626	0.683	0.719
PCRLv2 [37]	0.721	0.740	0.772	0.616	0.673	0.712
CAiD$_{MoCo-v2}$ [26]	0.740	0.768	0.783	0.634	0.691	0.719
IDQCE	**0.759**	**0.798**	**0.829**	**0.655**	**0.734**	**0.767**

and BYOL [8]. Specifically, we use the SSL pre-trained f_θ network as the feature extractor and append a liner layer on top of it. Throughout all downstream tasks, we update only the parameters of the linear layer while keeping the parameters of f_θ frozen. Unlike finetuning the entire backbone network, this ensures a fair evaluation of the discriminative ability of the representations learned by IDQCE. Table 1 presents the experimental results on test set samples in terms of AUC score, corresponding to the pre-training on the MIMIC dataset. In the MIMIC→NIH scenario involves pre-training on MIMIC-CXR images, while the downstream evaluation is performed on the NIH-Chest X-ray images. This experimental setup serves the evaluation under a transfer learning setting and aims to assess the generalization of the learned representations across datasets. Mainly, we evaluate the features extracted from f_θ by updating the parameters of the linear layer using different subsets (1%, 10%, and all) of training samples.

The results in Table 1 demonstrate the effectiveness of the proposed IDQCE framework compared to standard SOTA SSL and supervised methods. For the MIMIC→MIMIC scenario, IDQCE consistently outperforms all the considered baseline methods, including PCRLv2 and CAiD. Notably, IDQCE achieves an improvement of around 2% with the 1% subset of training samples, indicating its ability to learn informative representations with limited labeled data. As the size of the labeled training subset increases to 10%, IDQCE continues to perform better. In the MIMIC→NIH setting, IDQCE maintains its superiority, outperforming all baseline methods across different subsets of training samples with an average margin of more than 3%. The performance gain is particularly noticeable in scenarios with limited training samples (1% and 10%), where IDQCE pre-training exhibits a significant advantage. This indicates that the represen-

Table 2. Experimental test set results for linear classification tasks on the NIH dataset. The table presents the AUC score for different subsets of training samples (1%, 10%, and all) across in-distribution (NIH→NIH) and (NIH→Vinbig) under transfer learning evaluation settings.

Methods	NIH→NIH 1%	10%	all	NIH→Vinbig 1%	10%	all
Ran. Init.	0.514	0.551	0.597	0.530	0.564	0.696
Img. Init.	0.594	0.658	0.697	0.659	0.786	0.862
SimCLR [3]	0.613	0.663	0.703	0.610	0.709	0.763
MoCo [10]	0.607	0.654	0.694	0.606	0.698	0.758
BYOL [8]	0.609	0.649	0.676	0.587	0.669	0.748
SimSiam [4]	0.591	0.638	0.668	0.579	0.662	0.743
BT [35]	0.639	0.687	0.718	0.631	0.707	0.750
VICReg [2]	0.631	0.673	0.701	0.529	0.681	0.732
PCRLv2 [37]	0.638	0.720	0.723	0.594	0.700	0.796
CAiD$_{MoCo-v2}$ [26]	0.643	0.725	0.748	0.710	0.837	0.885
IDQCE	**0.679**	**0.774**	**0.810**	**0.722**	**0.883**	**0.925**

tations learned by the IDQCE framework on the MIMIC-CXR dataset are well generalized to the NIH dataset. We observe standard SSL methods like SimCLR, BYOL, SimSiam, VICReg, etc., which also perform well and demonstrate competitive performance, highlighting the general efficacy of these SSL frameworks for medical image representation learning.

Table 2 presents the experimental test set results in terms of AUC score for the NIH-Chext X-ray dataset by evaluating the features extracted from frozen f_θ. For this experimental setup, we pre-train the IDQCE framework on the NIH dataset, and the evaluation includes both in-distribution and transfer learning assessments. For in-distribution evaluation, the learned features are assessed on the NIH dataset, varying the subset of training samples (1%, 10%, and all), while the transfer learning evaluations are conducted using the VinBig-CXR dataset. For the subset with 1% of training samples, the proposed IDQCE framework outperforms all the baseline methods, including supervised and various SOTA SSL methods. IDQCE achieves a performance gain of more than 3% over both PCRLv2 and CAiD, with 1% training samples. As the size of training samples increases to 10%, IDQCE maintains its leading performance, with a performance gain of more than 5% compared to the best-performing baseline. The consistently high performance across different subsets, including all training samples, indicates the effectiveness of the proposed framework in learning more informative representations.

In the NIH→VinBig task, IDQCE again demonstrates its superior performance, outperforming all other methods with a significant margin across all subsets of training samples. It shows an AUC gain of more than 4% over CAiD in the case of 10% and all training samples. For the 1% subset also, IDQCE performs marginally better than the considered baseline methods. The performance gain in the transfer learning to the VinBig-CXR dataset indicates the generalization capability of the features learned by IDQCE on the NIH dataset.

Table 3. Experimental test set results for linear classification tasks on fundus images in terms of AUC score. The IDQCE framework, pre-trained on the EyePACS dataset, shows considerable performance gain over existing approaches on the MuReD and ODIR datasets for both 10% and all training samples. Bold indicates the best numbers, while underline is the second best.

Methods	EyePACS→MuReD		EyePACS→ODIR	
	10%	all	10%	all
Ran. Init.	0.653	0.750	0.602	0.661
Img. Init.	**0.824**	**0.907**	0.756	0.798
SimCLR [3]	0.763	0.852	0.732	0.783
MoCo [10]	0.729	0.819	0.714	0.751
BYOL [8]	0.712	0.775	0.652	0.688
SimSiam [4]	0.708	0.768	0.641	0.673
BT [35]	0.714	0.815	0.692	0.734
VICReg [2]	0.683	0.781	0.652	0.688
PCRLv2 [37]	0.735	0.839	0.703	0.764
CAiD$_{MoCo-v2}$ [26]	0.737	0.833	0.710	0.765
IDQCE	0.793	0.898	**0.768**	**0.806**

Table 3 exhibits the performance of various considered baseline methods and the proposed IDQCE framework on linear classification tasks for fundus image datasets. In the MuReD dataset, we observe that ImageNet-initialized weights show higher AUC values compared to other methods, highlighting the generalization capabilities of models pre-trained on a large-scale dataset for fundus images. Notably, IDQCE achieves superior performance among the SOTA standard SSL methods and PCRLv2, CAiD. The performance of IDQCE for 10% and all training samples reveals that it contributes to more discriminative and task-relevant features for fundus image classification. Similarly, on the ODIR dataset, IDQCE performs better, achieving the highest AUC values among all methods for both 10% and all training samples.

5 Discussion

In this section, we discuss the impact of different components of the IDQCE framework on learned representations and qualitative analysis of diagnostic heat maps.

Table 4. Ablation study on the impact of various key components in IDQCE framework. CB_size denotes the number of codebook vectors, and Dec ($f'\xi$) indicates the presence of the decoder network $f'\xi$. Fr and Fi denote frozen and fine-tuned states, respectively. The table reports the AUC score for different subsets of training samples in MIMIC→MIMIC and MIMIC→NIH settings.

Methods	CB_size	Dec (f'_ξ)	MIMIC→MIMIC 1%	MIMIC→MIMIC 10%	MIMIC→NIH 1%	MIMIC→NIH 10%
IDQCE	512	✗	**0.768**	**0.804**	**0.658**	**0.736**
	1024	✗	0.759	0.798	0.655	0.734
	4096	✗	0.760	0.800	0.657	0.735
IDQCE	1024	✓$_{Fr}$	0.760	0.801	0.657	0.732
	1024	✓$_{Fi}$	0.762	0.795	0.653	0.728
IDQCE+VQ-VAE	1024	-	0.756	0.791	0.653	0.723
SSL+Encoder	-	-	0.723	0.760	0.621	0.678

5.1 Ablation Study

To assess the importance of different components of the IDQCE framework, we systemically analyze the impact of vector quantization on the SSL pre-training in enhancing the representations. Specifically, we vary the number of codebook vectors during VQ-VAE pre-training to analyze its impact on the representations obtained during SSL pre-training at the later stage of the IDQCE framework. Additionally, we investigate the effects of integrating the decoder network f'_ξ into the SSL framework alongside the encoder f_ξ and codebook. Further, in an end-to-end setting, we simultaneously pre-train the SSL framework and VQ-VAE network to assess their combined impact on the learned representations. Lastly, we replace the vector quantized encoder with a regular CNN encoder during SSL pre-training to examine the impact of quantized embeddings during SSL pre-training. Table 4 shows the results of these experiments and provides valuable insights into the efficacy of the IDQCE framework under various ablation scenarios. The evaluation is conducted on MIMIC-CXR and NIH datasets, using the MIMIC-CXR dataset for pre-training, allowing for a comprehensive assessment across in-distribution and out-of-distribution settings.

The results in Table 4 demonstrate the impact of the number of codebook vectors (CB_size) during VQ-VAE pre-training on the representations learned by the IDQCE framework. Notably, we observe that increasing the number of codebook vectors from 512 to 4096 leads to a marginal degradation in the performance across all the tasks. Notably, for the MIMIC→MIMIC task, examining the 1% subset of training samples reveals that a CB_size of 512 achieves the highest performance among the configurations tested, indicating that a smaller codebook size is advantageous when dealing with limited data. When utilizing 10% and the entire training set, we observe comparative performance for both MIMIC and NIH datasets; however, the model with a CB_size of 512

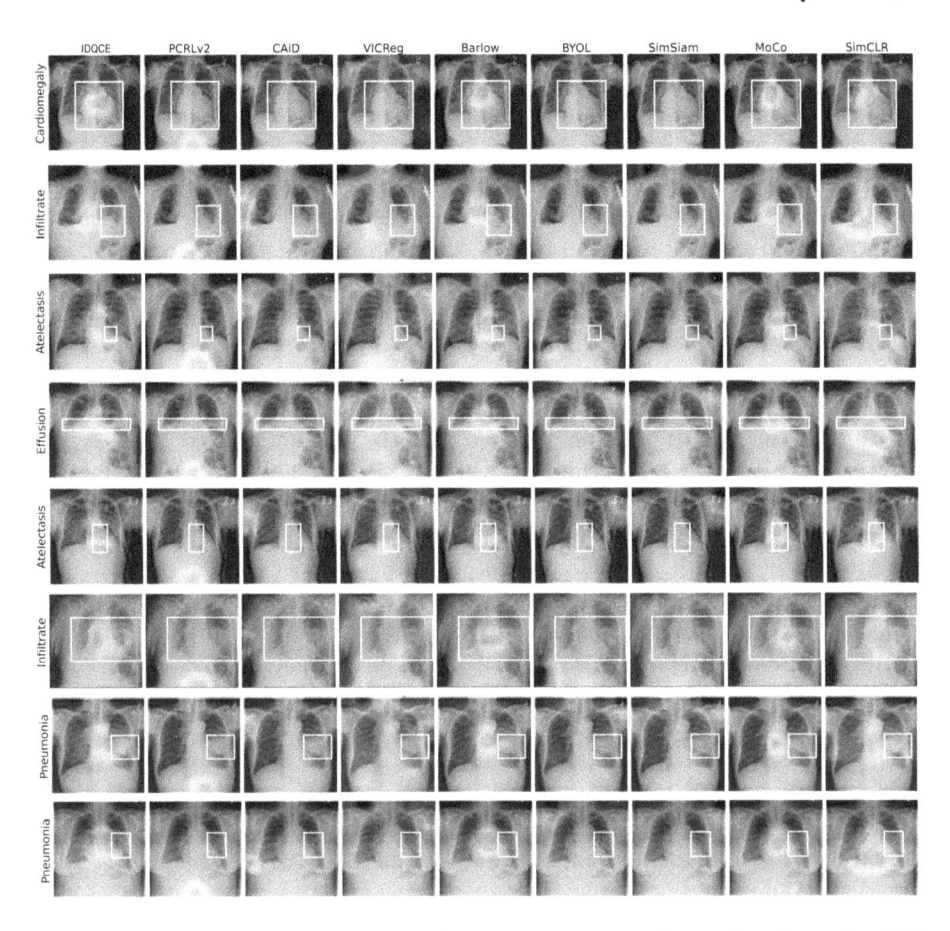

Fig. 2. Diagnostic heatmaps depict the interpretations of samples from the NIH dataset, corresponding to 1% of the labeled data. It provides insights into the regions of interest identified, offering a qualitative comparison of the diagnostic capabilities. Bounding boxes are the ground truth available in the public domain.

marginally outperforms other configurations across all subsets of training samples. This indicates that a smaller codebook size comparatively yields better results. Adopting a smaller codebook proves relatively advantageous by encouraging a higher probability of shared codebook vectors for anatomically similar structures. This facilitates effective compact information compression, allowing the IDQCE framework to distill essential anatomical features into SSL representations. Additionally, shared vectors enhance the model's generalization capabilities, promoting the capturing of more informative and discriminative features across different augmented views.

Table 4 further presents the results of integrating the decoder network (f'_ξ) pre-trained during VQ-VAE within the IDQCE framework. We keep the decoder frozen (Fr) and fine-tuned (Fi) state during IDQCE pre-training to do the comprehensive analysis. Surprisingly, no discernible improvement is observed upon integrating the decoder in both frozen and updated parameters during IDQCE pre-training. This indicates that the VQ-VAE encoder plays a crucial role in shaping informative SSL representations, while the decoder's contribution in the presence of the encoder may not be as influential.

Further, Table 4 represents the results corresponding to integrated end-to-end pre-training of IDQCE and VQ-VAE (IDQCE+VQ-VAE). We observe a slightly lower performance than the standalone IDQCE framework, suggesting that integrating the entire VQ-VAE during SSL pre-training may not offer significant benefits in this context. It implies that the information encoded by the VQ-VAE, such as the anatomically informed embeddings and the associated codebook, might not be fully exploited, necessary for enhancing the learned representations during SSL pre-training within the IDQCE framework. The last row of Table 4 explores the impact of SSL pre-training with a regular CNN encoder instead of the VQ-VAE encoder (SSL+Encoder). We pre-train the UNet model with ResNet-50 as the backbone encoder on the MIMIC-CXR dataset for this experimental setting. The results indicate a noticeable decrease in performance across all the tasks, emphasizing the importance of leveraging vector quantization and the anatomically informed quantized embeddings provided by the VQ-VAE encoder for effective SSL pre-training in medical imaging tasks.

5.2 Qualitative Analysis

Figure 2 presents the visual interpretations of pathological conditions exhibited in the chest X-ray diagnosis. The heatmaps correspond to the representations learned by IDQCE and the considered baseline methods for 1% training samples from the NIH dataset. Examining these heatmaps provides valuable insights into the model's decision-making process and the regions of interest identified in the chest X-ray images. Notably, we observe that IDQCE relatively identifies the more accurate diagnostic regions compared to the baseline methods for conditions like Cardiomegaly, Effusion, Atelectasis, etc. IDQCE performs relatively better than other SSL methods, especially non-contrastive ones like VICReg, BYOL, and SimSiam.

6 Conclusions

This work proposes an IDQCE framework to learn informative and discriminative features by leveraging the contextual quantized embeddings in the SSL model. IDQCE utilizes the VQ-VAE model to generate quantized embeddings that encapsulate valuable anatomical insights in a compact form to enhance the SSL model's capacity. Through comprehensive evaluations of multiple datasets from different modalities, we observe that quantized embeddings lead to better representations than existing SSL methods. Our empirical investigations, including experiments in linear evaluation, transfer learning, and ablation studies within the IDQCE framework, reveal the essential findings and the significance of information preservation in learning better representations of medical images. The future work includes expanding its application to a wider variety of medical imaging datasets, including 3D modalities such as CT, MRI, and ultrasound, to assess and enhance its generalizability across various medical contexts.

References

1. Azizi, S., Mustafa, B., Ryan, F., Beaver, Z., Freyberg, J., Deaton, J., Loh, A., Karthikesalingam, A., Kornblith, S., Chen, T., et al.: Big self-supervised models advance medical image classification. In: Proceedings of the IEEE/CVF international conference on computer vision. pp. 3478–3488 (2021)
2. Bardes, A., Ponce, J., LeCun, Y.: VICReg: Variance-invariance-covariance regularization for self-supervised learning. In: International Conference on Learning Representations (2022), https://openreview.net/forum?id=xm6YD62D1Ub
3. Chen, T., Kornblith, S., Norouzi, M., Hinton, G.: A simple framework for contrastive learning of visual representations. In: International conference on machine learning. pp. 1597–1607. PMLR (2020)
4. Chen, X., He, K.: Exploring simple siamese representation learning. In: Proceedings of the IEEE/CVF conference on computer vision and pattern recognition. pp. 15750–15758 (2021)
5. Chen, Y.J., Cheng, S.I., Chiu, W.C., Tseng, H.Y., Lee, H.Y.: Vector quantized image-to-image translation. In: European Conference on Computer Vision. pp. 440–456. Springer (2022)
6. Dugas, E., Jared, Jorge, Cukierski, W.: Diabetic retinopathy detection (2015), https://kaggle.com/competitions/diabetic-retinopathy-detection
7. Fu, Z., Jiao, J., Yasrab, R., Drukker, L., Papageorghiou, A.T., Noble, J.A.: Anatomy-aware contrastive representation learning for fetal ultrasound. In: European Conference on Computer Vision. pp. 422–436. Springer (2022)
8. Grill, J.B., Strub, F., Altché, F., Tallec, C., Richemond, P., Buchatskaya, E., Doersch, C., Avila Pires, B., Guo, Z., Gheshlaghi Azar, M., et al.: Bootstrap your own latent-a new approach to self-supervised learning. Adv. Neural. Inf. Process. Syst. **33**, 21271–21284 (2020)
9. Haghighi, F., Taher, M.R.H., Gotway, M.B., Liang, J.: Dira: Discriminative, restorative, and adversarial learning for self-supervised medical image analysis. In: Proceedings of the IEEE/CVF Conference on Computer Vision and Pattern Recognition. pp. 20824–20834 (2022)
10. He, K., Fan, H., Wu, Y., Xie, S., Girshick, R.: Momentum contrast for unsupervised visual representation learning. In: Proceedings of the IEEE/CVF conference on computer vision and pattern recognition. pp. 9729–9738 (2020)
11. Huang, M., Mao, Z., Chen, Z., Zhang, Y.: Towards accurate image coding: Improved autoregressive image generation with dynamic vector quantization. In: Proceedings of the IEEE/CVF Conference on Computer Vision and Pattern Recognition. pp. 22596–22605 (2023)
12. Huang, S.C., Pareek, A., Jensen, M., Lungren, M.P., Yeung, S., Chaudhari, A.S.: Self-supervised learning for medical image classification: a systematic review and implementation guidelines. NPJ Digital Medicine **6**(1), 74 (2023)
13. Jiang, Y., Sun, M., Guo, H., Bai, X., Yan, K., Lu, L., Xu, M.: Anatomical invariance modeling and semantic alignment for self-supervised learning in 3d medical image analysis. In: Proceedings of the IEEE/CVF International Conference on Computer Vision. pp. 15859–15869 (2023)

14. Johnson, A.E., Pollard, T.J., Greenbaum, N.R., Lungren, M.P., Deng, C.y., Peng, Y., Lu, Z., Mark, R.G., Berkowitz, S.J., Horng, S.: Mimic-cxr-jpg, a large publicly available database of labeled chest radiographs. arXiv preprint arXiv:1901.07042 (2019)
15. kaggle: Ocular disease recognition, https://www.kaggle.com/andrewmvd/ocular-disease-recognition-odir5k
16. Kobayashi, K., Hataya, R., Kurose, Y., Miyake, M., Takahashi, M., Nakagawa, A., Harada, T., Hamamoto, R.: Decomposing normal and abnormal features of medical images for content-based image retrieval of glioma imaging. Med. Image Anal. **74**, 102227 (2021)
17. Liu, X., Zhang, F., Hou, Z., Mian, L., Wang, Z., Zhang, J., Tang, J.: Self-supervised learning: Generative or contrastive. IEEE Trans. Knowl. Data Eng. **35**(1), 857–876 (2021)
18. Mao, C., Jiang, L., Dehghani, M., Vondrick, C., Sukthankar, R., Essa, I.: Discrete representations strengthen vision transformer robustness. In: International Conference on Learning Representations (2021)
19. Misra, I., Maaten, L.v.d.: Self-supervised learning of pretext-invariant representations. In: Proceedings of the IEEE/CVF Conference on Computer Vision and Pattern Recognition. pp. 6707–6717 (2020)
20. Nguyen, H.Q., Lam, K., Le, L.T., Pham, H.H., Tran, D.Q., Nguyen, D.B., Le, D.D., Pham, C.M., Tong, H.T., Dinh, D.H., et al.: Vindr-cxr: An open dataset of chest x-rays with radiologist's annotations. Scientific Data **9**(1), 429 (2022)
21. Nozawa, K., Sato, I.: Understanding negative samples in instance discriminative self-supervised representation learning. Adv. Neural. Inf. Process. Syst. **34**, 5784–5797 (2021)
22. Patel, A., Tudosiu, P.D., Pinaya, W.H., Graham, M.S., Adeleke, O., Cook, G., Goh, V., Ourselin, S., Cardoso, M.J.: Self-supervised anomaly detection from anomalous training data via iterative latent token masking. In: Proceedings of the IEEE/CVF International Conference on Computer Vision. pp. 2402–2410 (2023)
23. Pinaya, W.H., Tudosiu, P.D., Gray, R., Rees, G., Nachev, P., Ourselin, S., Cardoso, M.J.: Unsupervised brain imaging 3d anomaly detection and segmentation with transformers. Med. Image Anal. **79**, 102475 (2022)
24. Purushwalkam, S., Gupta, A.: Demystifying contrastive self-supervised learning: Invariances, augmentations and dataset biases. Adv. Neural. Inf. Process. Syst. **33**, 3407–3418 (2020)
25. Rodríguez, M.A., AlMarzouqi, H., Liatsis, P.: Multi-label retinal disease classification using transformers. IEEE Journal of Biomedical and Health Informatics (2022)
26. Taher, M.R.H., Haghighi, F., Gotway, M.B., Liang, J.: Caid: Context-aware instance discrimination for self-supervised learning in medical imaging. In: International Conference on Medical Imaging with Deep Learning. pp. 535–551. PMLR (2022)
27. Tudosiu, P.D., Varsavsky, T., Shaw, R., Graham, M., Nachev, P., Ourselin, S., Sudre, C.H., Cardoso, M.J.: Neuromorphologicaly-preserving volumetric data encoding using vq-vae. arXiv preprint arXiv:2002.05692 (2020)
28. Van Den Oord, A., Vinyals, O., et al.: Neural discrete representation learning. Advances in neural information processing systems **30** (2017)
29. Vu, Y.N.T., Wang, R., Balachandar, N., Liu, C., Ng, A.Y., Rajpurkar, P.: Medaug: Contrastive learning leveraging patient metadata improves representations for chest x-ray interpretation. In: Machine Learning for Healthcare Conference. pp. 755–769. PMLR (2021)

30. Wang, X., Peng, Y., Lu, L., Lu, Z., Bagheri, M., Summers, R.M.: Chestx-ray8: Hospital-scale chest x-ray database and benchmarks on weakly-supervised classification and localization of common thorax diseases. In: Proceedings of the IEEE conference on computer vision and pattern recognition. pp. 2097–2106 (2017)
31. Wang, X., Zhang, R., Shen, C., Kong, T., Li, L.: Dense contrastive learning for self-supervised visual pre-training. In: Proceedings of the IEEE/CVF Conference on Computer Vision and Pattern Recognition. pp. 3024–3033 (2021)
32. Yan, K., Cai, J., Jin, D., Miao, S., Guo, D., Harrison, A.P., Tang, Y., Xiao, J., Lu, J., Lu, L.: Sam: Self-supervised learning of pixel-wise anatomical embeddings in radiological images. IEEE Trans. Med. Imaging **41**(10), 2658–2669 (2022)
33. Yang, Z., Dong, W., Li, X., Huang, M., Sun, Y., Shi, G.: Vector quantization with self-attention for quality-independent representation learning. In: Proceedings of the IEEE/CVF Conference on Computer Vision and Pattern Recognition. pp. 24438–24448 (2023)
34. You, Y., Gitman, I., Ginsburg, B.: Large batch training of convolutional networks. arXiv preprint arXiv:1708.03888 (2017)
35. Zbontar, J., Jing, L., Misra, I., LeCun, Y., Deny, S.: Barlow twins: Self-supervised learning via redundancy reduction. In: International Conference on Machine Learning. pp. 12310–12320. PMLR (2021)
36. Zheng, C., Vedaldi, A.: Online clustered codebook. In: Proceedings of the IEEE/CVF International Conference on Computer Vision. pp. 22798–22807 (2023)
37. Zhou, H.Y., Lu, C., Chen, C., Yang, S., Yu, Y.: A unified visual information preservation framework for self-supervised pre-training in medical image analysis. IEEE Transactions on Pattern Analysis and Machine Intelligence (2023)
38. Zhou, H.Y., Lu, C., Yang, S., Han, X., Yu, Y.: Preservational learning improves self-supervised medical image models by reconstructing diverse contexts. In: Proceedings of the IEEE/CVF International Conference on Computer Vision. pp. 3499–3509 (2021)
39. Zhou, H.-Y., Yu, S., Bian, C., Hu, Y., Ma, K., Zheng, Y.: Comparing to Learn: Surpassing ImageNet Pretraining on Radiographs by Comparing Image Representations. In: Martel, A.L., Abolmaesumi, P., Stoyanov, D., Mateus, D., Zuluaga, M.A., Zhou, S.K., Racoceanu, D., Joskowicz, L. (eds.) MICCAI 2020. LNCS, vol. 12261, pp. 398–407. Springer, Cham (2020). https://doi.org/10.1007/978-3-030-59710-8_39
40. Zhou, Y., Wang, B., Huang, L., Cui, S., Shao, L.: A benchmark for studying diabetic retinopathy: segmentation, grading, and transferability. IEEE Trans. Med. Imaging **40**(3), 818–828 (2020)

Superpixel-Based Sparse Labeling for Efficient and Certain Medical Image Annotation

Somayeh Rezaei[1] and Xiaoyi Jiang[2](✉)

[1] Department of Information Engineering, University of Padova, Padua, Italy
[2] Faculty of Mathematics and Computer Science, University of Münster, Einsteinstrasse 62, 48149 Münster, Germany
xjiang@uni-muenster.de

Abstract. Supervised deep learning crucially depends on large amount of high-quality annotation data. While labeling for classification and grading tasks is rather efficient to achieve, labeling for segmentation is much more difficult and time-consuming, characterized by *pixel-based dense annotation* in practice. This approach suffers from two fundamental disadvantages: 1) *Lack of efficiency* because of the large number of pixels on region boundaries (other parts of an image can be easily labeled) and more importantly the need of precise positioning of boundary pixels. 2) *Lack of certainty*. The area around the boundaries is in fact the part of an image, where even medical experts are often uncertain and may make non-precise annotations, resulting in varying annotations by different experts (the serious problem of inter-observer variability). To overcome these disadvantages, we propose superpixel-based annotation instead of pixels. Importantly, we do not require to label in the area of boundaries with high uncertainty for medical experts. We automatically fill the unlabeled area (boundary gap). In addition to heuristic rules we also study the random walker. Experiments were conducted with three different medical segmentation tasks and two network models. Despite the easy-to-make sparse annotation we are able to achieve segmentation results that are comparable or even superior to those obtained by using pixel-based dense annotation. In addition to the high efficiency, our approach substantially reduces the inter-observer variability as a positive side effect.

Keywords: Medical image annotation · image segmentation · superpixel · inter-observer variability

1 Introduction

In medical imaging the state-of-the-art research has led to systems that match or even outperform the level of medical experts. One of the key elements for making this possible is the availability of large amount of high-quality annotation data. In contrast to other application fields, labeling medical imaging data typically requires significant expertise of medical experts. While global labeling for classification (e.g. designation of diseases) and grading (degree of deviation from normal case) can be achieved rather efficiently, labeling for image segmentation is much more demanding and time-consuming. In addition to efficiency, there is also an issue of labeling uncertainty.

Meanwhile, annotation tools have been developed to assist the labeling work [4]. But still the practice is characterized by *pixel-based dense* annotation in existing approaches. This dense annotation has several disadvantages, particularly:

- **Efficiency**. The use of pixels as basic processing unit causes a high amount of labeling work because of the large number of pixels on region boundaries (other parts of an image can be easily labeled) and more importantly the need of precise positioning of boundary pixels.
- **Certainty**. The area around the boundaries is in fact the part of an image, where even medical experts are often uncertain and may make non-precise annotations, which can be clearly observed in some of the sample images in Figure 2. Such situations typically result in varying annotations by different experts. This so-called inter-observer variability [30] is a serious problem in practice since the labeled training data and consequently the learned deep learning (DL) model are biased to the particular labeling expert and thus not objective. Multiple expert annotations can be combined to reduce the uncertainty [33]. But it would further increase the overall workload of labeling.

In our work we propose to use superpixels instead of pixels to overcome these disadvantages towards efficient and certain image annotation. Superpixels significantly reduce the quantity of basic processing units. More importantly, we do not require to label in the area of boundaries with high uncertainty. Instead, we automatically fill the unlabeled area (boundary gap). This approach thus helps to avoid the inter-observer variability towards increased objectivity. Despite this *superpixel-based sparse annotation* we envision to avoid loss of segmentation quality. That is, we want to achieve segmentation results that are comparable or even superior to those obtained by using pixel-based dense annotation. This requirement is absolutely essential to make the proposed approach practicable. The key challenge is thus how to automatically fill the boundary gap in such a way that the anticipated segmentation quality can be achieved using the sparse labeling only. This superpixel-based approach will significantly ease the annotation process for medical experts with regard to time and certainty, nevertheless with high-quality segmentation by the learned DL models. To our best knowledge, there is no published study that has demonstrated this potential before.

The remainder of the paper is organized as follows. In section 2 we discuss related work. In In section 3 we detail the scheme of superpixel-based sparse labeling and strategies for automatically filling the unlabeled area. Our approach is evaluated in In section 4 on three datasets using two DL models. Finally, we conclude the paper in In section 5.

2 Related work

2.1 Efficient image labeling

It is essential to reduce the workload for medical experts. Thus, the labeling can be done by non-experts first and then refined by medical experts [14]. In particular contexts it is possible to simplify the labeling task. For instance, instead of a full 3D labeling, 2D enface images (generated pixel-wise by aggregating, e.g. by averaging, maximum, or minimum, the flow information of the corresponding voxel stack in the specific retinal layer) are more easier to label for OCT Angiography volume data [18].

Weakly supervised learning attempts to use sparse annotations such as scribbles for effective training [27]. To overcome the inherent lacking structural information in such scribble supervision, superpixels are used to guide the scribbles walking towards unlabeled pixels [35].

2.2 Superpixel-based image analysis

Superpixel algorithms are a special kind of segmentation methods and partition an image into small irregularly shaped regions by grouping similar pixels together. A key requirement on superpixels is a good preservation of natural object boundaries. As a result, superpixels typically only contain pixels of the same semantic category (e.g. an organ) and do not cross the boundaries of different categories. Superpixels are attractive because of the considerably reduced quantity compared to pixels and their high level of semantics (information). They have been successfully used for image segmentation [22], depth image based rendering [23], and other image analysis tasks [25].

Superpixels have already found applications in DL model training. Generally, large parts of the segmentation targets are relatively easy to recognize and to label, but the area around the region boundaries is rather difficult to handle due to the inherent ambiguity and other factors (e.g. the partial volume effect). This labeling uncertainty results in unsatisfactory performance of the trained DL models. If a superpixel intersects with the annotation boundary, a high labeling uncertainty can be expected there. In [15] the pixel-level labels in this area are softened to probability values within [0, 1] based on signed distances to the annotation boundary. A similar soft label strategy was also used in [36]. Such approaches, however, still use a full pixel-based annotation for DL model training in contrast to our approach. Another application of superpixels is data augmentation. In [34] superpixelized images (The intensity values of all pixels within a superpixel are replaced by the average) are added to training data for augmentation. In [9] superpixels provide a context-aware and object-part-aware guidance to improve the popular cutmix-based data augmentation.

3 Method

We propose to use a sparse superpixel-based labeling to achieve the goals of efficient and certain image annotation.

3.1 Superpixel-based sparse labeling

Superpixels enable a light-weight annotation with each superpixel being annotated by one label only, thus achieving an efficient image annotation. The advantage of superpixels can be easily seen in Figure 1 that shows a X-ray chest image together with the pixel-based ground truth (GT) boundary and the computed superpixels.

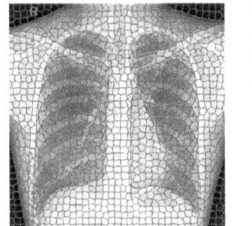

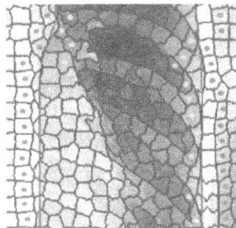

Fig. 1. A X-ray chest image: superimposed with pixel-based annotation in red and superpixels (left); illustration of superpixel-based labeling (right).

In terms of a perfect segmentation there are three different types of superpixels:

a) fully within the object of interest (foreground),
b) fully out of the object of interest (background),
c) crossing the boundary of foreground and background.

Superpixels of type a) and b) (marked yellow and green respectively in Figure 3) are easy to recognize and to label. Superpixels of type c) are the source of uncertainty for medical experts. Instead of soft labels [15,36], we do not require a pixel-based dense annotation. Instead, we leave uncertain superpixels of type c) unlabeled. This approach has two advantages. First, it avoids the pixel-based labeling in such boundary superpixels that would cause disproportionately high overhead. Second, the resulting annotation has a high certainty. Consequently, it can be expected that different medical experts will deliver very similar superpixel-based annotations, resulting a strong reduction of the inter-observer variability.

Modern annotation tools already provide functionality for superpixel-based labeling [4], but only as a means of supporting a full pixel-based annotation. Tools like Mask Editor [32] enable marking a small region by a single click only. A combined use of such annotation tools with our technique for handling unlabeled superpixels will deliver the foundation for generating large amount of annotations with high efficiency and certainty.

The main challenge here is how we can process this superpixel-based sparse annotation in such a way that we can avoid loss of segmentation quality. We will study different ways of automatically filling the boundary gap and thus generate pseudo GT (to be detailed below). Our experimental results show that that despite the sparse annotation it is possible to achieve segmentation results that are comparable or even superior to those obtained by using pixel-based dense annotation. As a positive side effect, leaving uncertain superpixels unlabeled and automatically filling the boundary gap also has the potential to significantly reduce the inter-observer variability that is inherent in pixel-based dense labeling.

Superpixel-based sparse labeling is common with weakly supervised learning such as scribbles [27,35] in their efficiency. However, superpixel labeling is rather dense apart from the boundary areas (the use of the wording sparse here is referred to the amount of superpixel labeling work). In contrast, scribbles are sparse *per se*, thus requesting the filling of large unlabeled areas. In addition, the use of superpixels and scribbles differs in the uncertainty for medical experts. Drawing scribbles has a high degree of freedom, which in turn may cause uncertainty for medical experts in terms of where, in which form, and how many to draw. This is particularly critical since interactive segmentation tends to be influenced by the input of scribbles or similar priors. Overall, these aspects, particularly that of the certainty, make superpixels an attractive choice compared to scribbles.

3.2 Gap filling strategies

We will study the following approaches to automatically fill the boundary gap and generate pseudo GT segmentations:

- No filling at all.
- A distance-based filling method.
- Use of the random walker segmentation algorithm.

In the first case there will be no labeling information available for boundary gap. Thus, the loss function used for training a DL model has to be adapted to exclude this part of the image.

The next two methods try to completely fill the gap in a meaningful manner. It thus transforms the superpixel-based sparse annotation into a pixel-based dense annotation. In distance-based filling we consider each pixel within an unlabeled superpixel. We determine its nearest distance to the foreground and background and assign it to one of them that is nearer to the pixel.

The random walker segmentation algorithm was first proposed by L. Grady [12]. Today, it remains a popular interactive segmentation method [21]. A user marks some seed pixels in each region. Then, the random walker method assigns all unseeded pixels to one of the regions using a probabilistic optimization scheme. This assignment is done by determining the region of the highest probability that a random walker starting from that pixel reaches the user provided scribbles (seeds). Random walker segmentation has found many applications [29]. In medical imaging, for instance, Biomedisa [17] is a recent platform for biomedical image segmentation that performs interpolation of sparsely pre-segmented slices based on simulation of random walker agents. Recent further development of random walker segmentation includes the nonlocal random walker [28], hierarchical random walker for segmenting large volumetric biomedical images [10], and end-to-end learned random walker [7]. In our case all pixels of the labeled superpixels of type a) and b) serve as seeds. Then, we use these seeds to obtain an assignment of the pixels of all unlabeled superpixels of type c) to obtain a pseudo dense GT segmentation.

3.3 Three-stage processing pipeline

In this work we follow a three-stage processing pipeline:

- A superpixel computation method is applied to obtain a partial GT segmentation.
- The gaps are filled to generate a full GT segmentation; except the option of keeping the gaps unfilled.
- Any backbone DL model that is suitable for the particular segmentation task under consideration can be used. The training is based on the pseudo GT generated in the last step.

This three-stage pipeline differs from unified approaches like [35], where a dedicated segmentation model is designed that cannot easily benefit from other advanced developments for image segmentation. Instead, it is modular and can be flexibly instantiated, making it possible to use advanced choices for each stage and to study different combinations. For instance, we have used general superpixel computation methods for the current experimental study (see Section 4). This element of the pipeline can be replaced by superpixel computation methods that are specially designed for particular medical imaging tasks [11,37]. The key issue of the pipeline is the gap filling. More sophisticated methods can be used, including the Segment Anything Model (SAM) [13].

4 Experimental evaluation

4.1 Experimental setting

Datasets. We use three datasets for the current study:

- Skin: 2018 ISIC Challenge on Skin Lesion Analysis Towards Melanoma Detection [8,19]. It is a comprehensive collection of dermoscopic images designed for the lesion segmentation challenge and comprises 2594 images, which are used for training and validation (80%, 20%). Additionally, the test dataset comprises 1,000 images.
- X-ray chest: It is from chest X-ray (CXR) imaging, valued for its diagnostic efficiency, affordability, and availability. This dataset encompasses a mix of tuberculosis-affected cases, normal conditions, and lung nodules (both benign and malignant). A total of 704 Kaggle CXR images are used in our study [20]. The data was divided into 70%, 20%, and 10% for training, validation, and testing, respectively.

- Ultrasound breast: It comprises 780 ultrasound breast images [3]. We divided the images into 70%, 20%, and 10% for training, validation, and testing purposes, respectively. The three used datasets have different resolutions. We uniformly normalize the size of the images to 256×256 pixels. Figure 2 shows some sample images from the three datasets.

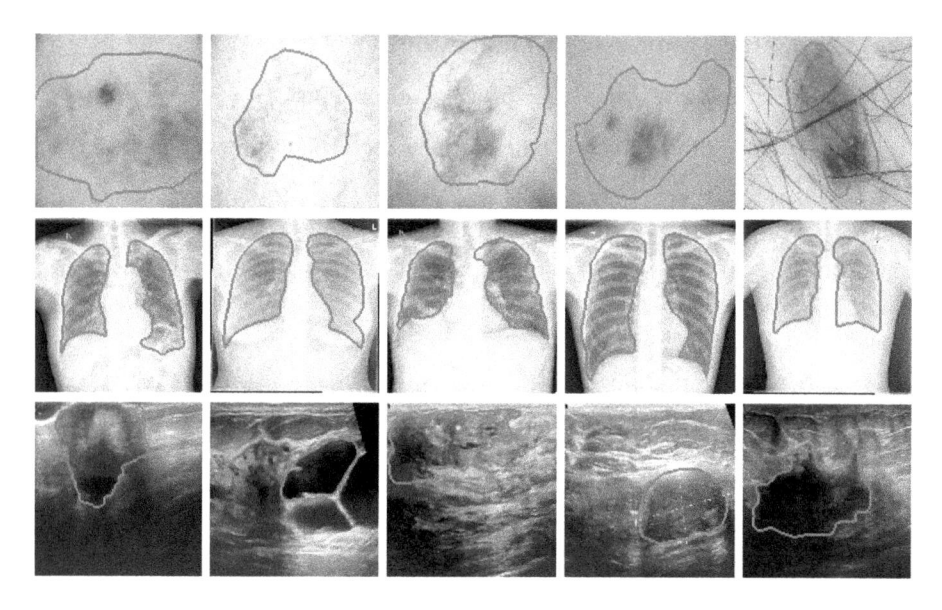

Fig. 2. Sample images with GT segmentation from the three datasets used in the study: Skin (top), X-ray chest (middle), ultrasound breast (bottom).

Experimental protocol. For the current study we simulate the superpixel-based sparse labeling. For this purpose we compute for each training image the superpixel representation. Then, each superpixel is classified into one of the three types described in Section 3.1 by using the existing original GT segmentation. Finally, all pixels of a classified superpixel takes the same related label. Thus, we have different GTs used for different purposes:

- One original GT.
- Three pseudo GTs resulting from the three different gap filling strategies.

We train four segmentation networks in total:

- The original GT is used for training a segmentation network, whose performance serves as a gold standard.
- Each of the pseudo GTs results in a segmentation network *using training data*. The performance of these networks is measured by means of the original GT *using test data*.

Superpixel generation. For the selection of superpixel computation methods, we first resort to [24]. It presents a comprehensive evaluation of 28 state-of-the-art superpixel algorithms utilizing a benchmark focusing on fair comparison and designed to provide new insights relevant for applications. The authors finally recommend six algorithms for use in practice. SLIC [1] and SEEDS [6] are among the recommended algorithms and will be used for our experiments. SLIC adapts the k-means clustering approach to efficiently generate superpixels. SEEDS performs superpixel computation within an energy optimization framework. In addition, we also consider the more recent superpixel algorithm SNIC [2]. Unlike SLIC, it is based on a non-iterative clustering and requires less space and computation resources.

We empirically fixed the parameters of the superpixel computation methods. For SLIC the compactness parameter is set to 10 and the number of segments is 800. For SEEDS the targeted number of superpixels is set to 800, the number of hierarchy levels is 4, prior is 2, and the number of histogram bins is 5. Since SNIC is a further development of SLIC, we set the two parameters (number of segments, compactness) the same as for SLIC.

Segmentation networks. We conducted experiments using two segmentation networks. Among the numerous segmentation models [31] we first chose the popular U-Net. The second one is the recent Dual-Aggregation Transformer Network called DuAT [26]. The reason for choosing DuAT is its superior performance over a number of state-of-the-art models for a variety of segmentation tasks.

Training details. Training was done in PyTorch and Keras libraries with AdamW optimizer. Key training parameters include a learning rate of 0.0001 and a dropout rate of 0.1. The models were trained with a batch size of 8 over 50 epochs, incorporating an early stopping mechanism triggered after 10 epochs without validation loss improvement.

Performance metrics. We apply the following widely used performance metrics: Dice score, Intersection over Union (IoU), precision (Prec), and sensitivity (Sens).

4.2 Experimental results

The segmentation performance is reported in Table 1. Note that our study only intends to demonstrate the relative performance of superpixel-based and full pixel labeling. Thus, we did not perform any preprocessing to achieve maximal possible performance. For instance, occlusion due to hairs in dermoscopic images (see Figure 2 for an example) may have negative impact on automatic segmentation. Hair removal is a preprocessing technique widely used in the literature [16].

Table 1. Performance on three datatsets. No_fill: No filling; Dist_fill: distance-based filling; RW_fill: random walker based filling. The best and second best result in each column are marked bold.

Skin:

Method		U-Net				DuAT			
Labeling	Superpixel	Dice	IOU	Prec	Sens	Dice	IOU	Prec	Sens
Original GT	–	0.8524	0.7623	0.8199	0.9312	0.8870	0.8109	0.8448	0.9620
SP: No_fill	SLIC	0.8406	0.7494	0.8587	0.8743	0.8071	0.6812	0.7240	0.9245
	SEEDS	0.8098	0.7055	**0.8637**	0.8187	0.7625	0.6180	0.6585	0.9229
	SNIC	0.8512	0.7640	**0.8823**	0.8665	0.8245	0.7054	0.7526	0.9234
SP: Dist_fill	SLIC	0.8568	0.7710	0.8348	0.9233	0.8798	0.7981	0.8354	0.9579
	SEEDS	**0.8818**	**0.8042**	0.8532	**0.9432**	0.8753	0.7922	0.8263	**0.9611**
	SNIC	0.8424	0.7461	0.8016	**0.9349**	**0.8855**	**0.8068**	0.8478	0.9546
SP: RW_fill	SLIC	**0.8594**	**0.7730**	0.8289	0.9321	**0.8814**	**0.8031**	0.8466	0.9503
	SEEDS	0.8356	0.7374	0.7871	0.9347	0.8580	0.7680	0.8163	0.9460
	SNIC	0.8540	0.7652	0.8224	0.9307	0.8537	0.7593	0.7884	**0.9668**

X-ray chest:

Method		U-Net				DuAT			
Labeling	Superpixel	Dice	IOU	Prec	Sens	Dice	IOU	Prec	Sens
Original GT	–	0.9537	0.9125	0.9523	0.9566	0.9537	0.9122	0.9339	0.9758
SP: No_fill	SLIC	0.7570	0.6151	**0.9853**	0.6269	0.7485	0.5987	0.8180	0.6988
	SEEDS	0.7988	0.6706	**0.9830**	0.6833	0.8254	0.6575	0.9164	0.7635
	SNIC	0.8644	0.7638	0.9486	0.8293	0.7747	0.6326	0.8228	0.7430
SP: Dist_fill	SLIC	**0.9434**	**0.8936**	0.9360	0.9524	0.9548	0.9143	**0.9521**	0.9590
	SEEDS	0.9247	0.8612	0.8795	**0.9775**	0.9568	0.9180	0.9509	**0.9642**
	SNIC	**0.9488**	**0.9033**	0.9371	0.9621	0.9597	0.9235	0.9593	0.9615
SP: RW_fill	SLIC	0.9238	0.8593	0.8947	0.9575	0.9476	0.9018	0.9370	0.9613
	SEEDS	0.9069	0.8309	0.8608	0.9619	0.9378	0.8851	0.9238	0.9576
	SNIC	0.8939	0.8093	0.8264	**0.9772**	0.9342	0.8781	0.9088	**0.9664**

Ultrasound breast:

Method		U-Net				DuAT			
Labeling	Superpixel	Dice	IOU	Prec	Sens	Dice	IOU	Prec	Sens
Original GT	–	0.8031	0.7145	0.8040	0.8594	0.9115	0.8386	0.8755	0.9117
SP: No_fill	SLIC	0.3593	0.2614	0.6309	0.2825	0.5279	0.3624	0.4290	0.7042
	SEEDS	0.2409	0.1702	0.5247	0.1826	0.4085	0.2595	0.2905	0.7175
	SNIC	0.3196	0.2056	0.2569	**0.8523**	0.5263	0.3609	0.4121	0.7575
SP: Dist_fill	SLIC	0.7859	0.6879	0.7906	0.8364	**0.8218**	**0.7220**	**0.7853**	0.8959
	SEEDS	0.7997	0.7015	**0.8326**	0.8246	**0.8296**	**0.7339**	**0.8132**	0.8872
	SNIC	**0.8078**	**0.7145**	**0.8230**	0.8480	0.8060	0.6976	0.7487	**0.9088**
SP: RW_fill	SLIC	0.7893	0.6991	0.7957	0.8241	0.7893	0.6933	0.8193	0.8134
	SEEDS	0.6804	0.5884	0.7109	0.7226	0.6864	0.5977	0.6825	0.7388
	SNIC	0.7830	0.6557	0.7842	0.7969	0.7642	0.6448	0.6743	**0.9307**

Segmentation performance. Overall, the random walker turns out to perform slightly worse than the simple distance-based filling. This is likely due to the fact

SLIC   SEEDS   SNIC

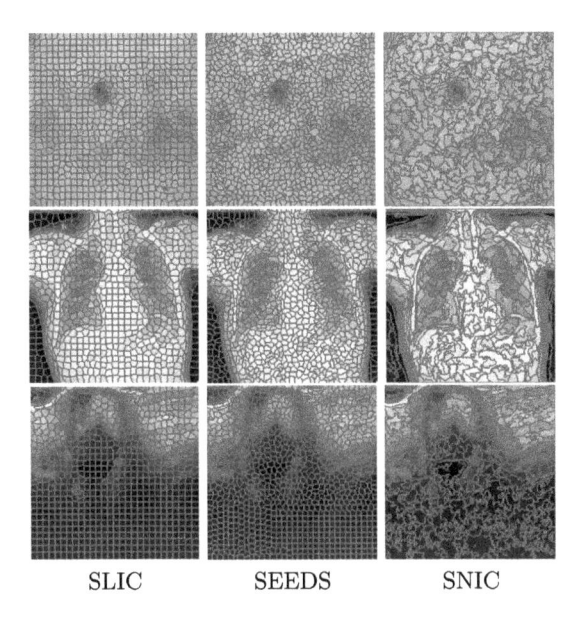

Fig. 3. Superpixel results for the sample images in the first column of Figure 2: Skin (top), X-ray chest (middle), ultrasound breast (bottom).

that the random walker is crucially controlled by an edge weight function dependent of the intensity/color differences of neighbor pixels. This implicitly assumes contours observable to some degree, which is not always available, particularly in skin and ultrasound images.

DuAT confirms the expectation of improved performance compared to the U-Net. Generally, we observe that superpixel-based labeling is able to produce segmentation performance on par or even superior to those from using the original pixel-based dense annotation. The only exception is DuAT on the Breast dataset. Among the three used datasets, the X-ray chest and the ultrasound breast dataset are more challenging, in the latter case due to the high level of noise and the low quality of the images [5]. Concerning the distance-based filling method, which seems to be a rather good compromise in terms of all performance metrics, there is no clear winner among the three tested superpixel algorithms.

Superpixel results. Figure 3 illustrates the results of the three used superpixel algorithms. While SLIC and SEEDS produce similar superpixels in term of geometric regularity, the superpixels from SNIC are rather irregular. Such elongated superpixels, however, may be advantageous in practice since they mean longer contours, thus leading to less superpixels to label.

Gap filling analysis. The gap filling is a key issue in our approach. Thus, we study to which extent the GT labeling can be reached by filling. For the samples images in Figure 2 the gap caused by superpixel-based labeling (SLIC) is illustrated in Figure 4. To quantify the filling effectivity we define:

- N: Total number of pixels in the gap area (black in Figure 4).

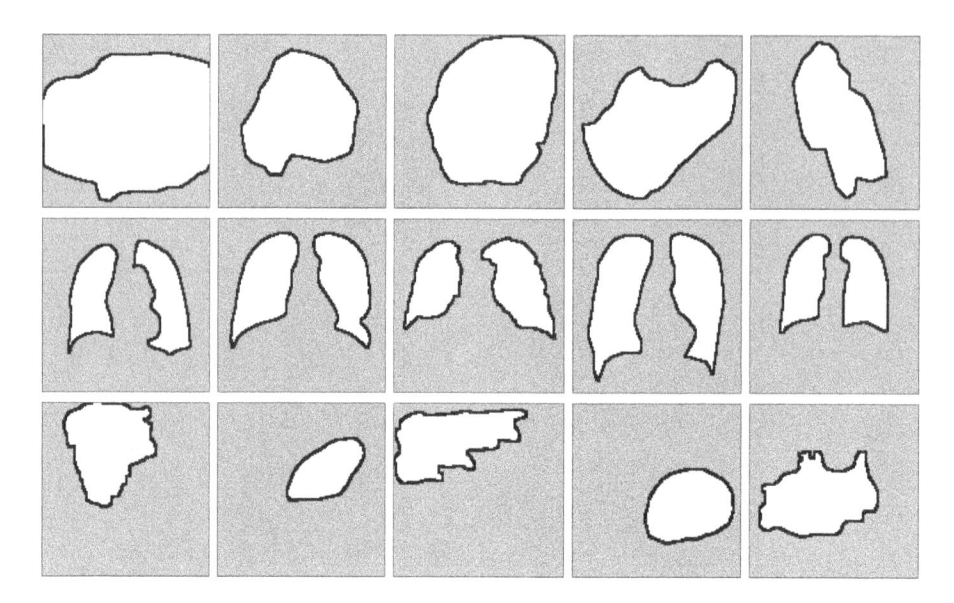

Fig. 4. Gap (in black) caused by superpixel-based labeling (SLIC) for the sample images in Figure 2: Skin (top), X-ray chest (middle), ultrasound breast (bottom).

Table 2. Gap filling analysis: Global for each dataset.

	Distance-based filling	Random walker filling
	skin dataset	
Average N	5162	5162
Average M	4503	4006
Average E_b	326	839
Average E_f	333	518
Ratio M/N	87.22%	77.58%
	chest dataset	
Average N	10183	10183
Average M	8811	7948
Average E_b	688	1607
Average E_f	684	625
Ratio of M/N	86.52%	78.05%
	breast dataset	
Average N	3661	3661
Average M	3260	2632
Average E_b	202	673
Average E_f	200	357
Ratio of M/N	89.03%	71.87%

Table 3. Gap filling analysis: Local for the samples images in Figure 2.

	Distance-based filling				
	skin dataset				
sample image	1	2	3	4	5
N	4568	9114	5986	6617	8722
M	3563	7018	4550	5103	6849
E_b	461	1036	734	758	979
E_f	544	1060	702	756	894
Ratio M/N	77.99%	77.00%	76.01%	77.11%	78.52%
	chest dataset				
N	10433	10249	9398	11714	8395
M	8914	8657	8105	9788	7189
E_b	799	778	710	945	567
E_f	720	814	583	981	629
Ratio M/N	85.44%	84.46%	86.24%	83.55%	85.75%
	breast dataset				
N	4262	5412	5357	6278	9277
M	3420	4505	4547	5350	7778
E_b	377	458	414	475	749
E_f	465	449	396	453	750
Ratio M/N	80.24%	83.24%	84.87%	85.21%	83.84%

- M: Those pixels in the gap area that are correctly assigned after gap filling.
- E_f: Foreground error (foreground pixels in the gap area are assigned to background after filling).
- E_b: Background error (background pixels in the gap area are assigned to foreground after filling).
- M/N: Ratio of correctly assigned pixels to total number of pixels in the gap area).

It holds that $N = M + E_f + E_b$. The related statistics for each dataset is presented in Table 2. Table 3 shows the same statistics for the samples images in Figure 2. The overwhelming majority (up to 90%) of the pixels in the gap area are corrected assigned by the distance-based filling strategies despite its simplicity, while random walker performs worse (see the reason discussed above). This explains the superior overall segmentation performance using the distance-based filling. Importantly, despite the remaining wrongly assigned pixels in the gap area (about 10%) we are still able to achieve final segmentation performance comparable or even superior to those obtained by using pixel-based dense annotation.

5 Conclusion

In this work we have demonstrated the fundamental potential of superpixel-based sparse labeling. In addition to high annotation efficiency, this approach reduces the inter-observer variability that is typically observed in medical imaging. Although the current experimental work is focused on medical image analysis, the proposed approach is general and can be applied to other domains as well.

Our three-stage pipeline is modular and can be flexibly instantiated. In real applications, its components need to be selected carefully considering the particular imaging modality and the object of interest. An important next step towards real applications is to explore the practical potential of our approach. This requires a user study involving medicine experts to do superpixel labeling (instead of the current simulation). For this purpose, there is also a need of developing a suitable user interface to support this novel form of labeling.

Acknowledgments. This work was supported by the Deutsche Forschungsgemeinschaft (DFG) - CRC 1450 - 431460824.

References

1. Achanta, R., Shaji, A., Smith, K., Lucchi, A., Fua, P., Süsstrunk, S.: SLIC superpixels compared to state-of-the-art superpixel methods. IEEE Trans. Pattern Anal. Mach. Intell. **34**(11), 2274–2282 (2012)
2. Achanta, R., Süsstrunk, S.: Superpixels and polygons using simple non-iterative clustering. In: CVPR. pp. 4895–4904 (2017)
3. Al-Dhabyani, W., Gomaa, M., Khaled, H., Fahmy, A.: Dataset of breast ultrasound images. Data Brief **28**, 104863 (2020)
4. Aljabri, M., AlAmir, M., Ghamdi, M.A., Abdel-Mottaleb, M., Collado-Mesa, F.: Towards a better understanding of annotation tools for medical imaging: a survey. Multimedia Tools and Applications **81**(18), 25877–25911 (2022)
5. Avola, D., Cinque, L., Fagioli, A., Foresti, G.L., Mecca, A.: Ultrasound medical imaging techniques: A survey. ACM Computing Surveys **54**(3), 67:1–67:38 (2022)
6. den Bergh, M.V., Boix, X., Roig, G., Gool, L.V.: SEEDS: superpixels extracted via energy-driven sampling. Int. J. Comput. Vision **111**(3), 298–314 (2015)
7. Cerrone, L., Zeilmann, A., Hamprecht, F.A.: End-to-end learned random walker for seeded image segmentation. In: IEEE Conference on Computer Vision and Pattern Recognition. pp. 12559–12568 (2019)
8. Codella, N.C.F., Gutman, D.A., Celebi, M.E., Helba, B., Marchetti, M.A., Dusza, S.W., Kalloo, A., Liopyris, K., Mishra, N.K., Kittler, H., Halpern, A.: Skin lesion analysis toward melanoma detection: A challenge at the 2017 international symposium on biomedical imaging (isbi), hosted by the international skin imaging collaboration (ISIC). In: 15th IEEE International Symposium on Biomedical Imaging (ISBI). pp. 168–172 (2018)
9. Dornaika, F., Sun, D.: LGCOAMix: Local and global context-and-object-part-aware superpixel-based data augmentation for deep visual recognition. IEEE Trans. Image Process. **33**, 205–215 (2024)

10. Drees, D., Eilers, F., Jiang, X.: Hierarchical random walker segmentation for large volumetric biomedical images. IEEE Trans. Image Process. **31**, 4431–4446 (2022)
11. Fang, L., Wang, X., Wang, M.: Superpixel/voxel medical image segmentation algorithm based on the regional interlinked value. Pattern Anal. Appl. **24**(4), 1685–1698 (2021)
12. Grady, L.J.: Random walks for image segmentation. IEEE Transactions on Pattern Analysis and Machince Intelligence **28**(11), 1768–1783 (2006)
13. Huang, Y., Yang, X., Liu, L., Zhou, H., Chang, A., Zhou, X., Chen, R., Yu, J., Chen, J., Chen, C., Liu, S., Chi, H., Hu, X., Yue, K., Li, L., Grau, V., Fan, D., Dong, F., Ni, D.: Segment anything model for medical images? Med. Image Anal. **92**, 103061 (2024)
14. Kuhlmann, J., Rothaus, K., Jiang, X., Faatz, H., Pauleikhoff, D., Gutfleisch, M.: 3d retinal vessel segmentation in octa volumes: Annotated dataset MORE3D and hybrid U-net with flattening transformation. In: DAGM German Conference on Pattern Recognition (GCPR) (2023)
15. Li, H., Wei, D., Cao, S., Ma, K., Wang, L., Zheng, Y.: Superpixel-guided label softening for medical image segmentation. In: MICCAI, Part IV. pp. 227–237 (2020)
16. Li, W., Raj, A.N.J., Tjahjadi, T., Zhuang, Z.: Digital hair removal by deep learning for skin lesion segmentation. Pattern Recogn. **117**, 107994 (2021)
17. Lösel, P.D., van de Kamp, T., Jayme, A., Ershov, A., Faragó, T., Pichler, O., Jerome, N.T., Aadepu, N., Bremer, S., Chilingaryan, S.A., Heethoff, M., Kopmann, A., Odar, J., Schmelzle, S., Zuber, M., Wittbrodt, J., Baumbach, T., Heuveline, V.: Introducing Biomedisa as an open-source onlineplatform for biomedical image segmentation. Nat. Commun. **11**, 5577 (2020)
18. Ma, Y., Hao, H., Xie, J., Fu, H., Zhang, J., Yang, J., Wang, Z., Liu, J., Zheng, Y., Zhao, Y.: ROSE: A retinal oct-angiography vessel segmentation dataset and new model. IEEE Trans. Med. Imaging **40**(3), 928–939 (2021)
19. Mirikharaji, Z., Abhishek, K., Bissoto, A., Barata, C., Avila, S., Valle, E., Celebi, M.E., Hamarneh, G.: A survey on deep learning for skin lesion segmentation. Med. Image Anal. **88**, 102863 (2023)
20. Rahman, T., Khandakar, A., Kadir, M.A., Islam, K.R., Islam, K.F., Mazhar, R.: Reliable tuberculosis detection using chest x-ray with deep learning, segmentation and visualization. IEEE Access **8**, 586–601 (2020)
21. Ramadan, H., Lachqar, C., Tairi, H.: A survey of recent interactive image segmentation methods. Computational Visual Media **6**(4), 355–384 (2020)
22. Sasmal, B., Dhal, K.G.: A survey on the utilization of superpixel image for clustering based image segmentation. Multimedia Tools and Applications **82**(23), 35493–35555 (2023)
23. Schmeing, M., Jiang, X.: Faithful disocclusion filling in depth image based rendering using superpixel-based inpainting. IEEE Trans. Multimedia **17**(12), 2160–2173 (2015)
24. Stutz, D., Hermans, A., Leibe, B.: Superpixels: An evaluation of the state-of-the-art. Comput. Vis. Image Underst. **166**, 1–27 (2018)
25. Subudhi, S., Patro, R.N., Biswal, P.K., Dell'Acqua, F.: A survey on superpixel segmentation as a preprocessing step in hyperspectral image analysis. IEEE Journal of Selected Topics in Applied Earth Observations and Remote Sensing **14**, 5015–5035 (2021)
26. Tang, F., Xu, Z., Huang, Q., Wang, J., Hou, X., Su, J., Liu, J.: DuAT: Dual-aggregation transformer network for medical image segmentation. In: Proc. of 6th Chinese Conference on Pattern Recognition and Computer Vision (PRCV). LNCS, vol. 14429, pp. 343–356. Springer (2023)

27. Valvano, G., Leo, A., Tsaftaris, S.A.: Learning to segment from scribbles using multi-scale adversarial attention gates. IEEE Trans. Med. Imaging **40**(8), 1990–2001 (2021)
28. Wang, H., Shen, J., Yin, J., Dong, X., Sun, H., Shao, L.: Adaptive nonlocal random walks for image superpixel segmentation. IEEE Transactions on Circuits Systems and Video Technology **30**(3), 822–834 (2020)
29. Wang, Z., Guo, L., Wang, S., Chen, L., Wang, H.: Review of random walk in image processing. Archives of Computational Methods in Engineering **26**, 17–34 (2019)
30. Webb, J.M., Adusei, S.A., Wang, Y., Samreen, N., Adler, K., Meixner, D.D., Fazzio, R.T., Fatemi, M., Alizad, A.: Comparing deep learning-based automatic segmentation of breast masses to expert interobserver variability in ultrasound imaging. Comput. Biol. Med. **139**, 104966 (2021)
31. Yao, X., Wang, X., Wang, S., Zhang, Y.: A comprehensive survey on convolutional neural network in medical image analysis. Multimedia Tools and Applications **81**(29), 41361–41405 (2022)
32. Zhang, C., Loken, K., Chen, Z., Xiao, Z., Kunkel, G.: Mask Editor: an image annotation tool for image segmentation tasks. CoRR **abs/1809.06461** (2018)
33. Zhang, L., Tanno, R., Xu, M., Huang, Y., Bronik, K., Jin, C., Jacob, J., Zheng, Y., Shao, L., Ciccarelli, O., Barkhof, F., Alexander, D.C.: Learning from multiple annotators for medical image segmentation. Pattern Recogn. **138**, 109400 (2023)
34. Zhang, Y., Yang, L., Zheng, H., Liang, P., Mangold, C., Loreto, R.G., Hughes, D.P., Chen, D.Z.: SPDA: superpixel-based data augmentation for biomedical image segmentation. In: International Conference on Medical Imaging with Deep Learning (MIDL). pp. 572–587 (2019)
35. Zhou, M., Xu, Z., Zhou, K., Tong, R.K.: Weakly supervised medical image segmentation via superpixel-guided scribble walking and class-wise contrastive regularization. In: MICCAI, Part II. pp. 137–147 (2023)
36. Zhou, Q., He, T., Zou, Y.: Superpixel-oriented label distribution learning for skin lesion segmentation. Diagnosis **12**(29), 938 (2022)
37. Zhuang, S., Li, F., Raj, A.N.J., Ding, W., Zhou, W., Zhuang, Z.: Automatic segmentation for ultrasound image of carotid intimal-media based on improved superpixel generation algorithm and fractal theory. Comput. Methods Programs Biomed. **205**, 106084 (2021)

Attention Seekers U-Net with Mamba for Sub-cellular Segmentation

Pratik Sinha[1(✉)] and Arif Ahmed Sekh[1,2]

[1] XIM University, Bhubaneswar, India
`pratiksinha.cs@gmail.com`
[2] UiT the Arctic University of Norway, Tromsø, Norway

Abstract. Accurate segmentation of subcellular structures from microscopy images is crucial for understanding cellular processes and functions, but it presents significant challenges due to factors such as noise, low signal-to-noise ratios, limited resolution, and complex spatial arrangements. To address these challenges, we introduce CMU-Net, a novel hybrid architecture that combines the strengths of U-Net, Mamba blocks (SSMs), and Convolutional Block Attention Modules (CBAM). U-Net provides a strong foundation for feature extraction, Mamba blocks efficiently capture long-range dependencies, and CBAM modules refine feature representations by selectively focusing on relevant information. We evaluated CMU-Net on three diverse datasets consisting both fluorescence and label-free microscopy images of mitochondria and endoplasmic reticulum (ER). The quantitative and qualitative results demonstrate that CMU-Net consistently outperforms various baseline methods, including established CNN-based and Transformer-based models, achieving improved segmentation accuracy and boundary representation. This study highlights the potential of our hybrid approach to significantly contribute to the field of subcellular image analysis, promoting a deeper understanding of cellular organization and function. Code is available at https://github.com/beasthunter758/CMU-Net.

Keywords: Subcellular Image Segmentation · Fluorescence Microscopy · Long-Range Dependencies · Mamba · Attention · Morphological analysis

1 Introduction

Microscopic subcellular segmentation has become crucial for understanding cellular processes and functions, especially in fields like disease diagnosis, drug development, and fundamental biological research. Precise segmentation of subcellular structures such as *mitochondria* and *endoplasmic reticulum (ER)* allows scientists to measure their shape, movement, and connections, offering a critical understanding of cellular well-being and operation [1,2]. However, the process of accurately dividing cells into smaller components is difficult because of the fundamental limitations of optical microscopy. These constraints include issues such as noise, low signal-to-noise ratios, restricted resolution due to diffraction, and the complex spatial arrangements of subcellular structures.

Several microscopy techniques offer essential tools for investigating subcellular structures, each possessing distinct advantages and limits. *Fluorescence microscopy*[3] utilizes fluorescent markers to precisely label molecules and structures of interest, enabling accurate studying and analysis of their distribution and movement within cells. This method is highly effective for analyzing individual proteins, organelles, or cellular processes. However, it necessitates the use of external markers, which might interfere with the natural cellular surroundings. In contrast, *label-free microscopy*[4] techniques, such as phase-contrast or differential interference contrast (DIC) microscopy, offer an easy way to examine cellular structures without demanding external markers. These approaches apply changes in refractive indices or optical path lengths to generate contrast and highlight cellular architecture. While label-free approaches offer advantages in retaining the original state of the cells, they sometimes result in poorer contrast and may require additional computational processing for precise segmentation.

Deep learning[5] techniques are becoming highly effective tools for image segmentation, showcasing exceptional achievements in many biological applications. Architectures such as U-Net [6] have been popular due to their encoder-decoder structure and skip connections, which allow for rapid extraction of features and preservation of complex details. U-Net++ [7] further improves upon this through implementing dense skip connections and convolution blocks on skip paths, boosting gradient flow and segmentation accuracy. Additionally, the inclusion of attention mechanisms, such as Convolutional Block Attention Modules (CBAM) [8], has shown helpful in refining feature representations by focusing on the most important information for certain tasks. Nevertheless, these models frequently encounter difficulties when faced with the specific challenges posed by subcellular segmentation in both fluorescence and label-free microscopy images.

Research Gap and Objectives: The current research gap resides in the lack of deep learning models specifically designed to address the challenges of subcellular segmentation while successfully taking advantage of the capabilities of existing architectures and developments in sequence modeling. Our objective is to develop a novel hybrid architecture that combines the advantages of U-Net[6], Mamba blocks[9], and CBAM[8] to achieve superior performance in subcellular segmentation tasks across multiple imaging modalities. We seek to assess our model on multiple datasets covering both fluorescence and label-free microscopy images[3,4], comparing its performance to various state-of-the-art models using evaluation metrics such as F1 score and Intersection over Union (IoU) to quantify segmentation accuracy. By closing this gap, we want to promote more efficient and precise study of subcellular structures, ultimately contributing to a deeper knowledge of cellular function and dysfunction.

Contribution: Our key contribution is the development of a novel hybrid architecture that effectively integrates U-Net[6], Mamba blocks[9], and CBAM[8] specifically for subcellular segmentation tasks. This integration attempts to take advantage of the capabilities of each component to tackle the particular challenges of this area while giving a fast yet efficient solution. The U-Net[6] framework provides an effective basis for feature extraction and multi-scale processing, adapting to the various properties of both fluorescence and label-free images. Mamba blocks[9], recognized for their computing speed and ability to capture long-range relationships, address limitations in resolution and the complex spatial arrangements of subcellular organelles. This enables the network to efficiently model the complex relationships between various parts of the structures, resulting to more accurate segmentation results. Finally, the addition of CBAM[8] modules allows the network to dynamically improve feature representations by focusing on the most discriminative information for precise segmentation, thus improving its ability to distinguish between foreground and background pixels. This combination of speed and precision makes our suggested methodology extremely suitable for evaluating large-scale microscopy datasets, supporting high-throughput research and quantitative study of subcellular structures.

2 Related Work

The field of medical image segmentation has undergone significant developments in recent years, driven by the fast development of deep learning methods. Convolutional neural networks (CNNs) and Transformers have emerged as the most prominent architectures, each with their unique strengths and limitations. Our proposed hybrid architecture expands upon the foundation built by these current approaches, including the U-Net framework, U-Mamba (with its SSM block)[10], and CBAM to solve the unique issues of subcellular segmentation.

2.1 U-Net and Its Variants

U-Net [6] has become the foundation in biomedical image segmentation due to its efficient encoder-decoder architecture and strong feature representation learning. The encoder progressively downsamples the input image through a series of convolution, activation, and max-pooling operations, capturing contextual information at various scales. The decoder mirrors this process by upsampling the feature maps and integrating them with similar feature maps from the encoder using skip connections. This allows the network to recover spatial information lost during downsampling, leading to precise localization and segmentation of objects.

The success of U-Net has led to various versions and extensions, each trying to improve upon its performance and handle unique issues in medical image segmentation. *U-Net++* [7] offers nested and dense skip routes, significantly improving the fusion of multi-scale information while improving gradient flow during training. *Attention U-Net* [11] includes an attention gate mechanism that learns to focus on target structures and ignore irrelevant background information. Other notable versions include *UNETR* [12], which integrates a Transformer encoder for capturing long-range dependencies, and *Swin-UNETR* [13], which utilizes a hierarchical Swin Transformer with shifted windows for efficient global context modeling.

Mathematically, the U-Net architecture can be represented as a series of nested functions. Let x be the input image and y be the output segmentation map. The encoder function, f_{enc}, maps the input image to a low-dimensional latent representation: $z = f_{enc}(x)$. The decoder function, f_{dec}, then upsamples the latent representation and combines it with features from the encoder to generate the segmentation map: $y = f_{dec}(z, f_{enc}(x))$. The skip connections ensure that fine-grained details from the encoder are incorporated into the upsampling process, contributing to accurate segmentation results.

2.2 U-Mamba and State Space Sequence Models

U-Mamba [10] enhances the U-Net architecture by placing *Mamba blocks* [9] into the encoder. Mamba blocks, based on the ideas of *State Space Sequence Models (SSMs)* [14], offer an efficient method for capturing long-range dependencies in sequential data. SSMs operate by maintaining an internal hidden state that expands over time, allowing each element in the sequence to interact with information from all previous elements. This property makes them particularly well-suited for positions where understanding long-range context is critical, such as subcellular segmentation.

Mamba implements a structured state space model, *S4* [14], which efficiently represents long-range dependencies through a precise parameterization of the state matrix. The model also integrates an input-dependent selection procedure, allowing it to selectively focus on relevant information based on the input sequence. This selected process sets Mamba apart from standard self-attention techniques found in Transformers, offering both efficiency and effectiveness in modeling long-range dependencies.

Recent publications have examined the usefulness of Mamba blocks in medical image segmentation. *VM-UNet* [15] utilizes Mamba blocks within a U-Net design, showing its potential for improving segmentation accuracy. *H-vmunet* [16] further extends this by introducing a high-order vision Mamba U-Net, progressively reducing duplicate information and improving local feature extraction. *SegMamba* [17] examines the use of Mamba for long-range sequential modeling in 3D medical image segmentation. These findings demonstrate the growing interest and promise of SSM-based methods in solving the issues of medical image analysis.

Mathematically, the SSM can be described by the following equations:

$$\frac{dh(t)}{dt} = Ah(t) + Bx(t) \text{ and } y(t) = Ch(t)$$

where $x(t)$ is the input sequence, $h(t)$ is the hidden state at time t, and A, B, and C are matrices that govern the evolution of the hidden state and the generation of the output sequence $y(t)$. The discretization of the continuous-time model into a discrete-time model is achieved using a timescale parameter and a discretization rule, such as the zero-order hold (ZOH) method.

2.3 Convolutional Block Attention Modules (CBAM)

Convolutional Block Attention Modules (CBAM) [8] are attention mechanisms meant to improve the representational power of convolutional neural networks. CBAMs refine intermediate feature maps by inferring attention maps along two separate dimensions: channel and spatial. The channel attention module focuses on "what" is meaningful in the given data, whereas the spatial attention module focuses on "where" the informative components are placed.

The channel attention module utilizes global average pooling and max pooling operations to collect spatial information, followed by a shared multi-layer perceptron (MLP) to generate channel attention weights. These weights are then applied to the input feature map, emphasizing informative channels and suppressing less significant ones. The spatial attention module utilizes average pooling and max pooling operations along the channel axis, concatenating the resulting feature maps. A convolution layer then processes this concatenated representation to provide a spatial attention map, highlighting relevant places in the feature map.

The integration of CBAM modules into various U-Net topologies has proven beneficial in improving segmentation performance. *MALUNet* [18] combines CBAM with a multi-attention mechanism and a lightweight U-Net for skin lesion segmentation. *MSNet* [19] implements CBAM within a multi-scale subtraction network for polyp segmentation. These experiments demonstrate the versatility and usefulness of CBAM in refining feature representations and improving segmentation accuracy across different tasks.

3 Methodology

3.1 Hybrid Architecture for Subcellular Segmentation

We propose a new hybrid architecture, called *CMU-Net*, which efficiently addresses the challenges of subcellular segmentation by using the capabilities of U-Net, Mamba blocks (SSMs), and Convolutional Block Attention Modules (CBAM). This architecture aims to achieve accurate and efficient segmentation by effectively capturing long-range dependencies and selectively focusing on relevant features within subcellular images. The overall architecture is depicted in Figure 1, and the individual components are shown in Figure 2.

3.2 Network Architecture

U-Net Backbone : *CMU-Net* utilizes a U-Net [6] architecture as its backbone, with an encoder-decoder structure with skip connections. The encoder consists of four downsampling levels, each consisting of two 3 × 3 convolutional layers with Instance Normalization[20] and LeakyReLU activation[21]. Max pooling operations with a stride of 2 downsample the feature maps at each level, while the number of feature channels doubles, starting from 32 in the first level and reaching 512 in the fourth level. The hierarchical structure of the network allows it to learn multi-scale feature representations, covering both local details and global context.

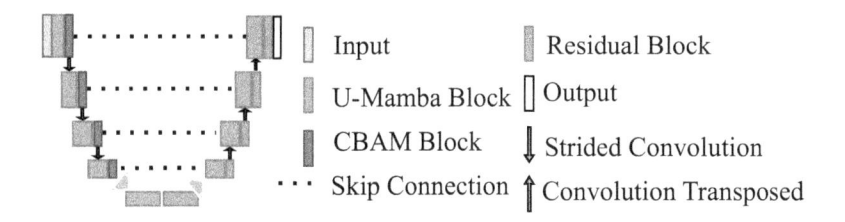

Fig. 1. The overall architecture of the CMU-NET Model.

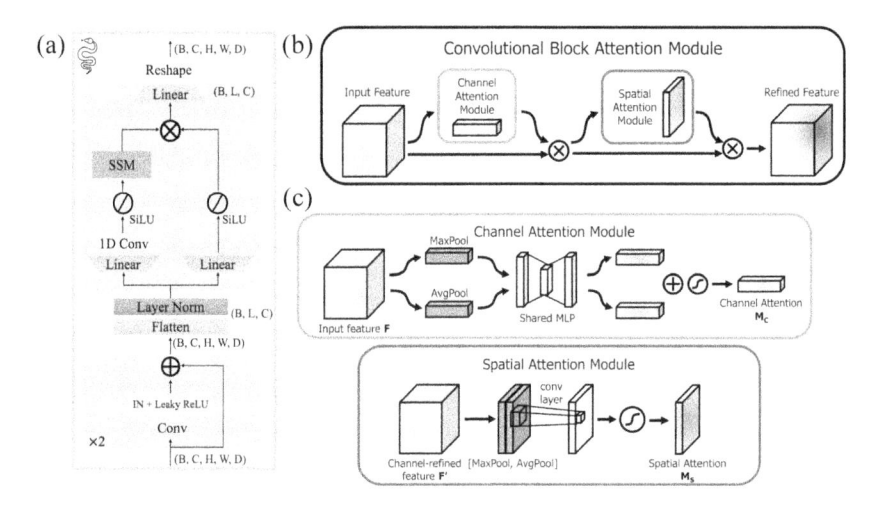

Fig. 2. Components of the CMU-Net architecture: (a) Structure of the U-Mamba block, Adapted from [10] (b) Overall CBAM block, Adapted from [8] (c) Breakdown of the CBAM block, Adapted from [8].

Mamba Block Integration : We integrate Mamba blocks [9,10] into the U-Net encoder to improve the network's ability to represent distant relationships that are important for understanding the complex spatial organization of subcellular structures. In addition, a Mamba layer is integrated following the two convolutional blocks at every downsampling level. The Mamba layer initially employs Layer Normalization on the input feature map x_l at level l and then feeds it into the Mamba block. The Mamba block leverages the Structured State Space Sequence Model (S4) [14], noted for its efficiency in addressing long-range dependencies. Let $h_l(t)$ represent the hidden state of the SSM at time step t. The state updating and output equations for the SSM can be stated as:

$$\frac{dh_l(t)}{dt} = A_l h_l(t) + B_l x_l(t) \; and \; y_l(t) = C_l h_l(t)$$

where A_l, B_l, and C_l are matrices guiding the state transition and the generation of output at level l. Discretization of this continuous-time model is accomplished utilizing a timescale parameter Δ_l and the zero-order hold (ZOH) approach.

In our implementation, the Mamba block consists of a linear projection layer (in_proj) that transforms the input features to a higher-dimensional space, followed by a depthwise convolution ($conv1d$) with a kernel size of 4 and SiLU activation. Further linear projections (x_proj and dt_proj) are utilized for state updates, and a final linear layer (out_proj) generates the output features y_l.

CBAM Incorporation : To further refine the feature representations learned by the Mamba blocks, we incorporate CBAM modules [8] after each Mamba layer in the encoder. The CBAM module applies channel and spatial attention mechanisms sequentially:

Channel Attention: Average pooling and max pooling operations are applied along the spatial dimensions of the Mamba block output y_l, resulting in $F_{avg}^{c,l}$ and $F_{max}^{c,l}$. Both pooled features pass through a shared multi-layer perceptron (MLP) with one hidden layer to generate channel attention weights:

$$M_c(y_l) = \sigma(MLP(AvgPool(y_l)) + MLP(MaxPool(y_l)))$$

The channel attention map $M_c(y_l)$ is then multiplied element-wise with the input feature map y_l to emphasize informative channels: $y_l' = M_c(y_l) \otimes y_l$.

Spatial Attention: Average pooling and max pooling operations are applied along the channel dimension of the channel-refined feature map y_l', resulting in $F_{avg}^{s,l}$ and $F_{max}^{s,l}$. The pooled features are concatenated and processed using a convolution layer with a kernel size of 7×7 to generate a spatial attention map:

$$M_s(y_l') = \sigma(f^{7\times 7}([AvgPool(y_l'); MaxPool(y_l')]))$$

4 Experiments and Results

4.1 Datasets

To evaluate the performance and adaptability of our proposed CMU-Net model, we conducted experiments on three diverse datasets containing images of subcellular structures, as summarized in Table 1 and visualized in Figure 3. The datasets consist of fluorescence and label-free microscopy techniques, including different cell types, imaging devices, and subcellular targets.

Dataset A (Rat Heart Cell, Mitochondria)[22]: This dataset contains 512 fluorescence microscopy images of mitochondria within rat heart cells captured using an OMEX microscope. The images display varying mitochondrial morphology, intensity, and background noise, providing a more realistic and challenging context to evaluate the model's performance.

Dataset B (ER)[23]: This dataset consists of 175 label-free microscopy pictures depicting the endoplasmic reticulum (ER) in cell types that are not identified. The images were captured using a Nikon Eclipse Ti-E microscope and display

Table 1. Datasets Used for Experimentation

Reference	Cell/Sub-cell Type	Imaging Device	Number of Samples
Sekh et al. [22] (A)	Rat heart cell, mitochondria	OMEX	512
Luo et al. [23] (B)	Unknown cell, ER	Nikon Eclipse Ti-E	175
Luo et al. [23] (C)	Unknown cell, mitochondria	Nikon Eclipse Ti-E	253

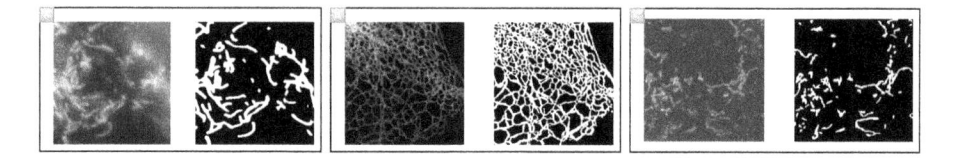

Fig. 3. Datasets used for evaluation.

the unique tubular network structure of the endoplasmic reticulum (ER). This dataset allows us to assess the model's ability to segment subcellular structures in label-free images and its generalizability to different organelles.

Dataset C (Unknown Cell, Mitochondria)[23]: This dataset includes 253 label-free microscopy images of mitochondria within unknown cell types, also taken with a Nikon Eclipse Ti-E microscope. Similar to Dataset C, this dataset allows us to examine the model's performance on label-free images and its ability to generalize to different cell types and imaging settings.

4.2 Implementation Details

We implemented the *CMU-Net* model using the popular nnU-Net framework [24] due to its modular design and self-configuring features. This framework allowed us to focus on the network architecture while maintaining consistency in image preprocessing, data augmentation, and training procedures. All networks, including our proposed model and baseline models used for comparison, were trained for 150 epochs on a single NVIDIA RTX A5000 GPU.

The nnU-Net[24] framework automatically configures hyperparameters based on the dataset attributes. The patch size, batch size, and network parameters such as the number of resolution levels and downsampling processes were determined automatically for each dataset. We utilized stochastic gradient descent (SGD)[25] as the optimizer, and the loss function was a combination of Dice loss and cross-entropy loss, as this has been demonstrated to provide stable performance across many segmentation tasks. The learning rate was initially set to 0.01 and decayed using a cosine annealing schedule with a minimum learning rate of $1e - 5$.

To enhance data diversity and improve model generalizability, we applied standard data augmentation techniques during training. These included random horizontal and vertical flips, random rotations, and elastic deformations. The specific augmentation parameters were determined by the nnU-Net framework based on the characteristics of each dataset.

4.3 Evaluation Metrics

To quantitatively assess the performance of our *CMU-Net* model and baseline methods, we employed two widely used evaluation metrics for image segmentation: F1 score and Intersection over Union (IoU).

F1 Score: The F1 score is the harmonic mean of precision and recall, providing a balanced evaluation of the model's accuracy in identifying both foreground (subcellular structures) and background pixels. It is particularly beneficial in situations with uneven class distributions, such as subcellular segmentation, where the background pixels often outnumber the foreground pixels.

Intersection over Union (IoU): IoU quantifies the overlap between the predicted segmentation mask and the ground truth mask, providing an indication of the model's capability to reliably identify the boundaries of subcellular structures. It is calculated as the ratio between the intersection and union of the predicted and ground truth masks.

Justification for Using F1 Score and IoU:

Binary Segmentation: Both F1 score and IoU are well-established metrics for evaluating binary segmentation tasks, where the goal is to categorize each pixel as belonging to either the foreground (object) class or the background class.

Imbalanced Class Distributions: The F1 score is particularly beneficial in scenarios with imbalanced class distributions, which is common in subcellular segmentation where background pixels frequently dominate. It provides a more reasonable estimate than measurements like accuracy, which can be misleading in such cases.

Boundary Representation: IoU directly assesses the overlap between predicted and ground truth masks, making it an appropriate metric for measuring the model's capacity to accurately outline the boundaries of subcellular structures. This is critical for subsequent assessments of morphology and dynamics.

By employing both F1 score and IoU, we obtain a comprehensive evaluation of the segmentation performance, surrounding both pixel-wise classification accuracy and the accuracy of boundary localization.

4.4 Quantitative and qualitative segmentation results

To evaluate the performance of our proposed CMU-Net model and baseline methods across different datasets and subcellular structures, we present the quantitative results for each dataset separately.

Table 2 presents the quantitative segmentation performance on Dataset A containing fluorescence microscopy images of mitochondria within rat heart cells [22]. CMU-Net achieves the highest F1 score and IoU compared to all baseline methods, demonstrating its effectiveness in accurately segmenting mitochondria.

Table 3 presents the quantitative results on Dataset B, which consists of label-free microscopy images of the endoplasmic reticulum (ER) [23]. CMU-Net

Table 2. Performance of Segmentation Methods on Dataset A (Rat, Mito)[22]

Method	F1/DSC	IoU
nnU-Net[24]	0.7593	0.7228
Seg ResNet[26]	0.7590	0.7214
UNETR[12]	0.7532	0.7118
Swin UNETR[13]	0.7573	0.7194
U-Mamba[10]	0.7598	0.7231
CMU-Net	**0.7883**	**0.7420**

Table 3. Performance of Segmentation Methods on Dataset B (ER)[23]

F1/DSC	IoU
0.8438	0.7312
0.8405	0.7267
0.8447	0.7330
0.8449	0.7328
0.8404	0.7259
0.8669	**0.7559**

Table 4. Performance of Segmentation Methods on Dataset C (Mito)[23]

F1/DSC	IoU
0.8175	0.6949
0.8163	0.6938
0.8039	0.6765
0.8203	0.6990
0.8232	0.7028
0.8358	**0.7255**

once again demonstrates improved performance, highlighting its ability to handle various subcellular structures and generalize to label-free microscopy images.

Table 4 presents the quantitative results on Dataset C, which comprises label-free microscopy images of mitochondria [23]. CMU-Net maintains its leading performance, proving its effectiveness across diverse cell types and imaging settings within the context of label-free microscopy.

Our CMU-Net model consistently outperforms all baseline methods across both evaluation metrics and all datasets. This indicates the effectiveness of our hybrid approach in precisely segmenting subcellular structures. Notably, the improvement in IoU suggests that CMU-Net is particularly excels at accurately defining the boundaries of these structures. The integration of Mamba blocks, capable of capturing long-range dependencies, likely contributes to this improved boundary localization by allowing the network to effectively model the spatial relationships and context within the images. Furthermore, the inclusion of CBAM modules enhances the network's ability to focus on the most relevant features for segmentation. The channel attention mechanism within CBAM helps identify and emphasize informative channels, while the spatial attention mechanism guides the network to focus on important regions within the feature maps. This attention-based refinement further contributes to the improved accuracy and boundary identification observed in the CMU-Net results.

Qualitative Results Visually, as shown in Figure 4, the segmentations produced by CMU-Net display better agreement with the ground truth masks, particularly in challenging regions with complex morphology or low contrast. The baseline methods, especially those based solely on convolutional operations (nnU-Net and SegResNet), tend to produce less accurate segmentations, with noticeable errors in boundary segmentation and fragmented structures. The improved performance of CMU-Net can be attributed to its ability to capture both local and global context through the combination of U-Net, Mamba blocks, and CBAM. Mamba blocks enhance the modeling of long-range dependencies, allowing the network to better understand the overall structure and spatial relationships within the subcellular images. Simultaneously, CBAM modules refine the feature representations by focusing on the most biased information, leading to more accurate and detailed segmentations.

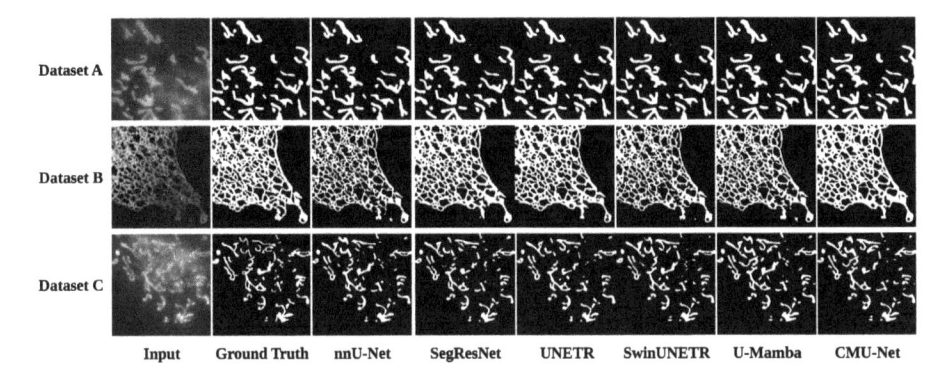

Fig. 4. Presents a qualitative comparison of segmentation results for our CMU-Net model and the baseline methods on representative images from Datasets A,B and C. Visually, the segmentations produced by CMU-Net exhibit better agreement with the ground truth masks, particularly in challenging regions with complex morphology or low contrast. The baseline methods, especially those based solely on convolutional operations (nnU-Net and SegResNet), tend to produce less accurate segmentations, with noticeable errors in boundary delineation and fragmented structures. The improved performance of CMU-Net can be attributed to its ability to capture both local and global context through the combination of U-Net, Mamba blocks, and CBAM. Mamba blocks facilitate the modeling of long-range dependencies, allowing the network to better understand the overall structure and spatial relationships within the subcellular images. Simultaneously, CBAM modules refine the feature representations by focusing on the most discriminative information, leading to more accurate and detailed segmentations.

Overall, both the quantitative and qualitative results demonstrate the improved performance of our CMU-Net model for subcellular segmentation tasks across different datasets and imaging modalities. The integration of Mamba blocks and CBAM modules within the U-Net framework allows the network to effectively capture complex spatial relationships and refine feature representations, leading to more accurate and visually appealing segmentations.

4.5 Morphological Analysis

Following the segmentation process, as illustrated in Figure 5 we conducted a detailed morphological analysis of the mitochondria to demonstrate the practical utility of our improved segmentation method. This analysis is crucial for understanding mitochondrial dynamics and their relevance to cellular health [27].

Classification Methodology We classified each disjoint mitochondrial structure into three categories: "Dot", "Rod", and "Network". The classification was based on size thresholds: $Dot < 100$ pixels, $100 \leq Rod \leq 500$ pixels, and $Network > 500$ pixels.

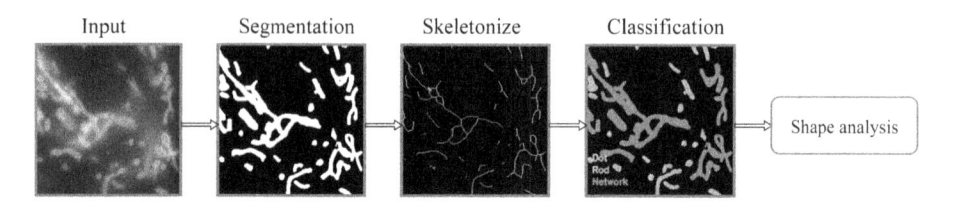

Fig. 5. Steps in morphological analysis: (a) Input image, (b) Segmentation, (c) Skeletonization, (d) Classification, (e) Shape analysis.

Statistical Analysis We hypothesized that an accurate segmentation method should correctly reflect the statistical information of different mitochondrial phenotypes as shown in Figure 6.

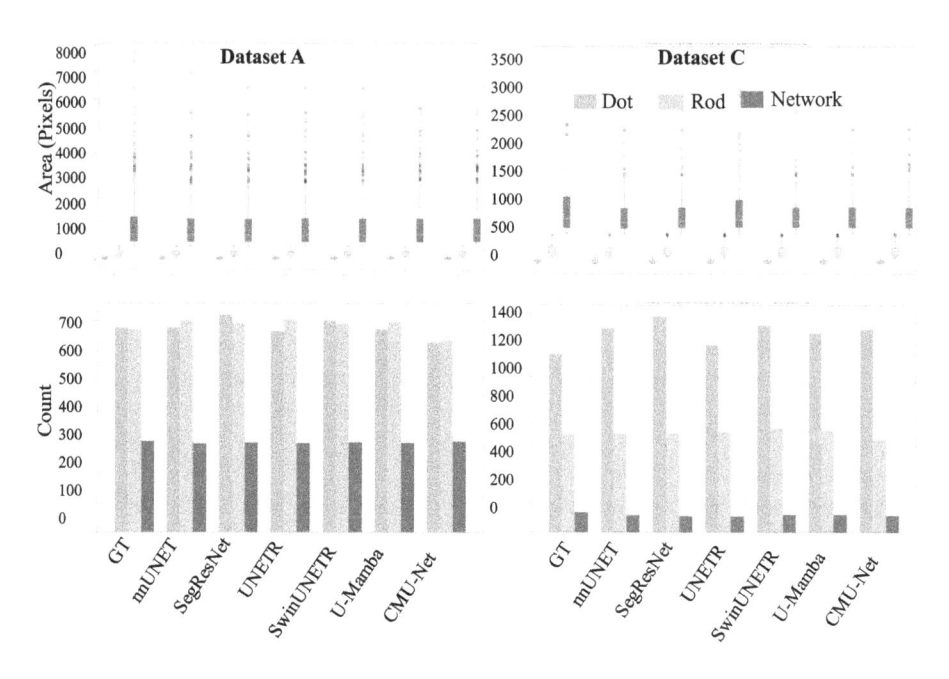

Fig. 6. Morphological analysis results.

The analysis focuses on two key metrics: the area occupied by each mitochondrial type (Dot, Rod, Network) and the count of each type.

For *Dataset A*, the graphs reveal that our CMU-Net method closely aligns with the ground truth in terms of both area distribution and count of mitochondrial structures. The U-Net baseline, while performing reasonably well, shows some differences, particularly in the network structures.

Similarly, for *Dataset C*, graphs reveal that CMU-Net outperforms the U-Net baseline, especially in accurately identifying and measuring rod-shaped mitochondria. This improvement is seen in both the area and count metrics.

The results consistently show that our proposed CMU-Net segmentation gives morphological statistics closely aligned with the ground truth, outperforming the baseline U-Net method across both datasets. This demonstrates the strength of our approach in handling different mitochondrial morphologies and datasets.

This fine-grained morphological analysis, enabled by our improved segmentation method, provides valuable insights into mitochondrial dynamics. The ability to accurately quantify and classify mitochondrial structures opens new avenues for investigating cellular health, disease progression, and the effects of various treatments on mitochondrial morphology. The stability of CMU-Net's performance across different datasets emphasizes its potential as a reliable tool for mitochondrial analysis in various biological contexts.

4.6 Ablation Study

In order to evaluate the specific contributions of each component in our proposed CMU-Net architecture, we conducted a thorough ablation study. We compared four configurations: (i) U-Net (baseline), (ii) U-Net combined with Mamba, (iii) U-Net combined with CBAM, and (iv) Full CMU-Net (U-Net + Mamba + CBAM).

We evaluated these configurations on all three datasets to determine the applicability and reliability of each architectural component across different subcellular segmentation tasks. The findings of this ablation investigation are displayed in Table 5, showcasing the F1 score and IoU for each configuration across the three datasets.

Table 5. Ablation study results on Datasets A, B, and C.

	Dataset A (Rat,Mito)		Dataset B (ER)		Dataset C (Mito)	
Model Variation	F1/DSC	IoU	F1/DSC	IoU	F1/DSC	IoU
U-Net	0.7593	0.7228	0.8438	0.7312	0.8175	0.6949
U-Net + Mamba	0.7598	0.7231	0.8404	0.7259	0.8232	0.7028
U-Net + CBAM	0.7557	0.7170	0.8412	0.7274	0.8142	0.6900
CMU-Net	**0.7883**	**0.7420**	**0.8669**	**0.7559**	**0.8358**	**0.7255**

The results of our ablation study present multiple crucial conclusions, particularly when evaluating the individual subcellular structures in each dataset:

1. **Baseline Performance:** The typical U-Net (configuration i) sets a solid baseline across all datasets, indicating its effectiveness in segmenting both mitochondria (Datasets A and C) and endoplasmic reticulum (Dataset B).

2. **Mamba Contribution:** The inclusion of Mamba blocks (configuration ii) demonstrates a consistent improvement compared to the baseline U-Net for the task of segmenting mitochondria in Datasets A and C. However, there's a minor drop in performance for ER segmentation (Dataset B). This suggests that Mamba's ability to capture long-range dependencies is particularly beneficial for mitochondrial structures, which often exhibit complex, branching morphologies.
3. **CBAM Contribution:** The addition of CBAM (configuration iii) demonstrates mixed outcomes. It increases performance for ER segmentation (Dataset B) but marginally affects performance for mitochondrial segmentation (Datasets A and C). This shows that CBAM's attention mechanism may be more effective in amplifying elements crucial to the tubular network structure of the ER, while potentially disregarding some subtle features important for mitochondrial segmentation.
4. **Collaborative Effect:** The entire CMU-Net architecture (configuration iv) consistently beats all other configurations across both mitochondrial and ER segmentation tasks. This exhibits a collaborative effect when integrating Mamba blocks with CBAM within the U-Net framework. The performance improvements over the baseline U-Net are notably noteworthy, with increases in F1 score of 3.82% and 2.24% for mitochondrial segmentation (Datasets A and C), and 2.74% for ER segmentation (Dataset B).

These results demonstrate the effectiveness of our hybrid approach in tackling the issues of subcellular segmentation across different organelles. The various effects of individual components (Mamba and CBAM) on different organelles highlights the significance of a hybrid approach. While Mamba blocks seem to be particularly effective for mitochondrial segmentation, likely due to their ability to capture the long-range spatial dependencies in mitochondrial networks, CBAM appears to offer advantages in ER segmentation, possibly by helping the network focus on the distinctive tubular structures of the ER.

The constant superior performance of the complete CMU-Net across both mitochondrial and ER segmentation tasks demonstrates its adaptability and effectiveness in handling various subcellular details. This demonstrates our architectural design decisions and emphasizes the potential of CMU-Net as an effective choice for various subcellular segmentation challenges as well.

5 Discussion and conclusion

This paper introduced CMU-Net, a novel hybrid architecture designed to address the challenges of subcellular segmentation. By integrating U-Net, Mamba blocks, and CBAM modules, our model effectively captures both local features and long-range dependencies crucial for accurately segmenting subcellular structures in microscopy images. The quantitative and qualitative results demonstrate the improved performance of CMU-Net compared to various baseline methods, including established CNN-based and Transformer-based segmentation models. While some baseline methods achieved comparable performance to

CMU-Net in specific cases, our model consistently outperformed them across different datasets and evaluation metrics. The integration of Mamba blocks, capable of efficiently modeling long-range dependencies, played a key contribution in improving segmentation accuracy, particularly in identifying the boundaries of subcellular structures. Moreover, the inclusion of CBAM modules further enhanced the model's ability to focus on key features and suppress irrelevant information, leading to the overall improvement in segmentation quality. The primary focus of this work was the development and evaluation of the CMU-Net architecture. However, there are various options for additional investigation and extension of this approach. One potential direction is to investigate the application of CMU-Net to other biomedical image segmentation tasks, such as cell membrane segmentation or nucleus segmentation. Additionally, exploring the integration of different attention mechanisms or exploring alternative SSM structures within the Mamba blocks could lead to further improvements in performance. Furthermore, the model can be adapted and optimized for 3D segmentation tasks to analyze volumetric microscopy data[28]. In conclusion, this paper presents a novel hybrid architecture, CMU-Net, that effectively addresses the challenges of subcellular segmentation. By integrating the strengths of CNNs, Mamba blocks, and the lightweight CBAM attention module, our model achieves accurate and efficient segmentation without incurring significant computational overhead. The promising results obtained in this study highlight the potential of CMU-Net to contribute significantly to the field of subcellular image analysis, providing a deeper understanding of cellular processes and functions. Future work will explore the generalizability of this approach to other segmentation tasks and investigate further enhancements to the model's performance and efficiency.

Acknowledgements. The Project funding including hardware resources(GPU) and other costs of the project are funded by the Science and Engineering Research Board(SERB), Govt. of India, Project No: SRG/2022/000122, executed in XIM University, Bhubaneswar, India, supervised by Arif Ahmed Sekh.

References

1. Juan C Caicedo, Jonathan Roth, Allen Goodman, Tim Becker, Kyle W Karhohs, Matthieu Broisin, Csaba Molnar, Claire McQuin, Shantanu Singh, Fabian J Theis, et al. Evaluation of deep learning strategies for nucleus segmentation in fluorescence images. *Cytometry Part A*, 95(9):952–965, 2019
2. Ismail Oztel, Gozde Yolcu, Ilker Ersoy, Tommi White, and Filiz Bunyak. Mitochondria segmentation in electron microscopy volumes using deep convolutional neural network. In *2017 IEEE International Conference on Bioinformatics and Biomedicine (BIBM)*, pages 1195–1200, 2017
3. Jeff W Lichtman and José-Angel Conchello. Fluorescence microscopy. *Nature methods*, 2(12):910–919, 2005
4. Chawin Ounkomol, Sharmishtaa Seshamani, Mary M Maleckar, Forrest Collman, and Gregory R Johnson. Label-free prediction of three-dimensional fluorescence images from transmitted-light microscopy. *Nature methods*, 15(11):917–920, 2018

5. Ian J. Goodfellow, Yoshua Bengio, and Aaron Courville. *Deep Learning*. MIT Press, Cambridge, MA, USA, 2016. http://www.deeplearningbook.org
6. O. Ronneberger, P.Fischer, and T. Brox. U-net: Convolutional networks for biomedical image segmentation. In *Medical Image Computing and Computer-Assisted Intervention (MICCAI)*, volume 9351 of *LNCS*, pages 234–241. Springer, 2015. (available on arXiv:1505.04597 [cs.CV])
7. Zongwei Zhou, Md Mahfuzur Rahman Siddiquee, Nima Tajbakhsh, and Jianming Liang. Unet++: A nested u-net architecture for medical image segmentation. In *Deep Learning in Medical Image Analysis and Multimodal Learning for Clinical Decision Support: 4th International Workshop, DLMIA 2018, and 8th International Workshop, ML-CDS 2018, Held in Conjunction with MICCAI 2018, Granada, Spain, September 20, 2018, Proceedings 4*, pages 3–11. Springer, 2018
8. Sanghyun Woo, Jongchan Park, Joon-Young Lee, and In So Kweon. Cbam: Convolutional block attention module. In *Proceedings of the European conference on computer vision (ECCV)*, pages 3–19, 2018
9. Albert Gu and Tri Dao. Mamba: Linear-time sequence modeling with selective state spaces. arXiv preprint arXiv:2312.00752, 2023
10. Jun Ma, Feifei Li, and Bo Wang. U-mamba: Enhancing long-range dependency for biomedical image segmentation. arXiv preprint arXiv:2401.04722, 2024
11. Ozan Oktay, Jo Schlemper, Loic Le Folgoc, Matthew Lee, Mattias Heinrich, Kazunari Misawa, Kensaku Mori, Steven McDonagh, Nils Y Hammerla, Bernhard Kainz, et al. Attention u-net: Learning where to look for the pancreas. arXiv preprint arXiv:1804.03999, 2018
12. Ali Hatamizadeh, Yucheng Tang, Vishwesh Nath, Dong Yang, Andriy Myronenko, Bennett Landman, Holger R Roth, and Daguang Xu. Unetr: Transformers for 3d medical image segmentation. In *Proceedings of the IEEE/CVF winter conference on applications of computer vision*, pages 574–584, 2022
13. Ali Hatamizadeh, Vishwesh Nath, Yucheng Tang, Dong Yang, Holger R Roth, and Daguang Xu. Swin unetr: Swin transformers for semantic segmentation of brain tumors in mri images. In *International MICCAI Brainlesion Workshop*, pages 272–284. Springer, 2021
14. Albert Gu, Karan Goel, and Christopher Ré. Efficiently modeling long sequences with structured state spaces. arXiv preprint arXiv:2111.00396, 2021
15. Jiacheng Ruan and Suncheng Xiang. Vm-unet: Vision mamba unet for medical image segmentation. arXiv preprint arXiv:2402.02491, 2024
16. Renkai Wu, Yinghao Liu, Pengchen Liang, and Qing Chang. H-vmunet: High-order vision mamba unet for medical image segmentation. arXiv preprint arXiv:2403.13642, 2024
17. Zhaohu Xing, Tian Ye, Yijun Yang, Guang Liu, and Lei Zhu. Segmamba: Long-range sequential modeling mamba for 3d medical image segmentation. arXiv preprint arXiv:2401.13560, 2024
18. Jiacheng Ruan, Suncheng Xiang, Mingye Xie, Ting Liu, and Yuzhuo Fu. Malunet: A multi-attention and light-weight unet for skin lesion segmentation. In *2022 IEEE International Conference on Bioinformatics and Biomedicine (BIBM)*, pages 1150–1156. IEEE, 2022
19. Xiaoqi Zhao, Lihe Zhang, and Huchuan Lu. Automatic polyp segmentation via multi-scale subtraction network. In *Medical Image Computing and Computer Assisted Intervention–MICCAI 2021: 24th International Conference, Strasbourg, France, September 27–October 1, 2021, Proceedings, Part I 24*, pages 120–130. Springer, 2021

20. Dmitry Ulyanov, Andrea Vedaldi, and Victor Lempitsky. Instance normalization: The missing ingredient for fast stylization. arXiv preprint arXiv:1607.08022, 2016
21. Andrew L Maas, Awni Y Hannun, Andrew Y Ng, et al. Rectifier nonlinearities improve neural network acoustic models. In *Proc. icml*, volume 30, page 3. Atlanta, GA, 2013
22. Arif Ahmed Sekh, Ida S Opstad, Gustav Godtliebsen, Åsa Birna Birgisdottir, Balpreet Singh Ahluwalia, Krishna Agarwal, and Dilip K Prasad. Physics-based machine learning for subcellular segmentation in living cells. *Nature Machine Intelligence*, 3(12):1071–1080, 2021
23. Yaoru Luo, Yuanhao Guo, Wenjing Li, Guole Liu, and Ge Yang. Fluorescence microscopy image datasets for deep learning segmentation of intracellular orgenelle networks, 2020
24. Fabian Isensee, Jens Petersen, Andre Klein, David Zimmerer, Paul F Jaeger, Simon Kohl, Jakob Wasserthal, Gregor Koehler, Tobias Norajitra, Sebastian Wirkert, et al. nnu-net: Self-adapting framework for u-net-based medical image segmentation. arXiv preprint arXiv:1809.10486, 2018
25. Sebastian Ruder. An overview of gradient descent optimization algorithms. arXiv preprint arXiv:1609.04747, 2016
26. Andriy Myronenko. 3d mri brain tumor segmentation using autoencoder regularization. In *Brainlesion: Glioma, Multiple Sclerosis, Stroke and Traumatic Brain Injuries: 4th International Workshop, BrainLes 2018, Held in Conjunction with MICCAI 2018, Granada, Spain, September 16, 2018, Revised Selected Papers, Part II 4*, pages 311–320. Springer, 2019
27. Li, Y., He, Y., Miao, K., Zheng, Y., Deng, C., Liu, T.-M.: Imaging of macrophage mitochondria dynamics in vivo reveals cellular activation phenotype for diagnosis. Theranostics **10**(7), 2897 (2020)
28. Boden, A., Pennacchietti, F., Coceano, G., Damenti, M., Ratz, M., Testa, I.: Volumetric live cell imaging with three-dimensional parallelized resolft microscopy. Nat. Biotechnol. **39**(5), 609–618 (2021)

Cross-Domain Multi-contrast MR Image Synthesis via Generative Adversarial Network

Guowen Wang[1], Silei Wang[2], Lu Wang[2], Congbo Cai[2], Shuhui Cai[2], and Zhong Chen[2(✉)]

[1] Institute of Artificial Intelligence, Xiamen University, Xiamen, China
[2] Department of Electronic Science, Fujian Provincial Key Laboratory of Plasma and Magnetic Resonance, Xiamen University, Xiamen, China
chenz@xmu.edu.cn

Abstract. Magnetic resonance imaging (MRI) simulations necessitate a substantial number of multi-contrast MR images, which can be time-consuming and costly to obtain. To overcome this challenge, synthetic data has emerged as a viable alternative. However, existing methods are limited in their ability to generate multiple modalities within a single dataset and struggle to produce modalities that are not present in the original dataset, resulting in poor domain transfer capability. To tackle this issue, we propose MCGAN for cross-domain multi-contrast MR image synthesis. The MCGAN framework employs a two-stage learning strategy. In the first stage, a domain adaptation module is utilized to align the source domain distribution with the target domain using unsupervised learning, effectively bridging domain gaps. Subsequently, the second stage involves an image-to-image module that empowers the model to generate additional modalities. By combining these two stages, MCGAN framework overcomes the limitations of single-stage generation methods, resulting in a model that synthesizes a comprehensive array of modalities within each dataset. Experimental results show that MCGAN method outperforms other transfer learning-based image-to-image methods and cross-dataset image synthesis methods in terms of both data distribution realism and texture details. Code is available at https://github.com/DropInOcean/MCGAN.

Keywords: Cross-Domian Synthesis · Generative Adversarial Network · Multi-contrast · MRI

1 Introduction

Magnetic resonance imaging (MRI) simulations are instrumental in optimizing, evaluating pulse sequences, and artifact tracing [1–3]. However, conducting MRI simulations necessitates a significant quantity of multi-contrast MR images that precisely capture the distribution characteristics of this modality and provide specific texture details of the human brain. Nevertheless, challenges such as slow imaging speed and high financial costs hinder the acquisition of multi-contrast MR images. In response to these challenges, synthetic data has emerged as a valuable solution, addressing the limitations posed by the scarcity of large datasets [4]. Synthetic data has demonstrated its utility in various applications, such as cardiac imaging and nuclei segmentation in histopathologic images [5, 6]. Moreover, the generation of synthetic data holds promise in addressing patient privacy concerns within the medical domain.

Previous research yielded impressive results on specific dataset such as IXI or BRATS [8–10], yet it has been constrained to generating only a limited number of specific modalities from each dataset, thus falling short of the demands for comprehensive MRI simulations. When the trained network is directly applied to other datasets, the outcomes have proven unsatisfactory, primarily due to MR scanner variations and individual subjects' movements. These discrepancies render data acquisition challenging and result in collected data that differs from the source domain, posing obstacles when converted into other contrast images. Our goal is to generate additional MR modalities that are not initially included in any existing dataset, thereby expanding the data quantity and enhancing the variety of image modalities available for MRI simulations.

The challenges inherent in this task include: 1) Handling the diversity of modalities present in different datasets to synthesize a comprehensive array of modalities within each dataset. 2) Overcoming domain gaps between various datasets to improve the transfer ability of the generative model across different datasets. 3) Establishing criteria for evaluating the reasonableness of synthesized images and ensuring they align with the requirements of MRI simulations, particularly in the absence of labeled comparisons for the generated data.

In this work, we propose the MCGAN framework to address these challenges. T1 images are widely used in disease assessment and positioning due to their fast-scanning time and clear anatomical structures, making them routinely acquired in hospitals and prevalent in various public datasets. Exploiting the advantages of T1 images, we propose to utilize them as a bridge to generate arbitrary contrast MR images. In order to solve the cross-dataset domain shift problem, we introduce a two-stage learning algorithm. In the first stage, the source domain image is transformed into the target domain image through an unsupervised method. Subsequently, the desired contrast image is generated through a supervised method, thereby generating the required various contrast MR images for MRI simulations.

To evaluate the suitability of the synthesized data for MRI simulation platforms, we propose a novel measurement indicator. This indicator involves segmenting cerebrospinal fluid (CSF), gray matter (GM), and white matter (WM) from the T1 image and then analyzing the distribution of each part. By comparing the distribution of these three regions between the synthesized data and the corresponding real data, and by calculating the weighted Kullback-Leibler ($KL_{average}$) divergence value of the three regions, the accuracy and reasonableness of the synthesized data can be assessed.

Experimental results demonstrate that the proposed method significantly outperforms existing image-to-image generation methods (Attention-U-Net [7], pix2pix [8], pGAN [9], cGAN [10], CUT [34], SRC [35]) and cross-dataset image generation methods (UNIT [11]) on three public datasets. The main contributions of this work are as follows:

1) MCGAN addresses the cross-dataset domain shift problem by building an image-to-image module and a domain adaptation module, employing a two-step training strategy to generate required MR image modality.

2) Leveraging T1 images, MCGAN synthesizes multi-contrast MR images that were previously unattainable through existing methods., This capability effectively addresses the challenge of generating a comprehensive array of MR modalities, particularly modalities absent from most public datasets (e.g., DWI, T2-FLAIR, and $T2^*$).

3) MCGAN introduces an attention gate module to the generator to enhance perception and generation of the foreground part, overcoming visual artifacts produced by pix2pix [8], pGAN [9], cGAN[10].

4) To assess the plausibility of synthesized images, we propose a new metric inspired by intra-modal connections, which offers reproducibility compared to subjective evaluation methods that suffer from time and cost constraints and inter-observer variability.

2 Related Work

2.1 Image-To-Image

Translating the synthesis of new modality images from available MR images has been a prominent research area in the field of medical imaging [12]. Traditional methods include patch regression [13], atlas-based methods [14, 15], and MR physics-based techniques [16]. In recent years, there has been a shift towards utilizing convolutional neural network (CNN)-based methods [17]. The introduction of generative adversarial networks aims to synthesize realistic images across different modalities [18]. For example, Salman et al. proposed pGAN [9], a conditional GAN-based approach that incorporates style transfer GAN for contrast-aware synthesis. Zhu et al. proposed unsupervised CycleGAN [19] for image transformation tasks. Additionally, Bing Cao et al. proposed Auto-GAN [20], which considers the complementary information from various modalities and employs a self-supervised method to synthesize a single modality from multiple source modalities. More recently, Park et al. introduced contrastive learning into image generation [34], replacing the cycle-consistency in unsupervised learning to achieve one-way conversion. Furthermore, Jung et al. proposed SRC [35], which effectively leverages heterogeneous semantics within the images.

2.2 Domain Adaptation

Domain adaptation (DA) techniques have been developed to transfer knowledge from a source domain to a target domain of interest [21, 22]. A large number of DA methods have been proposed, including instance re-weighting [23, 24], covariance alignment [25, 26], maximum mean difference [27, 28], pixel-level adaptation [29, 30], and more. Although the UNIT method [11] can be categorized within these DA techniques as it adjusts the model trained on the source domain to the target domain, it differs from existing methods because its focus is on generation tasks rather than understanding tasks like image classification or segmentation. Additionally, W. Bian et al. integrates variational models to realize MRI reconstruction and generation [36], particularly suited for generating one modality from multiple modalities.

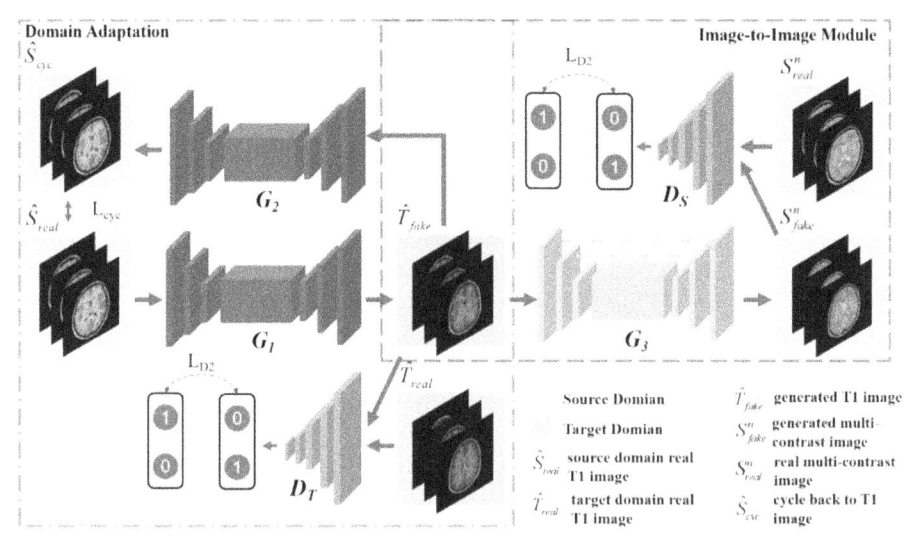

Fig. 1. The proposed MCGAN framework consists of a domain adaptation module and an image-to-image module. The domain adaptation module facilitates the conversion of the source domain to the target domain, leveraging a discriminator to ensure image authenticity. The image to the image module enables the conversion of T1 images to other contrast MR images.

3 Method

In this section, we present the architecture of MCGAN, shown in Fig. 1. To overcome the limitations of single-stage generation approaches, we have embraced a two-stage approach that amalgamates the advantages of various techniques, leading to the creation of models with enhanced performance. The first stage focused on translating the source domain T1 image ($\hat{S}_{real}$) into the target domain image ($\hat{T}_{fake}$). Subsequently, the second stage engages in image-to-image translation, converting the synthetic target domain image ($\hat{T}_{fake}$) into multi-contrast MR images (S^n_{real}). This dual-phase training regimen plays a pivotal role in equipping the image-to-image translation model with cross-dataset generation capabilities.

3.1 Domain Adaptive Module

Despite the T1 image obtained from different scanning machines being generally similar, there are still difference in image contrast, lightness and shading. Therefore, a domain shift problem exists between the source domain (original T1 image) and the target domain (unified T1 image). In this study, the source domain image is converted into the modality of the target domain through unsupervised learning to address this tissue. Here, S represents the source domain and T represents the target domain. The subscript "real" represents the real image, and "fake" refers to the synthesized image. The symbol "^" represents the T1 image, and the superscript $n = 1, 2, 3...$, represents other contrast MR images (such as T2, PD, DWI, T2*, T2-FLAIR, etc.).

Adversarial Loss. We implement an adversarial loss to enforce the translation of the source domain $\hat{S}_{real}$ to the target domain $\hat{T}_{real}$. The adversarial loss is defined as:

$$L_{adv_1}(G_1, D_T) = -E_T\left[\left(D_T(\hat{T}_{real}) - 1\right)^2\right] - E_S\left[D_T(G_1(\hat{S}_{real}))^2\right] \quad (1)$$

where $\hat{T}_{real}$ is the target domain T1 image, $\hat{S}_{real}$ is the source domain T1 image, E denotes the expected value, G_1 is the generator that converts source domain T1 image to target domain T1 image, and D_T is the discriminator that distinguishes between the generated target domain T1w images and the real target domain T1 images.

Cycle Consistency Loss. We propose the cycle consistency loss reduces the domain transformation mapping space and makes the reconstructed image ($\hat{S}_{cyc}$) consistent with the original image ($\hat{S}_{real}$). The adversarial loss is defined as:

$$L_{cycle}(G_1, G_2) = E\left[\left\|G_2(G_1(\hat{S}_{real})) - \hat{S}_{real}\right\|_1\right] \quad (2)$$

where $G_2(G_1(\hat{S}_{real}))$ tries to generate an image for consistency constraints

$$L_{DA} = \lambda_{adv}L_{adv_1}(G_1, D_T, S, T) + \lambda_{cyc}L_{cycle}(G_1, G_2) \quad (3)$$

Here, λ_{adv} and λ_{cyc} control the relative weighting of the adversarial loss and consistency loss, respectively.

3.2 Image-To-Image Module

Building upon the availability of paired data, the target domain T1 image ($\hat{T}_{fake}$) can be better converted to other multi-contrast MR images (S_{real}^n) through supervised learning. The generator G is trained to generate "real" images. To address artifact challenges similar to those encountered in pix2pix [8], pGAN [9] and cGAN [10], we draw inspiration from the Attention U-Net [32, 33] to refine the generator G. D is trained to distinguish between real and generated images.

Attention Gate. To enhance the network's receptive field and improve the translation of high-level semantic information, an attention mechanism is incorporated into the architecture. Inspired by the Attention U-Net, we introduce the attention gate into our generative network. The attention gate focus on foreground information while filtering out background details beyond the gate, as shown in Fig. 2. Convolutional layers progressively extract higher dimensional image representations (x) by processing local information layer by layer. $\hat{x}$ represents the attention output result. The gate utilizes upsampling information (g) and encoding information (x), calculates the attention value (α) and then multiplies it with each encoding information to derive the output formula of the attention gate:

$$\hat{x} = \alpha x \quad (4)$$

As shown in Fig. 6, our experiments demonstrate that the attention mechanism we introduced focuses primarily on internal brain information and does not attend to external information. This occurs because the external background consistently maintains

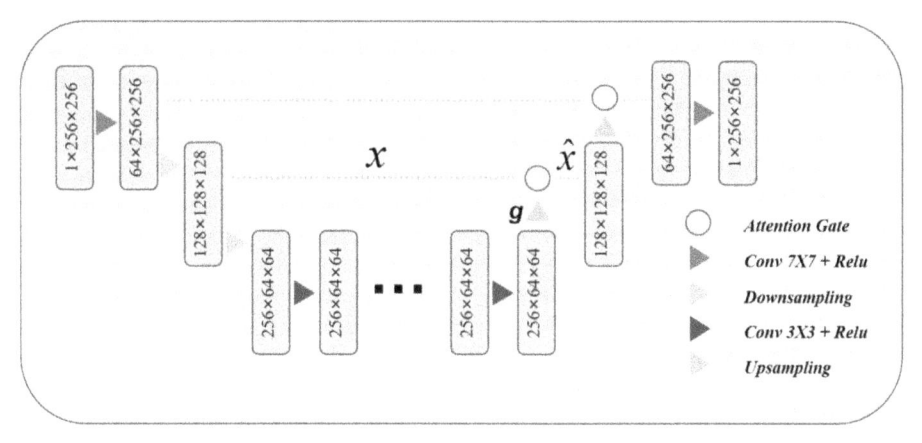

Cross-Domain Multi-contrast MR Image Synthesis   413

Fig. 2. The overall structure of the generator. Attention gate is added to enhance perception and generation of the foreground part, overcoming visual artifacts.

a value of 0 without variation, thus not attracting attention. In contrast, the values of different foreground modalities within the brain exhibit variability, leading the attention mechanism to consistently focus on these changing values. The attention gate integrates the concept of attention mechanism into convolutional network. The computation of the attention value is as follows:

$$A = \psi^\top \left(\sigma_1 \left(W_x^\top x + W_g^\top g + b_g \right) \right) + b_\psi \tag{5}$$

$$\alpha = \sigma_2(A(x, g; \Theta_{att})) \tag{6}$$

where W_x^T, W_g^T denote the linear transformations computed using channel-wise 1x1x1 convolutions for the input tensors. b_g and b_ψ are the bias terms. The rectified linear unit $\sigma_1(x) = \max(0, x)$ and $\sigma_2(x) = \frac{1}{1+\exp(-x)}$ denote sigmoid activation function.

Perceptual Loss Perceptual losses relies on differences in higher-level feature representations, typically extracted from networks pre-trained for more general tasks. A commonly used network for this purpose is the VGG-Net, trained on the ImageNet dataset for object classification. Here, feature maps are extracted before the second max-pooling operation following our pre-trained VGG16 on ImageNet. The resulting loss function can be written as:

$$L_{vgg}(G_3) = E[\| V(S_{real}^n) - V(G_3(\hat{T}_{fake})) \|_1] \tag{7}$$

Adversarial Loss We use an adversarial loss to enforce synthetic T1 image translation to other multi-contrast MR images. The adversarial loss is formulated as follows:

$$L_{adv2}(G_3, D_s) = -E_{\hat{T}_{fake}, S_{real}^n}\left[\left(D_s(\hat{T}_{fake}, S_{real}^n) - 1 \right)^2 \right] - E_{\hat{T}_{fake}}\left[D_s(\hat{T}_{fake}, G_3(\hat{T}_{fake}))^2 \right] \tag{8}$$

Pixel-wise Loss This loss focuses on minimizing pixel-wise differences between images:

$$L_{L_1}(G_3) = E\big[\big\|S_{real}^n - G_3\big(\hat{T}_{fake}\big)\big\|_1\big] \tag{9}$$

This yields the following aggregate loss function for training:

$$L_{I2I} = \lambda_{adv2}L_{adv2}(G_3, D_s) + \lambda_{L_1}L_{L_1}(G_3) + \lambda_{vgg}L_{vgg}(G_3) \tag{10}$$

4 Experiment

4.1 Experimental Setup

1) Data: Our experiment is based on 3 independent publicly available datasets: IXI, HCP, and Simons. The data contains T1, T2, and PD image of the IXI dataset; T1w, T2, and DWI image of the HCP dataset; T1, T2-FLAIR, and T2* image of the Simons dataset. Prior to processing, images excluding T1 images were registered to their respective T1 images. To ensure optimal model training and unbiased quantitative assessments, the datasets were normalized to standardize voxel intensity ranges across subjects. This normalization involved scaling image intensities within each subject to a range of [0,1] via division by the maximum intensity.

a. IXI dataset: From the 574 subjects, 459 were selected for training (T1- > PD, T1- > T2) and 115 for testing (T1- > T2-FLAIR, T1- > DWI, T1- > T2*).
b. HCP dataset: Out of 1113 subjects, 890 were selected for training (T1- > DWI) and 223 for testing (T1- > T2-FLAIR, T1- > PD, T1- > T2*).
c. Simons dataset: Among the 62 subjects, 46 were selected for training (T1- > T2*, T1- > T2-FLAIR) and 16 for testing (T1- > PD, T1- > DWI, T1- > T2).

2) Network: The generative and discriminative networks for domain adaption and image translation are built upon the generative network architecture proposed by Johnson et al., which achieved impressive results in neural style transfer and super-resolution tasks [31]. The two generator network structures of domain adaption are composed of multiple convolutional layers, including 3 up-sampled convolution layers (one with stride 7, two with stride 3), 9 residual blocks, and 3 down-sampled convolution layers (one with stride7, two with stride 3). The discriminator network employs PatchGAN [8], featuring 4 down-sampled convolutional layers with stride 4. The image translation generator differs from the domain adaption generator only by incorporating 2 additional residual blocks, while the discriminator remains the same.

3)Training details: Our method is implemented using Pytorch on an Nvidia 2080Ti GPU. We use the Adam optimizer with an initial learning rate of 0.0002, linearly reducing it after 100 epochs. The batch size is set to 4, and the total number of epoch is 200. In addition, different hyperparameters are set for different losses: λ_{vgg}, λ_{L1}, and λ_{cyc} are set to 100, while λ_{adv1} and λ_{adv2} is set to 1.

4.2 Metrics

In this section, we propose a new metric for cross-domain medical image synthesis. As commonly known, KL divergence measures the difference between two probability distributions. The brain can be categorized into 3 regions: CSF, GM, and WM. Each slice of each sample has a fixed partition and it occupies a fixed area. Leveraging FSL software, we segment the T1 image of each sample to obtain masks for the three regions and calculate the proportion of each mask to the entire brain (λ_{CSF}, λ_{GM}, λ_{WM}). These masks are then applied to the generated images to extract the corresponding three regions of the synthesized images. Given that the three-region distribution varies across different MR image modalities, we compare and analyze the synthesized modality's three-region distribution with that of the corresponding region in the real image, subsequently calculating the weighted KL divergence value of the three-region, as shown in the formula 11. A smaller KL divergence suggests greater consistency between the synthesized images and the real images, indicating the reasonableness of the synthesized images.

$$KL_{average} = \lambda_{CSF} KL_{CSF} + \lambda_{GM} KL_{GM} + \lambda_{WM} KL_{WM} \qquad (11)$$

4.3 Results

We applied and evaluated MCGAN on three datasets (IXI, HCP and Simons). As depicted in Fig. 3, MCGAN successfully synthesizes MR image modalities that were not originally present in the dataset and demonstrates strong generalization capabilities. The synthetic data are indistinguishable from real data, matching the contrast level of real modalities and accurately capturing the specific texture details of the human brain. MCGAN demonstrates its effectiveness in generating extensive datasets to serve as templates for MRI simulations.

4.4 Comparison

In our study, we conducted a comprehensive comparison between our proposed MCGAN and several existing image-to-image methods, including pix2pix [8], Attention U-Net [7], pGAN [9], cGAN [10], UNIT [11], CUT[34] and SRC[35]. These methods are implemented by officially provided code. We then proceeded to analyze the performance of all existing methods through both qualitative and quantitative assessments. As depicted in Fig. 4, the data generated from the MCGAN algorithm exhibits a more intricate texture structure compared to the results obtained from other comparative methods.

In Fig. 5, it is evident that the pixel distribution of the images generated by the MCGAN model closely aligns with the pixel distribution of real data. Furthermore, with the incorporation of the attention mechanism module, artifacts that were present in the pix2pix [8], pGAN [9], and cGAN [10] approaches are notably absent in the outputs of the MCGAN method. These artifacts typically manifest as values close to 0 and upward curves. Specific KL divergence values provided in Table 1 highlight that the images generated by the MCGAN outperform those created by the other comparative methods in terms of accuracy and reasonableness. The results underscore the superior performance of the MCGAN framework in producing high-quality images with enhanced texture details and reduced artifacts, making it a robust solution for image generation tasks.

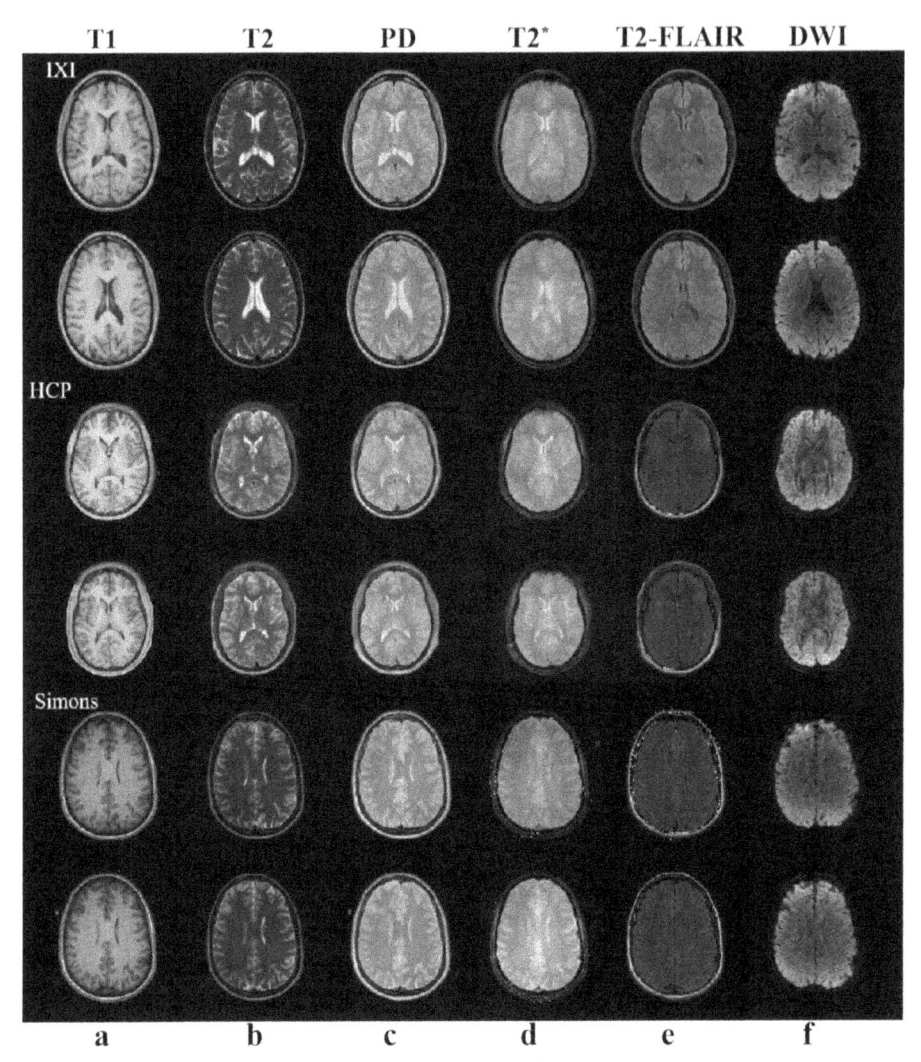

Fig. 3. Multi-contrast MR images synthesized from MCGAN. These images not only accurately depict the human brain, but also adhere to the distribution characteristics to each contrast. In the IXI dataset, a, b, and c are real images, and d, e, and f are synthetic images. In the HCP dataset, a, b, and f are real images, and c, d, and e are synthetic images. In the Simons dataset, a, d, and e are real images, and b, c, and f are synthetic images.

4.5 Ablation Study

In our ablation study, we conducted quantitative assessments to evaluate the impact of specific modules within our proposed MCGAN. The ablation experiments focused on two key factors: the attention module and the discriminator, with the analysis conducted using the IXI dataset.

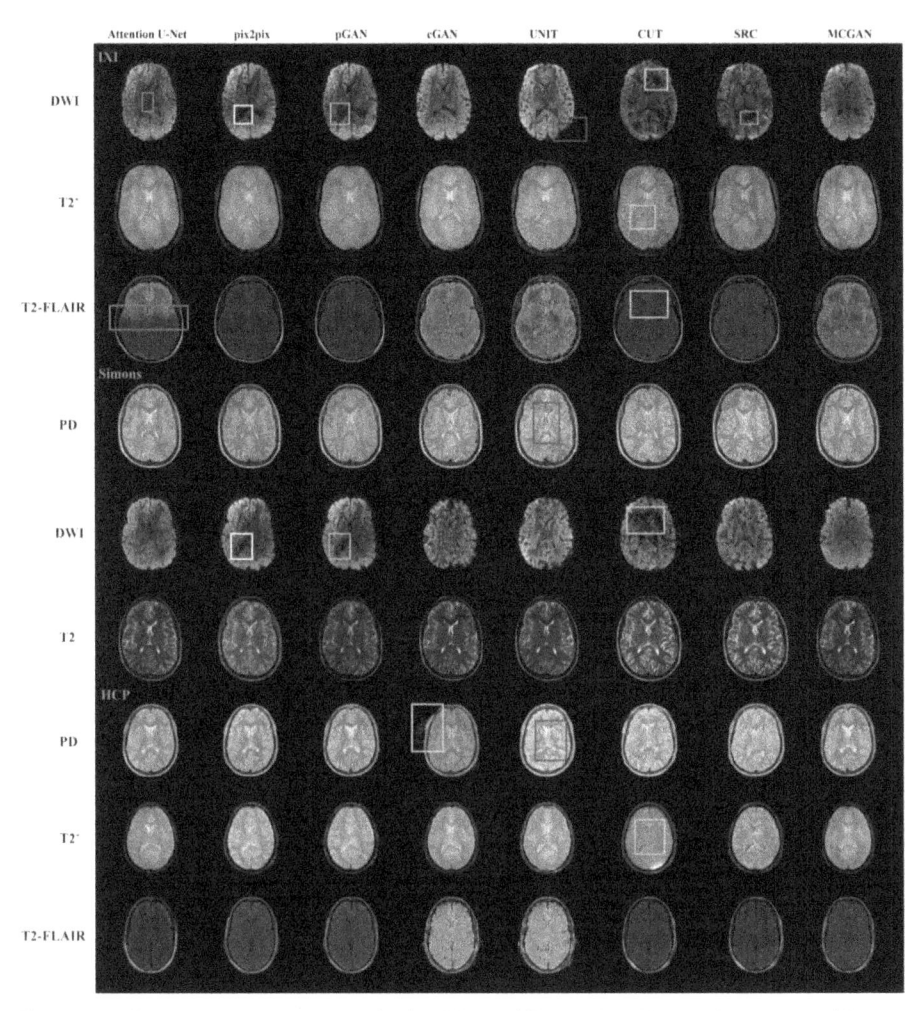

Fig. 4. Multi-contrast MR images synthesized from different algorithms. The unreasonable parts of the image generated by each model are marked with boxes. MCGAN generates more realistic data with clearer texture details and is proven to be better than other methods on the three datasets (IXI, Simons, and HCP).

As illustrated in Fig. 6, the attention coefficients vary significantly across different training epochs, indicating that the focus of the attention gate model is dynamically adjusted throughout the training process. The initial epochs show a broad and less specific pattern of attention, while later epochs show a more targeted approach, focusing on crucial components such as CSF, GM and WM. Our experiments demonstrate that the attention mechanism we introduced focuses primarily on internal brain information and does not attend to external information.

Table 2 presents quantitative analysis results of the ablation study. The metrics indicated a notable degradation in performance when either the attention module or

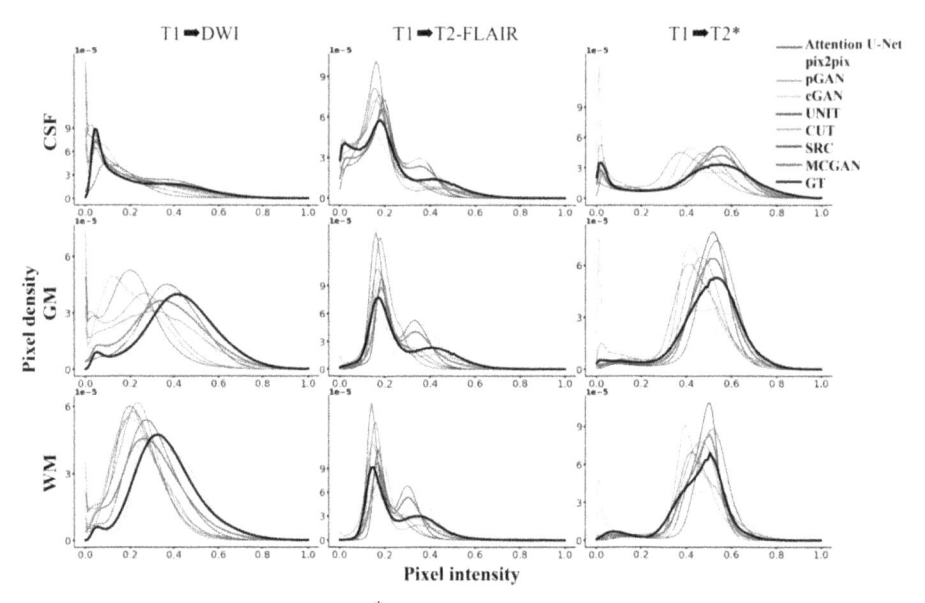

Fig. 5. Pixel distribution of DWI, T2*, T2-FLAIR images generated from different methods (pix2pix, Attention U-Net, pGAN, cGAN, UNIT, CUT, SRC and MCGAN) based on T1 images of IXI dataset.

the discriminator was removed from the MCGAN architecture, underscoring their importance.

Table 1. The KL divergence value of synthetic data and real data

Method	Modality	pGAN	pix2pix	cGAN	Attention U-Net	UNIT	CUT	SRC	MCGAN
IXI	DWI	0.20061	0.31953	0.27279	0.10748	0.33475	0.307963	0.29083	**0.08742**
	T2*	0.12043	0.17218	0.16708	0.10947	0.11617	0.17153	0.14981	**0.07711**
	T2-FLAIR	0.16062	0.25219	0.10502	0.24253	0.21754	0.15279	0.09775	**0.07396**
HCP	PD	0.10151	0.16712	0.16344	0.12014	**0.09295**	0.16042	0.14904	0.10558
	T2*	0.17962	0.22431	0.20484	0.12761	0.15360	0.19925	0.15274	**0.09679**
	T2-FLAIR	0.348776	0.370198	0.38077	0.27021	0.27344	0.27089	0.24177	**0.22464**
Simons	PD	0.14093	0.16566	0.03237	0.05918	0.10667	0.04982	0.03195	**0.03088**
	T2	0.32109	0.42118	0.23210	0.22698	0.29245	0.26210	0.19146	**0.18627**
	DWI	0.31693	0.37849	0.26814	0.11121	0.15299	0.19251	0.19439	**0.10828**

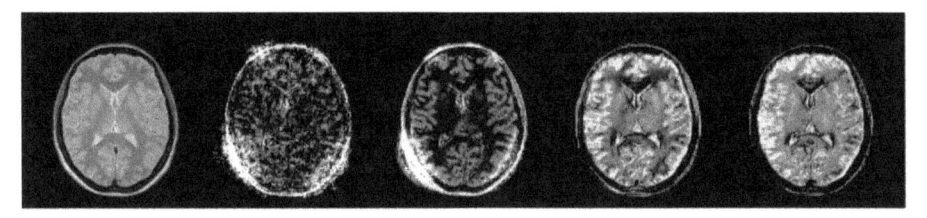

Fig. 6. Attention coefficients across different training epochs (2, 6, 10, 60) together with the corresponding PD image (the first column).

Table 2. Ablation study results

Method	IXI		
	DWI	$T2^*$	T2-FLAIR
W/o discriminator	0.08893	0.10842	0.11271
W/o attention gate	0.18798	0.09579	0.14731
W/o both	0.25438	0.16641	0.16239
Ours	0.08742	0.07711	0.07396

5 Conclusions

This paper proposes a novel framework for synthesizing diverse MRI simulation templates, with a specific focus on addressing cross-domain medical image synthesis challenges. The proposed MCGAN algorithm demonstrates strong generalization capabilities across datasets and effectively mitigates the domain shift problem through a two-stage training strategy. The first stage facilitates domain transformation from the source domain into the target domain, thus solving domain shift issues. Subsequently, the second stage enables the conversion of target domain images into other desired modality images. Experimental results show that our method outperforms other image-to-image algorithms and transfer learning methods across three public datasets. This indicates the potential of MCGAN as an effective algorithm for synthesizing multi-contrast MR images.

References

1. Huang, H. et al. High-efficient Bloch simulation of magnetic resonance imaging sequences based on deep learning. Phys. Med. Biol. **68**, 085002 (2023)
2. Calamante, F., Gadian, D.G., Connelly, A.: Delay and dispersion effects in dynamic susceptibility contrast MRI: simulations using singular value decomposition. Magn. Reson. Med. **44**(3), 466–473 (2000)
3. Li, T., Wang, J., Yang, Y., Glide-Hurst, C.K., Wen, N., Cai, J.: Multi-parametric MRI for radiotherapy simulation. Med. Phys. **50**, 5273–5293 (2023)
4. Frangi, A.F., Tsaftaris, S.A., Prince, J.L.: Simulation and Synthesis in Medical Imaging. IEEE Trans. Med. Imaging **37**, 673–679 (2018)

5. Gilbert, A., et al.: Generating Synthetic Labeled Data From Existing Anatomical Models: An Example With Echocardiography Segmentation. IEEE Trans. Med. Imaging **40**, 2783–2794 (2021)
6. Zhou, Y., et al.: A Framework for the Generation of Realistic Synthetic Cardiac Ultrasound and Magnetic Resonance Imaging Sequences From the Same Virtual Patients. IEEE Trans. Med. Imaging **37**, 741–754 (2018)
7. Oktay, O. et al. Attention U-Net: Learning Where to Look for the Pancreas. Preprint at http://arxiv.org/abs/1804.03999 (2018)
8. Isola, P., Zhu, J.-Y., Zhou, T. & Efros, A. A. Image-to-Image Translation with Conditional Adversarial Networks. (2018)
9. Dar et al. Image Synthesis in Multi-Contrast MRI With Conditional Generative Adversarial Networks. IEEE TRANSACTIONS ON MEDICAL IMAGING, VOL. 38, NO. 10, OCTOBER (2019)
10. Mirza, M. & Osindero, S. Conditional Generative Adversarial Nets. Preprint at http://arxiv.org/abs/1411.1784 (2014)
11. Liu, Ming-Yu, Thomas Breuel, and Jan Kautz. "Unsupervised image-to-image translation networks." Advances in neural information processing systems 30 (2017)
12. Meng, X., Sun, K., Xu, J., He, X. & Shen, D. Multi-modal Modality-masked Diffusion Network for Brain MRI Synthesis with Random Modality Missing. IEEE Trans. Med. Imaging 1–1 (2024)
13. A. Jog, S. Roy, A. Carass, and J. L. Prince, "Magnetic resonance image synthesis through patch regression," in 2013 IEEE 10th International Symposium on Biomedical Imaging, pp. 350–353, 2013
14. S. Roy, A. Carass, N. Shiee, D. L. Pham, and J. L. Prince, "MR contrast synthesis for lesion segmentation," in 2010 IEEE International Symposium on Biomedical Imaging: From Nano to Macro, pp. 932–935, 2010
15. Roy, S., Carass, A., Prince, J.L.: Magnetic resonance image examplebased contrast synthesis. IEEE Trans. Med. Imaging **32**(12), 2348–2363 (2013)
16. Blystad, I., Warntjes, J.B.M., Smedby, O., Landtblom, A.-M., Lundberg, P., Larsson, E.-M.: Synthetic MRI of the brain in a clinical setting. Acta Radiol. **53**(10), 1158–1163 (2012)
17. H. V. Nguyen, K. Zhou, and R. Vemulapalli, "Cross-domain synthesis of medical images using efficient location-sensitive deep network," in International Conference on Medical Image Computing and Computer Assisted Intervention, pp. 677–684, Springer, 2015
18. Jiang Liu, Ben Duffy, Keshav Datta, and Greg Zaharchuk, "One Model to Synthesize Them All: Multi-contrast Multi-scale Transformer for Missing Data Imputation," IEEE Transactions on Medical Imaging
19. Zhu, J.-Y., Park, T., Isola, P. & Efros, A. A. Unpaired Image-to-Image Translation Using Cycle-Consistent Adversarial Networks. in 2017 IEEE International Conference on Computer Vision (ICCV) 2242–2251 (IEEE, Venice, 2017)
20. Cao, B., Zhang, H., Wang, N., Gao, X., Shen, D.: Auto-gan: Self-supervised collaborative learning for medical image synthesis. Proceedings of the AAAI Conference on Artificial Intelligence **34**(07), 10486–10493 (Apr.2020)
21. K. Saenko, B. Kulis, M. Fritz, and T. Darrell. Adapting visual category models to new domains. In ECCV, 2010. 2
22. Y. Ganin, E. Ustinova, H. Ajakan, P. Germain, H. Larochelle, F. Laviolette, M. Marchand, and V. Lempitsky. Domainadversarial training of neural networks. JMLR, 2016
23. Huang, J., Smola, A. J., Gretton, A., Borgwardt, K. M. & Schölkopf, B. Correcting Sample Selection Bias by Unlabeled Data. in Advances in Neural Information Processing Systems 19 (eds. Schölkopf, B., Platt, J. & Hofmann, T.) 601–608 (The MIT Press, 2007)
24. Gong, Boqing, Kristen Grauman, and Fei Sha. "Correcting sample selection bias by unlabeled data." International conference on machine learning. (PMLR, 2013)

25. Sun, B. & Saenko, K. Deep CORAL: Correlation Alignment for Deep Domain Adaptation. in Computer Vision – ECCV 2016 Workshops (eds. Hua, G. & Jégou, H.) vol. 9915 443–450 (Springer International Publishing, Cham, 2016)
26. Sun, B., Feng, J. & Saenko, K. Return of Frustratingly Easy Domain Adaptation. (AAAI 30, 2016)
27. Long, Mingsheng, et al. "Learning transferable features with deep adaptation networks." International conference on machine learning. (PMLR, 2015)
28. Pan, S.J., Tsang, I.W., Kwok, J.T., Yang, Q.: Domain Adaptation via Transfer Component Analysis. IEEE Trans. Neural Netw. **22**, 199–210 (2011)
29. Hoffman, Judy, et al. "Cycada: Cycle-consistent adversarial domain adaptation." International conference on machine learning. (PMLR, 2018)
30. Murez, Zak, et al. "Image to image translation for domain adaptation." Proceedings of the IEEE conference on computer vision and pattern recognition. (2018)
31. Simonyan, K. & Zisserman, A. Very Deep Convolutional Networks for Large-Scale Image Recognition. Preprint at http://arxiv.org/abs/1409.1556 (2015)
32. Vaswani, Ashish, et al. "Attention is all you need." Advances in neural information processing systems 30 (2017)
33. O. Oktay, J. Schlemper, L. Le Folgoc, M. Lee, M. Heinrich, K. Misawa, K. Mori, S. McDonagh, N. Y Hammerla, B. Kainz, B. Glocker, and D. Rueckert, "Attention U-Net: Learning where to look for the pancreas," (2018)
34. Taesung Park, Alexei A.Efros, Richard Zhang, and Jun-Yan Zhu. Contrastive Learning for Unpaired Image-to-Image Translation. In European Conference on Computer Vision (ECCV), 2020
35. Chanyong Jung, Gihyun Kwon, and Jong Chul Ye. Exploring patch-wise semantic relation for contrastive learning in image-to-image translation tasks. In CVPR, 2022
36. Bian, W., Zhang, Q., Ye, X., Chen, Y.: A learnable variational model for joint multimodal MRI reconstruction and synthesis, pp. 354–364. In Medical Image Computing and Computer Assisted Intervention, MICCAI (2022)

Fusion of Machine Learning and Deep Neural Networks for Pulmonary Arteries and Veins Segmentation in Lung Cancer Surgery Planning

Hongyu Cheng[1], Limin Zheng[1], Zeyu Yan[2], Haoran Zhang[3], Bo Meng[1], and Xiaowei Xu[4(✉)]

[1] Shenzhen University, Shenzhen, China
[2] Easylinkin Technology Co., Ltd., Wuhan, China
[3] The Hong Kong Polytechnic University, Hongkong, China
[4] Guangdong Provincial People's Hospital, Guangzhou, China
`xiao.wei.xu@foxmail.com`

Abstract. Lung cancer, the second most common type of cancer worldwide, is primarily treated through surgery. During the operations, preserving pulmonary arteries and veins is a crucial problem. In recent years, 3D visualization techniques like virtual reality and 3D printing have been increasingly used in clinical practice for lung cancer surgery planning. Under the success of these techniques, automatic segmentation of pulmonary arteries and veins plays a key role. Particularly, the state-of-art approaches rely on two techniques, i.e. the deep neural networks (DNNs) or the traditional machine learning (ML) method, and both techniques have respective shortages. Basically, the ML-based methods generally demonstrate a limited performance, while the DNN-based methods lack sufficient annotation for accurate segmentation. In response to such a dilemma, this paper proposes a fusion method to combine the DNN-based and ML-based methods to segment pulmonary arteries and veins for lung cancer surgery planning. Particularly, the anatomy prior mask corresponding to pulmonary arteries and veins are identified using the marching cubes algorithm and Attention U-Net. Subsequently, an enhanced attention U-Net, is used to integrate the original CT scans with the anatomy prior mask to generate the refined segmentation results. Following this, an anatomy structure enhancement module is used to refine the segmentation further by refining disconnected vessel segments and correcting misclassified vessels based on anatomy prior masks. We experimented the proposed approach on a private dataset of 95 CT scans collected from patients after surgery, and then annotated by lung cancer experts. The results demonstrate that our approach outperforms the existing methods with an improvement of 5.1% to 16.2% in Dice score. The dataset and code have also been published [1] to facilitate further research in this field.

Keywords: Lung cancer · Surgery planning · Pulmonary arteries and veins · 3D Segmentation · Deep neural networks · Marching cubes

1 Introduction

Lung cancer is one of the leading causes of cancer-related morbidity and mortality worldwide, with 1.8 million deaths estimated annually [14]. Lung surgery especially segmentectomy has been adopted as the main treatment of lung cancer patients, in which precise cut of the affected lung part without affecting pulmonary arteries and veins is the key consideration [13]. If the anatomic structure of the patient is not well awared, bleeding and increased surgical resection may happen during lung surgery. Recently, 3D visualization including visual reality (VR) and 3D printing has been widely used for lung cancer surgery planning in which surgeons can obtain a vivid understanding of the anatomic structures [4]. However, in current clinical practice, pulmonary arteries and veins are manually segmented by experienced radiologists, which is time-consuming, costly, and lacks reproducibility [5].

Currently, there are two main approaches for the automatic segmentation of pulmonary arteries and veins: deep neural network (DNN) based approach and machine learning based approach. Most of the existing works can be categorized into the first approach which generally employs an end-to-end DNN with some domain-specific optimizations [8,12]. Nardelli et al.[8] first introduced DNNs to artery-vein segmentation. Qin et al.[12] incorporated anatomy prior of lung context map and distance transform to differentiate pulmonary arteries and veins. Another approach is based on traditional machine learning (ML) methods[2,10,11] with limited training data. Charbonnier et al.[2] constructed a graph representation of the segmented vessels to classify sub-structures of pulmonary arteries and veins. Payer et al.[10] classified each sub-structures via integer programming. Pu et al.[11] adopted a computational differential geometry method, marching cubes [7], to automatically identify the tubular-like structures in the lungs with high densities. Although both approaches have demonstrated applicability to specific use cases, however, they also feature respective drawbacks. The DNN-based approach cannot handle the thin vessels well due to the lack of annotations, while the traditional ML-based approach is of limited performance. In addition, the details of their evaluation, including the adopted datasets and comparison methods, vary significantly. For example, Qin et al. [12] used a combination of several public datasets with complement target labels from other works. Pu et al.[11] collected a private dataset for evaluation but no comparison with related methods are performed.

In this paper, we perform the fusion of DNNs and traditional ML to segment pulmonary arteries and veins for lung cancer surgery planning. To evaluate the proposed method, a collection of 95 computed tomography (CT) scans from surgery-treated patients is constructed. Results on this dataset show that our fusion approach achieves an improvement over existing works by 10.6% in Dice score. The dataset and code are also published [1] to facilitate further research in this field. Our contributions are as follows: (1) a new dataset of 95 CT scans in which all patients have underwent surgery, (2) a fusion method to combine the DNN-based and ML-based methods, and (3) improved performance by 5.1% to 16.2% compared with existing methods.

2 Method

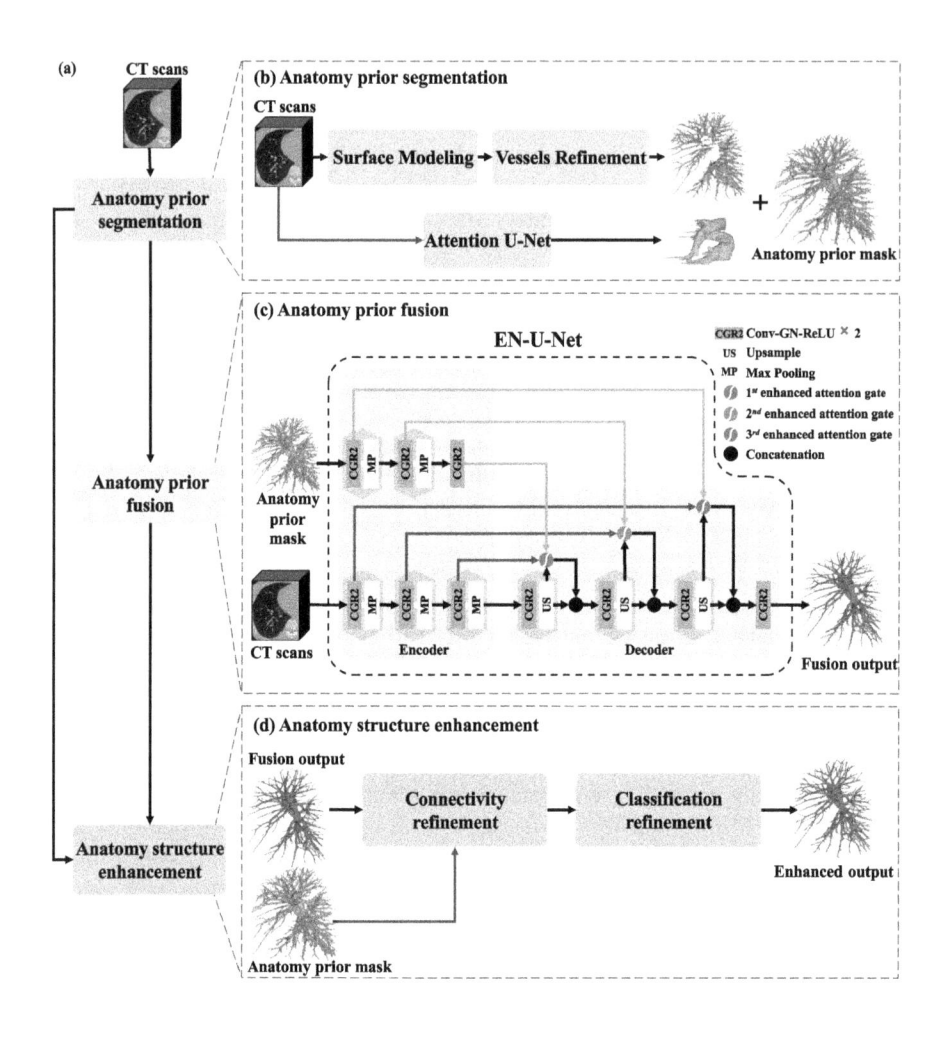

Fig. 1. The flow of our proposed fusion method: (a) the overall flow, (b) anatomy prior segmentation, (c) anatomy prior fusion, and (d) anatomy structure enhancement (ASE).

2.1 Overview

Our proposed approach for pulmonary arteries and veins segmentation is outlined in Fig. 1(a) and involves a systematic three-step process. Initially, potential regions of interest corresponding to the pulmonary arteries and veins are identified using the marching cubes algorithm and Attention U-Net [9] during the anatomy prior segmentation stage. Subsequently, in the anatomy prior fusion stage, an enhanced attention U-Net (EA-U-Net), is used to integrate the original CT scans with the anatomy prior mask to generate the refined segmentation results. Following this, the anatomy structure enhancement (ASE) stage refines the segmentation further by refining disconnected vessel segments and correcting misclassified vessels based on anatomy prior masks, if any. Note that the anatomy prior mask obtained in the first stage is added to the decoder of the network in the second stage, which could help guide the network to focus on related regions. More details of each stage are specified in the subsequent sections.

2.2 Anatomy prior segmentation

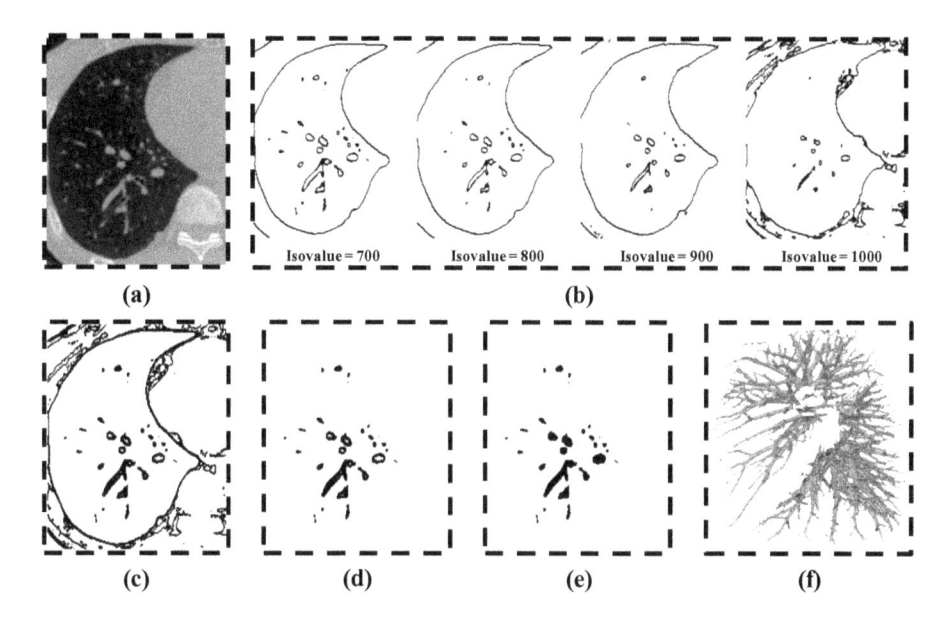

Fig. 2. Processing steps of marching cubes to get prior anatomy inside the lung including (a) the original CT image, (b) results of marching cubes with 2D projection of the equivalence surfaces from 700 to 1000, (c) image overlay of 4 isosurfaces, (d) intrapulmonary part reservation, (e) hole filling, and (f) 3D visualization.

Anatomy prior segmentation includes two regions corresponding to areas inside (thin vessels) and outside the lungs (thick vessels). The anatomy prior mask inside the lungs are obtained by the marching cubes algorithm [7], while that outside the lungs are obtained by an Attention U-Net. The final anatomy prior mask are obtained by combining the above two methods. Note that the anatomy prior mask does not differentiate between arteries and veins.

The marching cube algorithm process is depicted in Fig. 1(b), which can create rough surface models representing pulmonary arteries and veins by generating triangular meshes. For a comprehensive understanding of this process, please refer to [6]. The primary hyper-parameter for this step is the isovalue, which determines the threshold for surface creation. To specifically target pulmonary arteries and veins, pixel values within the range of [700 HU, 1000 HU] are analyzed. To ensure comprehensive coverage, isovalue settings ranging from 700 HU to 1000 HU at 100 HU intervals are selected, as depicted in Fig. 2(b). Note that 3D surface models, represented as triangular meshes, need to be mapped back to the original image space for fusion with CT images. The isosurfaces obtained from isovalue settings ranging from 700 to 1000 HU are aggregated to serve as boundary representations for pulmonary vessels in various dimensions, as illustrated in Fig. 2(c). However, the generation of lung vessel surface models may inadvertently include additional tissue structures such as lung walls and bones. Therefore, the results are refined to preserve regions exclusively within the lung window, with hole-filling techniques employed to identify potential artery and vein regions, as show in Fig. 2(d) and (e). The 3D visualization results of the prior anatomy inside the lungs are shown in Fig. 2(f).

2.3 Anatomy prior fusion

The process is illustrated in Fig. 1(c). Particularly, we introduce enhanced attention U-Net (EA-U-Net), for an accurate segmentation and differentiation of pulmonary arteries and veins. This framework operates with two primary inputs: original CT scans and the prior anatomy mask. Within the encoder, the CT scans and the prior anatomy mask undergo convolution and downsampling operations, resulting in the generation of feature maps of various dimensions. These feature maps serve as the foundation for subsequent fusion within the decoder.

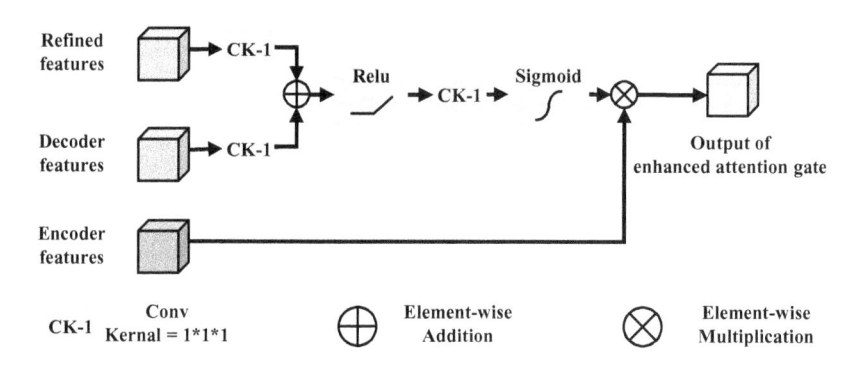

Fig. 3. Detailed structure of enhanced attention gate.

We propose an enhanced attention gate for efficient fusion. As depicted in Fig. 3, the enhanced attention gate integrates three key inputs: encoder features, decoder features, and refined features. The encoder features are derived from CT scans in the encoder process. The decoder features are produced by upsampling blocks in the decoder process. The refined features are generated from the anatomy prior mask in the encoder process. Traditionally, attention gates involve only two inputs, with encoder features acting as key and value vectors, and decoder features serving as query vectors. However, in the enhanced attention gate, the refined features are regarded as additional key vectors, marking a pivotal distinction from existing approaches. Within the enhanced attention gate, both refined features and decoder features undergo convolution operations before they are combined to assess their similarity. Attention scores for each channel are determined through the application of ReLU activation, a convolution operation, and a sigmoid function. The final feature representation is derived by multiplying the attention scores with the encoder features' mappings.

The rationale underneath this fusion approach has two points. First, it harnesses the effectiveness of multi-scale feature fusion, a technique substantiated by prior studies [8,12]. Second, it capitalizes on the potency of attention mechanisms, guiding the network's focus towards regions of interest. Given that the marching cubes stage has already identified potential pulmonary vessels regions, this specialized fusion method aims to refine and prioritize these areas, resulting in potentially more precise segmentation performance. Notably, the coarse segmentation achieved by the marching cubes stage encompasses distal small vessels, which may help address the challenge of gradient disappearance in peripheral regions.

2.4 Anatomy structure enhancement

Connectivity refinement Addressing disconnected vessels entails solving two key problems: spatial localization and spatial estimation of disconnected vessels.

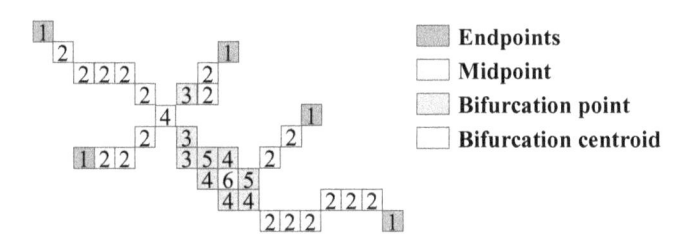

Fig. 4. Possible scenarios for the number of adjacent points to the center point.

(1) Spatial localization of disconnected vessels. First, we utilize the method proposed by Lee et al. [3] to extract the centerline of blood vessels, which yields a graph with a series of nodes. The possible scenarios for the number of neighboring points in a node, denoted as N, are depicted in Fig. 4. When $N = 1$, the node signifies the end of the vessel, referred to as the endpoint. When $N = 2$, the node represents the midpoint along the centerline. When $N \geq 3$, the node denotes a bifurcation point, indicating the vessel's division into multiple directions. When $N \geq 3$ and the number of neighboring nodes is maximal, the

node is termed as a bifurcation centroid. Nodes of disconnected vessels typically fall into the categories of endpoints and bifurcation centroids. Second, we include these endpoints or bifurcation centroids as candidate points but require further filtering to refine the selection. A continuous centerline exhibits the characteristic that nodes are essentially co-linear with adjacent nodes, allowing for the filtering and matching of candidate points based on this property. We can traverse all candidate points, with the notation of the candidate point as $P_0(x_0, y_0, z_0)$. The third node of the consecutive neighboring points to the point P_0 is $K(x_k, y_k, z_k)$. If point P_0 represents the center of the disconnected cross-section, then there should exist a node that is co-linear with point P_0 and point K in a close region. Consequently, we select the five closest candidate points $Q_m(x_m, y_m, z_m), m = 1, ..., 5$ to the point P_0, as illustrated in Fig. 5.

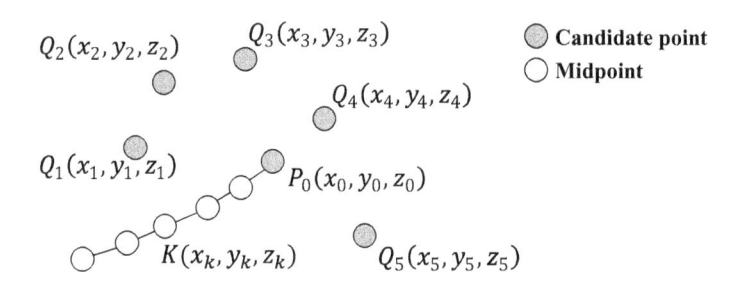

Fig. 5. Illustration of candidate points of disconnected vessels.

When points P_0, K, and Q_m are co-linear, the angle between $\overrightarrow{Q_m P_0}$ and $\overrightarrow{P_0 K}$ should be zero. In fact, P_0, K, and Q_m cannot completely co-linear. Therefore, we set a threshold T to constrain the selection of the center point of the disconnected cross-section. Points P_0 and Q_m are considered to be the centers of the bottom surface of the circular table when Equation 2 is satisfied.

$$\alpha = \arccos \frac{(\overrightarrow{Q_m P_0} * \overrightarrow{P_0 K})}{|\overrightarrow{Q_m P_0}| * |\overrightarrow{P_0 K}|} \quad (1)$$

$$\alpha < T \quad (2)$$

(2) Spatial estimation of disconnected vessels. We leverage the anatomical insight that the blood vessels have a tubular structure with a progressive decrease/increase in vessel radius. Thus, a round table can be employed to describe the largest possible spatial location of the disconnected vessels as illustrated in Fig. 6. The disconnected vessels are estimated as the intersection of the round table and the anatomy prior mask. More details are as follows. First, the upper and lower surfaces of the round table are determined. The centers of the end of the two vessels are denoted as P_0 and Q_m, and the normal vector of the

upper and lower surfaces of the round table is $\overrightarrow{Q_m P_0}$. With the normal vector, P_0 and Q_m, the planes D_1 and D_2 where the upper and lower bottom surfaces are located are determined using Equation 3 and 4. Second, the radius of the upper and lower surfaces of the round table is calculated. The distance between the segmentation results and the center point P_0/Q_m in the upper/lower plane are calculated, and the maximum distance in the two planes is taken as the radius of the upper and lower base planes, denoted as R_1 and R_2, respectively. Third, disconnected vessels are estimated, which are the intersection of the anatomy prior mask and the round table. Note that the anatomy prior mask are usually a good estimation of the the surface shape of the vessel. Forth, the label of the repaired vessel is set to the same as the one with a larger radius among the two terminal vessels with radii of R_1 and R_2.

$$(x_0 - x_m, y_0 - y_m, z_0 - z_m) \cdot (x - x_0, y - y_0, z - z_0) = 0 \quad (3)$$

$$(x_0 - x_m, y_0 - y_m, z_0 - z_m) \cdot (x - x_m, y - y_m, z - z_m) = 0 \quad (4)$$

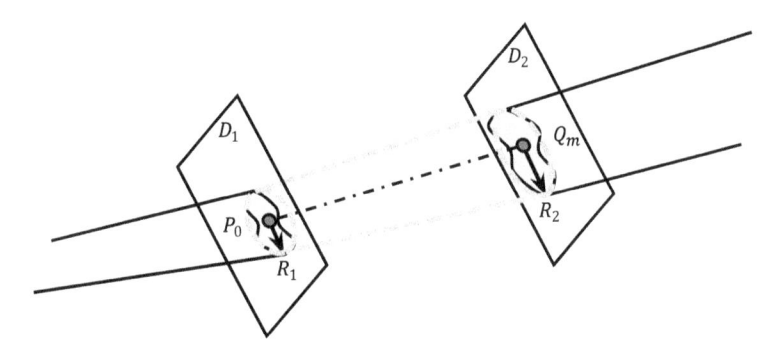

Fig. 6. Round table estimation of disconnected vessels.

Classification refinement Misclassification of pulmonary arteries and veins may lead to isolated vessels, as depicted in Fig. 7. To rectify this issue, we identify connected regions of the segmentation result in 3D space. Initially, we designate the largest volume of pulmonary arteries and veins within the connected region as trunk vessels. Subsequently, we examine whether there are isolated vessels connected to the trunk vessel but classified differently. If such isolated vessels exist, we correct the label of the isolated vessel accordingly. This iterative process proceeds until there are no remaining isolated vessels, signaling the completion of correcting the misclassified vessels.

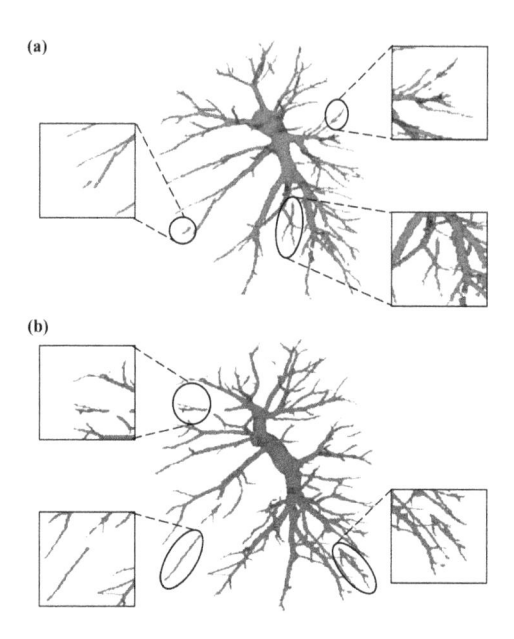

Fig. 7. Isolated vessels in pulmonary (a) arteries and (b) veins segmentation results.

3 Experiment

3.1 Dataset

Our dataset comprises 95 3D CT scans obtained from Siemens' SOMATOM Definition Flash machine. Patients are aged 29 to 82 years, with an average age of 58.4 years. The images contain $512 \times 512 \times (280-370)$ voxels, with a typical voxel size of $0.25 \times 0.25 \times 0.5 mm^3$. The annotations encompass intrapulmonary and extrapulmonary arteries and veins. All the annotations are performed by lung cancer expert surgeons with 2-3.5 invested hours for each image. It's important to note that **all patients in the dataset had undergone lung cancer surgeries, and the annotations had been effectively employed in clinical practice to support surgeons in planning their surgeries**. Due to the requirement of lung cancer surgeries, left and/or right lungs with nodes are considered to be labeled. Consequently, most images have annotations for only the left or right lung, while a few have labels for both lungs. In our experiments, we divided the dataset into two subsets, each representing half of the lung, and only the subset with annotations was utilized for training and testing. This led to a final dataset consisting of 106 subjects. In addition, we also labeled the lung areas for two purposes. On one hand, this helps distinguish pulmonary arteries and veins inside and outside the lung. On the other hand, some existing works [2,8,12,15–19] focus on pulmonary arteries and veins either inside the lung or outside the lung. The additional label of lung area can help make fair comparison for future studies. Fig. 8 illustrates examples of the images and corresponding labels including pulmonary arteries and veins and lung areas in our dataset.

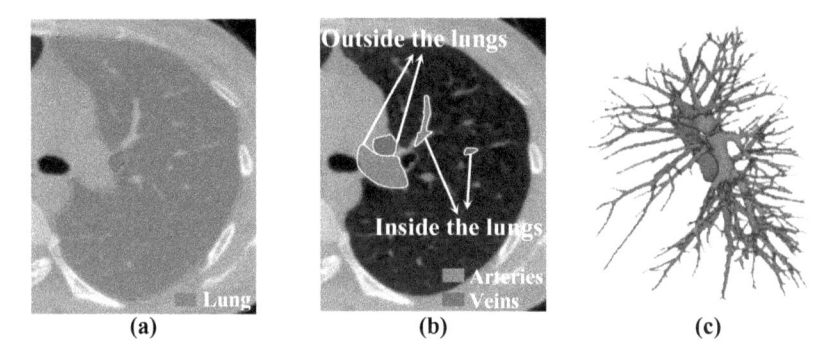

Fig. 8. Examples of the collected dataset including the labels of (a) lung areas, and (b) arteries and veins inside and outside the lungs, and (c) 3D visualization of labels.

3.2 Implementation Details

The experiments are conducted on a single Nvidia A40 GPU equipped with 48GB graphics memory. Since the size of CT scans varies, we divide the data into multiple patches with a fixed size of 176 17676 voxels and a stride of 64. Three-fold cross-validation is adopted, with a ratio of 13:1:7 for the training, validation, and test sets, respectively. To prevent overfitting and achieve good generalization, early stopping is applied during network training. Specifically, if the loss of the validation set stops decreasing for eight consecutive epochs, the training process will be terminated. Additionally, we impose constraints on the learning rate, starting with an initial value of 1e-4 and decaying down to a minimum value of 1e-6. If the Dice score does not decrease for 2 consecutive epochs on the validation set, the learning rate will be reduced by a factor of 10. Lastly, the hyperparametric threshold T in ASE is set to 10.

3.3 Results and discussion

Comparison with existing works Two prior works [16] [12] with publicly available code are selected for benchmarking the result of our method. The results are presented in Table 1. Compared with [16] and [12], our EA-U-Net achieved a significant improvement for pulmonary arteries and veins segmentation, namely 5.5% and 6.2% in DSC, respectively. Note that our EA-U-Net obtains optimal performance in all other evaluation metrics, including precision, recall, specificity, and Hausdorff distance. In addition, our method with ASE further boost the performance with an improvement of 9.4% and 5.1% in recall and

DSC over that without ASE. Therefore, our method have a significant improvement of 10.6% in DSC over the state-of-the-art work [12]. In addition, we have performed comparison with two recent methods [20] [11] as shown in Table I. We would like to highlight that our segmentation results obtain an improvement of 5.1% to 16.2% compared with existing works. We also discuss the robustness of our method using the variance of DSC and Hausdorff distances, and the results are presented in Table 2. The significant boost in the variance of the DSC and Hausdorff distance is mainly attributed to the correction of classification errors by ASE. Correcting these errors effectively eliminates inconsistencies between the results and the labels. In terms of the inference time for each CT image, it takes about 5 minutes for marching cubes on 11th Gen Intel(R) Core(TM)i7-11800H CPU, and it takes about 30 seconds for our EN-U-Net, and about 5 seconds for anatomy structure enhancement.

Table 1. Performance comparison of EA-U-Net with existing works.

Meth	Precision	Recall	Specificity	DSC	Hausdorff distance
Zulfiqar et al. [20]	0.722	0.669	0.998	0.684	9.941
Wu et al. [16]	0.693	0.775	0.998	0.733	14.735
Qin et al. [12]	0.702	0.778	0.998	0.740	7.574
Pu et al.[11]	0.762	0.769	0.997	0.759	**3.878**
Ours (EA-U-Net)	0.820	0.782	**0.999**	0.795	5.032
Ours (EA-U-Net with ASE)	**0.825**	**0.876**	**0.999**	**0.846**	4.162

Table 2. Variance comparison of Dice and Hausdorff of EA-U-Net with existing works.

Method	DSC	Hausdorff
Zulfiqar et al. [20]	0.00427	472.299
Wu et al. [16]	0.00429	1114.285
Qin et al. [12]	0.00575	285.882
Pu et al. [11]	0.00198	68.525
Ours (EA-U-Net)	0.00203	24.126
Ours (EA-U-Net with ASE)	**0.00151**	**3.992**

Result visualization and comparison of EA-U-Net Visual comparison of the segmentation results is shown in Fig 9. Note that for fair comparison, as a post-processing step, ASE is not adopted here. Overall, our segmentation results obtained the most complete structure of the pulmonary arteries and veins tree, with the least number of disconnected fine vessels. Most of the disconnections in the segmentation results of other methods occur at the end of the vessels. This is because the terminal vessels have smaller diameters. Thin vessels are difficult to detect in CT images, thus making segmentation more difficult. Our EA-U-Net method achieves better segmentation of small vessels by integrating the anatomy prior mask into the attention mechanism.

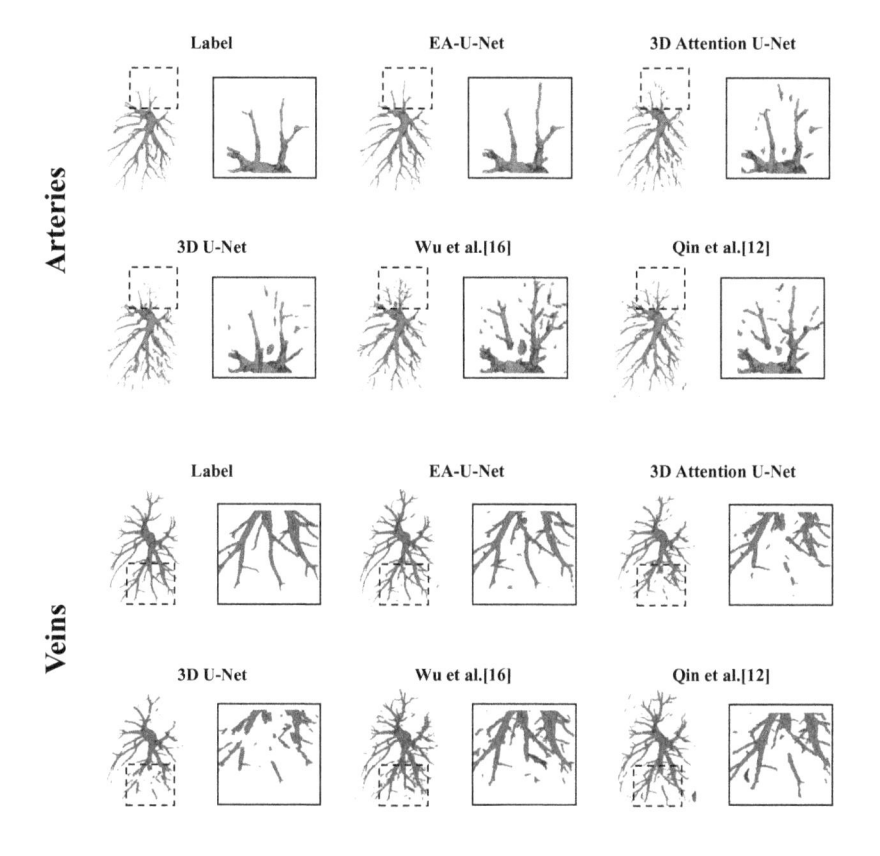

Fig. 9. Visual comparison of EA-U-Net results with existing methods. Note that for fair comparison, as a post-processing step, ASE is not adopted here.

Visual analysis of ASE The visual analysis of ASE is illustrated in Fig. 10. We can notice that visualization results depict the refinement of vessels at terminal disconnections more clearly. In addition, it can be observed that most of the refined vessels are correctly classified. We also define the *number of terminal vessels* as a metric to evaluate connectivity. Note that pulmonary arteries and veins can extend progressively within the lungs, leading to the branching of vessels, thus creating a complex vessel topology. Therefore, the terminal vessels

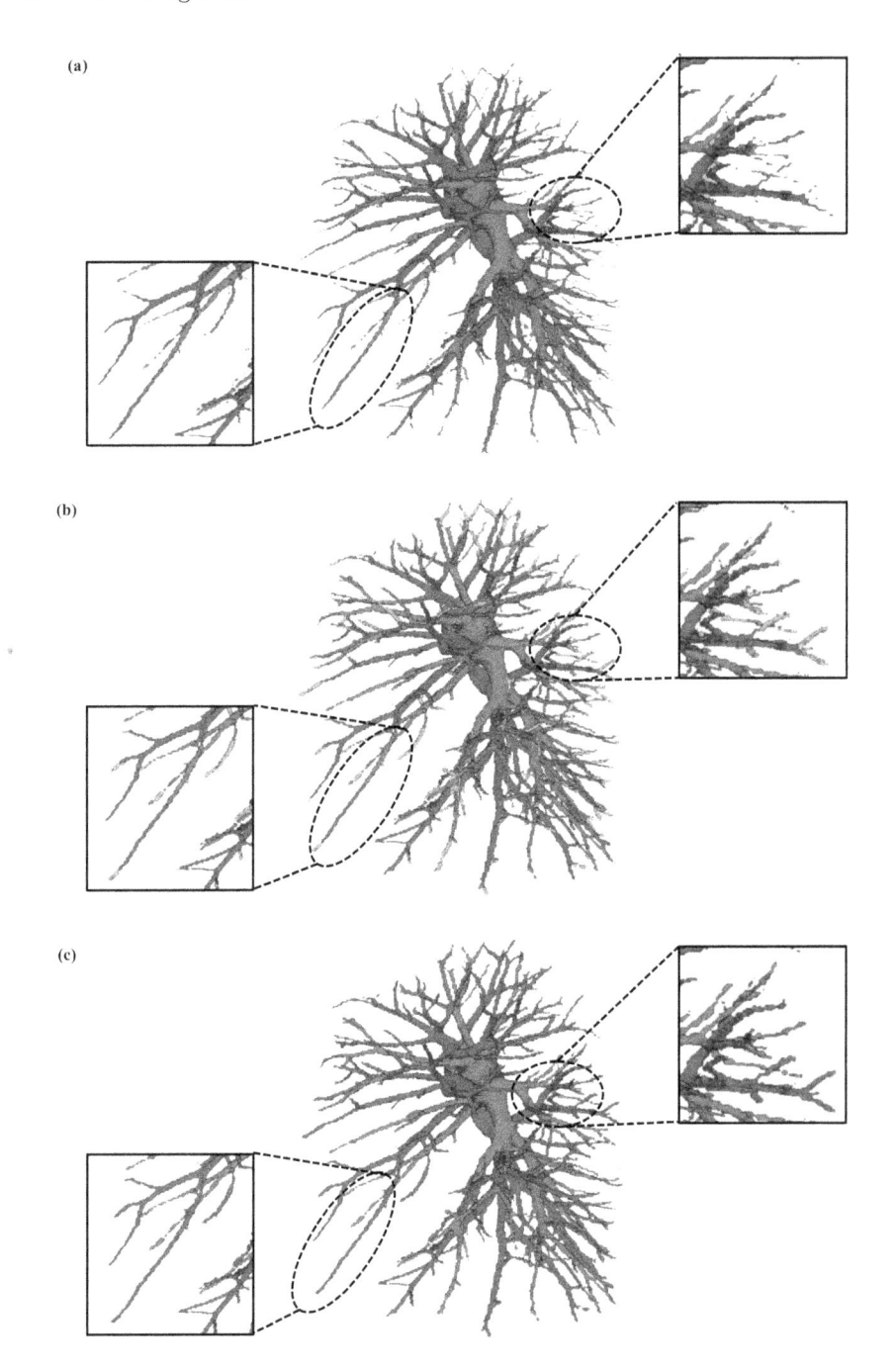

Fig. 10. Examples of segmentation results with and without ASE: (a) segmentation results without ASE, (b) segmentation results with ASE and highlighted difference, and (c) segmentation results with ASE.

represent the smallest terminal branches of the main pulmonary vessels. The *number of terminal vessels* objectively reflects the completeness and connectivity complexity of the vessel tree. Obviously, disconnected vessels will result in a larger number of terminal vessels. The results corresponding to Fig. 10 are shown in Table 3. The average number of refined terminal vessels increased by 59 after ASE, indicating a positive impact of our method on connectivity refinement. Segmentation results without ASE exhibited a decrease of 42 terminal vessels compared with labels, whereas they increased by 17 with ASE. This suggests that the segmentation results obtained by EA-U-Net actually covered the labeled terminal vessels but are not accounted for by annotators due to extreme thin vessels. We effectively addressed this issue by refining vessels at the disconnections. The refined segmentation results were able to identify more fine blood vessels than the labeled results.

Table 3. Comparison of mean number of terminal vessels.

Method	Number of terminal vessels
Ground truth	283
EA-U-Net	241
EA-U-Net with ASE	**300**

Table 4. Ablation analysis with EA-U-Net.

Meth	Precision	Recall	Specificity	DSC	Hausdorff distance
3D U-Net	0.767	0.725	0.998	0.739	12.88
Attention U-Net	0.773	0.742	0.998	0.752	12.17
Ours (EA-U-Net)	0.820	0.782	**0.999**	0.795	8.59
Ours (EA-U-Net with ASE)	**0.825**	**0.876**	**0.999**	**0.846**	**8.52**

Ablation analysis The ablation results is shown in Table 4 where we use precision, recall, specificity, Dice score (DSC), and Hausdorff distance as evaluation metrics. According to the results, EA-U-Net achieves superior results in all evaluation metrics, while Attention U-Net outperforms 3D U-Net in all the metrics. Particularly, EA-U-Net obtains an improvement of 3.8% and 3.62 over Attention U-Net in Dice and Hausdorff distance, respectively.

Visual comparison of the feature maps output by the attention gate between Attention U-Net and EA-U-Net is shown in Fig. 11. In the feature maps of the first attention gate, we observed a focus on the approximate region of the pulmonary vessels earlier than that in Attention U-Net. In the feature maps of the second attention gate, the region of attention shifted to the edges of the vessels, resulting in clearer boundaries compared with that in Attention U-Net. While for the third attention gate, our enhanced attention gate accurately identified the pulmonary vascular region. Particularly, the highly attentive region of our enhanced attention gate covers the labeled region more comprehensively, exhibiting a significant difference in attention compared with Attention U-Net.

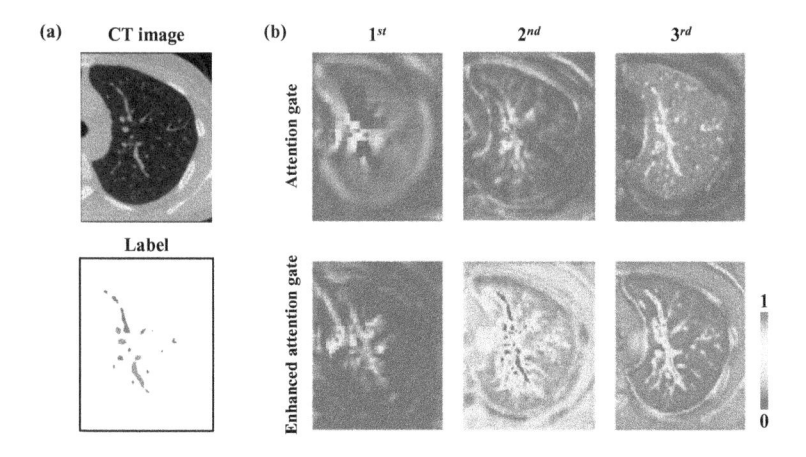

Fig. 11. Visual comparison of output feature maps of attention gates: (a) CT image and its label, and (b) corresponding feature maps. The positions of 1st, 2nd and 3rd attention gate can refer to Fig. 1.

4 Conclusion

This paper introduces a novel fusion approach that combines the marching cubes algorithm with EA-U-Net for the segmentation of the pulmonary arteries and veins for lung cancer surgery planning. A dataset comprised of 95 computed tomography (CT) scans from surgery-treated patients is collected for evaluating the proposed approach. The experimental results demonstrate that our proposed fusion approach achieves a significant improvement of 5.1% to 16.2% in DSC over existing methods. A visual comparison of the segmentation results further shows that our method can effectively focus on the region of interest and detect the disconnected vessels. To facilitate further research in this field, the dataset and code [1] are also published to the community.

Acknowledgement. This work was supported by the National Natural Science Foundation of China (No. 62276071), Guangdong Special Support Program-Science and Technology Innovation Talent Project (No. 0620220211), the Science and Technology Planning Project of Guangdong Province, China (No. 2019B020230003), Guangdong Peak Project (No. DFJH201802), Guangzhou Science and Technology Planning Project (No. 202206010049), Guangdong Basic and Applied Basic Research Foundation (No. 2022A1515010157).

References

1. Dataset. https://github.com/XiaoweiXu/PulmonaryVesselSegSurgicalPlanning
2. Charbonnier, J.P., Brink, M., Ciompi, F., Scholten, E.T., Schaefer-Prokop, C.M., Van Rikxoort, E.M. (eds.): Automatic pulmonary artery-vein separation and classification in computed tomography using tree partitioning and peripheral vessel matching, vol. 35. IEEE (2015)
3. Lee, T.C., Kashyap, R.L., Chu, C.N. (eds.): Building skeleton models via 3-D medial surface axis thinning algorithms, vol. 56. Elsevier (1994)
4. Li, C., Zheng, B., Yu, Q., Yang, B., Liang, C., Liu, Y. (eds.): Augmented reality and 3-dimensional printing technologies for guiding complex thoracoscopic surgery, vol. 112. Elsevier (2021)
5. Liu, X., Zhao, Y., Xuan, Y., Lan, X., Zhao, J., Lan, X., Han, B., Jiao, W.: Three-dimensional printing in the preoperative planning of thoracoscopic pulmonary segmentectomy. Translational Lung Cancer Research **8**(6), 929 (2019)
6. Lorensen, W.E., Cline, H.E. (eds.): Marching cubes: A high resolution 3D surface construction algorithm, vol. 21. ACM New York, NY, USA (1987)
7. Lorensen, W.E., Cline, H.E.: Marching cubes: A high resolution 3d surface construction algorithm. In: Seminal graphics: pioneering efforts that shaped the field, pp. 347–353 (1998)
8. Nardelli, P., Jimenez-Carretero, D., Bermejo-Pelaez, D., Washko, G.R., Rahaghi, F.N., Ledesma-Carbayo, M.J., Estépar, R.S.J. (eds.): Pulmonary artery–vein classification in CT images using deep learning, vol. 37. IEEE (2018)
9. Oktay, O., Schlemper, J., Folgoc, L.L., Lee, M., Heinrich, M., Misawa, K., Mori, K., McDonagh, S., Hammerla, N.Y., Kainz, B., et al.: Attention u-net: Learning where to look for the pancreas. arXiv preprint arXiv:1804.03999 (2018)
10. Payer, C., Pienn, M., Bálint, Z., Shekhovtsov, A., Talakic, E., Nagy, E., Olschewski, A., Olschewski, H., Urschler, M. (eds.): Automated integer programming based separation of arteries and veins from thoracic CT images, vol. 34. Elsevier (2016)
11. Pu, J., Leader, J.K., Sechrist, J., Beeche, C.A., Singh, J.P., Ocak, I.K., Risbano, M.G. (eds.): Automated identification of pulmonary arteries and veins depicted in non-contrast chest CT scans, vol. 77. Elsevier (2022)
12. Qin, Y., Zheng, H., Gu, Y., Huang, X., Yang, J., Wang, L., Yao, F., Zhu, Y.M., Yang, G.Z. (eds.): Learning tubule-sensitive cnns for pulmonary airway and artery-vein segmentation in ct, vol. 40. IEEE (2021)
13. Saji, H., Okada, M., Tsuboi, M., Nakajima, R., Suzuki, K., Aokage, K., Aoki, T., Okami, J., Yoshino, I., Ito, H., et al. (eds.): Segmentectomy versus lobectomy in small-sized peripheral non-small-cell lung cancer (JCOG0802/WJOG4607L): a multicentre, open-label, phase 3, randomised, controlled, non-inferiority trial, vol. 399. Elsevier (2022)
14. Sung, H., Ferlay, J., Siegel, R.L., Laversanne, M., Soerjomataram, I., Jemal, A., Bray, F. (eds.): Global cancer statistics 2020: GLOBOCAN estimates of incidence and mortality worldwide for 36 cancers in 185 countries, vol. 71. Wiley Online Library (2021)
15. Suzuki, H., Kawata, Y., Aokage, K., Matsumoto, Y., Sugiura, T., Tanabe, N., Nakano, Y., Tsuchida, T., Kusumoto, M., Marumo, K., Kaneko, M., Niki, N.: Aorta and main pulmonary artery segmentation using stacked u-net and localization on non-contrast-enhanced computed tomography images. MEDICAL PHYSICS (2023)

16. Wu, Y., Qi, S., Wang, M., Zhao, S., Pang, H., Xu, J., Bai, L., Ren, H.: Transformer-based 3d u-net for pulmonary vessel segmentation and artery-vein separation from ct images. Medical & Biological Engineering & Computing **61**(10), 2649–2663 (2023)
17. Xu, X., Jia, Q., Yuan, H., Qiu, H., Dong, Y., Xie, W., Yao, Z., Zhang, J., Nie, Z., Li, X., et al.: A clinically applicable ai system for diagnosis of congenital heart diseases based on computed tomography images. Med. Image Anal. **90**, 102953 (2023)
18. Xu, X., Wang, T., Shi, Y., Yuan, H., Jia, Q., Huang, M., Zhuang, J.: Whole heart and great vessel segmentation in congenital heart disease using deep neural networks and graph matching. In: MICCAI, Shenzhen, China, October 13–17, 2019, Proceedings, Part II 22. pp. 477–485. Springer (2019)
19. Xu, X., Wang, T., Zhuang, J., Yuan, H., Huang, M., Cen, J., Jia, Q., Dong, Y., Shi, Y.: Imagechd: A 3d computed tomography image dataset for classification of congenital heart disease. In: MICCAI, Lima, Peru, October 4–8, 2020, Proceedings, Part IV 23. pp. 77–87. Springer (2020)
20. Zulfiqar, M., Stanuch, M., Wodzinski, M., Skalski, A.: Dru-net: Pulmonary artery segmentation via dense residual u-network with hybrid loss function. Sensors **23**(12), 5427 (2023)

SEANet: Rethinking Skip-Connections Design in Encoder-Decoder Networks via Synergistic Spatial-Spectral Fusion for LDCT Denoising

Abhijit Das[1], Vandan Gorade[1], Dwarikanath Mahapatra[2], and Sudipta Roy[1](✉)

[1] Artificial Intelligence and Data Science, Jio Institute, Navi Mumbai 410206, India
sudiptaroy01@yahoo.com
[2] Inception Institute of AI, Abu Dhabi, UAE

Abstract. In medical imaging, low-dose CT (LDCT) is favored over normal-dose CT (NDCT) to reduce radiation risks, but LDCT suffers from increased noise, prompting the need for advanced denoising techniques. Traditional iterative reconstruction algorithms face challenges in balancing noise reduction and preserving fine details, whereas recent CNN or transformer based methods learn mappings from noisy to clean images incorporating encoder-decoder architectures with skip-connections. Conventional skip connections with vanilla convolutions lead to loss of semantic details in deeper layers, while transformer-based methods may struggle to focus on patch edges and exhibit limited efficacy capturing local representations precisely with limited data. To establish the balance between feature retention and noise reduction we introduce SEANet, a novel architecture integrating spectral residual units (SRUs) with encoder-decoder frameworks, synergizing complementary spatial and spectral spaces. In SEANet, SRUs guide the corresponding decoder blocks with synergistic local-to-global features via skip-connections. SRUs with encoding dilation modules (DMs) precisely separate noise from relevant features by distinct frequency bands in spectral space and attends patch-edges well. SEANet eliminates drawback of spatial pooling and felicitates precise denoising for low contrast regions in LDCT images. Extensive experiments and interpretable spectral maps demonstrate SEANet's superiority over state-of-the-art methods like RED-CNN and CTFormer with a margin of 1.92 in PSNR and 2.75 in SSIM. Find the code for SEANet at https://github.com/aj-das-research/SEANet.

Keywords: LDCT denoising · Spatial-spectral synergistic residuals · Learning in complementary spaces

Supplementary Information The online version contains supplementary material available at https://doi.org/10.1007/978-3-031-78198-8_29.

1 Introduction

Low-dose CT or LDCT denoising process has gain lot of attention over the pas few years as LDCT reduce radiation exposure while still providing effective imaging for early detection and monitoring of various medical conditions, particularly in cancer screening. However, LDCT is produced by using lower radiation doses during scanning, resulting in noisier images due to reduced signal-to-noise ratio inherent in lower radiation exposures. This largely impacts image clarity and quality with unnecessary artifacts and low contrast regions [17]. Moreover, CT images often contain intricate structural details that may be challenging to preserve during reconstruction, especially in the presence of noise. Also it is crucial to match the noise profile present in standard NDCT images.

Traditional approaches [4,26] to LDCT denoising involve methods such as filtering and iterative reconstruction algorithms [14]. These methods often struggle to balance noise reduction with preserving fine image details due to inherent trade-offs and require huge computational cost [27]. For example, filtered back projection (FBP) [25] applies a filter to raw projection data and then back-projects the filtered data to reconstruct the image, but it tends to yield images with higher noise levels and lower spatial resolution. In contrast, recent deep learning-based approaches [8,13], utilizing convolutional neural networks (CNNs), have demonstrated significant success. These approaches learn complex mappings directly from noisy input images to clean output images, effectively removing noise while preserving important structural details. They adapt to various noise patterns and capture intricate relationships between noisy and clean images, leading to superior denoising performance over the traditional methods.

Deep learning-based approaches, utilizing CNNs [9,22] or Transformers [23, 24,28], excel in learning intricate mappings from noisy input images to clean outputs, effectively removing noise while preserving essential structural details. These models adapt to various noise patterns and can capture nuanced relationships between noisy and clean images, outperforming traditional methods. Notably, incorporating skip-connections, as seen in transfer learning for LDCT image denoising, proves crucial, especially in scenarios with unknown noise levels. Works like RED-CNN [13] emphasize the efficacy of symmetric convolution and deconvolution networks with skip connections, achieving rapid denoising of LDCT images. Moreover, approaches like DRL-E-MP [15] leverage dilated residual learning and edge detection layers to better preserve structural details, particularly at image boundaries. Integrating ResNet with auto-encoder principles enables learning of both local and global image features, mitigating issues like vanishing gradients. Additionally, Transformer-based methods like TransUNet [1] and CTFormer [23] excel in capturing global dependencies, further enhancing denoising performance in LDCT images.

However, many existing approaches [2,12,19,20] tend to overlook either global or local information, presenting a challenge in achieving comprehensive

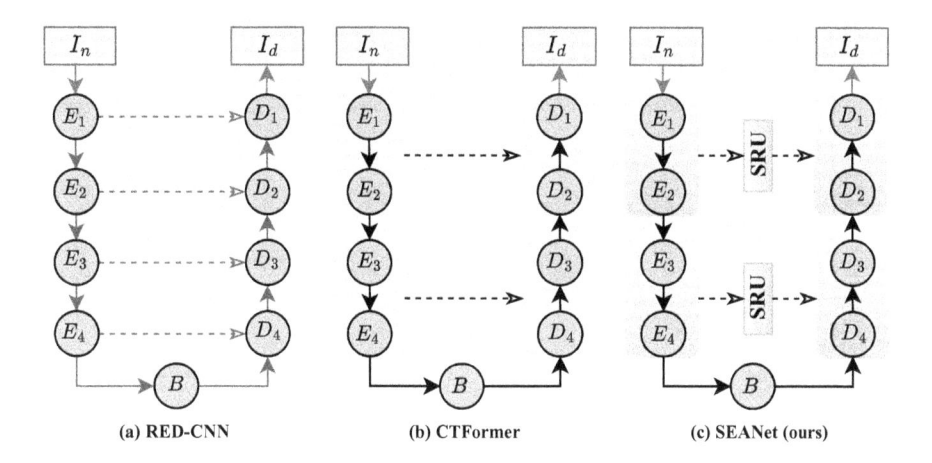

Fig. 1. An illustration showing the different skip-connection schemes implemented in existing denoising methods against proposed SEANet model with synergistic residual.

denoising. Furthermore, conventional skip connections, as employed in RED-CNN or UNet, often concatenate feature maps without considering potential loss of spatial or semantic details, particularly in deeper layers with reduced spatial resolution. Methods like CTFormer that employs transformers as the building blocks also suffers to focus on the edges of a given patch [18]. And for smaller datasets it has limited efficacy to extract the local representations [18].

Inspiring by recent advancements in the domain of synergizing complementary spaces [5–7], we rethink design on skip-connections in CNN and transformer-based encoder-decoder architectures and propose novel architecture called SEANet. As shown in Fig 1, integrating spectral and spatial features through SRU enables the model to combine the strengths of both domains resulting in local-to-global synergistic features in a unified framework. Spectral transformations in SEANet allow the model to allocate attention to relevant frequency bands, enabling multi-scale analysis that enables learning features in low-contrast regions also. Analyzing the frequency components in the spectral domain captures the orientation, edges and textures and complements the strong pixel-level representations learned by encoding dilation modules (DMs) (as shown in Fig 2). In SRU pointwise multiplication emphasizes high-frequency components without introducing spatial average filtering and results in fewer unwanted artifacts in denoised outputs. To this end, our contributions in this paper are summarized as the following:

1. We propose SEANet that utilizes spectral residual units (SRUs) with encoder-decoder architectures Synergizing spatial and spectral domains. Multiscale learning in spectral space complements the spatial features extracted from each encoding block.

2. SRU along with dilation modules (DMs) enables precise separation of noise from relevant anatomical features in spectral space by the distinct frequency signatures and attends edges for all patches. This felicitates precise denoising for low-contrast regions and mitigates the limitations of spatial pooling.
3. SEANet is evaluated on the largest publicly available Mayo LDCT dataset [16] across five different metrics with in-depth spectral map analysis. SEANet outperforms existing SOTA methods with a margin of 2.75 in structural similarity index (SSIM) and 8.93 in mean absolute percentage error (MAPE).

2 Method

2.1 Preliminaries

Problem Statement. The problem of LDCT denoising can be formulated as optimizing a loss function $\mathcal{L}(V, I_n, I_d)$, that can be formulated as:

$$\mathcal{L}(V; I_n, I_d) = \frac{1}{N} \sum_{i=1}^{N} \left(\mathcal{D}(V, I_n^{(i)}) - I_d^{(i)} \right)^2 \tag{1}$$

where, N represents the total number of samples in the dataset, $\mathcal{D}(\cdot)$ is the denoising function, $I_n^{(i)}$ and $I_d^{(i)}$ represent the i-th noisy LDCT image and corresponding denoised NDCT image, respectively. The goal is to adjust the parameters V to minimize the average squared difference between the predicted denoised image and the ground truth image over all samples in the dataset.

Fast Fourier Transformation (FFT). FFT [10] decomposes an image $f(h, w)$ into real and imaginary components, representing the image in a frequency domain. Let $\mathcal{F}(x, y)$ denote the Fourier Transform of 2D image $f(h, w)$ at spatial frequency coordinate (x, y). The Fourier Transform pair can be expressed as:

$$\mathcal{R}(\mathcal{F}(x,y)) = \frac{1}{HW} \sum_{h=0}^{H-1} \sum_{w=0}^{W-1} f(h, w) \cos\left(2\pi \left(\frac{xh}{H} + \frac{yw}{W} \right) \right) \tag{2}$$

$$\mathcal{I}(\mathcal{F}(x,y)) = \frac{1}{HW} \sum_{m=0}^{H-1} \sum_{w=0}^{W-1} f(h, w) \sin\left(2\pi \left(\frac{xh}{H} + \frac{yw}{W} \right) \right) \tag{3}$$

where, $f(h, w)$ represents the pixel at position (h, w) in the original image, while $\mathcal{F}(x, y)$ represents the function to represent the image in the frequency domain pertaining to the position (x, y). $H \times W$ represents the dimension of the image, and i is $\sqrt{-1}$.

Fourier Unit. A Fourier Unit primarily disassembles the spatial structure into image frequencies through a 2D FFT, followed by a convolution operation in the frequency domain, and ultimately reconstructs the structure using an 2D inverse FFT (IFFT) operation. For an input feature map $\mathcal{X} \in \mathbb{R}^{N \times C \times H \times W}$ fourier unit

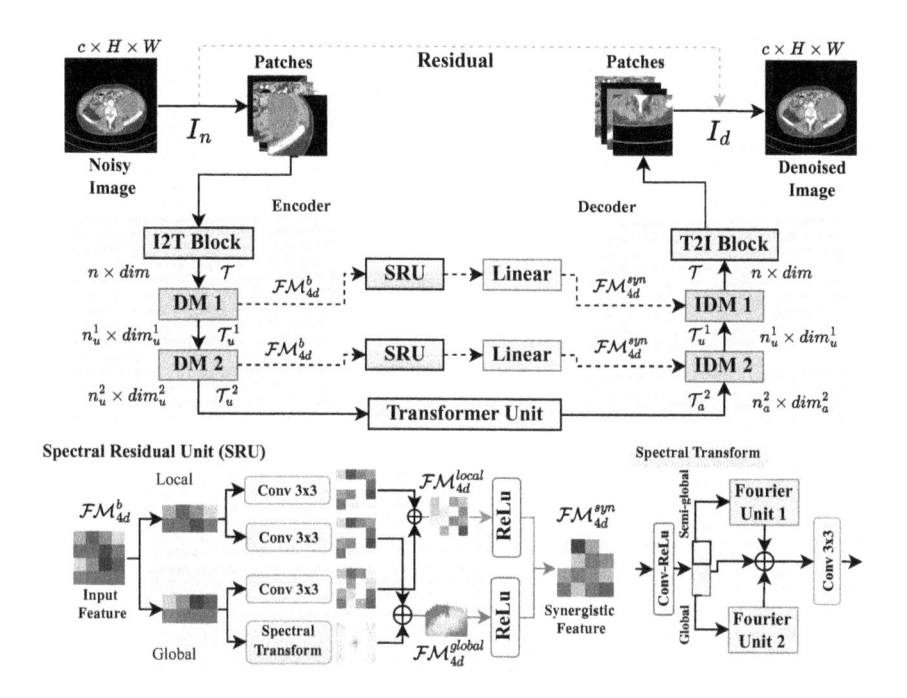

Fig. 2. Holistic architecture of proposed SEANet. SEANet consists of an encoder-decoder structure that takes an input noisy LDCT image in the form of patches. SEANet incorporates 2 image-token converters (I2T block and T2I block) along with 2 Dilation Modules (DMs) followed by a Transformer Unit and 2 consecutive Inverted Dilation Modules (IDMs). The Spectral Residual Units (SRUs) bridge the encoder-decoder with local to global synergistic feature maps $\mathcal{FM}_{4d}^{syn}$ guiding the boundary aware $\mathcal{FM}_{4d}^{b}$ during synthesis of denoised image.

outputs an enhanced feature map $\mathcal{Y} \in \mathbb{R}^{N \times D \times H' \times W'}$. First, $\mathcal{X}$ undergoes a spatial scale transformation to adjust its spatial resolution. This transformation is represented as $\mathcal{X}' = \text{Interpolate}(\mathcal{X}, \text{scale\_factor}, \text{mode})$. FFT computes the real $\mathcal{R}(\mathcal{F}(x,y))$ and imaginary $\mathcal{I}(\mathcal{F}(x,y))$ discrete fourier transforms using equations 2 and 3. The real and imaginary parts are stacked together along a new dimension to form a complex representation, $\mathcal{X}_c = [\mathcal{R}(\text{FT}(\mathcal{X})), \mathcal{I}(\text{FT}(\mathcal{X}))]$. With added spectral positional encoding the complex tensor undergoes a 1×1 spatial convolution operation followed by a rectified linear unit (ReLU) activation. The output real and imaginary components are converted back to the spatial domain using the inverse Fourier transform (IFFT), represented as:

$$\mathcal{Y} = f(x,y) = \frac{1}{HW} \sum_{h=0}^{H-1} \sum_{n=0}^{W-1} \left[\mathcal{R}(\mathcal{F}(h,w)) \cos\left(2\pi \left(\frac{xh}{H} + \frac{yw}{W}\right)\right) \right. \\ \left. - \mathcal{I}(\mathcal{F}(h,w)) \sin\left(2\pi \left(\frac{xh}{H} + \frac{yw}{W}\right)\right) \right] \quad (4)$$

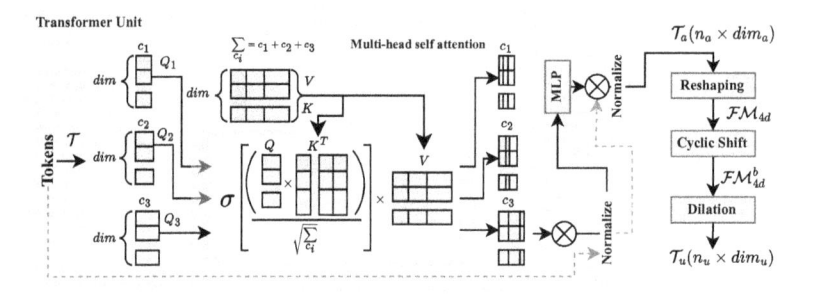

Fig. 3. Architecture of dilation modules (DMs). DMs contain a transformer unit followed by a cyclic shift and dilated unfolding operation. Transformer unit incorporates a multi-head self attention module and outputs attended 3D tokens $\mathcal{T}_a \in \mathbb{R}^{b \times n \times dim_a}$.

2.2 SEANet

Overview. As shown in Fig 2, SEANet employs a transformer based encoder-decoder architecture where SRUs output synergistic feature residuals mitigating the structural information loss. An input image $I_n \in \mathbb{R}^{b \times c \times H \times W}$ is patchified into two dimensional tokens $\mathcal{T} \in \mathbb{R}^{b \times n \times dim}$, where b, n and dim are the batch size, number of tokens and token dimension, respectively. Encoding starts with an image to token (I2T) block followed by couple of dilation modules, transformer unit and couple of inverted DMs with SRUs we get denoised tokens of the same shape. Then a similar approach is applied to revert back the tokens to image patches in final decoder layer followed by denoised patches and full image.

Transformer Unit. Transformer unit consists of a multi-head attention, a multi-layer perceptron (MLP) and residual connections, like a vanilla transformer block [3]. An input token sequence $\mathcal{T}$ is mapped into Query ($\mathcal{Q}$), Key ($\mathcal{K}$), and Value ($\mathcal{V}$) via self-attention, as shown in Fig 3. Additionally the softmax (σ) is approximated using a kernel method to reduce computational complexity.

Dilation and Inverted Dilation Modules. Dilation Module consists of a transformer unit followed by a progressive token-to-token (T2T) dilation. As shown in Fig 3 transformer unit outputs attended tokens $\mathcal{T}_a \in \mathbb{R}^{b \times n \times dim_a}$. T2T implements a progressive tokenization on $\mathcal{T}_a$ enhancing feature representation by capturing local structure information. T2T followed by reshaping, cyclic shift and dilation captures more spatial relationships across larger regions amongst tokens. Where vanilla transformer blocks fails to learn the inter-token dependencies at initial layers [18], proposed approach enhances the capabilities of learning the structural orientations with a broader receptive field. Sequence of attended tokens output by the transformer unit $\mathcal{T}_a \in \mathbb{R}^{b \times n \times dim_a}$ are transposed and reshaped to obtain four dimensional feature maps $\mathcal{FM}_{4d} \in \mathbb{R}^{b \times dim_a \times \sqrt{n} \times \sqrt{n}}$. The pixel intensities in $\mathcal{FM}_{4d}$ undergoes through a cyclic shift [21]. This forces the consequent transformer units to manipulate tokens from different feature maps. This reduces the tendency of poor attentiveness towards boundary regions in transformers. [18]. Boundary aware feature maps $\mathcal{FM}_{4d}^{b}$ are then unfolded to retokenize using overlapped splitting. These three dimensional unfolded tokens $\mathcal{T}_u \in \mathbb{R}^{b \times n_u \times dim_u}$ capture the correlations among surrounding patches. The unfolding operation is dilation based that helps to capture long range dependencies [11]. For a given perceptual field $P = \prod_{i=0}^{1} \exp\left(\log(2k_i + r_i - 1)\right)$ and desired unfolded dimension $dim_u = dim_a \times \exp\left(\sum_i \log(k_i)\right)$ dilation operation can be expressed as:

$$n_u = \left\lfloor \exp\left(\sum_{i=0}^{1} \log\left(\frac{\mathcal{S}(i) - \text{dilation} \times (K_i - 1) - 1}{\text{stride}} + 1\right)\right) \right\rfloor \quad (5)$$

where, $\mathcal{S}(i)$ denotes the size in the ith dimension, k_i and r_i are kernel size and dilation rate, respectively.

The inverted dilation modules (IDMs) are symmetric to the DMs. During decoding an inverse cyclic shift is applied. This transforms the tokens into 4D feature maps. Inside each IDM transformer unit takes these feature maps along with spectral residuals from the corresponding DMs.

Spectral Residual Unit. Inside spectral residual unit (SRU), we integrated fast fourier convolution followed by a linear layer. SRU splits channels of boundary aware residual feature maps $\mathcal{FM}_{4d}^{b}$ into local and global branches, as shown in Fig 2. Vanilla 3 × 3 convolutions are applied in local branch to learn the spatial features. And Spectral Transform module in the global branch learns global texture information and captures long-range contexts. Spectral transform uses two Fourier Units to capture the global and semi-global representations, as shown in Fig 2. Implemented fourier unit resembles operations discussed in equations 2, 3 and 4. Extracted feature maps $\mathcal{FM}_{4d}^{local}$ and $\mathcal{FM}_{4d}^{global}$ are concatenated to obtain a synergistic feature representation $\mathcal{FM}_{4d}^{syn}$. $\mathcal{FM}_{4d}^{syn}$ goes through a linear operation and reshaped to match the shape of the features output by $\mathcal{FM}_{4d}^{b}$ corresponding IDMs. Both the boundary aware and synergistic feature maps are concatenated channel-wise and enhanced feature maps $\mathcal{FM}_{4d}^{enh} = (\mathcal{FM}_{4d}^{syn} + \mathcal{FM}_{4d}^{b})$ are propagated to the consequent IDM modules.

3 Experimental Setup

Dataset and Experimental Settings. To validate the clinical efficiency of proposed SEANet, we utilized the *2016 NIH-AAPM-Mayo Clinic Low Dose CT Grand Challenge* dataset authorized by Mayo Clinics. This dataset consists of 2378 pairs of LDCT and NDCT images of size 512 × 512 and 3mm thickness for ten patients. The images of patient id *L506* are used for testing and the rest is used for training. Our models were implemented using PyTorch and trained on a 16GB NVIDIA RTX A4000 GPU. Only random rotation augmentation is used. Token dimension for all transformer units is 64 with depth=12 and attention heads=12. Kernel size is set to 7 with a stride 2 in both I2T and T2I blocks.

The kernel size in folding and unfolding is set to 3 with dilation rate r_i of 2,1,1,2 for DM1, DM2, IDM2 and IDM1, respectively. The no of tokens for DM1 and DM2 are $n_u^1 = 841$, $n_u^2 = 625$ and spatial dimensions $dim_u^1 = \sqrt{841} = 29$, $dim_u^2 = \sqrt{625} = 25$, respectively. We have trained SEANet for 200 epochs with a batch size of 16. Adam optimizer is utilized to optimize the MSE loss with initial learning rate of 1.0×10^{-5}.

Techniques of Comparison. SEANet is quantitatively evaluated using three standard metrics including peak signal to noise ratio (PSNR), structural similarity index measure (SSIM) and root mean squared error (RMSE). Additionally we have utilized two metrics- feature similarity index (FSI) and mean absolute percentage error (MAPE). FSI is calculated as: FSI = $w_1 \cdot$ PSNR$_{\text{norm}}$ + $w_2 \cdot$ SSIM$_{\text{norm}}$ − $w_3 \cdot$ RMSE$_{\text{norm}}$, where $w_1 = -1$, $w_2 = 1$, and $w_3 = -1$ are weights to balance the contributions of PSNR, SSIM, and RMSE, respectively. MAPE provides a relative measure of error, which can be useful for understanding the accuracy of denoising methods across different intensity levels. MAPE = $\frac{1}{n}\sum_{i=1}^{n}\left|\frac{I_o(i)-I_d(i)}{I_o(i)}\right| \times 100$, where, I_d and I_o are the denoised image and original LDCT image, respectively, and n is the total number of pixels in a single image. To demonstrate effectiveness of SEANet we compare against SOTA methods such as, RED-CNN, WGAN-VGG, CTFormer. We have analyzed the effectiveness of fourier convolutions, spectral residuals with different combinations and also qualitatively analyzed the denoising outputs by absolute errors, spectral maps and 3D frequency maps for all the baselines against SEANet.

4 Results

Table 1. Comparative results of LDCT denoising. **Blue** denotes original parameters of the test set. Green denotes the best of the existing SOTA methods and Red denotes the best of all.

Method	PSNR	ΔPSNR ↓	SSIM ↑	RMSE ↓	FSI ↑	MAPE ↓
	21.60	00.00	80.00	34.18	3.93	24.07
UNet	16.40	05.20	60.61	28.27	15.94	24.07
WGAN-VGG	16.76	04.84	64.98	28.70	19.52	21.98
MAP-NN	28.05	06.45	62.19	27.66	17.61	23.08
RedCNN	26.40	04.80	81.90	19.56	28.24	21.67
CTFormer	24.75	03.15	82.00	23.90	27.28	14.58
SEANet^2	23.06	**01.47**	83.42	19.34	**29.38**	06.76
SEANet	22.83	01.23	84.75	18.06	29.84	05.65

Quantitative results. From Table 1 it can be infered that our propose method demonstrate best performance compared to SOTA methods. To be precise, SEANet$_2$ when combined with CNN-based architecture RED-CNN improves its ability to capture global dependencies as well as pixel-level local features indicated by Higher SSIM and RMSE values respectively. This is highly important for preservation of structural details and low-level edge details. Noise in images often manifests as high-frequency components. When the initial RED-CNN model produces a undesirable smoothed output, it inadvertently removes some of these high-frequency details along with the noise. By incorporating spectral information through the synergistic SRU, the model can better preserve these high-frequency components. When denoising, the model utilizes the spectral information to guide the reconstruction process. Instead of overly smoothing the image, the model focuses on preserving high-frequency details, resulting in sharper denoised images. Further, SEANet also demonstrate better signal-to-noise ratio and pixel-level preservation demonstrating ability to effectively learn textures, edges indicated by higher FSI. However we notice marginal deterioration in term of local feature preservation indicated by lower RMSE and MAPE.

When integrated with CTFormer, SEANet demonstrate even better performance in terms of preserving global dependencies and surprisingly shows better local information preservation than RECNN and CTFormer. For instance, compared to CTformer, SEANet shows improvement of 8.56pp MAPE and 2.75pp SSIM. CTFormer, being a convolutional transformer architecture, excels in capturing long-range dependencies in the image through self-attention mechanisms. However, it struggles with preserving fine-grained local information or avoiding undesirable texture enhancements due to its global perspective. The synergy between the SRU and CTFormer ensures that both global and local information are preserved harmoniously, leading to superior performance in terms of preserving overall image structure and detail.

Qualitative results. Figure 4 provides qualitative comparisons, visually justifying the improved quantitative results. From the Regions of Interest (ROIs) depicted in Figure 5, it's evident that skip-connection-based networks like UNet excel in retaining local information. However, UNet tends to lack in preserving anatomical structures, likely due to its skip-connection design. On the other hand, RED-CNN captures anatomical structures effectively but may introduce blurry artifacts. Traditional spatial domain convolutions used in RED-CNN tend to blur edges because they average pixel values within the kernel's neighborhood. Meanwhile, CTFormer is prone to undesirable textures commonly associated with traditional spatial domain filtering, such as ringing or halo effects. Both ringing and halo effects degrade the visual quality as they operate directly on the pixel values of the image without considering the underlying frequency content or spatial relationships. In contrast, our proposed method demonstrates the preservation of anatomical patterns without excessive smoothing, offering a favorable balance between local detail preservation and structural integrity. Fast-fourier convolutions in SRU preserves edges better because it performs pointwise multiplication in the frequency domain, which emphasizes high-frequency components without spatial averaging. Overall SEANet demonstrate superior ability

448   A. Das et al.

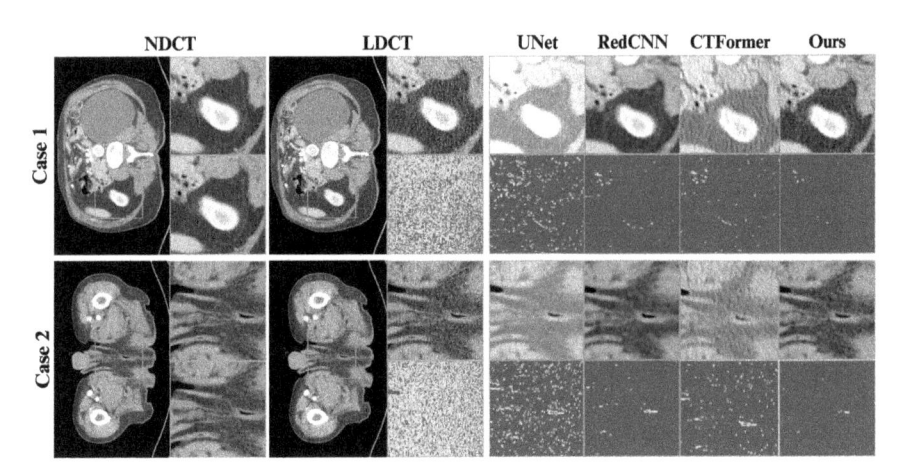

Fig. 4. Qualitative analysis of denoised images and corresponding error maps of the RoIs for two test cases. UNet shows inability to retain anatomical features during reconstruction showing more error points around organ-boundary. RED-CNN reduces noise and artifacts but CTFormer enhances the undesirable textures. SEANet (Ours) preserves the anatomical patterns well, prevents over-smoothness (unlike RED-CNN) and precisely removes the noise, resulting in lesser error.

to generate denoised output more clinically similar to NDCT images.

Spectral map analysis. The frequency domain is well-suited for separating signal from noise. By operating in the frequency domain, SRU can effectively suppress noise while preserving signal information. In Fig 6 spectral maps and 3d frequency maps are presented to understand how SRU contributes to the differentiating factor in SEANet's superior performance. In Fig 6 spectral maps are the normalized magnitude spectrum and 3d spectral frequencies represent $\mathcal{FM}_{4d}^{global}$ in normalized frequency space. Brighter regions in spectral map refers to higher energy or amplitude in those frequency bands. In NDCT images, the spectral distribution appears wide and spread out, with brighter regions covering a larger area. This indicates that NDCT images contain a diverse range of frequency components. Whereas, in LDCT the spectral distribution appears narrower and more concentrated with dark shades showing noisy appearances in whole image. Clustered noise or cross hatched patches in the UNet-denoised image produces irregularities in the spectral domain, leading to the formation of such white patches. While RED-CNN shows better performance but still has a lot of undesirable noise in the spectrum lacking global context. Spectrum spikes around the centre in CTFormer illustrates more pronounced discrepancy in the denoised output that cannot be identified in separate frequency bands, resulting in enhanced textures in output. On the other hand the separable bands of frequencies in the spectral map for SEANet helps the noises to be discarded precisely and restore the anatomical important features. The 3d frequency maps also

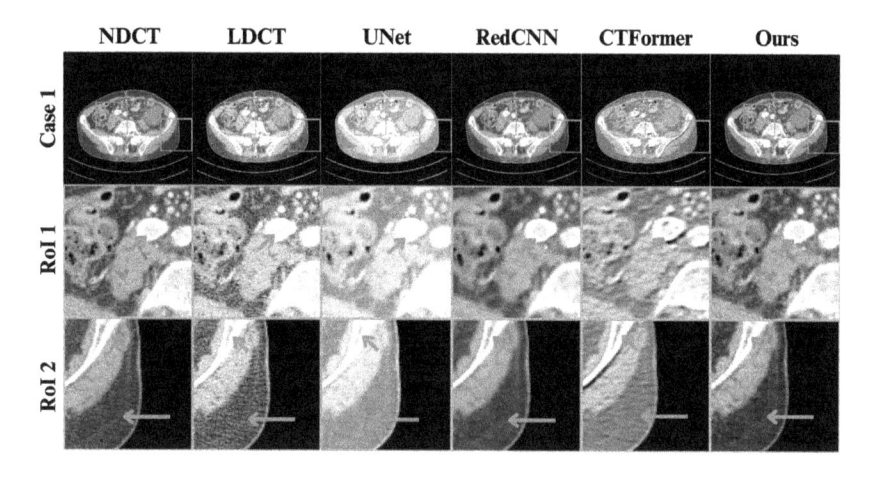

SEANet: Synergistic Spatial-Spectral Residual for LDCT Denoising   449

Fig. 5. Figure showing efficacy of different methods in retaining fine-grained lesions and controlling smoothness in peripheral regions. **Case 1:** In RoI-1 and RoI-2 the lesions are not retained by UNet, CTFormer distorts and produces dark spots in the boundary regions. Whereas SEANet (ours) outperforms RED-CNN by also capturing the fine details present inside the lesion. On the other hand the peripheral structures in RoI-2 are more accurately denoised by SEANet. (More example in supplementary.)

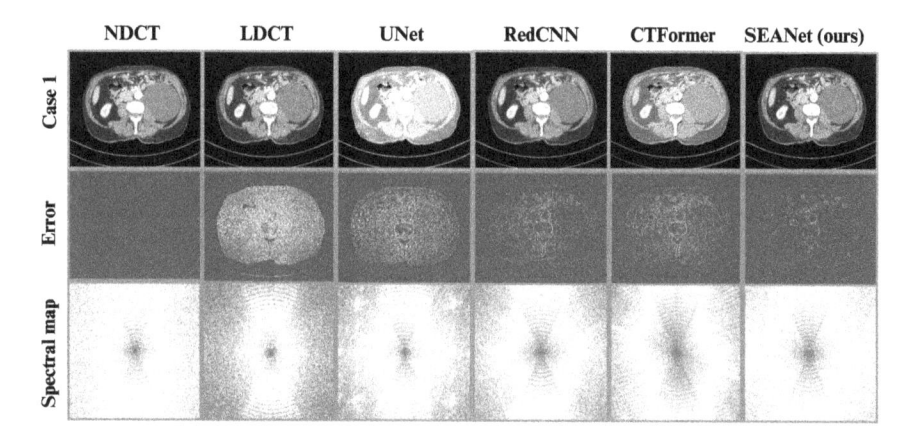

Fig. 6. In NDCT images, the spread out spectral map indicates that energy is distributed across various frequencies. In LDCT the spectral distribution appears more concentrated with dark shades for noisy appearances. Clustered noise or cross hatched patches in the UNet-denoised image leads to the formation of white patches. While RED-CNN shows better performance but still has a lot of undesirable noise in the spectrum. Spectrum spikes around the centre in CTFormer illustrates irregular structures and more pronounced discrepancy in the denoised output that cannot be identified in separate frequency bands. SEANet (ours) is very similar to the NDCT image, containing separable bands of frequencies for denoising. (More example in supplementary.)

signify a similar behaviour in frequency planes and patterns that qualitatively interprets SEANet as a superior algorithm in LDCT denoising task.

5 Ablation Studies

Table 2. Ablation study. **Top:** Effect of fourier convolutions into encoding I2T and decoding T2I block. **Middle:** Effect of different dilation rates. **Bottom:** Effect of different global (G) and local (L) split ratios in Spectral Transform module. Blue: Original values for the test cases. Green: Best of the category and Red: Best results.

Method	PSNR	ΔPSNR ↓	SSIM ↑	RMSE ↓	FSI ↑	MAPE ↓
	21.60	00.00	80.00	34.18	3.93	24.07
Vanilla E-D	24.75	03.15	82.00	23.90	27.28	14.58
Fourier E	24.02	2.42	60.61	23.02	28.23	12.33
Fourier D	23.85	2.25	64.98	22.66	28.40	10.45
Fourier E-D	23.25	01.25	82.35	21.38	28.98	08.70
r = [3, 1, 1, 3]	26.55	04.95	76.41	25.24	24.67	14.58
r = [3, 2, 2, 3]	24.75	03.15	78.55	23.90	26.08	14.58
r = [2, 1, 1, 2]	22.83	01.23	84.75	18.06	29.84	05.65
SEANet-5L5G	23.25	**01.25**	82.35	21.38	28.98	08.70
SEANet-4L6G	23.05	01.45	**83.42**	**19.54**	**28.24**	**06.44**
SEANet-3L7G	22.83	01.23	84.75	18.06	29.84	05.65

Effect of Fourier Convolutions. To study the effect of fourier convolutions over vanilla convolutions we have substituted the vanilla convolutions in the Transformer Unit with Fourier convolutions with different combinations. First, we have strictly employed fourier transformer unit in the I2T tokenizer and consequent Dilation Modules. This forms our fourier encoder (Fourier E) based SEANet. This integration results in more effective information flow from the encoder via conventional residuals and improves the PSNR while reducing RMSE loss, compared to the UNet alike vanilla encoder-decoder (Vanilla E-D) architecture. But due to the information loss around the edges (as shown in Fig 7) during decoding of each patch, the structural similarity decreases resulting in poor SSIM output. Similarly, substituting the decoder transformer with fourier convoluted units also suffers from lack of long range context propagation from the encoder block, resulting in marginal improvement in FSI metric, as shown in Table 2. Fully fourier encoder-decoder (Fourier E-D) results in lack of local information. This is due to the intrinsic nature of spectral transform's linearity and global frequency capturing efficacy. As shown in Fig 7, Fourier E-D induces undesired smoothness though effectively captures the global patterns with improved SSIM and FSI (as shown in Table 2).

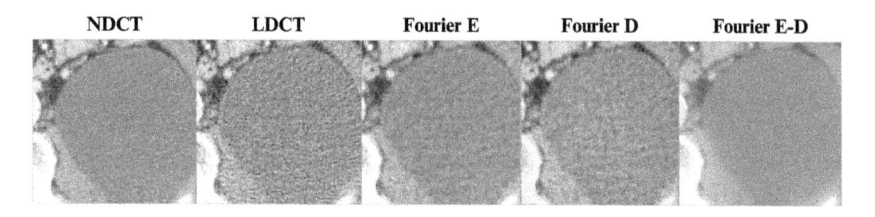

Fig. 7. Ablation study of the effectiveness of fourier transformers in SEANet. Fourier-E and Fourier-D fails to generate optimal denoised output due to the lack of local to global mixed feature flow with only one of the encoder or decoder containing transformer. Whereas, Fourier E-D contains spectral convolutions at each layer suppresses the noise or artifacts well, but fails to retain the global representations.

Effect of Dilation Rates in DMs and IDMs: Dilation rates control the spacing between the elements within the receptive field of a convolutional operation. A dilation rate of 1 represents the standard convolutional operation without gaps between elements, while dilation rates greater than 1 introduce gaps between elements, effectively expanding the receptive field. By judiciously choosing dilation rates, models can efficiently capture both local and global contextual information while maintaining spatial resolution. We have experimented SEANet with different dilation rates (r_i) during folding and unfolding between $\mathcal{FM}_{4d}^b$ and $\mathcal{T}_u$. For different combination of r_i inside DM1, DM2, IDM2 and IDM1 we have listed the performances of SEANet in Table 2. r = [2, 1, 1, 2] gives optimal output as this configuration provides a balanced mix of dilated and non-dilated convolutions, resulting in receptive fields that capture both local and global contextual information. While higher dilation rates results in disproportionately larger receptive field at the edges of the network, potentially sacrificing the ability to capture fine details (e.g., noise or lesions) within the input. Ultimately, the choice of dilation rates often depends on empirical performance and the specific requirements of the task or dataset.

Effect of Different Local to Global Feature Ratios in SRU: In SRU the split between local and global feature maps plays a crucial role in determining the model's ability to capture both local and global spatial information effectively. We have experimented with three different combination of local (L) to global (G) split ratios, as listed in Table 2. 3L7G configuration outputs optimal performance ensuring that the model can effectively capture both fine details and broader spatial relationships, leading to more robust feature representations. While using Transformer Units as a building block of our dilation modules, the maximal global context flow via the SRU's synergistic feature map $\mathcal{FM}_{4d}^{syn}$ that enriches the representations in $\mathcal{FM}_{4d}^{enh}$ during synthesis of denoised CT images.

6 Conclusion

In conclusion, we introduced SEANet, a novel architecture for LDCT denoising by synergizing spatial and spectral domains through SRUs and DMs. SEANet balances feature retention and noise reduction, outperforming state-of-the-art methods like RED-CNN and CTFormer. Extensive experiments and spectral map analysis demonstrate SEANet's superiority, achieving significant improvements across five different metrics. Our work emphasizes the importance of learning local to global synergistic features with complementary spaces. In future we will explore techniques for learning in complementary discrete and continuous spaces within SEANet to handle specific noise patterns and the integration of self-supervised learning to enhance model generalization and robustness.

References

1. Chen, J., Lu, Y., Yu, Q., Luo, X., Adeli, E., Wang, Y., Lu, L., Yuille, A.L., Zhou, Y.: Transunet: Transformers make strong encoders for medical image segmentation (2021)
2. Das, A., Jha, D., Gorade, V., Biswas, K., Pan, H., Zhang, Z., Ladner, D.P., Velichko, Y., Borhani, A., Bagci, U.: Pam-unet: Shifting attention on region of interest in medical images. arXiv preprint arXiv:2405.01503 (2024)
3. Dosovitskiy, A., Beyer, L., Kolesnikov, A., Weissenborn, D., Zhai, X., Unterthiner, T., Dehghani, M., Minderer, M., Heigold, G., Gelly, S., Uszkoreit, J., Houlsby, N.: An image is worth 16x16 words: Transformers for image recognition at scale (2021)
4. Fan, F., Shan, H., Kalra, M.K., Singh, R., Qian, G., Getzin, M., Teng, Y., Hahn, J., Wang, G.: Quadratic autoencoder (q-ae) for low-dose ct denoising. IEEE Trans. Med. Imaging **39**(6), 2035–2050 (2019)
5. Gorade, V., Mittal, S., Jha, D., Bagci, U.: Synergynet: Bridging the gap between discrete and continuous representations for precise medical image segmentation. In: Proceedings of the IEEE/CVF Winter Conference on Applications of Computer Vision. pp. 7768–7777 (2024)
6. Gorade, V., Mittal, S., Jha, D., Singhal, R., Bagci, U.: Harmonized spatial and spectral learning for robust and generalized medical image segmentation. arXiv preprint arXiv:2401.10373 (2024)
7. Gorade, V., Susladkar, O., Durak, G., Keles, E., Aktas, E., Cebeci, T., Medetalibeyoglu, A., Ladner, D., Jha, D., Bagci, U.: Towards synergistic deep learning models for volumetric cirrhotic liver segmentation in mris. arXiv preprint arXiv:2408.04491 (2024)
8. He, J., Yang, Y., Wang, Y., Zeng, D., Bian, Z., Zhang, H., Sun, J., Xu, Z., Ma, J.: Optimizing a parameterized plug-and-play admm for iterative low-dose ct reconstruction. IEEE Trans. Med. Imaging **38**(2), 371–382 (2018)
9. Huang, L., Jiang, H., Li, S., Bai, Z., Zhang, J.: Two stage residual cnn for texture denoising and structure enhancement on low dose ct image. Comput. Methods Programs Biomed. **184**, 105115 (2020)
10. Jia, X., Bartlett, J., Chen, W., Song, S., Zhang, T., Cheng, X., Lu, W., Qiu, Z., Duan, J.: Fourier-net: Fast image registration with band-limited deformation. In: Proceedings of the AAAI Conference on Artificial Intelligence. vol. 37, pp. 1015–1023 (2023)

11. Jiao, J., Tang, Y.M., Lin, K.Y., Gao, Y., Ma, J., Wang, Y., Zheng, W.S.: Dilateformer: Multi-scale dilated transformer for visual recognition. IEEE Transactions on Multimedia (2023)
12. Leuliet, T., Maxim, V., Peyrin, F., Sixou, B.: Combining conditional gan with vgg perceptual loss for bones ct image reconstruction. In: 16th International Meeting on Fully Three-Dimensional Image Reconstruction in Radiology and Nuclear Medicine (Fully3D). No. 281-284 (2021)
13. Li, Z., Shi, W., Xing, Q., Miao, Y., He, W., Yang, H., Jiang, Z., et al.: Low-dose ct image denoising with improving wgan and hybrid loss function. Computational and Mathematical Methods in Medicine **2021** (2021)
14. Liu, J., Zhang, Y., Zhao, Q., Lv, T., Wu, W., Cai, N., Quan, G., Yang, W., Chen, Y., Luo, L., et al.: Deep iterative reconstruction estimation (dire): approximate iterative reconstruction estimation for low dose ct imaging. Physics in Medicine & Biology **64**(13), 135007 (2019)
15. Marcos, L., Quint, F., Babyn, P., Alirezaie, J.: Dilated convolution resnet with boosting attention modules and combined loss functions for ldct image denoising. In: 2022 44th annual international conference of the IEEE engineering in medicine & biology society (EMBC). pp. 1548–1551. IEEE (2022)
16. McCollough, C.H., Bartley, A.C., Carter, R.E., Chen, B., Drees, T.A., Edwards, P., Holmes, D.R., III., Huang, A.E., Khan, F., Leng, S., et al.: Low-dose ct for the detection and classification of metastatic liver lesions: results of the 2016 low dose ct grand challenge. Med. Phys. **44**(10), e339–e352 (2017)
17. Power, S.P., Moloney, F., Twomey, M., James, K., O'Connor, O.J., Maher, M.M.: Computed tomography and patient risk: Facts, perceptions and uncertainties. World journal of radiology **8**(12), 902 (2016)
18. Raghu, M., Unterthiner, T., Kornblith, S., Zhang, C., Dosovitskiy, A.: Do vision transformers see like convolutional neural networks? Adv. Neural. Inf. Process. Syst. **34**, 12116–12128 (2021)
19. Ronneberger, O., Fischer, P., Brox, T.: U-net: Convolutional networks for biomedical image segmentation (2015)
20. Shan, H., Padole, A., Homayounieh, F., Kruger, U., Khera, R.D., Nitiwarangkul, C., Kalra, M.K., Wang, G.: Competitive performance of a modularized deep neural network compared to commercial algorithms for low-dose ct image reconstruction. Nature Machine Intelligence **1**(6), 269–276 (2019)
21. Song, Z., Yu, J., Chen, Y.P.P., Yang, W.: Transformer tracking with cyclic shifting window attention. In: Proceedings of the IEEE/CVF conference on computer vision and pattern recognition. pp. 8791–8800 (2022)
22. Wang, C., Shang, K., Zhang, H., Li, Q., Hui, Y., Zhou, S.K.: Dudotrans: dual-domain transformer provides more attention for sinogram restoration in sparse-view ct reconstruction. arXiv preprint arXiv:2111.10790 (2021)
23. Wang, D., Fan, F., Wu, Z., Liu, R., Wang, F., Yu, H.: Ctformer: convolution-free token2token dilated vision transformer for low-dose ct denoising. Physics in Medicine & Biology **68**(6), 065012 (2023)
24. Wang, D., Wu, Z., Yu, H.: Ted-net: Convolution-free t2t vision transformer-based encoder-decoder dilation network for low-dose ct denoising. In: Machine Learning in Medical Imaging: 12th International Workshop, MLMI 2021, Held in Conjunction with MICCAI 2021, Strasbourg, France, September 27, 2021, Proceedings 12. pp. 416–425. Springer (2021)
25. Willemink, M.J., Noël, P.B.: The evolution of image reconstruction for ct-from filtered back projection to artificial intelligence. Eur. Radiol. **29**, 2185–2195 (2019)

26. Wu, D., Kim, K., El Fakhri, G., Li, Q.: Iterative low-dose ct reconstruction with priors trained by artificial neural network. IEEE Trans. Med. Imaging **36**(12), 2479–2486 (2017)
27. Yin, X., Zhao, Q., Liu, J., Yang, W., Yang, J., Quan, G., Chen, Y., Shu, H., Luo, L., Coatrieux, J.L.: Domain progressive 3d residual convolution network to improve low-dose ct imaging. IEEE Trans. Med. Imaging **38**(12), 2903–2913 (2019)
28. Zhang, Z., Yu, L., Liang, X., Zhao, W., Xing, L.: Transct: dual-path transformer for low dose computed tomography. In: Medical Image Computing and Computer Assisted Intervention–MICCAI 2021: 24th International Conference, Strasbourg, France, September 27–October 1, 2021, Proceedings, Part VI 24. pp. 55–64. Springer (2021)

Extracting Vitals from ICU Monitor Images: An Insight from Analysis of 10K Patient Data

Akshat Rampuria, Kushagra Khare, Ayush Soni, and Debi Prosad Dogra[✉]

School of Electrical and Computer Sciences, Indian Institute of Technology, Bhubaneswar, Bhubaneswar 752050, India
{20cs02013,20cs02004,20ee01007,dpdogra}@iitbbs.ac.in

Abstract. In intensive care units (ICUs), continuous monitoring of patient vitals is crucial for timely interventions and optimal outcomes. While traditional methods rely on direct observation by healthcare professionals, technological advancements have led to a growing interest in image-based monitoring systems, particularly utilizing closed-circuit television (CCTV) cameras within ICU environments. However, integrating such systems poses challenges, notably in extracting vital information from monitor images efficiently. Current approaches, including manual interpretation and specialized algorithms, are often laborious and error-prone. In response, this paper presents a robust pipeline for automatic extraction of patient vitals from ICU monitor images. Our approach involves automatic detection and segmentation of monitor screens, followed by extraction of relevant vital signs. By enhancing existing ICU environments with this technology, we aim to improve patient care and resource utilization. Our contributions include a powerful pre-processing pipeline, a comprehensive study of optical character recognition (OCR) frameworks, and a geometry-based heuristic implemented in kornia for vital sign detection. Through this work, we lay the groundwork for innovative solutions that can revolutionize patient monitoring in ICU settings. The method has been tested on a large dataset comprises with 10K ICU monitor images.

Keywords: Vital extraction · ICU monitor segmentation · Keyword extraction · Medical image processing

1 Introduction

In intensive care units (ICUs), monitoring of patient vitals is paramount for ensuring timely interventions and optimizing patient outcomes. Traditionally, such monitoring is performed through visual observations done by healthcare professionals. However, there is a growing trend towards utilizing image-based automatic monitoring systems, particularly through closed-circuit television (CCTV) cameras installed within the ICU environments [1]. The paradigm shift towards CCTV-based monitoring is due to several factors [2]. Firstly, it offers continuous and non-intrusive means of monitoring the patients, allowing healthcare professionals to remotely observe multiple patients simultaneously. It also provides an opportunity for real-time data collection and analysis, facilitating early detection of abnormalities and timely interventions. However, integration of CCTV-based monitoring systems into existing ICU setups faces several challenges including extraction of vital information from the captured images. Our research endeavors to develop a robust and efficient pipeline for extracting patient vitals from ICU monitor images. The pipeline involves automatic detection and segmentation of the monitor's screen, followed by identification and extraction of relevant vital signs.

In this paper, we have made the following technical contributions: (i) A powerful pre-processing pipeline for processing ICU monitor images and video streams with acceptable latency and high accuracy. (ii) Integration of well-known OCR frameworks to extract information from the processed monitor images and to validate the effectiveness of the approach in detecting vital information from ICU monitor images. (iii) A screen geometry-based algorithm implemented in kornia to detect vital parameters from ICU monitor images by using proximity of keywords, distances, and color information. (iv) Validation of the proposed method on a large dataset comprising 10K ICU monitor images.

The rest of the paper is organized as follows. In the next section, we present the existing work available in the literature. Next, in Section 3, we present the proposed methodology through various subsections. Results and analysis are presented in Section 4. In Section 5, we present the discussion of challenges faced and in Section 6, we present the conclusion and future scopes of the present work.

2 Prior Art

Recent advancements in computer vision and deep learning have spurred a shift towards contactless monitoring techniques [4],[5],[6]. Despite these advancements, much of the existing research is confined to controlled laboratory settings [3] or specific clinical scenarios, thus limiting its scalability to real-world adult ICU environments [7]. Janssen et al. [8] and Rossol et al. [4] have proposed video-based respiration monitoring systems tailored for neonatal ICUs. Nevertheless, the method has been validated using only 2 newborns, thereby restricting the broader applicability of their findings. Similarly, Jorge et al. [9] have explored non-contact RGB camera-based respiration monitoring in neonatal ICUs, albeit necessitating lengthy 30-second video frames for rate computation, which may impede real-time monitoring capabilities. Antognoli et al. [10] have evaluated video-based heart rate and respiratory rate estimation against conventional patient monitors. However, their study exclusively focuses on newborns. For adults, Jorge et al. [5] have proposed a technique to monitor post-operative ICU patients through a non-contact approach, albeit with a validation study involving merely 15 patients. Liu et al. [6] have collected a large-scale real-world ICU patient dataset and proposed an unsupervised learning method to extract respiration wave-forms from the video data. They have used a neural network to estimate respiratory rates from the wave-forms. This method requires at least 10 seconds of respiration waveform data, which may still be inadequate for real-time monitoring. Researchers have also used video analysis-based techniques to automate similar tasks. For example, conventional motion-based approach such as the method proposed by Massaroni et al. [11] relies on analyzing pixel intensity variations in region-of-interest of the video frames to extract breathing signals and respiratory rate. These methods are susceptible to illumination changes and rigid motions, thereby compromising their robustness. Some studies have explored deep learning-based approaches for contactless vital sign monitoring. For example, Chen et al. [12] have estimated respiratory rates from face videos using a convolutional attention networks. The method assumes unobstructed and visible faces, which may not always be feasible in ICU settings.

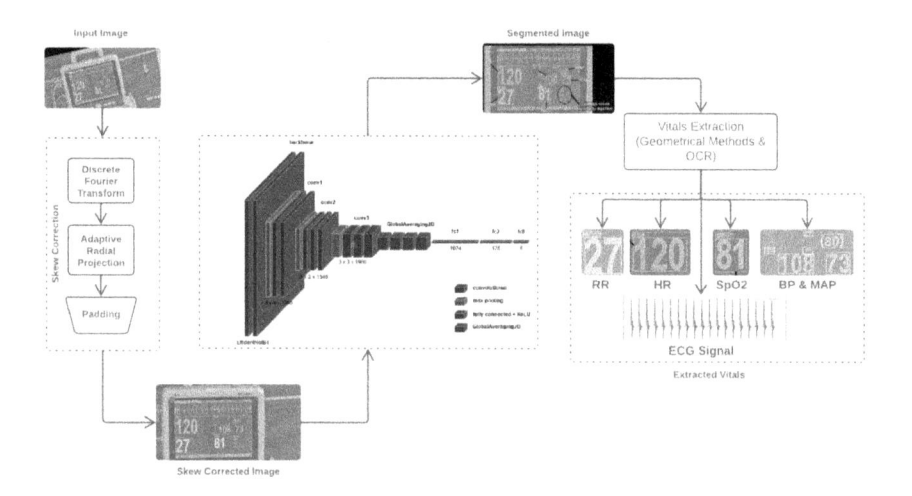

Fig. 1. The proposed methodology depicting various stages of the pipeline including pre-processing (skew correction), screen segmentation, and vital extraction.

3 Proposed Methodology

This section details the proposed methodology, where a computer vision-guided approach is used to extract features representing the patients' vitals from ICU monitor images to create a database. Such databases can easily be integrated with existing Electronic Medical Records (EMR) systems and used to train deep learning models for automatic ICU operations. A block diagram of the proposed methodology is presented in Fig. 1, which includes various stages of the pipeline, e.g., preprocessing (skew correction), screen segmentation, and vital extraction. The methodology is also presented in Algorithm 1.

3.1 Preprocessing

This step involves two stages, namely skew correction and padding, before the corrected images can be used as inputs in the monitor segmentation stage.

Algorithm 1 Extract Vitals From Images

1: **Input:** I_{input}(Image)
2: **Output:** $V_{\text{extracted}}$(Vitals)
3: **procedure** SKEW_CORRECTION(I_{input})
4:     $I_{\text{DFT}} \leftarrow \text{DFT}(I_{\text{input}})$, $I_{\text{ARP}} \leftarrow \text{Adaptive\_Radial\_Projection}(I_{\text{DFT}})$
5:     $I_{\text{padded}} \leftarrow \text{Padding}(I_{\text{ARP}})$
6:     **Return** $I_{\text{skew}} \leftarrow I_{\text{padded}}$
7: **end procedure**
8: **procedure** MONITOR_SEGMENTATION(I_{skew})
9:     Initialize CNN
10:    $I_{\text{seg}} \leftarrow \text{Feature Extraction and Pooling}(I_{\text{skew}})$
11:    **Return** I_{seg}
12: **end procedure**
13: **procedure** VITALS_EXTRACTION(I_{seg})
14:    $RR \leftarrow \text{Extract\_RR}(I_{\text{seg}})$, $HR \leftarrow \text{Extract\_HR}(I_{\text{seg}})$
15:    $SpO2 \leftarrow \text{Extract\_SpO2}(I_{\text{seg}})$, $BP, MAP \leftarrow \text{Extract\_BP\&MAP}(I_{\text{seg}})$
16:    $ECG_{\text{signal}} \leftarrow \text{Extract\_ECG}(I_{\text{seg}})$
17:    **Return** $V_{\text{extracted}} \leftarrow \{RR, HR, SpO2, BP, MAP, ECG_{\text{signal}}\}$
18: **end procedure**

Skew Correction and Padding: Raw monitor images may be skewed due to camera position. Such distortions affect the corner detection and OCR-based text extraction. Moreover, after detecting the four corners of the ICU monitor, it is necessary to crop the screen area for text extraction. In skewed images, top portion of the monitor or the background may be visible. This may result in detection of unimportant texts. To correct skew, we first calculate the skew angle using adaptive radial projection (ARP) on the Fourier magnitude spectrum [13]. The method has two steps: (i) Preprocessing and 2D Discrete Fourier Transformation of the input image. In this step, the image is first converted to grayscale. It is then passed through a 2D-DFT to obtain the magnitude spectrum. It is defined in Eq. (1), where x_{jk} represents the grayscale pixel values, N and M denote the height and width of the image. After obtaining the magnitude spectrum, it is normalized.

$$X_{lm} = \sum_{j=0}^{N-1} \sum k = 0^{M-1} x_{jk} e^{-2\pi i \left(\frac{jl}{N} + \frac{km}{M}\right)} \tag{1}$$

(ii) In the next step, adaptive radial projection (ARP) is performed on the spectrum extracted in the previous step. The projection radius $R = \min(H, W)$, angle list $[\theta_{\min}, \ldots, \theta_{\max}]$, and the initial projection value $A(\theta_i)$ at angle θ_i are calculated using Eqs. (2-5), where (c_x, c_y) is the center the spectrum, and $M[y, x]$ is the magnitude of spectrum at (x, y).

$$A(\theta_i) = \sum_{s=0}^{R} M[c_y + s \cdot \cos(\theta_i), c_x - s \cdot \sin(\theta_i)] \quad (2)$$

The correction projection value $B(\theta_j)$ at angle θ_j is computed using Eq. (3).

$$B(\theta_j) = \sum_{s=0}^{R} M[c_y + s \cdot \cos(\theta_j), c_x - s \cdot \sin(\theta_j)] \quad (3)$$

The candidate angles θ_a and θ_b are determined using Eq. (4).

$$\theta_a = \arg\max_{m} A(m), \quad \theta_b = \arg\max_{n} B(n) \quad (4)$$

And the final output angle θ_f is aggregated by the rule presented in Eq. (5). The original image is then rotated by the calculated skew angle.

$$\theta_f = \begin{cases} \theta_a, & \text{if } |\theta_a - \theta_b| > D \\ \theta_b, & \text{otherwise} \end{cases} \quad (5)$$

Adding a predetermined border to an image is referred to as padding. The dataset contains several images where a significant portion of the monitor may be missing. Such samples are corrected using padding (Fig. 2).

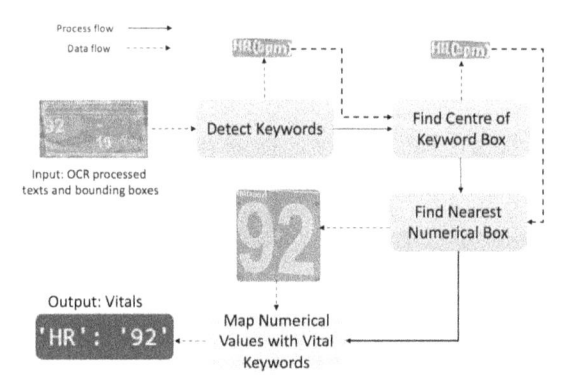

Fig. 2. The geometry-based heuristic used in the work to extract the vital signals from the segmented ICU monitor.

3.2 EfficientNet-guided Monitor Segmentation

We now eliminate unnecessary textual information from the image. We employ corner detection and cropped the desired portion containing the vitals. We have used a pretrained backbone model (EfficientNetB4) with CNN and Dense layers on top to detect the corner points. We have used (1280,3,3), (1540,3,3), and (1960,3,3) CNN layers, a global-average-2D pooling layer, and a (1024, 128, 8) dimension dense layer. The last layer produces the output as four coordinates of the desired segment of the monitor image. The EfficientNet [14] model, as depicted in Fig. 1, employs a method termed compound coefficient for the systematic scaling of models, ensuring a harmonious increase in width, depth, and resolution dimensions. Unlike random scaling approaches, compound scaling uniformly adjusts each dimension using predetermined scaling coefficients.

3.3 Extraction of Vitals

We now propose an algorithm to discern vital signs such as Heart Rate (HR), SpO2, BP, and ECG from the segmented monitor image. The process involves image segmentation, converting vitals into kornia's bboxes [17], after which OCR-guided text extraction is done, which is then followed by various geometrical methods to separate the segments from the monitor images and place them in the database. We adopt a nearest bounded bbox technique, implemented in kornia, that facilitates the association of vital values with the closest numerical representations.

Text Detection Using OCR: The first step in vital extraction involves applying OCR to identify and detect texts present in the image. We have tested EasyOCR [18], KerasOCR [19], and MMOCR [20] frameworks. Each framework offers unique features and capabilities, addressing various challenges in extracting text from images. In the results section, we provide a detailed evaluation of various OCR methods. We have used the PPOCR [16] model. In PPOCR, the text detection phase uses a pre-trained model (e.g., db_mv3_slim) that involves several steps. A lightweight backbone neural network is used to extract features from the segmented images, followed by a lightweight head to process these features. In the next step, a squeeze-and-excitation block is used to reduce computation time. The model uses Cosine learning rate decay as given in Eq. (6), where η_t is the learning rate at epoch t, η_{max} is the initial learning rate, η_{min} is the minimum learning rate, and T is the total number of epochs.

$$\eta_t = \eta_{min} + \frac{1}{2}(\eta_{max} - \eta_{min})\left(1 + \cos\left(\frac{t}{T}\pi\right)\right) \tag{6}$$

The learning rate is gradually increased from a small initial value. The FPGM pruner is used to prune the network based on the geometric median of the filters. In the next stage, bounding box detection is performed using a pre-defined model (e.g., dir_cls_mv3_slim), which involves the following operations: a separate lightweight backbone is used for feature extraction, and data augmentation (e.g., random cropping, flipping, etc.) is applied to improve the model's performance. The resolution of the input images is then adjusted. This process uses Parameterized Clipping Activation (PACT) for quantization, as given in Eq. (7), where α is a learnable parameter.

$$y = \max(\min(x, \alpha), 0) \qquad (7)$$

Finally, the text recognition phase is applied using another pre-trained model (e.g., `crnn_mv3_slim`). This phase also involves a lightweight backbone, data augmentation, Cosine learning rate decay, feature map resolution adjustment, and PACT quantization.

HR Extraction: Initially, we apply a filtering process among bboxes to eliminate detection below a predefined threshold, thereby mitigating noise interference. Subsequently, HR detection is executed. We perform a search among the present bboxes for the keywords directly associated with or closely related to 'HR' within the monitor's textual content, capturing the nearest integer to these instances. In cases where such keywords are absent, we identify and retrieve numerical values presented in green font. If there are no bboxes present with the 'HR' keyword, we resort to comparing respective values and prioritize the topmost result.

SPO2 and RR Extraction: Similarly, for the determination of Oxygen Saturation (SpO2), we pursue the identification of bboxes with keywords like 'SpO2', seeking the closest numerical representations. The process for Respiratory Rate (RR) detection involves scanning for bboxes with keywords such as 'rr' or 'resp'; failing which, we consider numerical values below 45 as potential RR indicators.

BP and MAP Extraction: The detection of Blood Pressure (BP) and Mean Arterial Pressure (MAP) involves a comprehensive strategy. Initially, we attempt BP detection by parsing all of the bboxes for the '/' symbol, utilizing the preceding and succeeding values as Systolic (SBP) and Diastolic (DBP) pressure, respectively. A secondary approach involves keyword-based search among the bboxes for terms like 'sys', 'dia', 'mmhg', among others, followed by proximity analysis to identify relevant bounding bboxes. MAP detection entails locating '(' or ')' symbols, extracting the numerical values adjacent to these symbols. Alternatively, we look for the keyword 'map' and identify the nearest bounding bbox. In our final endeavor to ascertain BP and MAP, we employ a clustering technique among the bboxes to identify three numerical values with the closest spatial proximity, subsequently determining SBP, DBP, and MAP based on some predefined thresholds. Upon completion of the extraction process, the gathered data is organized into a structured python dictionary format for comprehensive output.

ECG Graph Extraction: To detect the ECG graph, we first isolate the green color from the segmented image by masking the image, and then draw contours around the objects. We then identify the contour with the largest perimeter to extract the ECG graph. An example of the original ECG signal and the corresponding extracted graph is shown in Fig. 3.

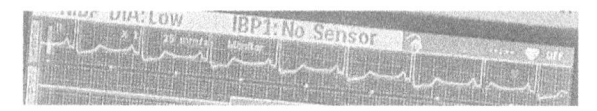

(a) An ECG graph taken from the original ICU monitor image

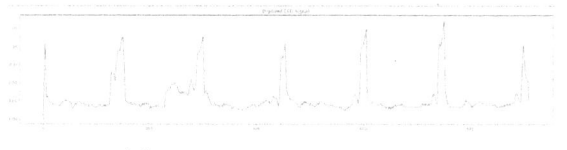

(b) The processed ECG graph

Fig. 3. Extraction of the EEG graph from the original ICU monitor image box.

4 Experiments and Results

In this section, we present the experiments and results obtained using publicly available ICU monitor image dataset. We have carried out various experiments to showcase the effectiveness of the proposed method.

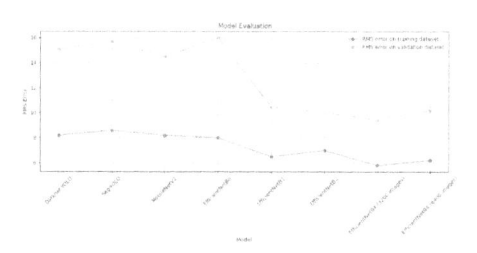

Fig. 4. Comparisons of various models used in the monitor segmentation stage of the proposed pipeline.

4.1 Dataset

The dataset [21] consists of labeled and unlabeled images of ICU monitors, captured across different hospitals. It includes ground truth bounding box information for all vital signs displayed on the monitors. A subset of the dataset is labeled with monitor variety, reflecting different brands and their unique orientations and positions for displaying vital information, which provided key insights for our methodology. The unlabeled portion includes images from various hospitals, including HCG Hospital in Bhavnagar, HCG Cancer Center in Nagpur, Kamalnayan Bajaj Hospital in Aurangabad, and Medical College Kolkata.

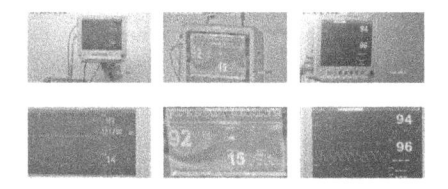

Fig. 5. Examples of segmented ICU monitor images using the proposed pipeline.

4.2 Results of Segmentation

This section presents the segmentation results using the EfficientNet framework, along with comparative assessments using other backbones. Fig. 4 shows the segmentation performance of various deep learning frameworks. The graph in Fig. 4 illustrates segmentation errors using different architectures, including MobileNetV2, Seg-YOLO, Darknet-YOLO, and various EfficientNet models (B0, B1, B2, and B4). The x-axis represents different models, while the y-axis shows the root mean square (RMS) error on training and validation datasets. As model complexity increases, generally characterized by a higher number of parameters, there is a noticeable decrease in RMS error, indicating better data pattern recognition. EfficientNet-B4, which performed the best, was selected for the final design. Sample segmentation results are shown in Fig. 5, demonstrating accurate segmentation. However, Fig. 6 highlights instances where the segmentation pipeline produced errors. In the first image, incorrect corner detection led to undesirable sections being included in the cropped image. In the second image, an incorrect skew angle resulted in a skewed output, while in the third image, poor corner detection caused vital information to be excluded from the active image area. Finally, Fig. 7 illustrates the impact of segmentation on the final output. When the input image bypasses the pre-processing and segmentation pipeline, detection performance degrades, leading to incorrect OCR and heuristic-based predictions of vital signs.

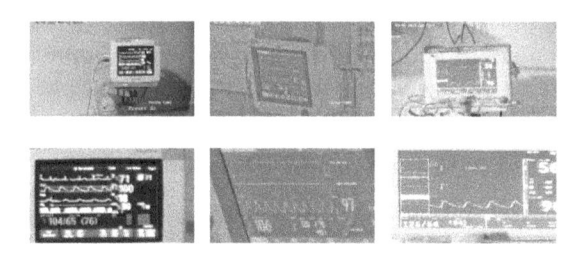

Fig. 6. Examples of segmentation errors where the proposed pipeline failed to detect all vitals correctly.

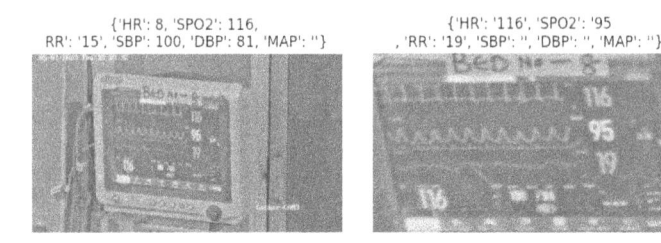

Fig. 7. Sample results with and without segmentation.

4.3 Ablation Study: Preprocessing

Skew correction adjusts the orientation of an image that is not perpendicular by an automatic rotation process. The binary image of an upright handwritten character has a distinct feature: the difference between the upper and lower limits of the histogram is relatively higher than in skewed samples. The test set includes three types of data: (i) samples with a 0° skew, (ii) samples with a 45° skew, and (iii) samples with a 90° skew.

We evaluated the selected CNN model on all test samples, comparing accuracy with and without the preprocessing stage, which included cleaning and skew correction. The results showed that the preprocessing stage significantly improved accuracy, particularly for images with a skew $\leq 45°$. For instance, with a 45° skew, accuracy increased from 23.89% to 44.11%, while for images with a $\leq 45°$ skew, accuracy improved from 84.56% to 88.96%. Random Padding also impacts results. Experiments were conducted on various CNN models trained using the CIFAR-10, CIFAR-100, and Fashion-MNIST datasets. Traditional padding was replaced with Random Padding in different layers, with results shown in Tables 1 and 2. For CIFAR-10, using AlexNet with Random Padding in the first three layers reduced the error rate to 12.75%. On VGG16, two Random Padding layers improved accuracy by 1.87%, achieving an error rate of 10.54%. Replacing traditional padding with Random Padding in GoogLeNet led to a new state-of-the-art test error of 10.20%, and on ResNet18, accuracy increased by 3.87%. Similar trends were observed on CIFAR-100 and Fashion-MNIST. In general, adding Random Padding to the first two layers improves deep learning models' performance, but increasing it further might reduce effectiveness.

4.4 Ablation Study: Segmentation Backbone

We have implemented and tested a few popular backbone networks for the segmentation model. EfficientNet [14] is a distinguished family of CNN architectures meticulously engineered to attain state-of-the-art performance while exhibiting remarkable efficiency in terms of parameter count and computational resources. Within this family, several variants have been developed, each denoted by a specific scaling factor, denoted by the symbol B. These variants, denoted as EfficientNet-Bx, are characterized by their respective parameter count and

Table 1. Results of applying Random Padding to various models on different datasets.

Dataset	Model	Random Padding Layers	Error Rate (%)
CIFAR-10	AlexNet	3	12.75
CIFAR-10	VGG16	2	10.54
CIFAR-10	GoogLeNet	3	10.20
CIFAR-10	ResNet18	1	3.87
CIFAR-100	AlexNet	3	40.66
CIFAR-100	VGG16	3	39.66
CIFAR-100	GoogLeNet	3	32.94
CIFAR-100	ResNet18	1	68.71
Fashion-MNIST	AlexNet	1	3.34
Fashion-MNIST	VGG16	2	5.49
Fashion-MNIST	GoogLeNet	2	5.82
Fashion-MNIST	ResNet18	1	0.16

Table 2. Effect of skew correction on handwriting recognition accuracy.

Skew	Accuracy before correction	Accuracy after correction
0 - 45 degrees	84.56%	88.96%
45 - 90 degrees	23.89%	44.11%

Floating Point Operations (FLOPs) requirements. EfficientNet-B0, the foundational member of this family, manifests with a parameter count of 5.3 million and necessitates 0.39 billion FLOPs for inference. Subsequent variants within the family, such as EfficientNet-B1, demonstrate incremental improvements in performance, albeit with a commensurate increase in model complexity. For instance, EfficientNet-B1 boasts a parameter count of 7.8 million and demands 0.70 billion FLOPs, while EfficientNet-B2, exhibiting further refinement, encompasses 9.2 million parameters and requires 1.0 billion FLOPs for computational operations. The progression continues with EfficientNet-B4, characterized by its extensive parameterization, comprising 19 million parameters and consuming 4.2 billion FLOPs. MobileNetV2 [15] presents a distinct architectural paradigm, structured around an inverted residual design. Central to its architecture are residual connections established between bottleneck layers, facilitating efficient information flow and feature extraction. Notably, MobileNetV2 incorporates lightweight depth-wise convolutions within its intermediate expansion layers, serving as a pivotal mechanism for introducing non-linearity while maintaining computational efficiency. Comprising an initial fully convolutional layer equipped with 32 filters, MobileNetV2 further integrates 19 residual bottleneck layers, each contributing to the network's capacity for feature abstraction and representation learning. The juxtaposition of EfficientNet and MobileNetV2 underscores the diverse architectural strategies employed in the pursuit of achieving optimal trade-offs between model complexity, computational efficiency, and performance across a spectrum of computer vision tasks. In terms of accuracy, EfficientNet outshines its predecessors by a marginal amount as depicted in Fig. ?? and Table 4. It may be observed that while traditional CNN-based backbones rely on 500+ million parameters, EfficientNet accomplishes the same with a mere 66 million parameters. This propels a remarkable increase in speed (eight folds) and it significantly enhances the network's usability in real-world industrial applications. The architecture's lightweight and robust design, combined with Mobile Inverted Bottleneck layers and Squeeze-and-Excitation optimization, consistently delivers high performance across several computer vision tasks.

4.5 Ablation Study: OCR Frameworks

In this section, we present a comprehensive comparison of three prominent optical character recognition (OCR) frameworks that we have tested with: EasyOCR, PaddleOCR, and MMOCR. Each framework offers unique features and capabilities, aiming to address various challenges in text extraction from images. We evaluate these frameworks based on their architecture, performance, and limitations. **EasyOCR**, built on Tesseract OCR engine and deep learning models, provides a straightforward interface for text recognition tasks. It utilizes a combination of the CRAFT algorithm for text detection and a CRNN model for text recognition. **PaddleOCR** offers a polished implementation of the CRNN architecture for text recognition, along with the CRAFT model for text detection. It simplifies the training pipeline with a high-level API, facilitating the development of text detection and OCR systems. **MMOCR** is part of the OpenMMLab project. It leverages PyTorch and MMdetection for text detection, recognition, and downstream tasks like key information extraction. It boasts a modular design, supporting various models for text-related tasks.

Table 3. Comparative performance of various OCRs

Word(s) Accuracy	Paddle OCR	MMOCR	EasyOCR
Correct	1141	270	1003
False	1185	2056	1323
Exact Match %	49.05%	11.61%	42.62%
Performance GPU (sec)	2.40	0.14	0.79
Performance CPU (sec)	3.22	1.85	37.76

Table 3 compares three OCR systems: Paddle OCR, MMOCR, and EasyOCR. It evaluates their word accuracy (based on 2326 sample size), exact match percentage, and performance on GPU and CPU platforms. Paddle OCR demonstrates the highest accuracy and efficiency, followed by EasyOCR, while MMOCR lags behind in both aspects. These insights inform the selection of the most suitable OCR system for specific applications.

Table 4. Results and comparisons of vital extraction using the proposed model and various combinations of the segmentation backbone and OCR

Model	HR accuracy	SpO2 accuracy	RR accuracy	BP accuracy	MAP accuracy
(PaddleOCR + EfficientNetB4)	87.74	67.19	85.02	89.23	64.81
(PaddleOCR + EfficientNetB2)	87.04	67.48	84.82	89.04	65.27
(EasyOCR + EfficientNetB4)	86.83	66.87	84.05	88.67	64.09
(PaddleOCR + EfficientNetB1)	85.41	66.32	83.77	88.36	63.74
(EasyOCR + EfficientNetB2)	84.19	65.96	83.29	88.02	63.38
(KerasOCR + EfficientNetB4)	86.78	65.605	82.83	87.73	62.94
(PaddleOCR + EfficientNetB0)	84.86	65.17	82.38	87.42	62.5
(PaddleOCR + MobileNetV2)	84.43	64.735	81.93	87.11	62.06
(MMOCR + EfficientNetB4)	84.01	64.3	81.48	86.8	61.62

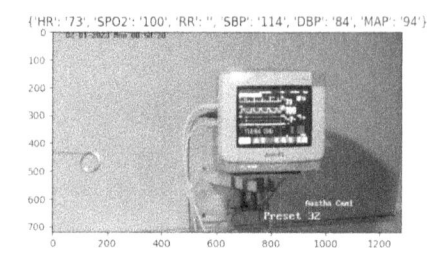

Fig. 8. Depiction of sample results obtained using the proposed pipeline.

4.6 Results of Vitals Extraction

We now present the results of vital parameter extraction. Table 4 presents the accuracy of each vital extracted, as compared with the ground truth. For this, a ground truth dataset has been created. The metric is in % accuracy. For each image, the vitals are extracted, and compared with the ground truth, to calculate accuracy. It can be observed from the table that the combination of PaddleOCR + EfficientNetB4 performs the best out of all other combinations tested in the experiments. It may be attributed to the better architecture of the B4 network and superior performance of the PaddleOCR in the present context. Finally, we present a snapshot of the final output. Fig. 8 presents the final outputs (vitals) extracted from the raw ICU monitor image with the help of the proposed pipeline. It may be observed that the extracted vitals are correctly matching with the ground truths.

5 Discussion

In this section, we discuss the challenges associated with existing ICU monitoring models and the rationale behind adopting the proposed computer vision (CV)-based approach.

5.1 Comparisons with Existing Approaches

Despite advancements in ICU monitoring technology, several challenges persist in the current models: **Manual Interpretation:** Traditional monitoring systems often rely on healthcare professionals to manually interpret data from physiological monitors. This process is labor-intensive, prone to human error, and can be inconsistent, especially under the high-stress conditions common in ICUs. The variability in interpretation can affect the timeliness of operations and the accuracy of diagnoses [22]. **Cost and Logistical Barriers:** Upgrading existing ICU equipment to newer models that support direct data integration with central systems can be prohibitively expensive. In resource-limited settings, such upgrades are often not feasible, making it challenging to implement state-of-the-art monitoring solutions that require specific hardware and software capabilities [23]. **Advantages of CV-Based Approach:** The proposed CV-based approach allows for continuous and non-intrusive monitoring of patient vitals by analyzing video feeds from existing CCTV cameras. This reduces the need for direct patient contact and minimizes disturbances, enhancing patient comfort and safety [24]. **Compatibility with Existing Equipment:** By focusing on image-based data extraction, our method circumvents the need for direct sensor integration. This makes it compatible with a wide range of existing ICU monitors, regardless of the manufacturer or model, thus avoiding the logistical and financial burdens associated with equipment upgrades [25]. **Enhancement of Remote Healthcare Provision:** Our approach supports remote healthcare by enabling the extraction of vital signs from video feeds. This is particularly beneficial in telemedicine scenarios, where healthcare providers need to monitor patients remotely without standardized equipment. It enhances the capabilities of remote healthcare systems by utilizing existing infrastructure effectively [26]. Given the aforementioned challenges, the proposed CV-based approach for extracting patient vitals from ICU monitor images offers a promising alternative to more expensive solutions.

6 Conclusion and Future Work

One potential improvement to the solution could involve the utilization of a broader spectrum of colors to enhance the detection of quantities. While the current system utilizes certain colors for specific associations such as blue for SpO2 and red/pink for BP and MAP, incorporating more color variations could help in capturing additional data. Despite encountering challenges with exceptions, leveraging this information could enhance overall accuracy. Additionally, implementing proper classification techniques on the unlabelled dataset could lead to more precise quantification of classes, particularly by utilizing location information for improved results. Furthermore, adopting a category segmentation-based pipeline could address the current method's limitation of disregarding the location information of numbers. This approach would involve integrating the quantities' location data into partial classification, potentially enhancing the accuracy of the binary classification model and allowing for the exploration of a third pipeline for further development and refinement.

Acknowledgement. We thank Himanshu Lal, C G Mitun Akil and Ramsundar Tanikella for their help during pipeline designing.

References

1. H. Wang, J. Huang, G. Wang, H. Lu and W. Wang, Contactless Patient Care Using Hospital IoT: CCTV-Camera-Based Physiological Monitoring in ICU, in *IEEE Internet of Things Journal*, 2023
2. H. Wang, W. Huang, J. Huang, G. Wang, H. Lu and W. Wang, Camera based Eye State Estimation for ICU Patients: A Pilot Clinical Study, in *2023 IEEE/CVF Conference on Computer Vision and Pattern Recognition Workshops*

3. Brown, J., Green, L., Taylor, M., White, P.: Autonomous Monitoring Systems in Intensive Care: A Review. Crit. Care Med. **49**(2), 345–352 (2021)
4. S.L. Rossol, J.K. Yang, C. Toney-Noland, J. Bergin, C. Basavaraju, P. Kumar, and H.C. Lee, Non-Contact Video-Based Neonatal Respiratory Monitoring, in *Children (Basel)*, 2020
5. J. Jorge, M. Villarroel, H. Tomlinson, O. Gibson, J.L. Darbyshire, J. Ede, M. Harford, J.D. Young, L. Tarassenko, and P. Watkinson, Non-contact physiological monitoring of post-operative patients in the intensive care unit, in *NPJ Digital Medicine*, 2023
6. Z. Liu et al., Contactless Respiratory Rate Monitoring For ICU Patients Based On Unsupervised Learning, in *2023 IEEE/CVF Conference on Computer Vision and Pattern Recognition Workshops*
7. B. Huang, S. Hu, Z. Liu, C.L. Lin, J. Su, C. Zhao, L. Wang, and W. Wang, Challenges and prospects of visual contactless physiological monitoring in clinical study, in *NPJ Digital Medicine*, 2023
8. R. Janssen, W. Wang, A. Moço, and G. de Haan, Video-based respiration monitoring with automatic region of interest detection, in *Physiological Measurement*, 2016
9. J. Jorge et al., Non-Contact Monitoring of Respiration in the Neonatal Intensive Care Unit, in *2017 12th IEEE International Conference on Automatic Face & Gesture Recognition*
10. L. Antognoli, P. Marchionni, S. Spinsante, S. Nobile, V.P. Carnielli and L. Scalise, Enhanced video heart rate and respiratory rate evaluation: standard multiparameter monitor vs clinical confrontation in newborn patients, in *2019 IEEE International Symposium on Medical Measurements and Applications*
11. C. Massaroni, E. Schena, S. Silvestri, F. Taffoni and M. Merone, Measurement system based on RGB camera signal for contactless breathing pattern and respiratory rate monitoring, in *2018 IEEE International Symposium on Medical Measurements and Applications*
12. M. Chen, Q. Zhu, H. Zhang, M. Wu and Q. Wang, Respiratory Rate Estimation from Face Videos, in *2019 IEEE EMBS International Conference on Biomedical & Health Informatics*
13. L. Pham, P.H. Hoang, X.T. Mai and T.A. Tran, Adaptive Radial Projection on Fourier Magnitude Spectrum for Document Image Skew Estimation, in *2022 IEEE International Conference on Image Processing*
14. M. Tan and Q.V. Le, EfficientNet: Rethinking Model Scaling for Convolutional Neural Networks, in arXiv:1905.11946 [cs.LG], 2019
15. A.G. Howard, M. Zhu, B. Chen, D. Kalenichenko, W. Wang, T. Weyand, M. Andreetto, and H. Adam, MobileNets: Efficient Convolutional Neural Networks for Mobile Vision Applications, in arXiv:1704.04861 [cs.CV], 2017
16. Y. Du et al., PP-OCR: A Practical Ultra Lightweight OCR System, in *CoRR abs/2009.09941*, 2020
17. Kornia's Bbox. Available online: https://kornia.readthedocs.io/en/stable/geometry.bbox.html
18. EasyOCR. Available online: https://github.com/JaidedAI/EasyOCR
19. KerasOCR. Available online: https://github.com/faustomorales/keras-ocr
20. MMOCR. Available online: https://github.com/open-mmlab/mmocr
21. Dataset. Available online: https://www.kaggle.com/datasets/
22. Davis, R., Wilson, S., Thompson, E., Rodriguez, M.: Challenges in Manual Interpretation of ICU Monitor Data. Healthcare Technology Letters **6**(4), 128–135 (2019)

23. P. Patel, S. Kumar, R. Desai and N. Shah, Cost and Logistical Barriers in Upgrading ICU Monitoring Equipment, in *International Journal of Medical Informatics*, vol. 130, pp. 104008, 2021
24. Roberts, E., Adams, J., Hall, M., Baker, S.: Non-Intrusive Patient Monitoring in ICUs Using Computer Vision. IEEE Trans. Med. Imaging **39**(12), 3903–3914 (2020)
25. Zhao, G., Chen, W., Zhang, X., Liu, J.: Ensuring Compatibility of ICU Monitoring Equipment with Central Systems. IEEE Access **8**, 19234–19244 (2020)
26. Gonzalez, S., Hernandez, F., Perez, L., Martinez, R.: Enhancing Remote Healthcare Provision through Video-Based Vital Sign Monitoring. Telemedicine and e-Health **27**(9), 951–958 (2021)

Author Index

A
Abdelhalim, Ibrahim 301
Agarwal, Snigdha 332
Ahmed, Sheraz 138
Aithal, Ninad 17

B
Baek, Hyeongboo 170
Balada, Christoph 138
Bappy, D. M. 170
Basuchowdhuri, Partha 60
Bhatt, Jignesh S. 122
Bhattacharya, Debanjali 17
Bodenhofer, Ulrich 106
Broti, Nawara Mahmood 235

C
Cai, Congbo 408
Cai, Shuhui 408
Chakraborti, Tapabrata 60
Chen, Zhong 408
Cheng, Hongyu 422

D
Dao, Duy-Phuong 267
Das, Abhijit 439
Das, Swagatam 283
Deng, Bowen 203
Dengel, Andreas 138
Dey, Samiran 60
Dhivyaa, S. P. 267
Dogra, Debi Prosad 455
Dworzak, Michael 316

E
Elazab, Ahmed 345
El-Baz, Ayman 301
Elsharkawy, Mohamed 301

F
Fierrez, Julian 75
Fröhler, Bernhard 106

G
Gall, Alexander 106
Gao, Feng 49
Gao, Li 49
Ghazal, Mohammed 301
Ghosh, Susmita 283
Gorade, Vandan 439
Gu, Yunrui 33
Gupta, Archit 1

H
Haq, Mohammad Z. 301
Haq, Rayan 301
He, Xiaohai 49
Heim, Anja 106
Hermosilla, Pedro 316
Hua, Yang 203

I
Iijima, Keiya 235
Issac, Thomas Gregor 17
Iwasaki, Masaki 235

J
Jiang, Xiaoyi 376

K
Kang, Donghwa 170
Khaled, Afifa 345
Khare, Kushagra 455
Kim, Jahae 267
Kong, Cong 33
Koo, Minsuk 170
Korra, Sathya Babu 90
Kumar, Sanidhya 122
Kundu, Bijon 60

L
Lee, Jinkyu 170
Lee, Youngmoon 170
Li, Qingli 33
Lucieri, Adriano 138

M
Macias-Fassio, Eric 75
Mahapatra, Dwarikanath 439
Mahmoud, Ali 301
Manna, Asim 251
Maurer-Granofszky, Margarita 316
Meng, Bo 422
Mishra, Deepak 360
Morales, Aythami 75

N
Nagaraju, K. 90

O
Ono, Yumie 235

P
Paranjape, Jay N. 187
Patel, Vishal M. 187
Peng, Yonghong 49
Phophalia, Ashish 122
Pruenza, Cristina 75

Q
Qing, Linbo 49

R
Rampuria, Akshat 455
Reiter, Michael 316
Rezaei, Somayeh 376
Roy, Santanu 1
Roy, Sudipta 439

S
Sablatnig, Robert 154
Saha, Sanjoy Kumar 60
Sahu, Palak 1
Sandhu, Harpal S. 301
Sawada, Masaki 235
Sekh, Arif Ahmed 391
Senck, Sascha 106

Sheet, Debdoot 251
Shen, Yiqing 220
Sikder, Shameema 187
Singh, Azad 360
Sinha, Neelam 17, 332
Sinha, Pratik 391
Song, Xiaoning 203
Soni, Ayush 455
Sterzinger, Rafael 154
Stippel, Christian 154

T
Takayama, Yutaro 235
Thanos, Aristomenis 301
Thota, Gokaramaiah 90
Tiwari, Shubhi 1

V
Varshney, Payal 138
Vedula, S. Swaroop 187
Vilvadrinath, Varun 122

W
Wang, Guowen 408
Wang, Lu 408
Wang, Silei 408
Wang, Yan 33
Weijler, Lisa 316
Weinberger, Patrick 106
Wu, Xiao-jun 203

X
Xu, Xiaowei 422

Y
Yan, Zeyu 422
Yang, Ge 49
Yang, Hyung-Jeong 267
Yin, Zhaoxia 33

Z
Zhang, Haoran 422
Zhang, Wenjie 203
Zhang, Yanteng 49
Zhang, Zhiwei 220
Zheng, Limin 422